Spon's External Works and Landscape Price Book

2010

Spon's External Works and Landscape Price Book

Edited by
DAVIS LANGDON LLP

in association with
LandPro Ltd
Landscape Surveyors

2010

Twenty-Ninth edition

Spon Press
an imprint of Taylor & Francis
LONDON AND NEW YORK

First edition 1978
Twenty-ninth edition published 2010
by Spon Press
2 Park Square, Milton Park, Abingdon, Oxon, OX14 4RN

Simultaneously published in the USA and Canada
by Spon Press
270 Madison Avenue, New York, NY 10016

Spon Press is an imprint of the Taylor & Francis Group, an informa business

© 2010 Spon Press

The right of Davis Langdon to be identified as the Author of this Work has been asserted by them in accordance with the Copyright, Designs and Patents Act 1988

Typeset in Arial by Taylor & Francis Books

Printed and bound in Great Britain by
TJ International Ltd, Padstow, Cornwall

All rights reserved. No part of this book may be reprinted or reproduced or utilised in any form or by any electronic, mechanical, or other means, now known or hereafter invented, including photocopying and recording, or in any information storage or retrieval system, without permission in writing from the publishers.

The publisher makes no representation, express or implied, with regard to the accuracy of the information contained in this book and cannot accept legal responsibility or liability for any errors or omissions that may be made.

British Library Cataloguing in Publication Data
A catalogue record for this book is available from the British Library

ISBN 13: 978-0-415-55260-8 (hardback)
ISBN 13: 978-0-203-87243-7 (ebook)

ISBN 10: 0-415-55260-5 (hardback)
ISBN 10: 0-203-87243-6 (ebook)

ISSN: 0267-4181

Contents

	page
Preface to the Twenty Ninth Edition	vii
Acknowledgements	ix
How to Use This Book	xxiii

PART 1: GENERAL — 1

Common Arrangement of Work Sections	3
Labour Rates Used in This Edition	5
Computation of Labour Rates Used in This Edition	6
Computation of the Cost of Materials	11
Computation of Mechanical Plant Costs	12
Landfill Tax	15

PART 2: APPROXIMATE ESTIMATES - MAJOR WORKS — 17

Cost Indices	18
Regional Variations	20
Approximate Estimates	21
Preliminaries	22
Demolition and Site Clearance	24
Groundwork	26
Ground Stabilization	28
In Situ Concrete	31
Brick/Block Walling	33
Roads and Pavings	35
Special Surfaces for Sport/Playgrounds	47
Preparation for Planting/Turfing	49
Seeding and Turfing	51
Planting	53
Landscape Maintenance	59
Fencing	61
Street Furniture	65
Drainage	66
Irrigation	72
Water Features	73
Timber Decking	74

PART 3: PRICES FOR MEASURED WORKS - MAJOR WORKS — 75

Introduction	75
New Items for This Edition	76
A Preliminaries	94
B Complete Buildings/Structures/Units	100
C Existing Site/Buildings/Services	102
D Groundwork	105
E In Situ Concrete/Large Precast Concrete	125
F Masonry	131
G Structural/Carcassing Metal/Timber	140
H Cladding/Covering	141

	page
J Waterproofing	142
M Surface Finishes	149
P Building Fabric Sundries	152
Q Paving/Planting/Fencing/Site Furniture	153
R Disposal Systems	289
S Piped Supply Systems	313
V Electrical Supply/Power/Lighting Systems	323

PART 4: APPROXIMATE ESTIMATES - MINOR WORKS — 331

Approximate Estimates	331
Demolition and Site Clearance	332
Groundwork	334
In Situ Concrete	336
Brick/Block Walling	337
Roads and Pavings	338
Preparation for Seeding/Turfing	342
Planting	343
Maintenance	346
Drainage	347

PART 5: PRICES FOR MEASURED WORKS - MINOR WORKS — 349

Introduction	349
New Items for This Edition	350
C Existing Site/Buildings/Services	353
D Groundwork	355
E In Situ Concrete/Large Precast Concrete	364
F Masonry	368
G Structural/Carcassing Metal/Timber	372
J Waterproofing	373
M Surface Finishes	374
Q Paving/Planting/Fencing/Site Furniture	375
R Disposal Systems	438

PART 6: FEES FOR PROFESSIONAL SERVICES — 449

Landscape Architects' Fees	449
Worked Examples of Percentage Fee Calculations	452

PART 7: RATES OF WAGES — 453

Building Industry Wages	453
Agricultural Wages, England and Wales	453
Agricultural Wages, Scotland	454

PART 8: DAYWORK AND PRIME COST — 457

PART 9: TABLES AND MEMORANDA — 467

INDEX — 537

Preface to the Twenty Ninth Edition

Market Conditions

This "credit crunch" edition of the price book reflects a surprising stability in the external works sector compared to other construction disciplines. Whilst all the data from our research at administrative or academic levels suggest that a 20% drop in prices should be in effect overall, this is not exactly the case. In summary, major suppliers have raised most of their prices, labour is stable but with some shorter working weeks in effect and plant prices may be negotiated keenly. The tender market is more competitive however and we may assume that overhead and profit are reduced. Most of the large to medium sized landscape contractors still report full order books, although smaller contractors reliant on the domestic market are definitely experiencing difficulties.

Supplier Prices

During the course of our price enquiries for this edition we have especially noted the following: the main industry drivers vary; steel is down; cement has increased and further increases are expected - this is driven by energy costs to a large extent; the cost of oil based products such as plastics and asphalt have risen; aggregates and builders merchant supplies are lower. Many of our other suppliers have requested that we apply minimal price changes or none at all. The most notable cost increases were implemented by the large precast suppliers who increased their material prices by 12%.

Labour Rates

Last year we reported that the labour rate was being stabilized by the broad use of eastern European labour. Unofficial reports indicate that this labour force is largely returning to their original homes as economic conditions are less harsh than those in the UK. The reality of the labour situation is that whilst wages cannot be reduced there have been redundancies. There is probably less overtime paid and there are certainly firms who have placed their staff on shorter weeks. The wage rate used this year is therefore reduced to £18.25 per hour for major works and £20.50 for minor works.

Plant Rates

Most of the plant suppliers referenced in this publication have advised us during our update to keep their prices the same. Some, such as site plant hire, have reflected a 10% reduction in rates. In truth we are aware that there is a lot of idle plant in suppliers' yards. It is general knowledge amongst hirers that a "squeeze" can be put onto suppliers for hire rates.

Tables and Memoranda

We remind readers of this section at the back of the book. It is often overlooked but contains a wealth of information useful to external works consultants and contractors.

Profit and Overhead

Spon's External Works and Landscape prices do not allow for profit or site overhead. Readers should evaluate the market conditions in the sector or environment in which they operate and apply a percentage to these rates. Company overhead is allowed for within the labour rates used. Please refer to 'Part 1 General' for an explanation of the build-up of this year's labour rates.

Prices for Suppliers and Services

We acknowledge the support afforded to us by the suppliers of products and services who issue us with the base information used. Their contact details are published in the directory section at the front of this book. We advise that readers wishing to evaluate the cost of a product or service should approach these suppliers directly should they wish to confirm prices prior to submission of tenders or quotations.

Whilst every effort is made to ensure the accuracy of the information given in this publication, neither the editors nor the publishers in any way accept liability for loss of any kind resulting from the use made by any person of such information.

We remind readers that we would be grateful to hear from them during the course of the year with suggestions or comments on any aspect of the contents of this year's book and suggestions for improvements. Our contact details are shown below.

DAVIS LANGDON
MidCity Place
71 High Holborn
London
WC1V 6QS

SAM HASSALL
LANDPRO LTD
Landscape Surveyors
14 Upper Bourne Lane
Farnham
Surrey
GU10 4RQ

Tel: 02070 617000
e-mail: spons@davislangdon.com

Tel: 01252 795030
e-mail: info@landpro.co.uk

Acknowledgements

This list has been compiled from the latest information available but as firms frequently change their names, addresses and telephone numbers due to reorganization, users are advised to check this information before placing orders.

ABG Environmental Geosynthetics Ltd
Unit E7
Meltham Mills Road
Meltham
West Yorkshire HD9 4DS
Geosynthetics
Tel: 01484 852096
Fax: 01484 851562
Website: www.abg-geosynthetics.com
E-mail: sales@abgltd.com

Agripower Ltd
Broomfield Farm
Rignall Road
Great Missenden
Bucks HP16 9PE
Drainage
Tel: 01494 866776
Fax: 01494 866779
Website: www.agripower.co.uk
E-mail: info@agripower.co.uk

Aco Technologies Plc
Aco Business Park
Hitchin Road
Shefford
Bedfordshire SG17 5TE
Linear drainage
Tel: 01462 816666
Fax: 01462 815895
Website: www.aco-online.co.uk
E-mail: technologies@aco.co.uk

Alumasc Exterior Building Products Ltd
White House Works
Bold Road
Sutton
St Helens
Merseyside WA9 4JG
Green roof systems
Tel: 01744 648400
Fax: 01744 648401
Website: www.alumasc-exterior-building-products.co.uk
E-mail: info@alumasc-exteriors.co.uk

Addagrip Surface Treatments UK Ltd
Addagrip House
Bell Lane Industrial Estate
Uckfield
East Sussex TN22 1QL
Epoxy bound surfaces
Tel: 01825 761333
Fax: 01825 768566
Website: www.addagrip.co.uk
E-mail: sales@addagrip.co.uk

Amberol Ltd
The Plantation
King Street
Alfreton
Derbyshire DE55 7TT
Street furniture
Tel: 01773 830930
Fax: 01773 834191
Website: www.amberol.co.uk
E-mail: info@amberol.co.uk

Amenity and Horticultural Services Ltd
Coppards Lane
Northiam
East Sussex TN31 6QP
Horticultural composts and fertilizers
Tel: 01797 252728
Fax: 01797 252724
Website: www.ahsdirect.co.uk
E-mail: info@ahsdirect.co.uk

Amenity Land Services Ltd (ALS)
Unit 2/3
Allscott
Telford
Shropshire TF6 5DY
Horticultural products supplier
Tel: 01952 641949
Fax: 01952 247369
Website: www.amenity.co.uk
E-mail: sales@amenity.co.uk

Anderton Concrete Products Ltd
Anderton Wharf
Soot Hill
Anderton
Northwich
Cheshire CW9 6AA
Concrete fencing
Tel: 01606 79436
Fax: 01606 871590
Website: www.andertonconcrete.co.uk
E-mail: sales@andertonconcrete.co.uk

Anglo Aquarium Plant Co Ltd
Strayfield Road
Enfield
Middlesex EN2 9JE
Aquatic plants
Tel: 020 8363 8548
Fax: 020 8363 8547
Website: www.anglo-aquarium.co.uk
E-mail: sales@anglo-aquarium.co.uk

Architectural Heritage Ltd
Taddington Manor
Cutsdean
Cheltenham
Gloucestershire GL54 5RY
Ornamental stone buildings and water features
Tel: 01386 584414
Fax: 01386 584236
Website: www.architectural-heritage.co.uk
E-mail: puddy@architectural-heritage.co.uk

Artificial Lawn Co
Hartshill Nursery
Thong Lane
Gravesend
Kent DA12 4AD
Artificial grass suppliers
Tel: 01474 364320
Fax: 01474 321587
Website: www.artificiallawn.co.uk
E-mail: sales@artificiallawn.co.uk

Autopa Ltd
Cottage Leap off Butlers Leap
Rugby
Warwickshire CV21 3XP
Street furniture
Tel: 01788 550556
Fax: 01788 550265
Website: www.autopa.co.uk
E-mail: info@autopa.co.uk

AVS Fencing Supplies Ltd
Unit 1 AVS Trading Park
Chapel Lane
Milford
Surrey GU8 5HU
Fencing
Tel: 01483 410960
Fax: 01483 860867
Website: www.avsfencing.co.uk
E-mail: sales@avsfencing.co.uk

Bauder Ltd
Broughton House
Broughton Road
Ipswich
Suffolk IP1 3QR
Green roof systems
Tel: 01473 257671
Fax: 01473 230761
Website: www.bauder.co.uk
E-mail: info@bauder.co.uk

Baylis Landscape Contractors Ltd
Hartshill Nursery
Thong Lane
Gravesend
Kent DA12 4AD
Sportsfield construction
Tel: 01474 569576
Fax: 01474 321587
Website: www.baylislandscapes.co.uk
E-mail: Baylis@baylislandscapecontractors.co.uk

Bituchem
Laymore
Forest Vale Industrial Estate
Cinderford
Gloucestershire GL14 2YH
Macadam contractor
Tel: 01594 826768
Fax: 01594 826948
Website: www.bituchem.com
E-mail: info@bituchem.com

Boddingtons Ltd
Blackwater Trading Estate
The Causeway
Maldon
Essex
Erosion control, soil stabilization, plant protection
Tel: 01621 874200
Fax: 01621 874299
Website: www.boddintons-ltd.com
E-mail: sales@boddingtons-ltd.com

Breedon Special Aggregates
(A division of Ennstone Johnston Ltd)
Breedon-on-the-Hill
Derby
Derbyshire DE8 1AP
Specialist gravels
Tel: 01332 694001
Fax: 01332 695159
Website: www.ennstone.co.uk
E-mail: sales@breedon.co.uk

Britannia Rainwater Recycling Systems
Studio House
Delamere Road
Cheshunt
Herts EN8 9SH
Rainwater tanks
Tel: 01992 633211
Fax: 01992 633212
E-mail: britanniaservices@btopenworld.com

British Seed Houses Ltd
Camp Road
Witham St Hughs
Lincoln LN6 9QJ
Grass and wildflower seed
Tel: 01522 868714
Fax: 01522 868095
Website: www.bshamenity.com
E-mail: seeds@bshlincoln.co.uk

Browse Bion
Unit 19, Lakeside Park
Medway City Estate
Rochester ME2 4LT
Park signage
Tel: 01634 710063
Fax: 01634 290112
E-mail: sales@browsebion.com

Broxap Street Furniture
Rowhurst Industrial Estate
Chesterton
NewcastleunderLyme
Staffordshire ST5 6BD
Street furniture
Tel: 0844 800 4085
Fax: 01782 565357
Website: www.broxap.com
E-mail: sales@broxap.com

Capital Garden Products Ltd
Gibbs Reed Barn
Pashley Road
Ticehurst
East Sussex TN5 7HE
Plant containers
Tel: 01580 201092
Fax: 01580 201093
Website: www.capital-garden.com
E-mail: sales@capital-garden.com

CED Ltd
728 London Road
West Thurrock
Grays
Essex RM20 3LU
Natural stone
Tel: 01708 867237
Fax: 01708 867230
Website: www.ced.ltd.uk
E-mail: sales@ced.ltd.uk

Charcon Hard Landscaping
Hulland Ward
Ashbourne
Derbyshire DE6 3ET
Paving and street furniture
Tel: 01335 372222
Fax: 01335 370074
Website: www.aggregate.com
E-mail: ukenquiries@aggregate.com

City Electrical Factors
Eastern Road
North Lane
Aldershot
Hampshire GU12 4YD
Electrical wholesaler
Tel: 01252 327661
Fax: 01252 343089
Website: www.cef.co.uk

Cooper Clarke Civils and Lintels
Special Products Division
Bloomfield Road
Farnworth
Bolton BL4 9LP
Wall drainage and erosion control
Tel: 01204 862222
Fax: 01204 793856
Website: www.heitonuk.com
E-mail: marketing@cooperclarke.co.uk

Crowders Nurseries
Lincoln Road
Horncastle
Lincolnshire LN9 5LZ
Plant protection
Tel: 01507 525000
Fax: 01507 524000
Website: www.crowders.co.uk
E-mail: sales@crowders.co.uk

CU Phosco Ltd
Charles House
Lower Road
Great Amwell
Ware
Hertfordshire SG12 9TA
Lighting
Tel: 01920 860600
Fax: 01920 860635
Website: www.cuphosco.co.uk
E-mail: sales@cuphosco.co.uk

DeeOrgan Ltd
5 Sandyford Road
Paisley
Renfrewshire PA3 4HP
Street furniture
Tel: 01418 897000
Fax: 01418 897764
Website: www.dee-organ.co.uk
E-mail: signs@dee-organ.co.uk

Deepdale Trees Ltd
Tithe Farm
Hatley Road
Sandy
Bedfordshire SG19 2DX
Trees, multi-stem and hedging
Tel: 01767 262636
Fax: 01767 262288
Website: www.deepdale-trees.co.uk
E-mail: mail@deepdale-trees.co.uk

DLF Trifolium Ltd
Thorn Farm
Evesham Road
Inkberrow
Worcestershire WR7 4LJ
Grass and wildflower seed
Tel: 01386 791102
Fax: 01386 792715
Website: www.dlf.co.uk
E-mail: amenity@dlf.co.uk

Duracourt (Spadeoak) Ltd
Town Lane
Wooburn Green
High Wycombe
Buckinghamshire HP10 0PD
Tennis courts
Tel: 01628 529421
Fax: 01628 810509
Website: www.duracourt.co.uk
E-mail: info@duracourt.co.uk

E.T. Clay Products
7-9 Fowler Road
Hainault
Essex IG6 3UT
Brick supplier
Tel: 0208 5012100
Fax: 0208 5009990
Website: www.etbricks.co.uk
E-mail: sales@etbricks.co.uk

Earth Anchors Ltd
15 Campbell Road
Croydon
Surrey CR0 2SQ
Anchors for site furniture
Tel: 020 8684 9601
Fax: 020 8684 2230
Website: www.earth-anchors.com
E-mail: sales@earth-anchors.com

EasyMix Concrete Ltd
31 Burrowfields Industrial Estate
Welwyn Garden City
Hertfordshire AL7 4SS
Concrete
Tel: 01707 338700
Fax: 01707 338701
Website: www.easymic-concrete.co.uk
E-mail: info@easymix-concrete.co.uk

Edwards Sports Products Ltd
North Mills Trading Estate
Bridport
Dorset
DT6 3AH
Sports equipment
Tel: 01308 424111
Fax: 01308 455800
Website: www.edsports.co.uk
E-mail: sales@edsports.co.uk

Elliott Hire
The Fen
Baston
Peterborough
Cambridgeshire PE6 9PT
Site offices
Tel: 01778 560891
Fax: 01778 560881
Website: www.elliotthire.co.uk
E-mail: hirediv@elliott-algeco.com

English Woodlands
Burrow Nursery
Cross in Hand
Heathfield
East Sussex TN21 0UG
Plant protection
Tel: 01435 862992
Fax: 01435 867742
Website: www.ewburrownursery.co.uk
E-mail: sales@ewburrownursery.co.uk

Eura Conservation Ltd
Unit H10, Halesfield 19
Telford
Shropshire TF7 4QT
Metalwork restoration
Tel: 01952 680 218
Fax: 01952 585 044
E-mail: enquiries@eura.co.uk

Eve Trakway
Bramley Vale
Chesterfield
Derbyshire S44 5GA
Portable roads
Tel: 08700 767676
Fax: 08700 737373
Website: www.evetrakway.co.uk
E-mail: marketing@evetrakway.co.uk

Exclusive Leisure Ltd
28 Cannock Street
Leicester LE4 9HR
Artificial sports surfaces
Tel: 0116 233 2255
Fax: 0116 246 1561
Website: www.exclusiveleisure.co.uk
E-mail: info@exclusiveleisure.co.uk

Fairwater Water Garden Design Consultants Ltd
Lodge Farm
Malthouse Lane
Ashington
West Sussex RH20 3BU
Water feature contractors
Tel: 01903 892228
Fax: 01903 892522
Website: www.fairwater.co.uk
E-mail: info@fairwater.co.uk

Farmura Environmental Products Ltd
Stone Hill
Egerton
Ashford
Kent TN27 9DU
Organic fertilizer suppliers
Tel: 01233 756241
Fax: 01233 756419
Website: www.farmura.com
E-mail: info@farmura.com

Fleet (Line Markers) Ltd
Spring Lane
Malvern
Worcestershire WR14 1AT
Sports line marking
Tel: 01684 573535
Fax: 01684 892784
Website: www.fleetlinemarkers.com
E-mail: sales@fleetlinemarkers.com

Forticrete Ltd
Hillhead Quarry
Harpur Hill
Buxton
Derbyshire SK17 9PS
Retaining wall systems
Tel: 01298 23333
Fax: 01298 23000
Website: www.forticrete.com
E-mail: enq@forticrete.com

FP McCann Ltd
Whitehill Road
Coalville
Leicester LE67 1ET
Precast concrete pipes
Tel: 01530 240056
Fax: 01530 240015
Website: www.fpmccann.co.uk
E-mail: info@fpmccann.co.uk

Furnitubes International Ltd
Meridian House
Royal Hill
Greenwich
London SE10 8RT
Street furniture
Tel: 020 8378 3200
Fax: 020 8378 3250
Website: www.furnitubes.com
E-mail: sales@furnitubes.com

The Garden Trellis Company
Unit 1 Brunel Road
Gorse Lane Industrial Estate
Great Clacton
Essex CO15 4LU
Garden joinery specialists
Tel: 01255 688361
Fax: 01255 688362
Website: www.gardentrellis.co.uk
E-mail: info@gardentrellis.co.uk

Grace Construction Products
Ajax Avenue
Slough
Berkshire SL1 4BH
Bituthene tanking
Tel: 01753 692929
Fax: 01753 691623
Website: www.uk.graceconstruction.com
E-mail: uksales@grace.com

Grass Concrete Ltd
Duncan House,
142 Thornes Lane,
Thornes,
West Yorkshire WF2 7RE,
Grass block paving
Tel: 01924 379443
Fax: 01924 290289
Website: www.grasscrete.com
E-mail: info@grasscrete.com

Greenfix Soil Stabilisation and Erosion Control Ltd
Allens West
Durham Lane
Eaglescliffe
Stockton On Tees TS16 ORW
Seeded erosion control mats
Tel: 01642 888693
Fax: 01642 888699
Website: www.greenfix.co.uk
E-mail: stockton@greenfix.co.uk

Greenleaf Horticulture
Ivyhouse Industrial Estate
Haywood Way
Hastings TN35 4PL
Root directors
Tel: 01424 717797
Fax: 01424 205240
E-mail: info@greenleafhorticulture.net

Grundon Waste Management Ltd
Goulds Grove
Ewelme
Wallingford OX10 6PJ
Gravel
Tel: 0870 4438278
Website: www.grundon.com
E-mail: sales@grundon.com

HSS Hire
Unit 3
14 Wates Way
Mitcham
Surrey CR4 4HR
Tool and plant hire
Tel: 020 8685 9500
Fax: 020 8685 9600
Website: www.hss.com
E-mail: hire@hss.com

Haddonstone Ltd
The Forge House
Church Lane
East Haddon
Northampton NN6 8DB
Architectural stonework
Tel: 01604 770711
Fax: 01604 770027
Website: www.haddonstone.co.uk
E-mail: info@haddonstone.co.uk

Harrison External Display Systems
Borough Road
Darlington
Co Durham DL1 1SW
Flagpoles
Tel: 01325 355433
Fax: 01325 461726
Website: www.harrisoneds.com
E-mail: sales@harrisoneds.com

Havells Sylvania Fixtures UK Ltd
Avis Way
Newhaven
East Sussex BN9 0ED
Lighting
Tel: 0870 6062030
Fax: 01273 512688
Website: www.havells-sylvania.com
E-mail: info.concord@havells-sylvania.com

Headland Amenity Ltd
1010 Cambourne Business Park
Cambourne
Cambridgeshire CB23 6DP
Tel: 01223 597834
Fax: 01223 598052
Chemicals
Website: www.headlandamenity.com
E-mail: info@headlandamenity.com

Heicom UK
4 Frog Lane
Tunbridge Wells
Kent TN1 1YT
Tel: 01892 522360
Fax: 01892 522767
Tree sand
Website: www.treesand.co.uk
E-mail: mike.q@treesand.co.uk

Hepworth
Hazelhead
Crow Edge
Sheffield S36 4HG
Drainage
Tel: 0870 4436000
Fax: 0870 4438000
Website: www.hepworthdrainage.co.uk
E-mail: info@hepworthdrainage.co.uk

Icopal UK Ltd
Lyon Way
St Albans
Hertfordshire AL4 0LB
Lake liner
Tel: 01727 830116
Fax: 01727 868045
Website: www.monarflex.icopal.co.uk
E-mail: enq@monarflex.co.uk

Inturf
The Chestnuts
Wilberfoss
York YO41 5NT
Turf
Tel: 01759 321000
Fax: 01759 380130
Website: www.inturf.com
E-mail: info@inturf.co.uk

J Toms Ltd
7 Marley Farm
Headcorn Road
Smarden
Kent TN27 8PJ
Plant protection
Tel: 01233 770066
Fax: 01233 770055
Website: www.jtoms.co.uk
E-mail: jtoms@btopenworld.com

Jacksons Fencing
Stowting Common
Ashford
Kent TN25 6BN
Fencing
Tel: 01233 750393
Fax: 01233 750403
Website: www.jacksons-fencing.co.uk
E-mail: sales@jacksons-fencing.co.uk

James Coles & Sons (Nurseries) Ltd
The Nurseries
Uppingham Road
Thurnby
Leicester LE7 9QB
Nurseries
Tel: 01162 412115
Fax: 01162 432311
Website: www.colesnurseries.co.uk
E-mail: sales@colesnurseries.co.uk

John Anderson Hire
Unit 5 Smallford Works
Smallford Lane
St Albans
Herts AL4 0SA
Mobile toilet facilities
Tel: 01727 822485
Fax: 01727 822886
Website: www.superloo.co.uk
E-mail: sales@superloo.co.uk

Johnsons Wellfield Quarries Ltd
Crosland Hill
Huddersfield
West Yorkshire HD4 7AB
Natural Yorkstone pavings
Tel: 01484 652311
Fax: 01484 460007
Website: www.johnsons-wellfield.co.uk
E-mail: sales@johnsons-wellfield.co.uk

Jones of Oswestry
Whittington Road
Oswestry
Shropshire SY11 1HZ
Channels, gulleys, manhole covers
Tel: 01691 653251
Fax: 01691 658222
Website: www.jonesofoswestry.com
E-mail: sales@jonesofoswestry.com

KAR UK
Invar Road,
Swinton,
Manchester M27 9HF
Irrigation
Tel: 01617 939703
Fax: 01617 945145
E-mail: sales@karuk.com
Website: www.karuk.co.uk

Keller Ground Engineering
Mereworth Business Centre
Danns Lane
Mereworth
Kent ME18 5LW
Hydro seeding specialist
Tel: 01622 816780
Fax: 01622 816791
Website: www.keller-ge.co.uk
E-mail: retaining@keller-ge.co.uk

Kompan Ltd
20 Denbigh Hall
Bletchley
Milton Keynes
Bucks MK3 7QT
Play equipment
Tel: 01908 642466
Fax: 01908 270137
Website: www.kompan.com
E-mail: kompan.uk@kompan.com

Land & Water Group Ltd
3 Weston Yard
The Street
Albury
Guildford
Surrey GU5 9AF
Specialist plant hire
Tel: 01483 202733
Fax: 01483 202510
Website: www.land-water.co.uk
E-mail: enquiries@land-water.co.uk

Landline Ltd
1 Bluebridge Industrial Estate
Halstead
Essex CO9 2EX
Pond and lake installation
Tel: 01787 476699
Fax: 01787 472507
Website: www.landline.co.uk
E-mail: sales@landline.co.uk

Lappset UK Ltd
Lappset House
Henson Way
Kettering
Northants NN16 8PX
Play equipment
Tel: 01536 412612
Fax: 01536 521703
Website: www.lappset.com
E-mail: customerservicesuk@lappset.com

LDC Limited
The Charcoal House
Blacksmith Lane
Guildford
Surrey GU4 8NQ
Willow walling
Tel: 01483 573817
Fax: 01483 532078
Website: www.ldclandscape.co.uk
E-mail: info@ldc.co.uk

Leaky Pipe Systems Ltd
Frith Farm
Dean Street
East Farleigh
Maidstone
Kent ME15 0PR
Irrigation systems
Tel: 01622 746495
Fax: 01622 745118
Website: www.leakypipe.co.uk
E-mail: sales@leakypipe.co.uk

Lister Lutyens Co Ltd
6 Alder Close
Eastbourne
East Sussex BN23 6QF
Street furniture
Tel: 01323 431177
Fax: 01323 639314
Website: www.listerteak.com
E-mail: sales@listerteak.com

Lorenz von Ehren
Maldfeldstrasse 4
D-21077 Hamburg
Germany
Topiary
Tel: 0049 40 761080
Fax: 0049 40 761081
Website: www.lve.de
E-mail: sales@lve.de

Maccaferri Ltd
7400 The Quorum
Oxford Business Park North
Garsington Road
Oxford OX4 2JZ
Gabions
Tel: 01865 770555
Fax: 01865 774550
Website: www.maccaferri.co.uk
E-mail: oxford@maccaferri.co.uk

Marshalls
Landscape House
Premier Way
Lowfields Business Park
Elland HX5 9HT
Hard landscape materials and street furniture
Tel: 01422 312000
Fax: 01422 330185
Website: www.marshalls.co.uk

Maxit Building Products Ltd
Heath Business Park
Runcorn
Cheshire WA7 4QX
Expanded clay aggregate
Tel: 01928 565656
Website: www.maxit-uk.co.uk/2367
E- mail: sales@maxit-uk.co.uk

McArthur Group Ltd
Foundry Lane
Bristol BS5 7UE
Security fencing
Tel: 0117 943 0500
Fax: 0117 943 0577
Website: www.mcarthur-group.com
E-mail: marketing@mcarthur-group.com

Melcourt Industries Ltd
Boldridge Brake
Long Newnton
Tetbury
Gloucestershire GL8 8RT
Mulch and compost
Tel: 01666 502711
Fax: 01666 504398
Website: www.melcourt.co.uk
E-mail: mail@melcourt.co.uk

Milton Pipes Ltd
Cooks Lane
Milton Regis
Sittingbourne
Kent ME10 2QF
Soakaway rings
Tel: 01795 425191
Fax: 01795 478232
Website: www.miltonpipes.com
E-mail: sales@miltonpipes.com

Neptune Outdoor Furniture Ltd
Thompsons Lane
Marwell
Winchester
Hampshire SO21 1JH
Street furniture
Tel: 01962 777799
Fax: 01962 777723
Website: www.nofl.co.uk
E-mail: info@nofl.co.uk

Norris and Gardiner Ltd
Lime Croft Road
Knaphill
Woking
Surrey GU21 2TH
Grounds maintenance
Tel: 01483 289111
Fax: 01483 289112
E-mail: rich@norg.co.uk

Oakover Nurseries Ltd
Maidstone Road
Hothfield
Ashford
Kent TN26 1AR
Native plant nurseries
Tel: 01233 713016
Fax: 01233 713510
Website: www.oakovernurseries.co.uk
E-mail: enquiries@oakovernurseries.co.uk

Orchard Street Furniture Ltd
Whistler House
51 The Green North
Warborough
Oxfordshire OX10 7DW
Street furniture
Tel: 01491 642123
Fax: 01491 642126
Website: www.orchardstreet.co.uk
E-mail: sales@orchardstreet.co.uk

PBA Solutions
Bryn
Rake Road
Liss
Hants GU33 7HB
Japanese Knotweed consultants
Tel: 01730 893460
Website: www.pba-solutions.com
E-mail: info@pba-solutions.com

Platipus Anchors Ltd
Kingsfield Business Centre
Philanthropic Road
Redhill
Surrey RH1 4DP
Tree anchors
Tel: 01737 762300
Fax: 01737 773395
Website: www.platipus-anchors.com
E-mail: info@platipus-anchors.com

Recycled Materials Ltd
PO Box 519
Surbiton
Surrey KT6 4YL
Disposal contractors
Tel: 020 8390 7010
Fax: 020 8390 7020
Website: www.recycledmaterialsltd.co.uk
E-mail: mail@recycledmaterialsltd.co.uk

Rigby Taylor Ltd
The Riverway Estate
Portsmouth Road
Peasmarsh
Guildford
Surrey GU3 1LZ
Horticultural supply
Tel: 01483 446900
Fax: 01483 534058
Website: www.rigbytaylor.com
E-mail: sales@rigbytaylor.com

RIW Ltd
Arc House
Terrace Road South
Binfield
Bracknell
Berkshire RG42 4PZ
Waterproofing products
Tel: 01344 397777
Fax: 01344 862010
Website: www.riw.co.uk
E-mail: enquiries@riw.co.uk

Road Equipment Ltd
28-34 Feltham Road
Ashford
Middlesex TW15 1DL
Plant hire
Tel: 01784 256565
Fax: 01784 240398
E-mail: roadequipment@aol.com

Rolawn Ltd
York Road
Elvington
York YO41 4XR
Industrial turf
Tel: 01904 608661
Fax: 01904 608272
Website: www.rolawn.co.uk
E-mail: info@rolawn.co.uk

RTS Ltd
UK Sales
Daisy Dene
Inglewhite Road
Goosnargh
Preston PR3 2EB
Permaloc edging
Tel: 01772 780234
Fax: 01772 780234
Website: www.rtslimited.uk.com
E-mail: rtslimited@hotmail.co.uk

Scotts UK Ltd
Paper Mill Lane
Bramford
Ipswich
Suffolk IP8 4BZ
Fertilizers and chemicals
Tel: 01473 830492
Fax: 01473 830046
Website: www.scottsprofessional.co.uk
E-mail: prof.sales@scotts.com

Sleeper Supplies Ltd
PO Box 1377
Kirk Sandall
Doncaster DN3 1XT
Sleeper supplier
Tel: 0845 230 8866
Fax: 0845 230 8877
Website: www.sleeper-supplies.co.uk
E-mail: sales@sleeper-supplies.co.uk

SMP Playgrounds Ltd
Ten Acre Lane
Thorpe
Egham
Surrey TW20 8RJ
Playground equipment
Tel: 01784 489100
Fax: 01784 431079
Website: www.smp.co.uk
E-mail: sales@smp.co.uk

Southern Conveyors
Unit 2, Denton Slipways Site
Wharf Road
Gravesend DAl2 2RU
Conveyors
Tel: 01474 564145
Fax: 01474 568036

Spadeoak Construction Co Ltd
Town Lane
Wooburn Green
High Wycombe
Bucks HP10 0PD
Macadam contractors
Tel: 01628 529421
Fax: 01628 810509
Website: www.spadeoak.co.uk
E-mail: email@spadeoak.co.uk

Steelway Brickhouse
Brickhouse Lane
West Bromwich
West Midlands B70 0DY
Access covers
Tel: 01215 214500
Fax: 01215 214551
Website: www.steelway.co.uk
E-mail: sales@steelwaybrickhose.co.uk

Steelway Fensecure
Queensgate Works
Bilston Road
Wolverhampton
West Midlands WV2 2NJ
Fencing
Tel: 01902 490919
Fax: 01902 4909296
Website: www.steelway.co.uk
E-mail: sales@steelway.co.uk

SteinTec UK Ltd
PO Box 381
Grays
Essex RM17 94B
Specialized mortars
Tel: 08707 555448
Fax: 08707 500218
Website: www.steintec.co.uk
E-mail: info@steintec.co.uk

Sugg Lighting Ltd
Sussex Manor Business Park
Gatwick Road
Crawley
West Sussex RH10 9GD
Lighting
Tel: 01293 540111
Fax: 01293 540114
Website: www.sugglighting.co.uk
E-mail: sales@sugglighting.co.uk

Swan Plant Ltd
New Barn Farm
Forest Road
Huncote
Leicestershire LE9 3LC
Landscape maintenance plant hire
Tel: 01455 888000
Fax: 01455 888009
Website: www.swan-services.co.uk
E-mail: info@swan-services.co.uk

Targetti Poulsen UK Ltd
Unit C 44 Barwell Business Park
Leatherhead Road
Chessington KT9 2NY
Outdoor lighting
Tel: 020 8397 4400
Fax: 020 8397 4455
Website: www.targetti.com
Or: www.louispoulsen.com
E-mail: nak-uk@lpmail.com

Tarmac Precast Concrete (RCC)
Barholm Road
Tallington
Stamford
Lincolnshire PE9 4RL
Precast concrete retaining units
Tel: 01778 381000
Fax: 01778 348041
Website: www.tarmac.co.uk
E-mail: enquiries@tarmac.co.uk

Tarmac Southern Ltd
Bell House Pit
Warren Lane
Colchester CO3 5NH
Ready mixed concrete
Tel: 020 8965 1864
Fax: 020 8965 1863

Tensar International
Cunningham Court
Shadsworth Business Park
Blackburn
Lancashire BB2 4PJ
Erosion control, soil stabilization
Tel: 01254 262431
Fax: 01254 266868
Website: www.tensar-international.com
E-mail: sales@tensar.co.uk

Terranova Lifting Ltd
Terranova House
Bennett Road
Reading
Berks RG2 0QX
Crane hire
Tel: 0118 931 2345
Fax: 0118 931 4114
Website: www.terranova-lifting.co.uk
E-mail: cranes@terranovagroup.co.uk

Thompson Landscapes
610 Goffs Lane
Goffs Oak
Herts EN7 5EP
Swimming pools
Tel: 01707 873444
Fax: 01707 873447
Website: www.thompsonlandscapes.com
E-mail: info@thompsonlandscapes.com

Townscape Products Ltd
Fulwood Road South
SuttoninAshfield
Nottinghamshire NG17 2JZ
Hard landscaping
Tel: 01623 513355
Fax: 01623 440267
Website: www.townscape-products.co.uk
E-mail: sales@townscape-products.co.uk

Tubex Ltd
Aberaman Park
Aberaman
South Wales CF44 6DA
Plant protection, tree guards
Tel: 01685 888000
Fax: 01685 888001
Website: www.tubex.com
E-mail: plantcare@tubex.com

Wade International Ltd
Third Avenue
Halstead
Essex CO9 2SX
Stainless steel drainage
Tel: 01787 475151
Fax: 01787 475579
Website: www.wadedrainage.co.uk
E-mail: sales@wade.eu

Wavin Plastics Ltd
Parsonage Way
Chippenham
Wiltshire SN15 5PN
Drainage products
Tel: 01249 766600
Fax: 01249 443286
Website: www.wavin.co.uk
E-mail: info@wavin.co.uk

Wicksteed Leisure Ltd
Digby Street
Kettering
Northamptonshire NN16 8YJ
Play equipment
Tel: 01536 517028
Fax: 01536 410633
Website: www.wicksteed.co.uk
E-mail: sales@wicksteed.co.uk

Woodscape Ltd
Church Works
Church Street
Church
Lancashire BB5 4JT
Street furniture
Tel: 01254 383322
Fax: 01254 381166
Website: www.woodscape.co.uk
E-mail: sales@woodscape.co.uk

Wybone Ltd
Mason Way
Platts Common Industrial Estate
Barnsley
South Yorkshire S74 9TF
Street furniture
Tel: 01226 744010
Fax: 01226 350105
Website: www.wybone.co.uk
E-mail: sales@wybone.co.uk

Yeoman Aggregates Ltd
Stone Terminal
Horn Lane
Acton
London W3 9EH
Aggregates
Tel: 020 8896 6820
Fax: 020 8896 6829
Website: www.foster-yeoman.co.uk
E-mail: sales@yeoman-aggregates.co.uk

How to Use This Book

INTRODUCTION

First-time users of *Spon's External Works and Landscape Price Book* and others who may not be familiar with the way in which prices are compiled may find it helpful to read this section before starting to calculate the costs of landscape works.

The cost of an item of landscape construction or planting is made up of many components:

- the cost of the product
- the labour and additional materials needed to carry out the job
- the cost of running the contractor's business

These are described more fully below.

IMPORTANT NOTES ON THE PROFIT ELEMENT OF RATES IN THIS BOOK

The rates shown in the Approximate Estimates and Measured Works sections of this book do not contain a profit element unless the rate has been provided by a subcontractor. Prices are shown at cost. This is the cost including the costs of company overhead required to perform this task or project. For reference please see the tables on pages 6–8.

Analysed Rates Versus Subcontractor Rates

As a general rule if a rate is shown as an analysed rate in the Measured Works section, i.e. it has figures shown in the columns other than the "Unit" and "Total Rate" column, it can be assumed that this has no profit or overhead element and that this calculation is shown as a direct labour/material supply/plant task being performed by a contractor.

On the other hand, if a rate is shown as a Total Rate only, this would normally be a subcontractor's rate and would contain the profit and overhead for the subcontractor.

The foregoing applies for the most part to the Approximate Estimates section, however in some items there may be an element of subcontractor rates within direct works build-ups.

As an example of this, to excavate, lay a base and place a macadam surface, a general landscape contractor would normally perform the earthworks and base installation. The macadam surfacing would be performed by a specialist subcontractor to the landscape contractor.

The Approximate Estimate for this item uses rates from the Measured Works section to combine the excavation, disposal, base material supply and installation with all associated plant. There is no profit included on these elements. The cost of the surfacing, however, as supplied to us in the course of our annual price enquiry from the macadam subcontractor, would include the subcontractor's profit element.

The landscape contractor would add this to his section of the work but would normally apply a mark up at a lower percentage to the subcontract element of the composite task. Users of this book should therefore allow for the profit element at the prevailing rates. Please see the worked examples below and the notes on overheads for further clarification.

INTRODUCTORY NOTES ON PRICING CALCULATIONS USED IN THIS BOOK

There are two pricing sections to this book:

Major Works - Average contract value £100,000.00 - £300,000.00

Minor Works - Average contract value £10,000.00 - £70,000.00

Typical Project Profiles Used in this Book

	Major Works	**Minor Works**
Contract value	100,000.00 - £300,000.00	£10,000.00 - £70,000.00
Labour rate (see page 7)	£18.25 per hour	£20.50 per hour
Labour rate for maintenance contracts	£15.25 per hour	-
Number of site staff	30	6–9
Project area	6000 m^2	1200 m^2
Project location	Outer London	Outer London
Project components	50% hard landscape 50% soft landscape and planting	20% hard landscape 80% soft landscape and planting
Access to works areas	Very good	Very good
Contract	Main contract	Main contract
Delivery of materials	Full loads	Part loads

As explained in more detail later the prices are generally based on wage rates and material costs current at spring 2009. They do not allow for preliminary items, which are dealt with in the Preliminaries section of this book, or for any Value Added Tax (VAT) which may be payable.

Adjustments should be made to standard rates for time, location, local conditions, site constraints and any other factors likely to affect the costs of a specific scheme.

Term contracts for general maintenance of large areas should be executed at rates somewhat lower than those given in this section.

There is now a facility available to readers that enables a comparison to be made between the level of prices given and those for projects carried out in regions other than outer London; this is dealt with on page 20.

Units of Measurement

The units of measurement have been varied to suit the type of work and care should be taken when using any prices to ascertain the basis of measurement adopted.

The prices per unit of area for executing various mechanical operations are for work in the following areas and under the following conditions:

Prices per m^2 relate to areas not exceeding 100 m^2
 (any plan configuration)

Prices per 100 m^2 relate to areas exceeding 100 m^2 but not exceeding 1/4 ha
 (generally clear areas but with some subdivision)

Prices per ha   relate to areas over 1/4 ha
 (clear areas suitable for the use of tractors and tractoroperated equipment)

The prices per unit area for executing various operations by hand generally vary in direct proportion to the change in unit area.

Approximate Estimates

These are combined Measured Work prices which give an approximate cost for a complete section of landscape work. For example, the construction of a car park comprises excavation, levelling, roadbase, surfacing and marking parking bays. Each of these jobs is priced separately in the Measured Works section, but a comprehensive price is given in the Approximate Estimates section, which is intended to provide a quick guide to the cost of the job. It will be seen that the more items that go to make up an approximate estimate, the more possibilities there are for variations in the PC prices and the user should ensure that any PC price included in the estimate corresponds to his specification.

Worked example

In many instances a modular format has been use in order to enable readers to build up a rate for a required task. The following table describes the trench excavation, pipe laying and backfilling operations as contained in section R12 of this book.

Pipe laying 100 m

	£
Excavate for drain 150 x 450 mm; excavated material to spoil heaps on site by machine	428.92
Lay flexible plastic pipe 100 mm wide	149.25
Backfilling with gravel rejects, blinding with sand and topping with 150 mm topsoil	332.67
Total	**910.84**

The figures given in Spon's External Works and Landscape Price Book are intended for general guidance, and if a significant variation appears in any one of the cost groups, the Price for Measured Work should be recalculated.

Measured Works

Prime Cost: Commonly known as the 'PC'. Prime Cost is the actual price of the material item being addressed such as paving, shrubs, bollards or turves, as sold by the supplier. Prime Cost is given 'per square metre', 'per 100 bags' or 'each' according to the way the supplier sells his product. In researching the material prices for the book we requested that the suppliers price for full loads of their product delivered to a site close to the M25 in London. Spon's rates do not include VAT. Some companies may be able to obtain greater discounts on the list prices than those shown in the book. Prime Cost prices for those products and plants which have a wide cost range will be found under the heading of Market Prices in the main sections of this book, so that the user may select the product most closely related to his specification.

Materials: The PC material plus the additional materials required to fix the PC material. Every job needs materials for its completion besides the product bought from the supplier. Paving needs sand for bedding, expansion joint strips and cement pointing; fencing needs concrete for post setting and nails or bolts; tree planting needs manure or fertilizer in the pit, tree stakes, guards and ties. If these items were to be priced out separately, Spon's External Works and Landscape Price Book (and the Bill of Quantities) would be impossibly unwieldy, so they are put together under the heading of Materials.

Labour: This figure covers the cost of planting shrubs or trees, laying paving, erecting fencing etc. and is calculated on the wage rate (skilled or unskilled) and the time needed for the job. Extras such as highly skilled craft work, difficult access, intermittent working and the need for labourers to back up the craftsman all add to the cost. Large regular areas of planting or paving are cheaper to install than smaller intricate areas, since less labour time is wasted moving from one area to another.

Plant, consumable stores and services: This rather impressive heading covers all the work required to carry out the job which cannot be attributed exactly to any one item. It covers the use of machinery ranging from JCBs to shovels and compactors, fuel, static plant, water supply (which is metered on a construction site), electricity and rubbish disposal. The cost of transport to site is deemed to be included elsewhere and should be allowed for elsewhere or as a preliminary item. Hired plant is calculated on an average of 36 hours working time per week.

Subcontract rates: Where there is no analysis against an item, this is deemed to be a rate supplied by a subcontractor. In most cases these are specialist items where most external works contractors would not have the expertise or the equipment to carry out the task described. It should be assumed that subcontractor rates include for the subcontractor's profit. An example of this may be found for the tennis court rates in section Q26.

Labour Rates Used in this Edition

The rates for labour used in this edition have been based on surveys carried out on a cross section of external works contractors. These rates include for company overheads such as employee administration, transport insurance, and on costs such as National Insurance. Please see the calculation of these rates shown on page 6–8. The rates do not include for profit.

Overheads: An allowance for this is included in the labour rates which are described above. The general overheads of the contract such as insurance, site huts, security, temporary roads and the statutory health and welfare of the labour force are not directly assignable to each item, so they are distributed as a percentage on each, or as a separate preliminary cost item. The contractor's and subcontractor's profits are not included in this group of costs. Site overheads, which will vary from contract to contract according to the difficulties of the site, labour shortages, inclement weather or involvement with other contractors, have not been taken into account in the build up of these rates, while overhead (or profit) may have to take into account losses on other jobs and the cost to the contractor of unsuccessful tendering.

ADJUSTMENT AND VARIATION OF THE RATES IN THIS BOOK

It will be appreciated that a variation in any one item in any group will affect the final Measured Works price. Any cost variation must be weighed against the total cost of the contract. A small variation in Prime Cost where the items are ordered in thousands may have more effect on the total cost than a large variation on a few items. A change in design that necessitates the use of earth moving equipment which must be brought to the site for that one job will cause a dramatic rise in the contract cost. Similarly, a small saving on multiple items will provide a useful reserve to cover unforeseen extras.

Worked examples

A variation in the Prime Cost of an item can arise from a specific quotation.

For example:
The PC of 900 x 600 x 50 mm precast concrete flags is given as £3.78 each which equates to £7.00/m^2, and to which the costs of bedding and pointing are added to give a total material cost of £10.79/m^2.
Further costs for labour and mechanical plant give a resultant price of £18.09/m^2.

If a quotation of £8.00/m^2 was received from a supplier of the flags the resultant price would be calculated as £18.09 less the original cost (£7.00) plus the revised cost (£8.00) to give £19.09/m^2.

A variation will also occur if, for example, the specification changes from 50 light standard trees to 50 extra heavy standard trees. In this case the Prime Cost will increase due to having to buy extra heavy standard trees instead of light standard stock.

Original price: The prime cost of an Acer platanoides 8 -10 cm bare root tree is given as £11.25, to which the cost of handling and planting, etc. is added to give an overall cost of £18.55. Further costs for labour and mechanical plant for mechanical excavation of a 600 x 600 x 600 mm pit (£3.32), a single tree stake (£5.12), importing "Topgrow" compost in 75 litre bags (£1.80), backfilling with imported topsoil (£6.73) and disposing of the excavated material (£3.42), give a resultant price of £38.94. Therefore the total cost of 50 light standard trees is £1947.00

Revised price: The prime cost of an Acer platanoides root ball 14 -16 cm is given as £57.00, however, taking into account, extra labour for handling and planting (£21.90), a larger tree pit 900 x 900 x 600 mm (£7.45), double staking (£7.08), more topsoil (£15.16), compost (3.97) and the disposal (£7,71), this cost gives a total unit price of £120.27. This is significantly more than the light standard version. Therefore the total cost of 50 extra heavy trees is £6013.50 and the additional cost is £4066.50.

This example of the effects of changing the tree size illustrates that caution is needed when revising tender prices, as merely altering the prime cost of an item will not accurately reflect the total cost of the revised item.

HOW THIS BOOK IS UPDATED EACH YEAR

The basis for this book is a database of Material, Labour, Plant and Subcontractor resources each with its own area of the database.

Material, Plant and Subcontractor Resources

Each year the suppliers of each material, plant or subcontract item are approached and asked to update their prices to those that will prevail in September of that year.

These resource prices are individually updated in the database. Each resource is then linked to one or many tasks. The tasks in the task library section of the database is automatically updated by changes in the resource library. A quantity of the resource is calculated against the task. The calculation is generally performed once and the links remain in the database.

On occasions where new information or method or technology are discovered or suggested, these calculations would be revisited. A further source of information is simple time and production observations made during the course of the last year.

Labour Resource Update

Most tasks, except those shown as subcontractor rates (see above), employ an element of labour. The Data Department at Davis Langdon conducts ongoing research into the costs of labour in various parts of the country. Tasks or entire sections are then re-examined and recalculated. Comments on the rates published in this book are welcomed and may be submitted to the contact addresses shown in the Preface.

www.davislangdon.com

SPON'S PRICEBOOKS 2010

Spon's Architects' and Builders' Price Book 2010

Davis Langdon

The most detailed, professionally relevant source of UK construction price information currently available anywhere.

In the 2010 edition the overheads and profits have been revised sharply downwards in line with actual levels. Preliminaries have also been dropped, on a lower cost base. And labour rates have been adjusted to reflect today's economic conditions. It provides about 150 new Measured Works items, including: additional ACO drainage systems and an overhauled Kalzip roofing section.

Hb & electronic package*: 1000pp approx.
978-0-415-55256-1: £145
electronic package only
978-0-203-87245-1: £135
(inc. sales tax where appropriate)

Spon's External Works and Landscape Price Book 2010

Davis Langdon

The only comprehensive source of information for detailed external works and landscape costs.

The 2010 edition gives a number of new Measured Works items: sustainable urban drainage systems (SUDS); sustainable pavings, and bases for sustainable pavings; step construction (in brick, in-situ concrete, slab, timber and stone); pergolas; and planting transplants (preparations and planting). And the definitive listing of common landscape chemical applications has been revised to comply with recently introduced legislation.

Hb & electronic package*: 424pp approx.
978-0-415-55260-8: £115
electronic package only
978-0-203-87243-7: £105
(inc. sales tax where appropriate)

Spon's Mechanical and Electrical Services Price Book 2010

Davis Langdon Mott Green Wall

The only comprehensive and best annual services engineering price book available for the UK.

As well as an extensive overhaul of prices in line with a volatile market, updates to the 2010 edition include a new section on the Façade System in engineering features which explains the relationship between the building envelope and the provision of cooling.

Hb & electronic package*: 640pp approx
978-0-415-55263-9: £145
electronic package only
978-0-203-87242-0: £135
(inc. sales tax where appropriate)

Spon's Civil Engineering and Highway Works Price Book 2010

Davis Langdon

Materials prices are still rising for most products, subcontract prices are volatile, tender prices falling… The 24th edition, in its easy to read format, incorporates a general review throughout. Assumptions on overheads and profits have been revised downwards; preliminaries have been cut, on a lower cost base; and labour rates have been adjusted to reflect today's economic climate.

Hb & electronic package:* 800pp approx.
978-0-415-55258-5: £155
electronic package only
978-0-203-87244-4: £145
(inc. sales tax where appropriate)

---

* Receive our ebook free when you order any hard copy Spon 2010 Price Book, with free estimating software to help you produce tender documents, customise data, perform word searches and simple calculations.
Or buy just the ebook, with free estimating software.
Visit www.pricebooks.co.uk

To Order: Tel: +44 (0) 1235 400524 **Fax:** +44 (0) 1235 400525
or Post: Taylor and Francis Customer Services, Bookpoint Ltd, Unit T1, 200 Milton Park, Abingdon, Oxon, OX14 4TA UK
Email: book.orders@tandf.co.uk

For a complete listing of all our titles visit: www.tandf.co.uk

Spon Press
an imprint of Taylor & Francis

Get more from your Spon Price Libraries

✶FAST ESTIMATE

Everything you need to price work quickly and accurately using your own rates together with your Spon Price libraries.

- ✓ **Resource Library** - Keep all your costs up-to-date and in one place
- ✓ **Organiser** – To Do Lists. Link Contacts to jobs and resources
- ✓ **Price Lists** - Access to the latest product prices from your key suppliers
- ✓ **Integrated Spon Price Libraries** – Access to a range of industry standard Price libraries
- ✓ **Custom Rate Library** - Maintain your own item library (with full build ups)
- ✓ **Composite Items** - Save time by selecting a single item covering multiple tasks
- ✓ **BoQ Preparation** - Deliver professional documents to clients and colleagues
- ✓ **Project Comparison** – Compare similar jobs by cost; profit; type. Area rate comparisons
- ✓ **Live Data views** – Watch the effect of your changes in tables or easy to view charts
- ✓ **Import From MS Excel®** – Fast, accurate data import from an industry standard format
- ✓ **Output to Microsoft® Applications** - Excel®, Project®
- ✓ **Fully Customisable** – Each user can have their own screen layouts (saved)

Fast Estimate shares the same clear pricing processes as Spon Estimating, but with more pricing tools to give you outstanding control over your tender documents.

Download your FREE trial copy:

www.estek.co.uk

Or call 01764 655331

PART 1

General

This part of the book contains the following sections:

Common Arrangement of Work Sections, page 3
Labour Rates Used in this Edition, page 5
Computation of Labour Rates Used in this Edition, page 6
Computation of the Cost of Materials, page 11
Computation of Mechanical Plant Costs, page 12
Landfill Tax, page 15

SPON'S PRICEBOOKS 2010

Spon's Architects' and Builders' Price Book 2010
Davis Langdon

The most detailed, professionally relevant source of UK construction price information currently available anywhere.

In the 2010 edition the overheads and profits have been revised sharply downwards in line with actual levels. Preliminaries have also been dropped, on a lower cost base. And labour rates have been adjusted to reflect today's economic conditions. It provides about 150 new Measured Works items, including: additional ACO drainage systems and an overhauled Kalzip roofing section.

Hb & electronic package*: 1000pp approx.
978-0-415-55256-1: **£145**
electronic package only
978-0-203-87245-1: **£135**
(inc. sales tax where appropriate)

Spon's External Works and Landscape Price Book 2010
Davis Langdon

The only comprehensive source of information for detailed external works and landscape costs.

The 2010 edition gives a number of new Measured Works items: sustainable urban drainage systems (SUDS); sustainable pavings, and bases for sustainable pavings; step construction (in brick, in-situ concrete, slab, timber and stone); pergolas; and planting transplants (preparations and planting). And the definitive listing of common landscape chemical applications has been revised to comply with recently introduced legislation.

Hb & electronic package*: 424pp approx.
978-0-415-55260-8: **£115**
electronic package only
978-0-203-87243-7: **£105**
(inc. sales tax where appropriate)

Spon's Mechanical and Electrical Services Price Book 2010
Davis Langdon Mott Green Wall

The only comprehensive and best annual services engineering price book available for the UK.

As well as an extensive overhaul of prices in line with a volatile market, updates to the 2010 edition include a new section on the Façade System in engineering features which explains the relationship between the building envelope and the provision of cooling.

Hb & electronic package*: 640pp approx
978-0-415-55263-9: **£145**
electronic package only
978-0-203-87242-0: **£135**
(inc. sales tax where appropriate)

Spon's Civil Engineering and Highway Works Price Book 2010
Davis Langdon

Materials prices are still rising for most products, subcontract prices are volatile, tender prices falling... The 24th edition, in its easy to read format, incorporates a general review throughout. Assumptions on overheads and profits have been revised downwards; preliminaries have been cut, on a lower cost base; and labour rates have been adjusted to reflect today's economic climate.

Hb & electronic package:* 800pp approx.
978-0-415-55258-5: **£155**
electronic package only
978-0-203-87244-4: **£145**
(inc. sales tax where appropriate)

---

* Receive our ebook free when you order any hard copy Spon 2010 Price Book, with free estimating software to help you produce tender documents, customise data, perform word searches and simple calculations.
Or buy just the ebook, with free estimating software.
Visit www.pricebooks.co.uk

To Order: Tel: +44 (0) 1235 400524 **Fax:** +44 (0) 1235 400525
or Post: Taylor and Francis Customer Services, Bookpoint Ltd, Unit T1, 200 Milton Park, Abingdon, Oxon, OX14 4TA UK
Email: book.orders@tandf.co.uk
For a complete listing of all our titles visit: www.tandf.co.uk

Spon Press
an imprint of Taylor & Francis

Common Arrangement of Work Sections

The main work sections relevant to landscape work and their grouping:

A Preliminaries/general conditions

- A10 Project particulars
- A11 Tender and contract documents
- A12 The site/existing buildings
- A13 Description of the work
- A20 The contract/subcontract
- A30 Employer's requirements: Tendering/Subletting/Supply
- A31 Employer's requirements: Provision, content and use of documents
- A32 Employer's requirements: Management of the Works
- A33 Employer's requirements: Quality standards/control
- A34 Employer's requirements: Security/Safety/Protection
- A35 Employer's requirements: Specific limitations on method/sequence/timing/use of site
- A36 Employer's requirements: facilities/temporary works/services
- A37 Employer's requirements: operation/maintenance of the finished building
- A40 Contractor's general cost items: Management and staff
- A41 Contractor's general cost items: Site accommodation
- A42 Contractor's general cost items: Services and facilities
- A43 Contractor's general cost items: Mechanical plant
- A44 Contractor's general cost items: Temporary works
- A50 Works/Products by/on behalf of the Employer
- A51 Nominated subcontractors
- A52 Nominated suppliers
- A53 Work by statutory authorities/undertakers
- A54 Provisional work
- A55 Dayworks
- A60 Preliminaries/General conditions for demolition contract
- A61 Preliminaries/General conditions for investigation/survey contract
- A62 Preliminaries/General conditions for piling/embedded retaining wall contract
- A63 Preliminaries/General conditions for landscape contract
- A70 General specification requirements for work package

B Complete buildings/structures/units

- B10 Prefabricated buildings/structures
- B11 Prefabricated building units

C Existing site/buildings/services

- C10 Site survey
- C11 Ground investigation
- C12 Underground services survey
- C13 Building fabric survey
- C20 Demolition
- C50 Repairing/Renovating/Conserving metal

D Groundwork

- D11 Soil stabilization
- D20 Excavating and filling
- D41 Crib walls/gabions/reinforced earth

E In situ concrete/large precast concrete

- E05 In situ concrete construction generally
- E10 Mixing/Casting/Curing in situ concrete
- E20 Formwork for in situ concrete
- E30 Reinforcement for in situ concrete

F Masonry

- F10 Brick/Block walling
- F20 Natural stone rubble walling
- F21 Natural stone ashlar walling/dressings
- F22 Cast stone ashlar walling/dressings

	F30	Accessories/Sundry items for brick/block/stone walling
	F31	Precast concrete sills/lintels/copings/features

G Structural/carcassing metal/timber

	G31	Prefabricated timber unit decking

H Cladding/covering

	H51	Natural stone slab cladding/features
	H52	Cast stone slab cladding/features

J Waterproofing

	J10	Specialist waterproof rendering
	J20	Mastic asphalt tanking/damp proofing
	J21	Mastic asphalt roofing/insulation/finishes
	J22	Proprietary roof decking with asphalt finish
	J30	Liquid applied tanking/damp proofing
	J31	Liquid applied waterproof roof coatings
	J40	Flexible sheet tanking/damp proofing
	J44	Sheet linings for pools/lakes/waterways
	J50	Green Roof systems

M Surface finishes

	M10	Cement:sand/Concrete screeds
	M20	Plastered/Rendered/Roughcast coatings
	M40	Stone/Concrete/Quarry/Ceramic tiling/Mosaic
	M60	Painting/clear finishing

P Building fabric sundries

	P30	Trenches/Pipeways/Pits for buried engineering services

Q Paving/planting/fencing/site furniture

	Q10	Kerbs/Edgings/Channels/paving accessories
	Q20	Granular sub-bases to roads/pavings
	Q21	In situ concrete roads/pavings
	Q22	Coated macadam/Asphalt roads/pavings
	Q23	Gravel/Hoggin/Woodchip roads/pavings
	Q24	Interlocking brick/block roads/pavings
	Q25	Slab/Brick/Sett/Cobble pavings
	Q26	Special surfacings/pavings for sport/general amenity
	Q30	Seeding/Turfing
	Q31	Planting
	Q32	Planting in Special environments
	Q35	Landscape maintenance
	Q40	Fencing
	Q50	Site/Street furniture/equipment

R Disposal systems

	R12	Drainage below ground
	R13	Land drainage

S Piped supply systems

	S10	Cold water
	S14	Irrigation
	S15	Fountains/Water features

V Electrical supply/power/lighting systems

	V41	Street/Area/Flood lighting

Labour Rates Used in This Edition

Based on surveys carried out on a cross section of external works contractors, the following rates for labour have been used in this edition.
These rates include for company overheads such as employee administration, transport insurance, and on-costs such as National Insurance.
The rates do not include for profit.

The rates for all labour used in this edition are as follows:

Major Works

General contracting	£18.25/hour
Maintenance contracting	£15.25/hour

Minor Works

General contracting	£20.50/hour

Computation of Labour Rates Used in This Edition

Different organizations will have varying views on rates and costs, which will in any event be affected by the type of job, availability of labour and the extent to which mechanical plant can be used. However this information should assist the reader to:

(1)   compare the prices to those used in his own organization
(2)   calculate the effect of changes in wage rates or prices of materials
(3)   calculate prices for work similar to the examples given

From 30 June 2008 basic weekly rates of pay for craft and general operatives are £401.70 and £302.25 respectively; to these rates have been added allowances for the items below in accordance with the recommended procedure of the Chartered Institute of Building in its "Code of Estimating Practice". The resultant hourly rates are £13.48 and £10.08 for craft operatives and general operatives respectively.

The items for which allowances have been made are:

- Lost time
- Construction Industry Training Board Levy
- Holidays with pay
- Accidental injury, retirement and death benefits scheme
- Sick pay
- National Insurance
- Severance pay and sundry costs
- Employer's liability and third party insurance

The tables that follow illustrate how the above hourly rates have been calculated. Productive time has been based on a total of 1801.8 hours worked per year for daywork calculations above and for 1953.54 hours worked per year for the all-in labour rates (including 5 hours per week average overtime) for the all in labour rates below.

How the Labour Rate has been Calculated

A survey of typical landscape/external works companies indicates that they are currently paying above average wages for multi-skilled operatives regardless of specialist or supervisory capability. In our current overhaul of labour constants and costs we have departed from our previous policy of labour rates based on national awards and used instead a gross rate per hour for all rate calculations. Estimators can readily adjust the rate if they feel it is inappropriate for their work.

The productive labour of any organization must return their salary plus the cost of the administration which supports the labour force.

The labour rate used in this edition is calculated as follows:

Team size

A three man team with annual salaries as shown and allowances for National Insurance, uniforms, site tools and a company vehicle, returns a basic labour rate of £14.79 per hour.

Basic labour rate calculation for Spon's 2010 Major and Minor Works

Standard Working Hours per year (2009/10) =1802
Less allowance for sick leave = -39
Less allowance for down time 2% = -35
Actual Working Hours per year (2009/10) = 1728

	Number of	Labour team	NI	Nett Cost	Uniforms	Site Tools	TOTAL 3 Man Team
Foreman	1	22000.00	2816.00	24816.00	200.00	150.00	
Craftsman	1	19500.00	2496.00	21996.00	200.00	150.00	
Labourer	1	15500.00	1984.00	17484.00	200.00	150.00	
	3.00	57000.00	7296.00	64296.00	600.00	450.00	£65346.00
Overtime Hours							
nominal hours	40.00	19.10	2.27	854.79			
/annum	40.00	16.93	2.01	757.66			
at 1.5 times normal standard rate	40.00	13.45	1.60	602.24			
Total staff in Team	3.00			66510.69	600.00	450.00	67560.69
Vehicle Costs Inclusive of Fuel Insurances etc	Working Days	£ /Day					
	240.00	38.00					£ 9,120.00
						TOTAL	£ 76,680.00

The basic average labour rate excluding overhead costs (below) is

$$\frac{£76680.00}{\text{(Total staff in team)} \times \text{(Working hours per year)}} = £14.79$$

Overhead costs

Add to the above basic rate the company overhead costs for a small company as per the table below. These costs are absorbed by the number of working men multiplied by the number of working hours supplied in the table below. This then generates an hourly overhead rate which is added to the Nett cost rate above.

Illustrative overhead for small company employing 12 – 30 landscape operatives

Cost Centre	Number of	Cost	Total
MD: Salary only: excludes profits; include vehicle	1	43000.00	45000.00
Senior contracts managers; include vehicle	1	37500.00	40000.00
Other contracts managers; include vehicle	1	28000.00	30000.00
Secretary	1.00	17000.00	17000.00
Book keeper	1.00	19000.00	19000.00
Rental	12.00	1000.00	12000.00
Insurances	1.00	5000.00	5000.00
Telephone and mobiles	12.00	250.00	3000.00
Office equipment	1.00	1000.00	1000.00
Stationary	12.00	50.00	600.00
Advertising	12.00	200.00	2400.00
Other vehicles not allocated to contract teams	1.00	9000.00	9000.00
Other consultants	1.00	6000.00	6000.00
Accountancy	1.00	2500.00	2500.00
Lights heating water	12.00	200.00	2400.00
Other expenses	1.00	8000.00	10000.00
TOTAL OFFICE OVERHEAD			**198400.00**

The final labour rate used is then generated as follows:

$$\text{Nett labour rate} + \frac{\text{Total Office Overhead}}{(\text{Working hours per year} \times \text{Labour resources employed})}$$

Total nr of site staff	Admin cost per hour	Total rate per man hour
12	9.57	24.36
15	7.65	22.45
18	6.38	21.17
20	5.74	20.53
25	4.59	19.38
30	3.83	18.62
40	2.87	17.66

Hourly labour rates used for Spon's External Works and Landscape 2010 (Major Works)

General contracting	£18.25
Maintenance contracting	£15.25

Minor Works

- The minor works labour calculation is based on the above but the administration cost of the organization is £96,840.00 per annum
- The owner of the business does not carry out site works and is assumed to be paid a salary of £38,000.00 per annum; this figure constitutes part of the £94,840.00 shown
- There are between 6 and 15 site operatives

Illustrative overhead for small company employing 6 – 15 landscape operatives

Cost Centre	Number of	Cost	Total
MD: Salary only: excludes profits; include vehicle	1	38000.00	40000.00
Secretary	1.00	17000.00	17000.00
Book keeper	1.00	8000.00	8000.00
Rental	12.00	300.00	3600.00
Insurances	1.00	4000.00	4000.00
Telephone and mobiles	12.00	200.00	2400.00
Office equipment	1.00	1000.00	1000.00
Stationary	12.00	20.00	240.00
Advertising	12.00	200.00	2400.00
Other vehicles not allocated to contract teams	1.00	4000.00	4000.00
Other consultants	1.00	3000.00	3000.00
Accountancy	1.00	2000.00	2000.00
Lights, heating, water	12.00	100.00	1200.00
Other expenses	1.00	8000.00	10000.00
TOTAL OFFICE OVERHEAD			94840.00

The final labour rate used is then generated as follows:

$$\text{Nett labour rate} + \frac{\text{Total Office Overhead}}{\text{(Working hours per year resources employed)}}$$

Total nr of site staff	Admin cost per hour	Total rate per man hour
3	18.06	32.96
6	9.03	23.82
9	6.02	20.81
12	4.52	19.31
15	3.61	18.40

Hourly labour rates used for Spon's External Works and Landscape 2010 (Minor Works)

General contracting	£20.50

Calculation of Annual Hours Worked

Normal hours worked

Normal hours worked per week		39 hours
Normal hours worked per year	39 hrs x 52 wks	2028 hours

Less holidays

Annual holidays	39 hrs x 4.2 wks	-163.8 hours
Public holidays	39 hrs x 1.6 wks	-62.4 hours

Less sick

	39 hrs x 1 wk	-39 hours

Less time lost for inclement weather @ 2%

	-35.26 hours

Actual hours worked per annum — **1727.54 hours**

Computation of the Cost of Materials

Percentages of default waste are placed against material resources within the supplier database. These range from 2.5% (for bricks) to 20% (for topsoil). An allowance for the cost of unloading, stacking etc. should be added to the cost of materials.

The following are typical hours of labour for unloading and stacking some of the more common building materials.

Material	Unit	Labourer hours
Cement	tonne	0.67
Lime	tonne	0.67
Common bricks	1000	1.70
Light engineering bricks	1000	2.00
Heavy engineering bricks	1000	2.40

Computation of Mechanical Plant Costs

Plant used within resource calculations in this book have been used according to the following principles:

1. Plant which is non-pedestrian is fuelled by the plant hire company.
2. Operators of excavators above the size of 5 tonnes are provided by the plant hire company and no labour calculation is allowed for these operators within the rate.
3. Three to five tonne excavators and dumpers are operated by the landscape contractor and additional labour is shown within the calculation of the rate.
4. Small plant is operated by the landscape contractor.
5. No allowance for delivery or collection has been made within the rates shown.
6. Downtime is shown against each plant type either for non-productive time or mechanical down time.

The following tables list the plant resources used in this year's book.

Prices shown reflect 32 working hours per week.

Plant Supplied by Road Equipment Ltd, Tel: 01784 256565

Description - Major Works	Unit	Supply Quantity	Rate Used £	Down Time %
Excavator 360 Tracked 21 tonne *fuelled and operated*	hour	1	41.25	20
Excavator 360 Tracked 5 tonne *fuelled only*	hour	1	10.93	20
Excavator 360 Tracked 7 tonne *fuelled only*	hour	1	11.88	20
Mini Excavator JCB 803 Rubber Tracks Self Drive *fuelled only*	hour	1	7.19	20
Mini Excavator JCB 803 Steel Tracks Self Drive *fuelled only*	hour	1	7.19	20
JCB 3Cx 4x4 Sitemaster *fuelled and operated*	hour	1	30.00	20
JCB 3Cx 4x4 Sitemaster + Breaker *fuelled and operated*	hour	1	31.25	20
Dumper 3 tonne, Thwaites Self Drive *fuelled only*	hour	1	3.13	20
Dumper 5 tonne, Thwaites Self Drive *fuelled only*	hour	1	3.75	20
Dumper 6 tonne, Thwaites Self Drive *fuelled only*	hour	1	4.06	20
Forklift Telehandler *fuelled only*	hour	1	14.58	40

Computation of Mechanical Plant Costs

Plant Supplied by Road Equipment Ltd, Tel: 01784 256565

Description - Minor Works	Unit	Supply quantity	Rate used £	Down time %
Excavator 360 Tracked 21 tonne *fuelled and operated*	hour	1	41.25	30
Excavator 360 Tracked 5 tonne *fuelled only*	hour	1	10.93	30
Excavator 360 Tracked 7 tonne *fuelled only*	hour	1	11.88	30
Mini Excavator JCB 803 Rubber Tracks Self Drive *fuelled only*	hour	1	7.19	30
Mini Excavator JCB 803 Steel Tracks Self Drive *fuelled only*	hour	1	7.19	30
JCB 3Cx 4x4 Sitemaster *fuelled and operated*	hour	1	30.00	30
JCB 3Cx 4x4 Sitemaster + Breaker *fuelled and operated*	hour	1	31.25	30
Dumper 3 tonne, Thwaites Self Drive *fuelled only*	hour	1	3.13	30
Dumper 5 tonne, Thwaites Self Drive *fuelled only*	hour	1	3.75	30
Dumper 6 tonne, Thwaites Self Drive *fuelled only*	hour	1	4.06	30
Forklift Telehandler *fuelled only*	hour	1	14.58	40

Plant Supplied by HSS Hire

Description - Major Works	Unit	Supply quantity	Supply cost (£)	Rate used (£)	Down time (%)
Post Hole Borer, 2 man, weekly rate (rate based on 25 productive hours)	hour	40	95.00	3.80	37.5
Petrol Poker Vibrator + 50 mm head (rate based on 25 productive hours)	hour	40	48.00	1.92	37.5
Vibrating Plate Compactor (rate based on 25 productive hours)	hour	40	69.00	2.76	37.5
Petrol Masonry Saw Bench, 350 mm (rate based on 25 productive hours)	hour	40	124.00	4.96	37.5
Diamond Blade Consumable, 450 mm	mm	1	46.00	46.00	-
Alloy Access Tower, 5.2 m	days	5	159.00	31.80	-
Oxy-acetylene Cutting Kit (rate based on 16 productive hours)	hour	40	102.00	6.38	80
Heavy-duty Petrol Breaker, 5 hrs/day (rate based on 25 productive hours)	hour	40	153.00	6.12	37.5
Cultivator, 110 kg (rate based on 25 productive hours)	hour	40	199.00	7.96	37.5
Rotavator + Tractor + Operator (rate based on 32 productive hours)	hour	40	394.50	12.33	-
Vibrating roller, 136 kg/10.1 Kn, 4 hrs/day (rate based on 20 productive hours)	hour	40 150.0	134.00	6.70	25

Computation of Mechanical Plant Costs

Description - Minor Works	Unit	Supply quantity	Supply cost £	Rate used £	Down time %
Post Hole Borer, 2 man, weekly rate (rate based on 20 productive hours)	hour	40	95.00	4.75	50.00
Petrol Poker Vibrator + 50 mm head (rate based on 20 productive hours)	hour	40	48.00	2.40	50.00
Vibrating Plate Compactor (rate based on 20 productive hours)	hour	40	69.00	3.45	50.00
Petrol Masonry Saw Bench, 350 mm (rate based on 20 productive hours)	Hour	40	124.00	6.20	50.00
Diamond Blade Consumable, 450 mm	mm	1	46.00	46.00	-
Alloy Access Tower, 5.2 m	days	5	159.00	31.80	-
Oxy-acetylene Cutting Kit (rate based on 16 productive hours)	hour	40	102.00	6.38	80
Heavy-duty Petrol Breaker, 5 hrs/day (rate based on 25 productive hours)	hour	40	153.00	6.12	50.00
Cultivator, 110 kg (rate based on 25 productive hours)	hour	40	199.00	7.96	50.00
Rotavator + Tractor + Operator (rate based on 32 productive hours)	hour	40	394.50	12.33	-
Vibrating Roller, 136 kg/10.1 Kn, 4 hrs/day (rate based on 20 productive hours)	hour	40	134.00	6.70	50

Landfill Tax

Inert Waste Liable at the Lower Rate

Group	Description of material	Conditions	Notes
1	Rocks and soils	Naturally occurring	Includes clay, sand, gravel, sandstone, limestone, crushed stone, china clay, construction stone, stone from the demolition of buildings or structures, slate, topsoil, peat, silt and dredgings. Glass includes fritted enamel, but excludes glass fibre and glass reinforced plastics.
2	Ceramic or concrete materials		Ceramics includes bricks, bricks and mortar, tiles, clayware, pottery, china and refractories. Concrete includes reinforced concrete, concrete blocks, breeze blocks and aircrete blocks, but excludes concrete plant washings.
3	Minerals	Processed or prepared, not used	Moulding sands excludes sands containing organic binders. Clays includes moulding clays and clay absorbents, including Fuller's earth and bentonite. Man-made mineral fibres includes glass fibres, but excludes glass-reinforced plastic and asbestos. Silica, mica and mineral abrasives.
4	Furnace slags		Vitrified wastes and residues from thermal processing of minerals where, in either case, the residue is both fused and insoluble slag from waste incineration.
5	Ash		Comprises only bottom ash and fly ash from wood, coal or waste combustion but excludes fly ash from municipal, clinical and hazardous waste incinerators and sewage sludge incinerators.
6	Low activity inorganic compound		Comprises only titanium dioxide, calcium carbonate, magnesium carbonate, magnesium oxide, magnesium hydroxide, iron oxide, ferric hydroxide, aluminium oxide, aluminium hydroxide and zirconium dioxide.
7	Calcium sulphate	Disposed of either at a site not licensed to take putrescible waste or in a containment cell which takes only calcium sulphate	Includes gypsum and calcium sulphate-based plasters, but excludes plasterboard.
8	Calcium hydroxide and brine	Deposited in brine cavity	
9	Water	Containing other qualifying material in suspension	

Volume to Weight Conversion Factors

Waste category	Typical waste types	Cubic metres to tonnes – multiply by:	Cubic yards to tonnes – multiply by:
Inactive or inert waste	Largely water insoluble and non or very slowly biodegradable: e.g. sand, subsoil, concrete, bricks, mineral fibres, fibreglass etc.	1.5	1.15

Notes:

The Landfill tax came into operation on 1 October 1996. It is levied on operators of licensed landfill sites at the following rates with effect from 1 April 2009:

£2.50 per tonne (frozen until 2010/11) - inactive or inert wastes
Included are soil, stones, brick, plain and reinforced concrete, plaster and glass.

£40.00 per tonne (increasing by £8.00 per tonne per year until at least 2013) - all other taxable wastes
Included are timber, paint and other organic wastes generally found in demolition work, builders skips, etc.

Mixtures containing wastes not classified as inactive or inert will not qualify for the lower rate of tax unless the amount of non-qualifying material is small and there is no potential for pollution. Water can be ignored and the weight discounted.

Calculating the Weight of Waste

There are two options:

- If licensed sites have a weighbridge, tax will be levied on the actual weight of waste.
- If licensed sites do not have a weighbridge, tax will be levied on the permitted weight of the lorry based on an alternative method of calculation based on volume to weight factors for various categories of waste.

For further information contact the National Advisory Service, Telephone: 0845 010 9000

ESSENTIAL READING FROM TAYLOR AND FRANCIS

Environmental Noise Barriers
2nd edition

Benz Kotzen and Colin English

Environmental Noise Barriers is a unique one-stop reference for practitioners, whether acoustical engineers, landscape architects or manufacturers, and for highways departments in local and central authorities. This extensively revised new edition is updated in line with UK and EU legislation and international provision of barriers.

Contents
Introduction;
Defining the Need for Barriers; Acoustic and Visual Impact Assessment – Environmental Impact Assessment Techniques and Mitigation;
Acoustic Performance of Barriers; Barrier Morphology and Design;
Types of Barrier and Barrier Materials; Planting and Bio-barriers;
Noise Amelioration and Roads; Tunnels;
Airports;
Environmental Noise Barriers in Other Locations – Industrial Uses

April 2009 246 x 189mm: 280pp
Hb: 978-0-415-43708-0 **£90.00**

To Order: Tel: +44 (0) 1235 400524 **Fax:** +44 (0) 1235 400525
or Post: Taylor and Francis Customer Services,
Bookpoint Ltd, Unit T1, 200 Milton Park, Abingdon, Oxon, OX14 4TA UK
Email: book.orders@tandf.co.uk

For a complete listing of all our titles visit:
www.tandf.co.uk

Taylor & Francis
Taylor & Francis Group

ESSENTIAL READING FROM TAYLOR AND FRANCIS

Spon's Irish Construction Price Book

Third Edition

Franklin + Andrews

This new edition of *Spon's Irish Construction Price Book*, edited by Franklin + Andrews, is the only complete and up-to-date source of cost data for this important market.

- All the materials costs, labour rates, labour constants and cost per square metre are based on current conditions in Ireland
- Structured according to the new Agreed Rules of Measurement (second edition)
- 30 pages of Approximate Estimating Rates for quick pricing

This price book is an essential aid to profitable contracting for all those operating in Ireland's booming construction industry.

Franklin + Andrews, Construction Economists, have offices in 100 countries and in-depth experience and expertise in all sectors of the construction industry.

April 2008: 246x174 mm: 510 pages
Hb: 978-0-415-45637-1: **£145.00**

To Order: Tel: +44 (0) 1235 400524 **Fax:** +44 (0) 1235 400525
or Post: Taylor and Francis Customer Services,
Bookpoint Ltd, Unit T1, 200 Milton Park, Abingdon, Oxon, OX14 4TA UK
Email: book.orders@tandf.co.uk

For a complete listing of all our titles visit:
www.tandf.co.uk

Taylor & Francis
Taylor & Francis Group

PART 2

Approximate Estimates – Major Works

This part of the book contains the following sections:

Cost Indices, page 18
Regional Variations, page 20
Approximate Estimates, page 21

Cost Indices

The purpose of this section is to show changes in the cost of carrying out landscape work (hard surfacing and planting) since 1990. It is important to distinguish between costs and tender prices: the following table reflects the change in cost to contractors but does not necessarily reflect changes in tender prices. In addition to changes in labour and material costs, which are reflected in the indices given below, tender prices are also affected by factors such as the degree of competition at the time of tender and in the particular area where the work is to be carried out, the availability of labour and materials, and the general economic situation. This can mean that in a period when work is scarce, tender prices may fall despite the fact that costs are rising, and when there is plenty of work available, tender prices may increase at a faster rate than costs.

The Constructed Cost Index

A Constructed Cost Index based on PSA Price Adjustment Formulae for Construction Contracts (Series 2). Cost indices for the various trades employed in a building contract are published monthly by HMSO and are reproduced in the technical press.

The indices comprise 49 Building Work indices plus seven "Appendices" and other specialist indices. The Building Work indices are compiled by monitoring the cost of labour and materials for each category and applying a weighting to these to calculate a single index.

Although the PSA indices are prepared for use with priceadjustment formulae for calculating reimbursement of increased costs during the course of a contract, they also present a time series of cost indices for the main components of landscaping projects. They can therefore be used as the basis of an index for landscaping costs.

The method used here is to construct a composite index by allocating weightings to the indices representing the usual work categories found in a landscaping contract, the weightings being established from an analysis of actual projects. These weightings totalled 100 in 1976 and the composite index is calculated by applying the appropriate weightings to the appropriate PSA indices on a monthly basis, which is then compiled into a quarterly index and rebased to 1976 = 100.

Constructed Landscaping (Hard Surfacing and Planting) Cost Index

Based on approximately 50% soft landscaping area and 50% hard external works.
1976 = 100

Year	First Quarter	Second Quarter	Third Quarter	Fourth Quarter	Annual Average
1990	330	334	353	357	344
1991	356	356	364	64	360
1992	364	365	376	378	371
1993	380	381	383	384	382
1994	387	388	394	397	391
1995	399	404	413	413	407
1996	417	419	428	430	424
1997	430	432	435	442	435
1998	442	445	466	466	455
1999	466	468	488	490	478
2000	493	496	513	514	504
2001	512	514	530	530	522
2002	531	540	573	574	555
2003	577	582	603	602	591
2004	602	610	639	639	623
2005	641	648	684	683	664
2006	688	692	710	710	710
2007	714	718	737	742	728
2008	752	771	803	794	780
2009	790*				

* Provisional

This index is updated every quarter in Spon's Price Book Update. The updating service is available, free of charge, to all purchasers of Spon's Price Books (see the back page of this book for details).

Regional Variations

Prices in Spon's External Works and Landscape Price Book are based upon conditions prevailing for a competitive tender in the outer London area. For the benefit of readers, this edition includes regional variation adjustment factors which can be used for an assessment of price levels in other regions.

Special further adjustment may be necessary when considering city centre or very isolated locations.

Region	Adjustment Factor
Outer London	1.00
Inner London	1.06
South East (excluding Greater London)	0.93
South West	0.90
East Midlands	0.85
West Midlands	0.86
East Anglia	0.87
Yorkshire & Humberside	0.87
North West	0.85
Northern	0.88
Scotland	0.93
Wales	0.90
Northern Ireland	0.72

The following example illustrates the adjustment of prices for regions other than Outer London, by use of regional variation adjustment factors.

A. Value of items priced using Spon's External Works and Landscape Price Book for Outer London £100,000

B. Adjustment to value of A. to reflect Northern Region price level 100,000 x 0.88 £ 88,000

Approximate Estimates

Prices in this section are based upon the Prices for Measured Works, but allow for incidentals which would normally be measured separately in a Bill of Quantities. They do not include for Preliminaries which are priced elsewhere in this book.

Items shown as subcontract or specialist rates would normally include the specialist's overhead and profit. All other items which could fall within the scope of works of general landscape and external works contractors would not include profit.

Based on current commercial rates, profits of 15% to 35% may be added to these rates to indicate the likely "with profit" values of the tasks below. The variation quoted above is dependent on the sector in which the works are taking place - domestic, public or commercial.

PRELIMINARIES

Item Excluding site overheads and profit	Unit	Total rate £
CONTRACT ADMINISTRATION AND MANAGEMENT		
Prepare tender bid for external works or landscape project; measured works contract; inclusive of bid preparation, provisional program and method statements		
Measured works bid; measured bills provided by the employer; project value		
£30,000.00	nr	366.75
£50,000.00	nr	637.25
£100,000.00	nr	1310.50
£200,000.00	nr	1943.00
£500,000.00	nr	2398.00
£1,000,000.00	nr	3238.00
Lump sum bid; scope document, drawings and specification provided by the employer; project value		
£30,000.00	nr	666.75
£50,000.00	nr	937.25
£100,000.00	nr	1790.50
£200,000.00	nr	2903.00
£500,000.00	nr	3838.00
£1,000,000.00	nr	5638.00
Prepare and maintain Health and Safety file; prepare risk and COSHH assessments, method statements, works programmes before the start of works on site; project value		
£35,000.00	nr	244.00
£75,000.00	nr	366.00
£100,000.00	nr	488.00
£200,000.00 - £500,000.00	nr	1220.00
Contract management		
Site management by site surveyor carrying out supervision and administration duties only; inclusive of on costs		
full time management	week	1200.00
managing one other contract	week	600.00
Parking		
Parking expenses where vehicles do not park on the site area; per vehicle		
metropolitan area; city centre	week	200.00
Metropolitan area; outer areas	week	160.00
suburban restricted parking areas	week	40.00
Congestion charging		
London only	week	40.00
SITE SETUP AND SITE ESTABLISHMENT		
Site fencing; supply and erect temporary protective fencing and remove at completion of works		
Cleft chestnut paling		
1.20 m high 75 mm larch posts at 3 m centres	100 m	678.00
1.50 m high 75 mm larch posts at 3 m centres	100 m	875.00
"Heras" fencing		
2 week hire period	100 m	700.00
4 week hire period	100 m	956.00
8 week hire period	100 m	1468.00
12 week hire period	100 m	1980.00
16 week hire period	100 m	2492.00
24 week hire period	100 m	3516.00

PRELIMINARIES

Item Excluding site overheads and profit	Unit	Total rate £
Site compounds Establish site administration area on reduced compacted Type 1 base; 150 mm thick; allow for initial setup of site office and welfare facilities for site manager and maximum 8 site operatives		
temporary base or hard standing; 20 x 10 m	200 m^2	1809.60
To hardstanding area above; erect office and welfare accommodation; allow for furniture, telephone connection and power; restrict access using temporary safety fencing to the perimeter of the site administration area Establishment; delivery and collection costs only		
site office and chemical toilet; 20 m of site fencing	nr	766.40
site office and chemical toilet; 40 m of site fencing	nr	830.80
site office and chemical toilet; 60 m of site fencing	nr	895.20
Weekly hire rates of site compound equipment		
site office 2.4 x 3.6 m; chemical toilet; 20 m of site fencing	week	97.35
site office 4.8 x 2.4 m; chemical toilet; 20 m of site fencing; armoured store 2.4 x 3.6 m	week	109.35
site office 2.4 x 3.6 m; chemical toilet; 40 m of site fencing	week	122.95
site office 2.4 x 3.6 m; chemical toilet; 60 m of site fencing	week	148.55
Removal of site compound Remove base; fill with topsoil and reinstate to turf on completion of site works	200 m^2	1568.26
Surveying and setting out Note: In all instances a quotation should be obtained from a surveyor. The following cost projections are based on the costs of a surveyor using EDM (electronic distance measuring equipment) where an average of 400 different points on site are recorded in a day. The surveyed points are drawn in a CAD application and the assumption used that there is one day of drawing up for each day of on-site surveying. Survey existing site using laser survey equipment and produce plans of existing site conditions		
site 1,000 to 10,000 m^2 with sparse detail and features; boundary and level information only; average five surveying stations	nr	750.00
site up to 1,000 m^2 with dense detail such as buildings, paths, walls and existing vegetation; average five survey stations	nr	1500.00
site up to 4000 m^2 with dense detail such as buildings, paths, walls and existing vegetation; average 10 survey stations	nr	2250.00

DEMOLITION AND SITE CLEARANCE

Item Excluding site overheads and profit	Unit	Total rate £
DEMOLITION AND SITE CLEARANCE		
Clear existing scrub vegetation including shrubs and hedges and burn on site; spread ash on specified area		
By machine		
light scrub and grasses	100 m^2	18.00
undergrowth brambles and heavy weed growth	100 m^2	39.00
small shrubs	100 m^2	39.00
As above but remove vegetation to licensed tip	100 m^2	32.00
By hand	100 m^2	81.00
As above but remove vegetation to licensed tip	100 m^2	53.00
Fell trees on site; grub up roots; all by machine; remove debris to licensed tip		
trees 600 mm girth	each	213.00
trees 1.5 - 3.00 m girth	each	606.00
trees over 3.00 m girth	each	1384.00
Demolish existing surfaces		
Break up plain concrete slab; remove to licensed tip		
150 mm thick	m^2	8.00
200 mm thick	m^2	9.70
300 mm thick	m^2	15.00
Break up plain concrete slab and preserve arisings for hardcore; break down to maximum size of 200 x 200 mm and transport to location; maximum distance 50 m		
150 mm thick	m^2	6.00
200 mm thick	m^2	7.00
300 mm thick	m^2	11.00
Break up reinforced concrete slab and remove to licensed tip		
150 mm thick	m^2	12.00
200 mm thick	m^2	15.00
300 mm thick	m^2	23.00
Break out existing surface and associated 150 mm thick granular base load to 20 tonne muck away vehicle for removal off site		
macadam surface; 70 mm thick	m^2	6.00
macadam surface; 150 mm thick	m^2	8.10
block paving; 50 mm thick	m^2	6.00
block paving; 80 mm thick	m^2	6.50
Demolish existing free standing walls; grub out foundations; remove arisings to tip		
Brick wall; 112 mm thick		
300 mm high	m	9.80
500 mm high	m	15.00
Brick wall; 225 mm thick		
300 mm high	m	13.00
500 mm high	m	16.00
1.00 m high	m	26.00
1.20 m high	m	28.00
1.50 m high	m	34.00
1.80 m high	m	38.00
Site clearance - General		
Clear away light fencing and gates (chain link, chestnut paling, light boarded fence or similar) and remove to licensed tip	100 m	107.00
Demolish existing structures		
Timber sheds on concrete base 150 mm thick inclusive of disposal and backfilling with excavated material; plan area of building		
10 m^2	nr	182.64
15 m^2	nr	258.53

DEMOLITION AND SITE CLEARANCE

Item Excluding site overheads and profit	Unit	Total rate £
Building of cavity wall construction; insulated with tiled roof; inclusive of foundations; allow for disconnection of electrical water and other services; grub out and remove 10 m of electrical cable, water and waste pipes		
10 m^2	nr	**1580.95**
20 m^2	nr	**2103.72**

GROUNDWORK

Item Excluding site overheads and profit	Unit	Total rate £
GROUNDWORK		
Cut and strip by machine turves 50 mm thick		
Load to dumper or palette by hand; stack on site not exceeding 100 m travel to stack	100 m²	50.00
As above but all by hand; including barrowing	100 m²	426.00
Prices for excavating and reducing to levels are for work in light or medium soils; multiplying factors for other soils are as follows		
clay	1.5	-
compact gravel	1.2	-
soft chalk	2.0	-
hard rock	3.0	-
Excavating topsoil for preservation		
Remove topsoil average depth 300 mm; deposit in spoil heaps not exceeding 100 m travel to heap; turn heap once during storage		
by machine	m³	11.00
by hand	m³	112.00
Excavate to reduce levels; remove spoil to dump not exceeding 100 m travel; all by machine		
Average 0.25 m deep	m²	2.00
Average 0.30 m deep	m²	3.00
Average 1.0 m deep; using 21 tonne 360 tracked excavator	m²	5.00
Average 1.0 m deep; using JCB	m²	7.00
Excavate to reduce levels; remove spoil to dump not exceeding 100 m travel; all by hand		
Average 0.10 m deep	m²	9.90
Average 0.20 m deep	m²	20.00
Average 0.30 m deep	m²	29.60
Average 1.0 m deep	m²	98.60
Extra for carting spoil to licensed tip offsite (20 tonnes)	m³	16.00
Extra for carting spoil to licensed tip offsite (by skip; machine loaded)	m³	49.00
Extra for carting spoil to licensed tip offsite (by skip; hand loaded)	m³	110.00
Spread excavated material to levels in layers not exceeding 150 mm; using scraper blade or bucket		
Average thickness 100 mm	m²	1.00
Average thickness 100 mm but with imported topsoil	m²	4.00
Average thickness 200 mm	m²	1.00
Average thickness 200 mm but with imported topsoil	m²	7.00
Average thickness 250 mm	m²	1.00
Average thickness 250 mm but with imported topsoil	m²	9.00
Extra for work to banks exceeding 30° slope	30%	-
Rip subsoil using approved subsoiling machine to a depth of 600 mm below topsoil at 600 mm centres; all tree roots and debris over 150 x 150 mm to be removed; cultivate ground where shown on drawings to depths as shown		
100 mm	100 m²	8.00
200 mm	100 m²	8.00
300 mm	100 m²	9.00
400 mm	100 m²	9.00

Approximate Estimates – Major Works 27

GROUNDWORK

Item Excluding site overheads and profit	Unit	Total rate £
Form landform from imported material; mounds to falls and grades to receive turf or seeding treatments; material dumped by 20 tonne lorry to location; volume 100 m³ spread over a 200 m² area; thickness varies from 150 to 500 mm Imported recycled topsoil		
by large excavator	100 m³	2197.12
by 5 tonne excavator	100 m³	2386.12
by 3 tonne excavator	100 m³	2507.12
TRENCHES		
Excavate trenches for foundations; trenches 225 mm deeper than specified foundation thickness; pour plain concrete foundations GEN 1 10 N/mm and to thickness as described; disposal of excavated material off site; dimensions of concrete foundations		
250 mm wide x 250 mm deep	m	12.00
300 mm wide x 300 mm deep	m	14.00
400 mm wide x 600 mm deep	m	35.00
600 mm wide x 400 mm deep	m	36.00
Excavate trenches for foundations; trenches 225 mm deeper than specified foundation thickness; pour plain concrete foundations GEN 1 10 N/mm and to thickness as described; disposal of excavated material off site; dimensions of concrete foundations		
250 mm wide x 250 mm deep	m³	211.00
300 mm wide x 300 mm deep	m³	154.00
600 mm wide x 400 mm deep	m³	149.00
400 mm wide x 600 mm deep	m³	136.00
Excavate trenches for services; grade bottoms of excavations to required falls; remove 20% of excavated material off site Trenches 300 mm wide		
600 mm deep	m	5.00
750 mm deep	m	6.42
900 mm deep	m	7.66
1.00 m deep	m	8.45
Ha-ha Excavate ditch 1200 mm deep x 900 mm wide at bottom battered one side 45° slope; excavate for foundation to wall and place GEN 1 concrete foundation 150 x 500 mm; construct one and a half brick wall (brick PC £300.00/1000) battered 10° from vertical; laid in cement:lime:sand (1:1:6) mortar in English garden wall bond; precast concrete coping weathered and throated set 150 mm above ground on high side; rake bottom and sides of ditch and seed with low maintenance grass at 35 g/m²		
wall 600 mm high	m	248.00
wall 900 mm high	m	307.00
wall 1200 mm high	m	380.00
Excavate ditch 1200 mm deep x 900 mm wide at bottom battered one side 45° slope; place deer or cattle fence 1.20 m high at centre of excavated trench; rake bottom and sides of ditch and seed with low maintenance grass at 35 g/m²		
fence 1200 mm high	m	19.00

GROUND STABILIZATION

Item Excluding site overheads and profit	Unit	Total rate £
GROUND STABILIZATION		
Excavate and grade banks to grade to receive stabilization treatments below; removal of excavated material not included		
By 21 tonne excavator	m^3	0.28
By 7 tonne excavator	m^3	2.01
By 5 tonne excavator	m^3	2.17
By 3 tonne excavator	m^3	3.38
Excavate sloped bank to vertical to receive retaining or stabilization treatment priced below; allow 1.00 m working space at top of bank; backfill working space and remove balance of arisings off site		
Bank height 1.00 m; excavation by 5 tonne excavator		
10°	m	75.00
30°	m	27.00
45°	m	17.00
60°	m	20.00
Bank height 1.00 m; excavation by 5 tonne excavator; arisings moved to stockpile on site		
10°	m	19.00
30°	m	10.00
45°	m	7.00
60°	m	6.00
Bank height 2.00 m; excavation by 5 tonne excavator		
10°	m	288.00
30°	m	96.00
45°	m	60.00
60°	m	39.00
Bank height 2.00 m; excavation by 5 tonne excavator but arisings moved to stockpile on site		
10°	m	63.00
30°	m	27.00
45°	m	21.00
60°	m	17.00
Bank height 3.00 m; excavation by 21 tonne excavator		
20°	m	285.00
30°	m	181.00
45°	m	102.00
60°	m	60.00
As above but arisings moved to stockpile on site		
20°	m	39.00
30°	m	27.00
45°	m	13.00
60°	m	8.00
Bank stabilization; erosion control mats		
Excavate fixing trench for erosion control mat 300 x 300 mm; backfill after placing mat selected below	m	6.00
To anchor trench above; place mat to slope to be retained; allow extra over to the slope for the area of matting required to the anchor trench		
Rake only bank to final grade; lay erosion control solution; sow with low maintenance grass at 35 g/m^2; spread imported topsoil 25 mm thick incorporating medium grade sedge peat at 3 kg/m^2 and fertilizer at 30 g/m^2; water lightly		
unseeded "Eromat Light"	m^2	3.00
seeded "Covamat Standard"	m^2	3.00
lay Greenfix biodegradable pre-seeded erosion control mat; fix with 6 x 300 mm steel pegs at 1.0 m centres	m^2	4.00
"Tensar" open textured erosion mat	m^2	5.00

GROUND STABILIZATION

Item Excluding site overheads and profit	Unit	Total rate £
Excavate trench and lay foundation concrete 1:3:6; 600 x 300 mm deep		
By machine	m	28.00
By hand	m	38.00
Concrete block retaining walls; Stepoc		
Excavate trench 750 mm deep and lay concrete foundation 600 mm wide x 600 mm deep; construct Forticrete precast hollow concrete block wall with 450 mm below ground laid all in accordance with manufacturer's instructions; fix reinforcing bar 12 mm as work proceeds; fill blocks with concrete 1:3:6 as work proceeds		
Walls 1.00 m high		
type 190; 400 x 225 x 190 mm	m	176.00
type 256; 400 x 225 x 256 mm	m	201.00
As above but 1.50 m high		
type 190; 400 x 225 x 190 mm	m	225.00
type 256; 400 x 225 x 256 mm	m	259.00
Walls 1.50 m high as above but with foundation 1.80 m wide x 600 mm deep		
type 190; 400 x 225 x 190 mm	m	274.00
type 256; 400 x 225 x 256 mm	m	311.00
Walls 1.80 m high		
type 190; 400 x 225 x 190 mm	m	297.00
type 256; 400 x 225 x 256 mm	m	336.00
On foundation measured above, supply and RCC precast concrete "L" shaped units, constructed all in accordance with manufacturer's instructions; backfill with approved excavated material compacted as the work proceeds		
1250 mm high x 1000 mm wide	m	202.00
2400 mm high x 1000 mm wide	m	331.00
3750 mm high x 1000 mm wide	m	628.00
Grass concrete; bank revetments; excluding bulk earthworks		
Bring bank to final grade by machine; lay regulating layer of Type 1; lay filter membrane; lay 100 mm drainage layer of broken stone or approved hardcore 28 - 10 mm size; blind with 25 mm sharp sand; lay grass concrete surface (price does not include edgings or toe beams); fill with approved topsoil and fertilizer at 35 g/m^2; seed with dwarf rye based grass seed mix		
Grasscrete in situ reinforced concrete surfacing GC 1; 100 mm thick	m^2	43.00
Grasscrete in situ reinforced concrete surfacing GC 2; 150 mm thick	m^2	50.00
Grassblock 103 mm open matrix blocks; 406 x 406 x 103 mm	m^2	43.00
Gabion walls		
Construct revetment of Maccaferri Ltd "Reno" mattress gabions laid on firm level ground; tightly packed with broken stone or concrete and securely wired; all in accordance with manufacturer's instructions		
gabions 6 x 2 x 0.17 m; one course	m^2	33.00
gabions 6 x 2 x 0.17 m; two courses	m^2	59.00

GROUND STABILIZATION

Item Excluding site overheads and profit	Unit	Total rate £
GROUND STABILIZATION - cont'd		
Timber log retaining walls		
Excavate trench 300 mm wide to one third of the finished height of the retaining walls below; lay 100 mm hardcore; fix machine rounded logs set in concrete 1:3:6; remove excavated material from site; fix geofabric to rear of timber logs; backfill with previously excavated material set aside in position; all works by machine		
100 mm diameter logs		
500 mm high (constructed from 1.80 m lengths)	m	14.00
1.20 mm high (constructed from 1.80 m lengths)	m	27.20
1.60 mm high (constructed from 2.40 m lengths)	m	35.90
2.00 mm high (constructed from 3.00 m lengths)	m	44.70
As above but 150 mm diameter logs		
1.20 mm high (constructed from 1.80 m lengths)	m	27.20
1.60 mm high (constructed from 2.40 m lengths)	m	35.90
2.00 mm high (constructed from 3.00 m lengths)	m	44.70
2.40 mm high (constructed from 3.60 m lengths)	m	53.50
Timber crib wall		
Excavate trench to receive foundation 300 mm deep; place plain concrete foundation 150 thick in 11.50 N/mm^2 concrete (sulphate-resisting cement); construct timber crib retaining wall and backfill with excavated spoil behind units		
timber crib wall system; average 4.2 m high	m	230.00
timber crib wall system; average 2.2 m high	m	190.00
Geogrid soil reinforcement (Note: measured per metre run at the top of the bank)		
Backfill excavated bank; lay Tensar geogrid; cover with excavated material as work proceeds		
2.00 m high bank; geogrid at 1.00 m vertical lifts		
10° slope	m	158.00
20° slope	m	86.00
30° slope	m	61.00
2.00 m high bank; geogrid at 0.50 m vertical lifts		
45° slope	m	57.00
60° slope	m	46.00
3.00 m high bank		
10° slope	m	158.00
20° slope	m	86.00
30° slope	m	61.00
45° slope	m	144.00

IN SITU CONCRETE

Item Excluding site overheads and profit	Unit	Total rate £
IN SITU CONCRETE		
Mix concrete on site; aggregates delivered in 20 tonne loads; deliver mixed concrete to location by mechanical dumper distance 25 m		
1:3:6	m³	82.30
1:2:4	m³	94.30
As above but ready mixed concrete		
10 N/mm²	m³	108.00
15 N/mm²	m³	114.00
Mix concrete on site; aggregates delivered in 20 tonne loads; deliver mixed concrete to location by barrow distance 25 m		
1:3:6	m³	125.00
1:2:4	m³	137.00
As above but aggregates delivered in 1 tonne bags		
1:3:6	m³	155.00
1:2:4	m³	167.00
As above but ready mixed concrete		
10 N/mm²	m³	150.00
15 N/mm²	m³	157.00
As above but concrete discharged directly from ready mix lorry to required location		
10 N/mm²	m³	110.00
15 N/mm²	m³	113.00
Excavate foundation trench mechanically; remove spoil offsite; lay 1:3:6 site mixed concrete foundations; distance from mixer 25 m; depth of trench to be 225 mm deeper than foundation to allow for three underground brick courses priced separately		
Foundation size		
200 mm deep x 400 mm wide	m	18.00
300 mm deep x 500 mm wide	m	26.00
400 mm deep x 400 mm wide	m	26.00
400 mm deep x 600 mm wide	m	38.00
600 mm deep x 600 mm wide	m	55.00
As above but hand excavation and disposal to spoil heap 25 m by barrow and off site by grab lorry		
200 mm deep x 400 mm wide	m	39.00
300 mm deep x 500 mm wide	m	63.00
400 mm deep x 400 mm wide	m	63.00
400 mm deep x 600 mm wide	m	71.00
600 mm deep x 600 mm wide	m	131.00
Excavate foundation trench mechanically; remove spoil offsite; lay ready mixed concrete GEN1 discharged directly from delivery lorry to location; depth of trench to be 225 mm deeper than foundation to allow for three underground brick courses priced separately; foundation size		
200 mm deep x 400 mm wide	m	17.00
300 mm deep x 500 mm wide	m	26.00
400 mm deep x 400 mm wide	m	25.00
400 mm deep x 600 mm wide	m	38.00
600 mm deep x 600 mm wide	m	54.00
Reinforced concrete wall to foundations above (site mixed concrete)		
Up to 1.00 m high x 200 mm thick	m²	91.20
Up to 1.00 m high x 300 mm thick	m²	91.20
Reinforced concrete wall to foundations above (ready mix concrete RC35)		
1.00 m high x 200 mm thick	m²	121.00
1.00 m high x 300 mm thick	m²	132.00

IN SITU CONCRETE

Item Excluding site overheads and profit	Unit	Total rate £
IN SITU CONCRETE - cont'd		
Grade ground to levels and falls; excavate to reduce levels for concrete slab; lay waterproof membrane, 100 mm hardcore and form concrete slab reinforced with A 142 mesh in 1:2:4 site mixed concrete to thickness; remove excavated material off site		
Concrete 1:2:4 site mixed		
100 mm thick	m²	**763.96**
150 mm thick	m²	**770.55**
250 mm thick	m²	**783.71**
300 mm thick	m²	**790.30**
Concrete ready mixed GEN 2		
100 mm thick	m²	**763.57**
150 mm thick	m²	**769.96**
250 mm thick	m²	**782.74**
300 mm thick	m²	**789.13**

BRICK/BLOCK WALLING

Item Excluding site overheads and profit	Unit	Total rate £
BRICK/BLOCK WALLING		
Excavate foundation trench 500 mm deep; remove spoil to dump off site; lay site mixed concrete foundations 1:3:6 350 x 150 mm thick; construct half brick wall with one brick piers at 2.0 m centres; laid in cement:lime:sand (1:1:6) mortar with flush joints; fair face one side; DPC two courses underground; engineering brick in cement:sand (1:3) mortar; coping of headers on end		
Wall 900 mm high above DPC		
in engineering brick (class B) - PC £239.00/1000	m	153.00
in sandfaced facings - PC £300.00/1000	m	177.00
in reclaimed bricks - PC £800.00/1000	m	210.00
Excavate foundation trench 400 mm deep; remove spoil to dump off site; lay GEN 1 concrete foundations 450 mm wide x 250 mm thick; construct one brick wall with one and a half brick piers at 3.0 m centres; all in English Garden Wall bond; laid in cement:lime:sand (1:1:6) mortar with flush joints, fair face one side; DPC two courses engineering brick in cement:sand (1:3) mortar; precast concrete coping 152 x 75 mm		
Wall 900 mm high above DPC		
in engineering brick (class B) - PC £239.00/1000	m	256.00
in sandfaced facings - PC £300.00/1000	m	269.00
in reclaimed bricks - PC £800.00/1000	m	394.00
Wall 1200 mm high above DPC		
in engineering brick (class B) - PC £239.00/1000	m	286.00
in sandfaced facings - PC £300.00/1000	m	301.00
in reclaimed bricks - PC £800.00/1000	m	393.00
Wall 1800 mm high above DPC		
in engineering brick (class B) - PC £239.00/1000	m	415.00
in sandfaced facings - PC £300.00/1000	m	438.00
in reclaimed bricks - PC £800.00/1000	m	576.00
Excavate foundation trench 450 mm deep; remove spoil to dump off site; lay GEN 1 concrete foundations 600 x 300 mm thick; construct one and a half brick wall with two thick brick piers at 3.0 m centres; all in English Garden Wall bond; laid in cement:lime:sand (1:1:6) mortar with flush joints; fair face one side; DPC two courses engineering brick in cement:sand (1:3) mortar; coping of headers on edge		
Wall 900 mm high above DPC		
in engineering brick (class B) - PC £239.00/1000	m	522.00
in sandfaced facings - PC £300.00/1000	m	276.00
Wall 1200 mm high above DPC		
in engineering brick (class B) - PC £239.00/1000	m	294.00
in sandfaced facings - PC £300.00/1000	m	308.00
Wall 1800 mm high above DPC		
in engineering brick (class B) - PC £239.00/1000	m	364.00
in sandfaced facings - PC £300.00/1000	m	470.00
in reclaimed bricks - PC £800.00/1000	m	901.00

BRICK/BLOCK WALLING

Item Excluding site overheads and profit	Unit	Total rate £
BRICK/BLOCK WALLING - cont'd		
Excavate foundation trench 450 mm deep; remove spoil to dump off site; lay GEN 1 concrete foundations 600 x 300 mm thick; construct wall of concrete block		
Solid blocks 7 N/mm^2		
100 mm thick	m^2	55.00
140 mm thick	m^2	61.00
100 mm blocks laid "on flat"		
215 mm thick	m^2	110.00
Hollow blocks filled with concrete		
215 mm thick	m^2	88.00
Hollow blocks but with steel bar cast into the foundation		
215 mm thick	m^2	110.00
Excavate foundation trench 450 mm deep; remove spoil to dump off site; lay GEN 1 concrete foundations 600 x 300 mm thick; construct wall of concrete block with brick face 112.5 mm thick to stretcher bond; place stainless steel ties at 4 nr/m^2 of wall face		
Solid blocks 7 N/mm^2 with stretcher bond brick face; reclaimed bricks PC £800.00/1000		
100 mm thick	m^2	170.00
140 mm thick	m^2	176.00

Approximate Estimates – Major Works 35

ROADS AND PAVINGS

Item Excluding site overheads and profit	Unit	Total rate £
ROADS AND PAVINGS		
BASES FOR PAVING		
Excavate ground and reduce levels to receive 38 mm thick slab and 25 mm mortar bed; dispose of excavated material off site		
Lay granular fill Type 1; Limestone aggregate 150 mm thick laid to falls and compacted		
all by machine	m²	11.00
all by hand except disposal by grab	m²	46.00
Lay granular fill Type 1; recycled material;150 mm thick laid to falls and compacted		
all by machine	m²	8.20
all by hand except disposal by grab	m²	44.00
Lay 1:2:4 concrete base 150 mm thick laid to falls		
all by machine	m²	22.00
all by hand except disposal by grab	m²	59.00
150 mm hardcore base with concrete base 150 mm deep		
concrete 1:2:4 site mixed	m²	29.00
concrete PAV 1 35 ready mixed	m²	28.00
Hardcore base 150 mm deep; concrete reinforced with A142 mesh		
site mixed concrete 1:2:4; 150 mm deep	m²	41.00
site mixed concrete 1:2:4; 250 mm deep	m²	57.00
concrete PAV 1 35 N/mm ready mixed; 150 mm deep	m²	33.00
concrete PAV 1 35 N/mm ready mixed; 250 mm deep	m²	49.00
KERBS AND EDGINGS		
Note: excavation is by machine unless otherwise mentioned		
Excavate trench and construct concrete foundation 300 mm wide x 150 mm deep; lay precast concrete kerb units bedded in semi-dry concrete; slump 35 mm maximum; haunching one side; disposal of arisings off site		
Kerbs laid straight		
125 mm high x 125 mm thick; bullnosed; type BN	m	24.00
125 x 255 mm; ref HB2; SP	m	27.00
150 x 305 mm; ref HB1	m	32.00
Excavate and construct concrete foundation 450 mm wide x 150 mm deep; lay precast concrete kerb units bedded in semi-dry concrete; slump 35 mm maximum; haunching one side; lay channel units bedded in 1:3 sand mortar; jointed in cement: sand mortar 1:3		
Kerbs; 125 mm high x 125 mm thick; bullnosed; type BN		
dished channel; 125 x 225 mm; Ref CS	m	44.00
square channel; 125 x 150 mm; Ref CS2	m	42.00
bullnosed channel; 305 x 150 mm; Ref CBN	m	52.00
Granite kerbs		
straight; 125 x 250 mm	m	47.00
curved; 125 x 250 mm	m	51.00
Brick channel; class B engineering bricks		
Excavate and construct concrete foundation 600 mm wide x 200 mm deep; lay channel to depths and falls bricks to be laid as headers along the channel; bedded in 1:3 lime:sand mortar bricks close jointed in cement:sand mortar 1:3		
3 courses wide	m	60.00
Precast concrete edging on concrete foundation 100 x 150 mm deep and haunching one side 1:2:4 including all necessary excavation disposal and formwork		
Rectangular, chamfered or bullnosed; to one side of straight path		
50 x 150 mm	m	20.00
50 x 200 mm	m	25.00

ROADS AND PAVINGS

Item Excluding site overheads and profit	Unit	Total rate £
KERBS AND EDGINGS - cont'd		
Precast concrete edging on concrete foundation 100 x 150 mm deep and haunching one side 1:2:4 including all necessary excavation disposal and formwork - cont'd		
50 x 250 mm	m	26.00
Rectangular, chamfered or bullnosed as above but to both sides of straight paths		
50 x 150 mm	m	43.00
50 x 200 mm	m	50.00
50 x 250 mm	m	51.00
Timber edgings; softwood		
Straight		
150 x 38 mm	m	7.00
150 x 50 mm	m	7.00
Curved; over 5 m radius		
150 x 38 mm	m	9.00
150 x 50 mm	m	9.00
Curved; 4 - 5 m radius		
150 x 38 mm	m	9.00
150 x 50 mm	m	10.00
Curved; 3 - 4 m radius		
150 x 38 mm	m	10.00
150 x 50 mm	m	11.00
Curved; 1 - 3 m radius		
150 x 38 mm	m	12.00
150 x 50 mm	m	13.00
Timber edgings; hardwood		
150 x 38 mm	m	12.00
150 x 50 mm	m	14.00
Brick or concrete block edge restraint; excavate for foundation; lay concrete 1:2:4 200 mm wide x 150 mm deep; on 50 mm thick sharp sand bed; lay blocks or bricks inclusive of haunching one side		
Blocks 200 x 100 x 60 mm; PC £7.63/m^2; butt jointed		
header course	m	15.00
stretcher course	m	12.00
Bricks 215 x 112.5 x 50 mm; PC £300.00/1000; with mortar joints		
header course	m	17.00
stretcher course	m	16.00
Sawn yorkstone edgings; excavate for foundations; lay concrete 1:2:4 150 mm deep x 33.3% wider than the edging; on 35 mm thick mortar bed; inclusive of haunching one side		
Yorkstone 50 mm thick		
100 mm wide x random lengths	m	23.00
100 x 100 mm	m	27.00
100 mm wide x 200 mm long	m	35.00
250 mm wide x random lengths	m	36.00
500 mm wide x random lengths	m	59.00
Granite edgings; excavate for footings; lay concrete 1:2:4 150 mm deep x 33.3% wider than the edging; on 35 mm thick mortar bed; inclusive of haunching one side		
Granite 50 mm thick		
100 mm wide x random lengths	m	23.00
100 x 100 mm	m	27.00
100 mm wide x 200 mm long	m	35.00
250 mm wide x random lengths	m	36.00
500 mm wide x random lengths	m	59.00

ROADS AND PAVINGS

Item Excluding site overheads and profit	Unit	Total rate £
ROADS		
Excavate weak points of excavation by hand and fill with well rammed Type 1 granular material		
100 mm thick	m^2	9.00
200 mm thick	m^2	12.00
Reinforced concrete roadbed 150 mm thick		
Excavate road bed and dispose excavated material off site; bring to grade; lay reinforcing mesh A142 lapped and joined; lay 150 mm thick in situ reinforced concrete roadbed 21 N/mm^2; with 15 impregnated fibreboard expansion joints and polysulphide based sealant at 50 m centres; on 150 mm hardcore blinded with ash or sand; kerbs 155 x 255 mm to both sides; including foundations haunched one side in 11.5 N/mm^2 concrete with all necessary formwork; falls, crossfalls and cambers not exceeding 15°		
4.90 m wide	m	206.00
6.10 m wide	m	278.00
7.32 m wide	m	293.00
Macadam roadway over 1000 m^2		
Excavate 350 mm for pathways or roadbed; bring to grade; lay 100 mm well-rolled hardcore; lay 150 mm Type 1 granular material; lay precast edgings kerbs 50 x 150 mm on both sides; including foundations haunched one side in 11.5 N/mm^2 concrete with all necessary formwork; machine lay surface of 90 mm macadam 60 mm base course and 30 mm wearing course; all disposal off site		
1.50 m wide	m	87.00
2.00 m wide	m	91.00
3.00 m wide	m	130.00
4.00 m wide	m	160.00
5.00 m wide	m	190.00
6.00 m wide	m	220.00
7.00 m wide	m	240.00
Macadam roadway 400 - 1000 m^2		
Excavate 350 mm for pathways or roadbed; bring to grade; lay 100 mm well-rolled hardcore; lay 150 mm Type 1 granular material; lay precast edgings kerbs 50 x 150 mm on both sides; including foundations haunched one side in 11.5 N/mm^2 concrete with all necessary formwork; machine lay surface of 90 mm macadam 60 mm base course and 30 mm wearing course; all disposal off site		
1.50 m wide	m	91.50
2.00 m wide	m	105.00
3.00 m wide	m	136.00
4.00 m wide	m	168.00
5.00 m wide	m	199.00
6.00 m wide	m	231.00
7.00 m wide	m	262.00

ROADS AND PAVINGS

Item Excluding site overheads and profit	Unit	Total rate £
CAR PARKS		
Excavate 350 mm for pathways or roadbed to receive surface 100 mm thick priced separately; bring to grade; lay 100 mm well-rolled hardcore; lay 150 mm Type 1 granular material; lay kerbs 125 x 255 mm on both sides; including foundations haunched one side in 11.5 N/mm² concrete with all necessary formwork		
Work to falls, crossfalls and cambers not exceeding 15°		
1.50 m wide	m	68.00
2.00 m wide	m	76.00
3.00 m wide	m	89.00
4.00 m wide	m	110.00
5.00 m wide	m	120.00
6.00 m wide	m	130.00
7.00 m wide	m	150.00
As above but excavation 450 mm deep and base of Type 1 at 250 mm thick		
1.50 m wide	m	129.00
2.00 m wide	m	156.00
3.00 m wide	m	209.00
4.00 m wide	m	261.00
5.00 m wide	m	314.00
6.00 m wide	m	367.00
7.00 m wide	m	420.00
To excavated and prepared base above, lay roadbase of 40 mm size dense bitumen macadam 70 mm thick; lay wearing course of 10 mm size dense bitumen macadam 30 mm thick; mark out car parking bays 5.0 x 2.4 m with thermoplastic road paint; surfaces all mechanically laid		
Per bay; 5.0 x 2.4 m	each	211.00
Gangway	m²	15.00
As above but stainless steel road studs 100 x 100 mm; two per bay in lieu of thermoplastic paint		
per bay; 5.0 x 2.4 m	each	214.00
Car park as above but with interlocking concrete blocks 200 x 100 x 80 mm; grey		
per bay; 5.0 x 2.4 m	each	457.00
gangway	m²	39.00
Car park as above but with interlocking concrete blocks 200 x 100 x 80 mm; colours		
per bay; 5.0 x 2.4 m	each	471.00
gangway	m²	40.00
Car park as above but with interlocking concrete blocks 200 x 100 x 60 mm; grey		
per bay; 5.0 x 2.4 m	each	446.00
gangway	m²	38.00
Car park as above but with interlocking concrete blocks 200 x 100 x 60 mm; colours		
per bay; 5.0 x 2.4 m	each	454.00
gangway	m²	38.00
Car park as above but with laying of Grass Concrete Grasscrete in situ continuously reinforced cellular surfacing; including expansion joints at 10 m centres; fill with topsoil and peat (5:1) and fertilizer at 35 g/m²; seed with dwarf rye grass at 35 g/m²		
GC1; 100 mm thick for cars and light traffic		
per bay; 5.0 x 2.4 m	each	376.00
gangway	m²	32.00
GC2; 150 mm thick for HGV traffic including dust carts		
per bay; 5.0 x 2.4 m	each	465.00
gangway	m²	39.00

ROADS AND PAVINGS

Item Excluding site overheads and profit	Unit	Total rate £
CAR PARKING FOR DISABLED PEOPLE		
To excavated and prepared base above, lay roadbase of 20 mm size dense bitumen macadam 80 mm thick; lay wearing course of 10 mm size dense bitumen macadam 30 mm thick; mark out car parking bays with thermoplastic road paint		
per bay; 5.80 x 3.25 m; ambulant	each	273.00
per bay; 6.55 x 3.80 m; wheelchair	each	360.00
gangway	m^2	15.00
Car park as above but with interlocking concrete blocks 200 x 100 x 60 mm; grey; but mark out bays		
per bay; 5.80 x 3.25 m; ambulant	each	3600.00
per bay; 6.55 x 3.80 m; wheelchair	each	4200.00
gangway	m^2	38.00
PLAY AREA FOR BALL GAMES		
Excavate for playground; bring to grade; lay 225 mm consolidated hardcore; blind with 100 mm type 1; lay macadam base of 40 mm size dense bitumen macadam to 75 mm hick; lay wearing course 30 mm thick		
Over 1000 m^2	100 m^2	3296.00
400 - 1000 m^2	100 m^2	3401.00
Excavate for playground; bring to grade; lay 100 mm consolidated hardcore; blind with 100 mm type 1; lay macadam base of dense bitumen macadam to 50 mm thick; lay wearing course 20 mm thick		
Over 1000 m^2	100 m^2	2552.00
400 - 1000 m^2	100 m^2	3039.00
Excavate for playground; bring to grade; lay 100 mm consolidated hardcore; blind with 100 mm type 1; lay macadam base of dense bitumen macadam to 50 mm thick; lay wearing course of Addagrip resin coated aggregate 3 mm diameter x 6 mm thick		
Over 1000 m^2	100 m^2	4686.00
400 - 1000 m^2	100 m^2	5030.00
Safety surfaced play area Excavate for playground; bring to grade; lay with 150 mm type 1; lay macadam base of dense bitumen macadam to 50 mm thick		
lay safety surface 35 mm; coloured	100 m^2	9052.00
INTERLOCKING BLOCK PAVING		
Edge restraint to block paving; excavate for groundbeam; lay concrete 1:2:4 200 mm wide x 150 mm deep; on 50 mm thick sharp sand bed; inclusive of haunching one side		
Blocks 200 x 100 x 60 mm; PC £7.63/m^2; butt jointed		
header course	m	15.00
stretcher course	m	12.00
Bricks 215 x 112.5 x 50 mm; PC £300.00/1000; with mortar joints		
header course	m	17.00
stretcher course	m	16.00
Excavate ground; supply and lay granular fill Type 1 150 mm thick laid to falls and compacted; supply and lay block pavers; laid on 50 mm compacted sharp sand; vibrated; joints filled with loose sand excluding edgings or kerbs measured separately		
Concrete blocks		
200 x 100 x 60 mm	m^2	49.00
200 x 100 x 80 mm	m^2	50.00

ROADS AND PAVINGS

Item Excluding site overheads and profit	Unit	Total rate £
INTERLOCKING BLOCK PAVING - cont'd		
Reduce levels; lay 150 mm granular material Type 1; lay vehicular block paving to 90° herringbone pattern; on 50 mm compacted sand bed; vibrated; jointed in sand and vibrated; excavate and lay precast concrete edging 50 x 150 mm; on concrete foundation 1:2:4		
Blocks 200 x 100 x 60 mm		
1.0 m wide clear width between edgings	m	89.00
1.5 m wide clear width between edgings	m	113.00
2.0 m wide clear width between edgings	m	137.00
Blocks 200 x 100 x 60 mm but blocks laid 45° herringbone pattern including cutting edging blocks		
1.0 m wide clear width between edgings	m	95.00
1.5 m wide clear width between edgings	m	119.00
2.0 m wide clear width between edgings	m	143.00
Blocks 200 x 100 x 60 mm but all excavation by hand disposal off site by grab		
1.0 m wide clear width between edgings	m	184.00
1.5 m wide clear width between edgings	m	221.00
2.0 m wide clear width between edgings	m	258.00
BRICK PAVING		
WORKS BY MACHINE		
Excavate and lay base Type 1 150 mm thick remove arisings; all by machine; lay clay brick paving		
200 x 100 x 50 mm thick; butt jointed on 50 mm sharp sand bed		
PC £300.00/1000	m^2	59.00
PC £600.00/1000	m^2	75.00
200 x 100 x 50 mm thick; 10 mm mortar joints on 35 mm mortar bed		
PC £300.00/1000	m^2	73.00
PC £600.00/1000	m^2	86.00
Excavate and lay base Type 1 250 mm thick all by machine; remove arisings; lay clay brick paving		
200 x 100 x 50 mm thick; 10 mm mortar joints on 35 mm mortar bed		
PC £300.00/1000	m^2	79.00
Excavate and lay 150 mm readymix concrete base reinforced with A393 mesh; all by machine; remove arisings; lay clay brick paving		
200 x 100 x 50 mm thick; 10 mm mortar joints on 35 mm mortar bed; running or stretcher bond		
PC £300.00/1000	m^2	92.00
PC £600.00/1000	m^2	105.00
200 x 100 x 50 mm thick; 10 mm mortar joints on 35 mm mortar bed; butt jointed; herringbone bond		
PC £300.00/1000	m^2	78.00
PC £600.00/1000	m^2	94.00
Excavate and lay base readymix concrete base 150 mm thick reinforced with A393 mesh; all by machine; remove arisings; lay clay brick paving		
215 x 102.5 x 50 mm thick; 10 mm mortar joints on 35 mm mortar bed		
PC £300.00/1000; herringbone	m^2	94.00
PC £600.00/1000; herringbone	m^2	107.00

ROADS AND PAVINGS

Item Excluding site overheads and profit	Unit	Total rate £
WORKS BY HAND		
Excavate and lay base Type 1 150 mm thick by hand; arisings barrowed to spoil heap maximum distance 25 m and removal off site by grab; lay clay brick paving 200 x 100 x 50 mm thick; butt jointed on 50 mm sharp sand bed		
PC £300.00/1000	m^2	96.60
PC £600.00/1000	m^2	124.00
Excavate and lay 150 mm concrete base; 1:3:6: site mixed concrete reinforced with A393 mesh; remove arisings to stockpile and then off site by grab; lay clay brick paving 215 x 102.5 x 50 mm thick; 10 mm mortar joints on 35 mm mortar bed		
PC £300.00/1000	m^2	95.30
PC £600.00/1000	m^2	122.00
FLAG PAVING TO PEDESTRIAN AREAS		
Prices are inclusive of all mechanical excavation and disposal.		
Supply and lay precast concrete flags; excavate ground and reduce levels; treat substrate with total herbicide; supply and lay granular fill Type 1 150 mm thick laid to falls and compacted Standard precast concrete flags bedded and jointed in lime:sand mortar (1:3)		
450 x 450 x 70 mm; chamfered	m^2	39.00
450 x 450 x 50 mm; chamfered	m^2	35.00
600 x 300 x 50 mm	m^2	33.00
600 x 450 x 50 mm	m^2	33.00
600 x 600 x 50 mm	m^2	30.00
750 x 600 x 50 mm	m^2	30.00
900 x 600 x 50 mm	m^2	29.00
Coloured flags bedded and jointed in lime:sand mortar (1:3)		
600 x 600 x 50 mm	m^2	32.00
450 x 450 x 70 mm; chamfered	m^2	44.00
400 x 400 x 65 mm	m^2	44.00
750 x 600 x 50 mm	m^2	32.00
900 x 600 x 50 mm	m^2	30.00
Marshalls Saxon; textured concrete flags; reconstituted yorkstone in colours; butt jointed bedded in lime:sand mortar (1:3)		
300 x 300 x 35 mm	m^2	64.00
450 x 450 x 50 mm	m^2	54.00
600 x 300 x 35 mm	m^2	50.00
600 x 600 x 50 mm	m^2	47.00
Tactile flags; Marshalls blister tactile pavings; red or buff; for blind pedestrian guidance laid to designed pattern		
450 x 450 mm	m^2	46.00
400 x 400 mm	m^2	49.00
PEDESTRIAN DETERRENT PAVING		
Excavate ground and bring to levels; treat substrate with total herbicide; supply and lay granular fill Type 1 150 mm thick laid to falls and compacted; supply and lay precast deterrent paving units bedded in lime:sand mortar (1:3) and jointed in lime: sand mortar (1:3) Marshalls Mono		
Lambeth pyramidal paving; 600 x 600 x 75 mm	m^2	37.00
Townscape		
Abbey square cobble pattern pavings; reinforced; 600 x 600 x 60 mm	m^2	56.00
Geoset chamfered studs; 600 x 600 x 60 mm	m^2	52.00

ROADS AND PAVINGS

Item Excluding site overheads and profit	Unit	Total rate £
IMITATION YORKSTONE PAVINGS		
Excavate ground and bring to levels; treat substrate with total herbicide; supply and lay granular fill Type 1 150 mm thick laid to falls and compacted; supply and lay imitation yorkstone paving laid to coursed patterns bedded in lime:sand mortar (1:3) and jointed in lime:sand mortar (1:3)		
Marshalls Heritage; square or rectangular		
300 x 300 x 38 mm	m²	68.00
600 x 300 x 30 mm	m²	55.00
600 x 450 x 38 mm	m²	52.00
450 x 450 x 38 mm	m²	50.00
600 x 600 x 38 mm	m²	46.00
As above but laid to random rectangular patterns		
various sizes selected from the above	m²	58.00
Imitation yorkstone laid random rectangular as above but on concrete base 150 mm thick		
by machine	m²	70.00
by hand	m²	110.00
NATURAL STONE SLAB PAVING		
WORKS BY MACHINE		
Excavate ground by machine and reduce levels; to receive 65 mm thick slab and 35 mm mortar bed; dispose of excavated material off site; treat substrate with total herbicide; lay granular fill Type 1 150 mm thick laid to falls and compacted; lay to random rectangular pattern on 35 mm mortar bed		
New riven slabs		
laid random rectangular	m²	120.00
New riven slabs; but to 150 mm plain concrete base		
laid random rectangular	m²	129.00
New riven slabs; disposal by grab		
laid random rectangular	m²	130.00
Reclaimed Cathedral grade riven slabs		
laid random rectangular	m²	153.00
Reclaimed Cathedral grade riven slabs; but to 150 mm plain concrete base		
laid random rectangular	m²	164.00
Reclaimed Cathedral grade riven slabs; disposal by grab		
laid random rectangular	m²	159.00
New slabs sawn 6 sides		
laid random rectangular	m²	110.00
three sizes; laid to coursed pattern	m²	110.00
New slabs sawn 6 sides; but to 150 mm plain concrete base		
laid random rectangular	m²	117.00
three sizes; laid to coursed pattern	m²	122.00
New slabs sawn 6 sides; disposal by grab		
laid random rectangular	m²	111.00
3 sizes, laid to coursed pattern	m²	116.00
WORKS BY HAND		
Excavate ground by hand and reduce levels, to receive 65 mm thick slab and 35 mm mortar bed; barrow all materials and arisings 25 m; dispose of excavated material off site by grab; treat substrate with total herbicide; lay granular fill Type 1 150 thick laid to falls and compacted; lay to random rectangular pattern on 35 mm mortar bed		
New riven slabs		
laid random rectangular	m²	154.00
New riven slabs laid random rectangular; but to 150 mm plain concrete base		
laid random rectangular	m²	168.00

ROADS AND PAVINGS

Item Excluding site overheads and profit	Unit	Total rate £
New riven slabs; but disposal to skip		
laid random rectangular	m²	169.00
Reclaimed Cathedral grade riven slabs		
laid random rectangular	m²	191.00
Reclaimed Cathedral grade riven slabs; but to 150 mm plain concrete base		
laid random rectangular	m²	205.00
Reclaimed Cathedral grade riven slabs; but disposal to skip		
laid random rectangular	m²	206.00
New slabs sawn 6 sides		
laid random rectangular	m²	142.00
three sizes; sawn 6 sides laid to coursed pattern	m²	147.00
New slabs sawn 6 sides; but to 150 mm plain concrete base		
laid random rectangular	m²	156.00
three sizes; sawn 6 sides laid to coursed pattern	m²	161.00
GRANITE SETT PAVING - PEDESTRIAN		
Excavate ground and bring to levels; lay 100 mm hardcore to falls; compacted with 5 tonne roller; blind with compacted Type 1 50 mm thick; lay 100 mm concrete 1:2:4; supply and lay granite setts 100 x 100 x 100 mm bedded in cement:sand mortar (1:3) 25 mm thick minimum; close butted and jointed in fine sand; all excluding edgings or kerbs measured separately		
Setts laid to bonded pattern and jointed		
new setts 100 x 100 x 100 mm	m²	97.10
second-hand cleaned 100 x 100 x 100 mm	m²	108.00
Setts laid in curved pattern		
new setts 100 x 100 x 100 mm	m²	104.00
second-hand cleaned 100 x 100 x 100 mm	m²	114.00
GRANITE SETT PAVING - TRAFFICKED AREAS		
Excavate ground and bring to levels; lay 100 mm hardcore to falls; compacted with 5 tonne roller; blind with compacted Type 1 50 mm thick; lay 150 mm site mixed concrete 1:2:4 reinforced with steel fabric; supply and lay granite setts 100 x 100 x 100 mm bedded in cement:sand mortar (1:3) 25 mm thick minimum; close butted and jointed in fine sand; all excluding edgings or kerbs measured separately		
Site mixed concrete		
new setts 100 x 100 x 100 mm	m²	105.00
second-hand cleaned 100 x 100 x 100 mm	m²	115.00
Ready mixed concrete		
new setts 100 x 100 x 100 mm	m²	103.00
second-hand cleaned 100 x 100 x 100 mm	m²	114.00
CONCRETE PAVING		
Pedestrian areas		
Excavate to reduce levels; lay 100 mm Type 1; lay PAV 1 air entrained concrete; joints at maximum width of 6.0 m cut out and sealed with sealant; inclusive of all formwork and stripping		
100 mm thick	m²	67.00
Trafficked areas		
Excavate to reduce levels; lay 150 mm Type 1 lay PAV 2 40 N/mm² air entrained concrete; reinforced with steel mesh 200 x 200 mm square at 2.22kg/m²; joints at max width of 6.0 m cut out and sealed with sealant		
150 mm thick	m²	85.00

ROADS AND PAVINGS

Item Excluding site overheads and profit	Unit	Total rate £
BEACH COBBLE PAVING		
Excavate ground and bring to levels and fill with compacted Type 1 fill 100 mm thick; lay GEN 1 concrete base 100 mm thick; supply and lay cobbles individually laid by hand bedded in cement:sand mortar (1:3) 25 mm thick minimum; dry grout with 1:3 cement:sand grout; brush off surplus grout and water in; sponge off cobbles as work proceeds; all excluding formwork edgings or kerbs measured separately		
Scottish beach cobbles 50 - 75 mm	m^2	113.00
Scottish beach cobbles 100 - 200 mm	m^2	120.00
Kidney flint cobbles 75 - 100 mm	m^2	120.00
CONCRETE SETT PAVING		
Excavate ground and bring to levels; supply and lay 150 mm granular fill Type 1 laid to falls and compacted; supply and lay setts bedded in 50 mm sand and vibrated; joints filled with dry sand and vibrated; all excluding edgings or kerbs measured separately		
Marshalls Mono; Tegula precast concrete setts		
random sizes 60 mm thick	m^2	45.00
single size 60 mm thick	m^2	43.00
random size 80 mm thick	m^2	49.00
single size 80 mm thick	m^2	60.00
GRASS CONCRETE PAVING		
Reduce levels; lay 150 mm Type 1 granular fill compacted; supply and lay precast grass concrete blocks on 20 sand and level by hand; fill blocks with 3 mm sifted topsoil and pre-seeding fertilizer at 50 g/m^2; sow with perennial ryegrass/chewings fescue seed at 35 g/m^2		
Grass concrete		
GB103 406 x 406 x 103 mm	m^2	36.00
GB83 406 x 406 x 83 mm	m^2	33.00
extra for geotextile fabric underlayer	m^2	1.00
Firepaths		
Excavate to reduce levels; lay 300 mm well rammed hardcore; blinded with 100 mm type 1; supply and lay Marshalls Mono "Grassguard 180" precast grass concrete blocks on 50 mm sand and level by hand; fill blocks with sifted topsoil and pre-seeding fertilizer at 50 g/m^2		
firepath 3.8 m wide	m	220.00
firepath 4.4 m wide	m	255.00
firepath 5.0 m wide	m	290.00
turning areas	m^2	58.00
Charcon Hard Landscaping Grassgrid; 366 x 274 x 100 mm	m^2	32.00
SLAB/BRICK PATHS		
Stepping stone path inclusive of hand excavation and mechanical disposal, 100 mm Type 1 and sand blinding; slabs laid 100 mm apart to existing turf		
600 mm wide; 600 x 600 x 50 mm slabs		
natural finish	m	48.00
coloured	m	48.00
exposed aggregate	m	57.00
900 mm wide; 600 x 900 x 50 mm slabs		
natural finish	m	52.00
coloured	m	53.00
exposed aggregate	m	68.00

Approximate Estimates – Major Works

ROADS AND PAVINGS

Item Excluding site overheads and profit	Unit	Total rate £
Pathway inclusive of mechanical excavation and disposal; 100 mm Type 1 and sand blinding; slabs close butted		
Straight butted path 900 mm wide; 600 x 900 x 50 mm slabs		
natural finish	m	19.00
coloured	m	20.00
exposed aggregate	m	29.00
Straight butted path 1200 mm wide; double row; 600 x 900 x 50 mm slabs; laid stretcher bond		
natural finish	m	33.00
coloured	m	35.00
exposed aggregate	m	56.00
Straight butted path 1200 mm wide; one row of 600 x 600 x 50 mm; one row 600 x 900 x 50 mm slabs		
natural finish	m	32.00
coloured	m	34.00
exposed aggregate	m	62.00
Straight butted path 1500 mm wide; slabs of 600 x 900 x 50 mm and 600 x 600 x 50 mm; laid to bond		
natural finish	m	40.00
coloured	m	42.00
exposed aggregate	m	62.00
Straight butted path 1800 wide; two rows of 600 x 900 x 50 slabs; laid bonded		
natural finish	m	44.00
coloured	m	53.00
exposed aggregate	m	84.00
Straight butted path 1800 mm wide; three rows of 600 x 900 x 50 mm slabs; laid stretcher bond		
natural finish	m	48.00
coloured	m	57.00
exposed aggregate	m	89.00
Brick paved paths; bricks 215 x 112.5 x 65 mm with mortar joints; prices are inclusive of excavation, disposal off site, 100 mm hardcore, 100 mm 1:2:4 concrete bed; edgings of brick 215 mm wide, haunched; all jointed in cement:lime:sand mortar (1:1:6)		
Path 1015 mm wide laid stretcher bond; edging course of headers		
rough stocks - PC £450.00/1000	m	81.00
engineering brick - PC £239.00/1000	m	76.00
Path 1015 mm wide laid stack bond		
rough stocks - PC £450.00/1000	m	85.00
engineering brick - PC £239.00/1000	m	81.00
Path 1115 mm wide laid header bond		
rough stocks - PC £450.00/1000	m	89.00
engineering brick - PC £239.00/1000	m	84.00
Path 1330 mm wide laid basketweave bond		
rough stocks - PC £450.00/1000	m	110.00
engineering brick - PC £239.00/1000	m	96.00
Path 1790 mm wide laid basketweave bond		
rough stocks - PC £450.00/1000	m	140.00
engineering brick - PC £239.00/1000	m	97.00

ROADS AND PAVINGS

Item Excluding site overheads and profit	Unit	Total rate £
SLAB/BRICK PATHS - cont'd		
Brick paved paths; brick paviors 200 x 100 x 50 mm chamfered edge with butt joints; prices are inclusive of excavation, 100 mm Type 1 and 50 mm sharp sand; jointing in kiln dried sand brushed in; exclusive of edge restraints		
Path 1000 mm wide laid stretcher bond; edging course of headers		
rough stocks - PC £450.00/1000	m	74.20
engineering brick - PC £239.00/1000	m	77.20
Path 1330 mm wide laid basketweave bond		
rough stocks - PC £450.00/1000	m	105.00
engineering brick - PC £239.00/1000	m	97.70
Path 1790 mm wide laid basketweave bond		
rough stocks - PC £450.00/1000	m	137.00
engineering brick - PC £239.00/1000	m	99.70
GRAVEL PATHS		
Reduce levels and remove spoil to dump on site; lay 150 mm hardcore well rolled; lay 25 mm sand blinding and geofabric; fix timber edge 150 x 38 mm to both sides of straight paths; lay 50 mm Cedec gravel watered and rolled		
Lay Cedec gravel 50 mm thick; watered and rolled		
1.0 m wide	m	28.00
1.5 m wide	m	38.00
2.0 m wide	m	47.00
Lay Breedon gravel 50 mm thick; watered and rolled		
1.0 m wide	m	28.00
1.5 m wide	m	37.00
2.0 m wide	m	47.00
BARK PAVING		
Excavate to reduce levels; remove all topsoil to dump on site; treat area with herbicide; lay 150 mm clean hardcore; blind with sand; lay 0.7 mm geotextile filter fabric water flow 50 l/m^2/sec; supply and fix treated softwood edging boards 50 x 150 mm to bark area on hardcore base extended 150 mm beyond the bark area; boards fixed with galvanized nails to treated softwood posts 750 x 50 x 75 mm driven into firm ground at 1.0 m centres; edging boards to finish 25 mm above finished bark surface; tops of posts to be flush with edging boards and once weathered		
Supply and lay 100 mm Melcourt conifer walk chips 10 - 40 mm size		
1.00 m wide	m	20.00
2.00 m wide	m	30.20
3.00 m wide	m	41.50
4.00 m wide	m	50.20
Extra over for wood fibre	m^2	-1.49
Extra over for hardwood chips	m^2	-1.31
FOOTPATHS		
Excavate footpath to reduce level; remove arisings to tip on site maximum distance 25 m; lay Type 1 granular fill 100 mm thick; lay base course of 28 mm size dense bitumen macadam 50 mm thick; wearing course of 10 mm size dense bitumen macadam 20 mm thick; timber edge 150 x 38 mm		
Areas over 1000 m^2		
1.0 m wide	m	30.00
1.5 m wide	m	42.00
2.0 m wide	m	52.00
Areas 400 - 1000 m^2		
1.0 m wide	m	33.00
1.5 m wide	m	47.00
2.0 m wide	m	71.00

Approximate Estimates – Major Works

SPECIAL SURFACES FOR SPORT/PLAYGROUNDS

Item Excluding site overheads and profit	Unit	Total rate £
BARK PLAY AREA		
Excavate playground area to 450 mm depth; lay 150 mm broken stone or clean hardcore; lay filter membrane; lay bark surface Melcourt Industries		
Playbark 10/50 - 300 mm thick	100 m²	2924.00
Playbark 8/25 - 300 mm thick	100 m²	3472.00
BOWLING GREEN CONSTRUCTION; BAYLIS LANDS		
Bowling green; complete Excavate 300 mm deep and grade to level; excavate and install 100 mm land drain to perimeter and backfill with shingle; install 60 mm land drain to surface at 4.5 m centres backfilled with shingle; install 50 m non perforated pipe 50 m long; install 100 mm compacted and levelled drainage stone, blind with grit and sand; spread 150 mm imported 70:30 sand:soil accurately compacted and levelled; lay bowling green turf and top dress twice luted into surface; exclusive of perimeter ditches and bowls protection		
6 rink green 38.4 x 38.4 m	each	56250.00
install "Toro" automatic bowling green irrigation system with pump, tank, controller, electrics, pipework and 8 nr "Toro 780" sprinklers; excluding pumphouse (optional)	each	8600.00
Supply and install to perimeter of green; "Sportsmark" preformed bowling green ditch channels		
"Ultimate Design 99" steel reinforced concrete channel; 600 mm long section	each	7950.00
Supply and fit bowls protection material to rear hitting face of "Ultimate" channels 1 and 2 above		
"Curl Grass" artificial grass; 0.45 m wide	each	2451.00
"Astroturf" artificial grass; 0.45 m wide	each	2060.00
bowls protection ditch liner laid loose; 300 mm wide	each	1250.00
JOGGING TRACK		
Excavate track 250 mm deep; treat substrate with Casoron G4; lay filter membrane; lay 100 mm depth gravel waste or similar; lay 100 mm compacted gravel	100 m²	5631.00
Extra for treated softwood edging 50 x 150 mm on 50 x 50 mm posts		
both sides	m	9.00
PLAYGROUNDS		
Excavate playground area and dispose of arisings to tip; lay Type 1 granular fill; lay macadam surface two coat work 80 mm thick; base course of 28 mm size dense bitumen macadam 50 mm thick; wearing course of 10 mm size dense bitumen macadam 30 mm thick Areas over 1000 m²		
excavation 180 mm base 100 mm thick	m²	23.00
excavation 225 mm base 150 mm thick	m²	26.00
Areas 400 - 1000 m²		
excavation 180 mm base 100 mm thick	m²	26.00
excavation 225 mm base 150 mm thick	m²	29.00
Excavate playground area to given levels and falls; remove soil off site and backfill with compacted Type 1 granular fill; lay ready mixed concrete to fall 2% in all direction		
Base 150 mm thick; surface 100 mm thick	m²	23.00
Base 150 mm thick; surface 150 mm thick	m²	29.00

SPECIAL SURFACES FOR SPORT/PLAYGROUNDS

Item Excluding site overheads and profit	Unit	Total rate £
SAFETY SURFACING; BAYLIS LANDSCAPE CONSTRUCTION		
Excavate ground and reduce levels to receive safety surface; dispose of excavated material off site; treat substrate with total herbicide; lay granular fill Type 1 150 mm thick laid to falls and compacted; lay macadam base 40 mm thick; supply and lay "Ruberflex" wet pour safety system to thicknesses as specified		
All by machine except macadam by hand black		
15 mm thick	100 m²	4810.00
35 mm thick	100 m²	6280.00
60 mm thick	100 m²	7640.00
coloured		
15 mm thick	100 m²	8540.00
35 mm thick	100 m²	9110.00
60 mm thick	100 m²	10100.00
All by hand except disposal by grab black		
15 mm thick	100 m²	8270.00
35 mm thick	100 m²	9820.00
60 mm thick	100 m²	11300.00
coloured		
15 mm thick	100 m²	12000.00
35 mm thick	100 m²	12700.00
60 mm thick	100 m²	13700.00
SPORTSGROUND CONSTRUCTION; BAYLIS LANDSCAPE CONSTRUCTION		
Plain sports pitches; site clearance, grading and drainage not included		
Cultivate ground and grade to levels; apply pre-seeding fertilizer at 900 kg/ha; apply pre-seeding selective weedkiller; seed in two operations with sports pitch type grass seed at 350 kg/ha; harrow and roll lightly; including initial cut; size		
association football; senior 114 x 72 m	each	3613.00
association football; junior 106 x 58 m	each	2914.00
rugby union pitch; 156 x 81 m	each	5303.00
rugby league pitch; 134 x 60 m	each	3496.00
hockey pitch; 95 x 60 m	each	2681.00
shinty pitch; 186 x 96 m	each	7459.00
men's lacrosse pitch; 100 x 55 m	each	2564.00
women's lacrosse pitch; 110 x 73 m	each	3730.00
target archery ground; 150 x 50 m	each	3030.00
cricket outfield; 160 x 142 m	each	9440.00
cycle track outfield; 160 x 80 m	each	5827.00
polo ground; 330 x 220 m	each	27852.00
Cricket square; excavate to depth of 150 mm; pass topsoil through 6 mm screen; return and mix evenly with imported marl or clay loam; bring to accurate levels; apply pre-seeding fertilizer at 50 g/m; apply selective weedkiller; seed with cricket square type g grass seed at 50 g/m²; rake in and roll lightly; erect and remove temporary protective chestnut fencing; allow for initial cut and watering three times		
22.8 x 22.8 m	each	15150.00

PREPARATION FOR PLANTING/TURFING

Item Excluding site overheads and profit	Unit	Total rate £
SURFACE PREPARATION BY MACHINE		
Treat area with systemic non selective herbicide one month before starting cultivation operations; rip up subsoil using subsoiling machine to a depth of 250 mm below topsoil at 1.20 m centres in light to medium soils; rotavate to 200 mm deep in two passes; cultivate with chain harrow; roll lightly; clear stones over 50 mm		
By tractor	100 m²	8.90
As above but carrying out operations in clay or compacted gravel	100 m²	9.50
As above but ripping by tractor rotavation by pedestrian operated rotavator; clearance and raking by hand; herbicide application by knapsack sprayer	100 m²	40.00
As above but carrying out operations in clay or compacted gravel	100 m²	54.00
Spread and lightly consolidate topsoil brought from spoil heap not exceeding 100 m; in layers not exceeding 150 mm; grade to specified levels; remove stones over 25 mm; all by machine		
100 mm thick	100 m²	65.00
150 mm thick	100 m²	98.00
300 mm thick	100 m²	196.00
450 mm thick	100 m²	293.00
Extra to above for imported topsoil PC £26.00 m³ allowing for 20% settlement		
100 mm thick	100 m²	312.00
150 mm thick	100 m²	468.00
300 mm thick	100 m²	936.00
450 mm thick	100 m²	1410.00
500 mm thick	100 m²	1560.00
600 mm thick	100 m²	1872.00
750 mm thick	100 m²	2340.00
1.00 m thick	100 m²	3120.00
Extra for incorporating mushroom compost at 50 mm/m² into the top 150 mm of topsoil (compost delivered in 20 m³) loads		
manually spread; mechanically rotavated	100 m²	160.00
mechanically spread and rotavated	100 m²	120.00
Extra for incorporating manure at 50 mm/m² into the top 150 mm of topsoil loads		
manually spread; mechanically rotavated; 20 m³ loads	100 m²	279.00
mechanically spread and rotavated; 60 m³ loads	100 m²	169.00
SURFACE PREPARATION BY HAND		
Spread only and lightly consolidate topsoil brought from spoil heap in layers not exceeding 150 mm; grade to specified levels; remove stones over 25 mm; all by hand		
100 mm thick	100 m²	366.00
150 mm thick	100 m²	548.00
300 mm thick	100 m²	1096.00
450 mm thick	100 m²	1643.00
As above but inclusive of loading to location by barrow maximum distance 25 m; finished topsoil depth		
100 mm thick	100 m²	804.00
150 mm thick	100 m²	1205.00
300 mm thick	100 m²	2410.00
450 mm thick	100 m²	3614.00
500 mm thick	100 m²	4014.00
600 mm thick	100 m²	4813.00

PREPARATION FOR PLANTING/TURFING

Item Excluding site overheads and profit	Unit	Total rate £
SURFACE PREPARATION BY HAND - cont'd		
Spread only and lightly consolidate topsoil brought from spoil heap in layers not exceeding 150 mm; grade to specified levels; remove stones over 25 mm; all by hand - cont'd		
As above but loading to location by barrow maximum distance 100 m; finished topsoil depth		
100 mm thick	100 m^2	822.00
150 mm thick	100 m^2	1233.00
300 mm thick	100 m^2	2465.00
450 mm thick	100 m^2	3697.00
500 mm thick	100 m^2	4106.00
600 mm thick	100 m^2	4923.00
Extra to the above for incorporating mushroom compost at 50 mm/m^2 into the top 150 mm of topsoil; by hand	100 m^2	187.00
Extra to above for imported topsoil PC £26.00 m^3 allowing for 20% settlement		
100 mm thick	100 m^2	312.00
150 mm thick	100 m^2	468.00
300 mm thick	100 m^2	936.00
450 mm thick	100 m^2	1410.00
500 mm thick	100 m^2	1560.00
600 mm thick	100 m^2	1872.00
750 mm thick	100 m^2	2340.00
1.00 m thick	100 m^2	3120.00

SEEDING AND TURFING

Item Excluding site overheads and profit	Unit	Total rate £
SEEDING AND TURFING		
Bring top 200 mm of topsoil to a fine tilth using tractor drawn implements; remove stones over 25 mm by mechanical stone rake and bring to final tilth by harrow; apply pre-seeding fertilizer at 50 g/m² and work into top 50 mm during final cultivation; seed with certified grass seed in two operations; roll seedbed lightly after sowing		
General amenity grass at 35 g/m²; BSH A3	100 m²	43.00
General amenity grass at 35 g/m²; Johnsons Taskmaster	100 m²	32.00
Shaded areas; BSH A6 at 50 g/m²	100 m²	52.00
Motorway and road verges; Perryfields Pro 120 25-35 g/m²	100 m²	34.00
Bring top 200 mm of topsoil to a fine tilth using pedestrian operated rotavator; remove stones over 25 mm; apply pre-seeding fertilizer at 50 mm g/m² and work into top 50 mm during final hand cultivation; seed with certified grass seed in two operations; rake and roll seedbed lightly after sowing		
General amenity grass at 35 g/m²; BSH A3	100 m²	110.00
General amenity grass at 35 g/m²; Johnsons Taskmaster	100 m²	99.00
Shaded areas; BSH A6 at 50 g/m²	100 m²	119.00
Motorway and road verges; Perryfields Pro 120 25-35 g/m²	100 m²	110.00
Extra to above for using imported topsoil spread by machine		
100 mm minimum depth	100 m²	340.00
150 mm minimum depth	100 m²	510.00
extra for slopes over 30°	50%	-
Extra to above for using imported topsoil spread by hand; maximum distance for transporting soil 100 m		
100 mm minimum depth	m²	678.00
150 mm minimum depth	m²	1020.00
extra for slopes over 30°	50%	-
Extra for mechanically screening top 25 mm of topsoil through 6 mm screen and spreading on seedbed; debris carted to dump on site not exceeding 100 m	m³	6.41
Bring top 200 mm of topsoil to a fine tilth; remove stones over 50 mm; apply pre-seeding fertilizer at 50 g/m² and work into top 50 mm of topsoil during final cultivation; seed with certified grass seed in two operations; harrow and roll seedbed lightly after sowing		
Areas inclusive of fine levelling by specialist machinery		
outfield grass at 350 kg/ha; Perryfields Pro 40	ha	4256.00
outfield grass at 350 kg/ha; Perryfields Pro 70	ha	3966.00
sportsfield grass at 300 kg/ha; Johnsons Sportsmaster	ha	3986.00
Seeded areas prepared by chain harrow		
low maintenance grass at 350 kg/ha	ha	4977.00
verge mixture grass at 150 kg/ha	ha	4278.00
Extra for wild flora mixture at 30 kg/ha		
BSH WSF 75 kg /ha	ha	3500.00
extra for slopes over 30°	50%	-
Cut existing turf to 1.0 x 1.0 x 0.5 m turves; roll up and move to stack not exceeding 100 m		
by pedestrian operated machine; roll up and stack by hand	100 m²	70.00
all works by hand	100 m²	198.00
Extra for boxing and cutting turves	100 m²	4.00

SEEDING AND TURFING

Item Excluding site overheads and profit	Unit	Total rate £
SEEDING AND TURFING - cont'd		
Bring top 200 mm of topsoil to a fine tilth in two passes; remove stones over 25 mm and bring to final tilth; apply pre-seeding fertilizer at 50 g/m^2 and work into top 50 mm during final cultivation; roll turf bed lightly		
Using tractor drawn implements and mechanical stone rake	m^2	**1.00**
Cultivation by pedestrian rotavator; all other operations by hand	m^2	**1.00**
As above but bring turf from stack not exceeding 100 m; lay turves to stretcher bond using plank barrow runs; firm turves using wooden turf beater		
using tractor drawn implements and mechanical stone rake	m^2	**2.00**
cultivation by pedestrian rotavator; all other operations by hand	m^2	**3.00**
As above but including imported turf; Rolawn Medallion		
using tractor drawn implements and mechanical stone rake	m^2	**4.00**
cultivation by pedestrian rotavator; all other operations by hand	m^2	**4.00**
Extra over to all of the above for watering on two occasions and carrying out initial cut		
by ride on triple mower	100 m^2	**2.00**
by pedestrian mower	100 m^2	**5.00**
by pedestrian mower; box cutting	100 m^2	**6.00**
Extra for using imported topsoil spread and graded by machine		
25 mm minimum depth	m^2	**1.00**
75 mm minimum depth	m^2	**3.00**
100 mm minimum depth	m^2	**4.00**
150 mm minimum depth	m^2	**6.00**
Extra for using imported topsoil spread and graded by hand; distance of barrow run 25 m		
25 mm minimum depth	m^2	**2.00**
75 mm minimum depth	m^2	**5.00**
100 mm minimum depth	m^2	**6.00**
150 mm minimum depth	m^2	**9.00**
Extra for work on slopes over 30° including pegging with 200 galvanized wire pins	m^2	**2.00**
Inturf Big Roll		
Supply, deliver in one consignment, fully prepare the area and install in Big Roll format Inturf 553, a turfgrass comprising dwarf perennial ryegrass, smooth stalked rneadowgrass and fescues; installation by tracked machine		
Preparation by tractor drawn rotavator	m^2	**4.00**
Cultivation by pedestrian rotavator; all other operations by hand	m^2	**4.00**
Erosion control		
On ground previously cultivated, bring area to level, treat with herbicide; lay 20 mm thick open texture erosion control mat with 100 mm laps, fixed with 8 x 400 mm steel pegs at 1.0 m centres; sow with low maintenance grass suitable for erosion control on slopes at 35 g/m^2; spread imported topsoil 25 mm thick and fertilizer at 35 g/m^2; water lightly using sprinklers on two occasions	m^2	**7.00**
as above but hand-watering by hose pipe maximum distance from mains supply 50 m	m^2	**7.00**

PLANTING

Item Excluding site overheads and profit	Unit	Total rate £
HEDGE PLANTING		
Works by machine; excavate trench for hedge 300 mm wide x 450 mm deep; deposit spoil alongside and plant hedging plants in single row at 200 mm centres; backfill with excavated material incorporating organic manure at 1 m^3 per 5 m^3; carry out initial cut; including delivery of plants from nursery		
Bare root hedging plants		
PC £0.30	100 m	610.00
PC £0.60	100 m	760.00
PC £1.00	100 m	960.00
PC £1.20	100 m	1060.00
PC £1.50	100 m	1210.00
As above but two rows of hedging plants at 300 mm centres staggered rows		
PC £0.30	100 m	794.00
PC £0.60	100 m	994.00
PC £1.00	100 m	1261.00
PC £1.50	100 m	1594.00
Works by hand; excavate trench for hedge 300 mm wide x 450 mm deep; deposit spoil alongside and plant hedging plants in single row at 200 mm centres; backfill with excavated material incorporating organic manure at 1 m^3 per 5 m^3; carry out initial cut; including delivery of plants from nursery		
Two rows of hedging plants at 300 mm centres staggered rows		
PC £0.30	100 m	996.00
PC £0.60	100 m	1196.00
PC £1.00	100 m	1463.00
PC £1.50	100 m	1796.00
TREE PLANTING		
Excavate tree pit by hand; fork over bottom of pit; plant tree with roots well spread out; backfill with excavated material, incorporating treeplanting compost at 1 m^3 per 3 m^3 of soil, one tree stake and two ties; tree pits square in sizes shown		
Light standard bare root tree in pit; PC £9.75		
600 x 600 mm deep	each	34.00
900 x 600 mm deep	each	48.00
Standard bare root tree in pit; PC £16.00		
600 x 600 mm deep	each	45.00
900 x 600 mm deep	each	53.00
Standard root balled tree in pit; PC £23.50		
600 x 600 mm deep	each	54.00
900 x 600 mm deep	each	62.00
1.00 x 1.00 m deep	each	83.00
Selected standard bare root tree in pit; PC £22.50		
900 x 900 mm deep	each	58.00
1.00 x 1.00 m deep	each	84.00
Selected standard root ball tree in pit; PC £32.50		
900 x 900 mm deep	each	84.00
1.00 x 1.00 m deep	each	130.00
Heavy standard bare root tree in pit; PC £39.25		
900 x 900 mm deep	each	95.00
1.00 x 1.00 m deep	each	110.00
1.50 m x 750 mm deep	each	150.00
Heavy standard root ball tree in pit; PC £54.25		
900 x 900 mm deep	each	110.00
1.00 x 1.00 m deep	each	130.00
1.50 m x 750 mm deep	each	155.00

PLANTING

Item Excluding site overheads and profit	Unit	Total rate £
TREE PLANTING - cont'd		
Excavate tree pit by hand; fork over bottom of pit; plant tree with roots well spread out; backfill with excavated material, incorporating treeplanting compost at 1 m³ per 3 m³ of soil, one tree stake and two ties; tree pits square in sizes shown - cont'd		
Extra heavy standard bare root tree in pit; PC £63.00		
1.00 x 1.00 m deep	each	**140.00**
1.20 x 1.00 m deep	each	**136.00**
1.50 m x 750 mm deep	each	**178.00**
Extra heavy standard root ball tree in pit; PC £62.50		
1.00 x 1.00 m deep	each	**141.00**
1.50 m x 750 mm deep	each	**171.00**
1.50 x 1.00 m deep	each	**203.00**
Excavate tree pit by machine; fork over bottom of pit; plant tree with roots well spread out; backfill with excavated material, incorporating organic manure at 1 m³ per 3 m³ of soil; one tree stake and two ties; tree pits square in sizes shown		
Light standard bare root tree in pit; PC £9.75		
600 x 600 mm deep	each	**29.00**
900 x 900 mm deep	each	**41.00**
Standard bare root tree in pit; PC £16.00		
600 x 600 mm deep	each	**40.00**
900 x 600 mm deep	each	**42.00**
900 x 900 mm deep	each	**48.00**
Standard root balled tree in pit; PC £23.50		
600 x 600 mm deep	each	**49.00**
900 x 600 mm deep	each	**51.00**
900 x 900 mm deep	each	**57.00**
Selected standard bare root tree in pit; PC £22.50		
900 x 900 mm deep	each	**58.00**
1.00 x 1.00 m deep	each	**68.00**
Selected standard root ball tree in pit; PC £32.50		
900 x 900 mm deep	each	**84.00**
1.00 x 1.00 m deep	each	**110.00**
Heavy standard bare root tree in pit; PC £39.25		
900 x 900 mm deep	each	**95.00**
1.00 x 1.00 m deep	each	**88.00**
1.20 x 1.00 m deep	each	**110.00**
Heavy standard root ball tree in pit; PC £54.25		
900 x 900 mm deep	each	**110.00**
1.00 x 1.00 m deep	each	**110.00**
1.20 x 1.00 m deep	each	**123.00**
Extra heavy standard bare root tree in pit; PC £63.00		
1.00 x 1.00 m deep	each	**130.00**
1.20 x 1.00 m deep	each	**136.00**
1.50 x 1.00 m deep	each	**163.00**
Extra heavy standard root ball tree in pit; PC £62.50		
1.00 x 1.00 m deep	each	**125.00**
1.20 x 1.00 m deep	each	**139.00**
1.50 x 1.00 m deep	each	**167.00**

PLANTING

Item Excluding site overheads and profit	Unit	Total rate £
SEMI MATURE TREE PLANTING		
Excavate tree pit deep by machine; fork over bottom of pit; plant rootballed tree Acer platanoides "Emerald Queen" using telehandler where necessary; backfill with excavated material, incorporating Melcourt Topgrow bark/manure mixture at 1 m³ per 3 m³ of soil; Platipus underground guying system; tree pits 1500 x 1500 x 1500 mm deep inclusive of Platimats; excavated material not backfilled to treepit spread to surrounding area		
16 - 18 cm girth; PC £85.00	each	247.00
18 - 20 cm girth; PC £100.00	each	267.00
20 - 25 cm girth; PC £125.00	each	358.00
25 - 30 cm girth; PC £165.00	each	455.00
30 - 35 cm girth; PC £300.00	each	688.00
As above but treepits 2.00 x 2.00 x 1.5 m deep		
40 - 45 cm girth; PC £500.00	each	1030.00
45 - 50 cm girth; PC £700.00	each	1244.00
55 - 60 cm girth; PC £1,200.00	each	1798.00
67 - 70 cm girth; PC £2,200.00	each	2982.00
75 - 80 cm girth; PC £4,500.00	each	5329.00
Extra to the above for imported topsoil moved 25 m from tipping area and disposal off site of excavated material Tree pits 1500 x 1500 x 1500 mm deep		
16 - 18 cm girth	each	177.00
18 - 20 cm girth	each	166.00
20 - 25 cm girth	each	156.00
25 - 30 cm girth	each	147.00
30 - 35 cm girth	each	139.00
TREE PLANTING WITH MOBILE CRANES		
Excavate treepit 1.50 x 1.50 x 1.00 m deep; supply and plant semi mature trees delivered in full loads; trees lifted by crane; inclusive of backfilling tree pit with imported topsoil, compost, fertilizers and underground guying using "Platipus" anchors		
Self managed lift; local authority applications, health and safety, traffic management or road closures not included; tree size and distance of lift; 35 tonne crane		
25 - 30 cm; maximum 25 m distance	each	623.00
30 - 35 cm; maximum 25 m distance	each	768.00
35 - 40 cm; maximum 25 m distance	each	988.00
55 - 60 cm; maximum 15 m distance	each	1784.00
80 - 90 cm; maximum 10 m distance	each	6803.00
Managed lift; inclusive of all local authority applications, health and safety, traffic management or road closures all by crane hire company; tree size and distance of lift; 35 tonne crane		
25 - 30 cm; maximum 25 m distance	each	633.00
30 - 35 cm; maximum 25 m distance	each	786.00
35 - 40 cm; maximum 25 m distance	each	1010.00
55 - 60 cm; maximum 15 m distance	each	1828.00
80 - 90 cm; maximum 10 m distance	each	6954.00
Self managed lift; local authority applications, health and safety, traffic management or road closures not included; tree size and distance of lift; 80 tonne crane		
25 - 30 cm; maximum 40 m distance	each	627.00
30 - 35 cm; maximum 40 m distance	each	776.00
35 - 40 cm; maximum 40 m distance	each	999.00
55 - 60 cm; maximum 33 m distance	each	1803.00
80 - 90 cm; maximum 23 m distance	each	6868.00

PLANTING

Item Excluding site overheads and profit	Unit	Total rate £
TREE PLANTING WITH MOBILE CRANES - cont'd		
Excavate treepit 1.50 x 1.50 x 1.00 m deep; supply and plant semi mature trees delivered in full loads; trees lifted by crane; inclusive of backfilling tree pit with imported topsoil, compost, fertilizers and underground guying using "Platipus" anchors - cont'd		
Managed lift; inclusive of all local authority applications, health and safety, traffic management or road closures all by crane hire company; tree size and distance of lift; 35 tonne crane		
25 - 30 cm; maximum 40 m distance	each	637.00
30 - 35 cm; maximum 40 m distance	each	794.00
35 - 40 cm; maximum 40 m distance	each	1030.00
55 - 60 cm; maximum 33 m distance	each	1848.00
80 - 90 cm; maximum 23 m distance	each	7023.00
SHRUBS GROUND COVERS AND BULBS		
Excavate planting holes 250 x 250 x 300 mm deep to area previously ripped and rotavated; excavated material left alongside planting hole		
By mechanical auger		
250 mm centres (16 plants per m^2)	m^2	11.00
300 mm centres (11.11 plants per m^2)	m^2	8.00
400 mm centres (6.26 plants per m^2)	m^2	5.00
450 mm centres (4.93 plants per m^2)	m^2	4.00
500 mm centres (4 plants per m^2)	m^2	3.00
600 mm centres (2.77 plants per m^2)	m^2	2.00
750 mm centres (1.77 plants per m^2)	m^2	2.00
900 mm centres (1.23 plants per m^2)	m^2	1.00
1.00 m centres (1 plant per m^2)	m^2	1.00
1.50 m centres (0.44 plants per m^2)	m^2	1.00
As above but excavation by hand		
250 mm centres (16 plants per m^2)	m^2	11.00
300 mm centres (11.11 plants per m^2)	m^2	8.00
400 mm centres (6.26 plants per m^2)	m^2	5.00
450 mm centres (4.93 plants per m^2)	m^2	4.00
500 mm centres (4 plants per m^2)	m^2	3.00
600 mm centres (2.77 plants per m^2)	m^2	2.00
750 mm centres (1.77 plants per m^2)	m^2	2.00
900 mm centres (1.23 plants per m^2)	m^2	1.00
1.00 m centres (1 plant per m^2)	m^2	1.00
1.50 m centres (0.44 plants per m^2)	m^2	1.00
Clear light vegetation from planting area and remove to dump on site; dig planting holes; plant whips with roots well spread out; backfill with excavated topsoil; including one 38 x 38 mm treated softwood stake, two tree ties and mesh guard 1.20 m high; planting matrix 1.5 x 1.5 m; allow for beating up once at 10% of original planting, cleaning and weeding round whips once, applying fertilizer once at 35 gm/m^2; using the following mix of whips, bare rooted		
Plant bare root plants average PC £0.27 each to a required matrix plant mix as above	100 m^2	252.00
Plant bare root plants average PC £0.75 each to a required matrix plant mix as above	100 m^2	263.00

PLANTING

Item Excluding site overheads and profit	Unit	Total rate £
Cultivate and grade shrub bed; bring top 300 mm of topsoil to a fine tilth, incorporating mushroom compost at 50 mm and Enmag slow release fertilizer; rake and bring to given levels; remove all stones and debris over 50 mm; dig planting holes average 300 x 300 x 300 mm deep; supply and plant specified shrubs in quantities as shown below; backfill with excavated material as above; water to field capacity and mulch 50 mm bark chips 20 - 40 mm size; water and weed regularly for 12 months and replace failed plants		
Shrubs 3 l PC £2.80; ground covers 9 cm PC £1.50 100% shrub area		
300 mm centres	100 m²	5752.00
400 mm centres	100 m²	3445.00
500 mm centres	100 m²	2377.00
600 mm centres	100 m²	1797.00
100% groundcovers		
200 mm centres	100 m²	7199.00
300 mm centres	100 m²	3464.00
400 mm centres	100 m²	2159.00
500 mm centres	100 m²	1554.00
groundcover 30%/shrubs 70% at the distances shown below		
200/300 mm	100 m²	6123.00
300/400 mm	100 m²	3421.00
300/500 mm	100 m²	2692.00
400/500 mm	100 m²	2292.00
groundcover 50%/shrubs 50% at the distances shown below		
200/300 mm	100 m²	6480.00
300/400 mm	100 m²	3450.00
300/500 mm	100 m²	2929.00
400/500 mm	100 m²	2263.00
Cultivate ground by machine and rake to level; plant bulbs as shown; bulbs PC £25.00/100		
15 bulbs per m²	100 m²	653.00
25 bulbs per m²	100 m²	1068.00
50 bulbs per m²	100 m²	2108.00
Cultivate ground by machine and rake to level; plant bulbs as shown; bulbs PC £13.00/100		
15 bulbs per m²	100 m²	452.00
25 bulbs per m²	100 m²	734.00
50 bulbs per m²	100 m²	1439.00
Form holes in grass areas and plant bulbs using bulb planter, backfill with organic manure and turf plug; bulbs PC £13.00/100		
15 bulbs per m²	100 m²	652.00
25 bulbs per m²	100 m²	1086.00
50 bulbs per m²	100 m²	2171.00
BEDDING		
Spray surface with glyphosate; lift and dispose of turf when herbicide action is complete; cultivate new area for bedding plants to 400 mm deep; spread compost 100 mm deep and chemical fertilizer "Enmag" and rake to fine tilth to receive new bedding plants; remove all arisings to skip		
Existing turf area		
disposal to skip	m²	4.87
disposal to compost area on site; distance 25 m	m²	5.24

PLANTING

Item Excluding site overheads and profit	Unit	Total rate £
BEDDING - cont'd		
Plant bedding to existing planting area; bedding planting PC £0.25 each Clear existing bedding; cultivate soil to 230 mm deep; incorporate compost 75 mm and rake to fine tilth; collect bedding from nursery and plant at 100 mm centres; irrigate on completion; maintain weekly for 12 weeks		
mass planted; 100 mm centres	m^2	33.00
to patterns; 100 mm centres	m^2	36.00
mass planted; 150 mm centres	m^2	19.00
to patterns; 150 mm centres	m^2	22.00
mass planted; 200 mm centres	m^2	19.00
to patterns; 200 mm centres	m^2	22.00
Extra for watering by hand held hose pipe		
Flow rate 25 litres/minute		
10 litres/m^2	100 m^2	0.11
15 litres/m^2	100 m^2	0.17
20 litres/m^2	100 m^2	0.22
25 litres/m^2	100 m^2	0.28
Flow rate 40 litres/minute		
10 litres/m^2	100 m^2	0.07
15 litres/m^2	100 m^2	0.11
20 litres/m^2	100 m^2	0.14
25 litres/m^2	100 m^2	0.17
PLANTING PLANTERS		
To brick planter; coat insides with 2 coats RIW liquid asphaltic composition; fill with 50 mm shingle and cover with geofabric; fill with screened topsoil incorporating 25% Topgrow compost and Enmag		
Planters 1.00 m deep		
1.00 x 1.00 m	each	130.00
1.00 x 2.00 m	each	208.00
1.00 x 3.00 m	each	295.00
Planters 1.50 m deep		
1.00 x 1.00 m	each	182.00
1.00 x 2.00 m	each	271.00
1.00 x 3.00 m	each	443.00
Container planting; fill with 50 mm shingle and cover with geofabric; fill with screened topsoil incorporating 25% Topgrow compost and Enmag		
Planters 1.00 m deep		
400 x 400 x 400 mm deep	each	9.00
400 x 400 x 600 mm deep	each	20.00
1.00 m x 400 mm wide x 400 mm deep	each	19.00
1.00 m x 600 mm wide x 600 mm deep	each	31.00
1.00 m x 100 mm wide x 400 mm deep	each	32.00
1.00 m x 100 mm wide x 600 mm deep	each	47.00
1.00 m x 100 mm wide x 1.00 m deep	each	77.00
1.00 m diameter x 400 mm deep	each	26.00
1.00 m diameter x 1.00 m deep	each	61.00
2.00 m diameter x 1.00 m deep	each	238.00

Approximate Estimates – Major Works

LANDSCAPE MAINTENANCE

Item Excluding site overheads and profit	Unit	Total rate £
LANDSCAPE MAINTENANCE		
MAINTENANCE OF GRASSED AREAS		
Maintenance executed as part of a landscape construction contract		
Grass cutting		
Grass cutting; fine turf; using pedestrian guided machinery; arisings boxed and disposed of off site		
per occasion	100 m^2	6.00
per annum 26 cuts	m^2	2.00
per annum 18 cuts	m^2	2.00
Grass cutting; standard turf; using self propelled three gang machinery		
per occasion	100 m^2	1.00
per annum 26 cuts	m^2	1.00
per annum 18 cuts	m^2	1.00
Maintenance for one year; recreation areas, parks, amenity grass areas; using tractor drawn machinery		
per occasion	ha	30.00
per annum 26 cuts	ha	755.00
per annum 18 cuts	ha	523.00
Aeration of turfed areas		
Aerate ground with spiked aerator; apply spring/summer fertilizer once; apply autumn/winter fertilizer once; cut grass, 16 cuts; sweep up leaves twice		
as part of a landscape contract, defects liability	ha	1832.00
as part of a long term maintenance contract	ha	1501.00
PLANTED AREA MAINTENANCE		
Vegetation control; native planting; roadside railway or forestry planted areas.		
Post planting maintenance; control of weeds and grass; herbicide spray applications; maintain weed free circles 1.00 m diameter to planting less than 5 years old in roadside, rail or forestry planting environments and the like; strim grass to 50 - 75 mm; prices per occasion (three applications of each operation normally required)		
Knapsack spray application; glyphosate; planting at		
1.50 mm centres	ha	984.00
1.75 mm centres	ha	645.00
2.00 mm centres	ha	678.00
Maintain planted areas; control of weeds and grass; maintain weed free circles 1.00 m diameter to planting less than 5 years old in roadside, rail or forestry planting environments and the like; strim surrounding grass to 50 - 75 mm; prices per occasion (three applications of each operation normally required)		
Herbicide spray applications; CDA (controlled droplet application) glyphosate and strimming; plants planted at the following centres		
1.50 mm centres	ha	594.00
1.75 mm centres	ha	645.00
2.00 mm centres	ha	678.00

LANDSCAPE MAINTENANCE

Item Excluding site overheads and profit	Unit	Total rate £
PLANTED AREA MAINTENANCE - cont'd		
Post planting maintenance; control of weeds and grass; herbicide spray applications; CDA (controlled droplet application); "Xanadu" glyphosate/diuron; maintain weed free circles 1.00 m diameter to planting less than 5 years old in roadside, rail or forestry planting environments and the like; strim grass to 50 - 75 mm; prices per occasion (1.5 applications of herbicide and three strim operations normally required)		
Plants planted at the following centres		
1.50 mm centres	ha	**543.00**
1.75 mm centres	ha	**608.00**
2.00 mm centres	ha	**655.00**
Ornamental shrub beds		
Hand weed ornamental shrub bed during the growing season; planting less than 2 years old		
Mulched beds; weekly visits; planting centres		
600 mm centres	100 m^2	**10.00**
400 mm centres	100 m^2	**11.00**
300 mm centres	100 m^2	**15.00**
ground covers	100 m^2	**19.00**
Mulched beds; monthly visits; planting centres		
600 mm centres	100 m^2	**14.00**
400 mm centres	100 m^2	**19.00**
300 mm centres	100 m^2	**22.00**
ground covers	100 m^2	**28.00**
Non-mulched beds; weekly visits; planting centres		
600 mm centres	100 m^2	**14.00**
400 mm centres	100 m^2	**15.00**
300 mm centres	100 m^2	**28.00**
ground covers	100 m^2	**37.00**
Non-mulched beds; monthly visits; planting centres		
600 mm centres	100 m^2	**19.00**
400 mm centres	100 m^2	**22.00**
300 mm centres	100 m^2	**28.00**
ground covers	100 m^2	**46.00**
Remulch planting bed at the start of the planting season; top up mulch 25 mm thick; Melcourt Ltd		
Larger areas maximum distance 25 m; 80 m^3 loads		
Ornamental bark mulch	100 m^2	**137.00**
Melcourt Bark Nuggets	100 m^2	**130.00**
Amenity Bark	100 m^2	**94.00**
Forest biomulch	100 m^2	**86.00**
Smaller areas; maximum distance 25 m; 25 m^3 loads		
Ornamental bark mulch	100 m^2	**169.00**
Melcourt Bark Nuggets	100 m^2	**162.00**
Amenity Bark	100 m^2	**126.00**
Forest biomulch	100 m^2	**118.00**

Approximate Estimates – Major Works

FENCING

Item Excluding site overheads and profit	Unit	Total rate £
TEMPORARY FENCING		
Site fencing; supply and erect temporary protective fencing and remove at completion of works		
Cleft chestnut paling		
1.20 m high 75 mm larch posts at 3 m centres	100 m	678.00
1.50 m high 75 mm larch posts at 3 m centres	100 m	875.00
"Heras" fencing		
2 week hire period	100 m	700.00
4 week hire period	100 m	956.00
8 week hire period	100 m	1468.00
12 week hire period	100 m	1980.00
16 week hire period	100 m	2492.00
24 week hire period	100 m	3516.00
CHAIN LINK AND WIRE FENCING		
Chain link fencing; supply and erect chain link fencing; form post holes and erect concrete posts and straining posts with struts at 50 m centres all set in 1:3:6 concrete; fix line wires		
3 mm galvanized wire 50 mm chainlink fencing		
900 mm high	m	23.00
1200 mm high	m	25.00
1800 mm high	m	35.00
plastic coated 3.15 gauge galvanized wire mesh		
900 mm high	m	22.00
1200 mm high	m	24.00
1800 mm high	m	34.00
Extra for additional concrete straining posts with 1 strut set in concrete		
900 mm high	each	64.00
1200 mm high	each	67.00
1800 mm high	each	81.00
Extra for additional concrete straining posts with 2 struts set in concrete		
900 mm high	each	92.00
1200 mm high	each	94.00
1800 mm high	each	122.80
Extra for additional angle iron straining posts with 2 struts set in concrete		
900 mm high	each	68.00
1200 mm high	each	77.60
1400 mm high	each	91.80
1800 mm high	each	94.40
2400 mm high	each	105.00
TIMBER FENCING		
Clear fenceline of existing evergreen shrubs 3 m high average; grub out roots by machine chip on site and remove off site; erect close boarded timber fence in treated softwood, pales 100 x 22 mm lapped, 150 x 22 mm gravel boards		
Concrete posts 100 x 100 mm at 3.0 m centres set into ground in 1:3:6 concrete		
900 mm high	m	66.00
1500 mm high	m	71.00
1800 mm high	m	77.00
As above but with softwood posts; three arris rails		
1350 mm high	m	69.00
1650 mm high	m	76.00
1800 mm high	m	78.00

FENCING

Item Excluding site overheads and profit	Unit	Total rate £
TIMBER FENCING - cont'd		
Erect chestnut pale fencing; cleft chestnut pales; two lines galvanized wire, galvanized tying wire, treated softwood posts at 3.0 m centres and straining posts and struts at 50 m centres driven into firm ground		
900 mm high; posts 75 mm diameter x 1200 mm long	m	6.00
1200 mm high; posts 75 mm diameter x 1500 mm long	m	7.00
Construct timber rail; horizontal hit and miss type; rails 150 x 25 mm; posts 100 x 100 mm at 1.8 m centres; twice stained with coloured wood preservative; including excavation for posts and concreting into ground (C7P) In treated softwood		
1800 mm high	m	55.00
Construct cleft oak rail fence with rails 300 mm minimum girth tennoned both ends; 125 x 100 mm treated softwood posts double mortised for rails; corner posts 125 x 125 mm, driven into firm ground at 2.5 m centres		
3 rails	m	17.00
4 rails	m	21.00
DEER STOCK RABBIT FENCING		
Construct rabbit-stop fencing; erect galvanized wire netting; mesh 31 mm; 900 mm above ground, 150 below ground turned out and buried; on 75 mm diameter treated timber posts 1.8 m long driven 700 mm into firm ground at 4.0 m centres; netting clipped to top and bottom straining wires 2.63 mm diameter; straining post 150 mm diameter x 2.3 m long driven into firm ground at 50 m intervals		
turned in 150 mm	100 m	901.00
buried 150 mm	100 m	980.00
Deer fence Construct deer-stop fencing; erect five 4 mm diameter plain galvanized wires and five 2 ply galvanized barbed wires at 150 mm spacing on 45 x 45 x 5 mm angle iron posts 2.4 m long driven into firm ground at 3.0 m centres, driven into firm ground at 3.0 m centres, with timber droppers 25 x 38 mm x 1.0 m long at 1.5 m centres	100 m	1544.00
Forestry fencing Supply and erect forestry fencing of three lines of 3 mm plain galvanized wire tied to 1700 x 65 mm diameter angle iron posts at 2750 m centres with 1850 x 100 mm diameter straining posts and 1600 x 80 mm diameter struts at 50.0 m centres driven into firm ground		
1800 mm high; three wires	100 m	877.80
1800 mm high; three wires including cattle fencing	100 m	1156.00
CONCRETE FENCES		
Supply and erect precast concrete post and panel fence in 2 m bays; panels to be shiplap profile, aggregate faced one side; posts set 600 mm in ground in concrete		
1500 mm high	m	29.00
1800 mm high	m	33.00
2100 mm high	m	38.00

FENCING

Item Excluding site overheads and profit	Unit	Total rate £
SECURITY FENCING		
Supply and erect chainlink fence; 51 x 3 mm mesh; with line wires and stretcher bars bolted to concrete posts at 3.0 m centres and straining posts at 10 m centres; posts set in concrete 450 x 450 mm x 33% of height of post deep; fit straight extension arms of 45 x 45 x 5 mm steel angle with three lines of barbed wire and droppers; all metalwork to be factory hot-dip galvanized for painting on site		
Galvanized 3 mm mesh		
900 mm high	m	27.00
1200 mm high	m	31.00
1800 mm high	m	41.00
As above but with PVC coated 3.15 mm mesh (diameter of wire 2.5 mm)		
900 mm high	m	26.00
1200 mm high	m	29.00
1800 mm high	m	41.00
Add to fences above for base of fence to be fixed with hairpin staples cast into concrete ground beam 1:3:6 site mixed concrete; mechanical excavation disposal to on site spoil heaps		
125 x 225 mm deep	m	5.00
Add to fences above for straight extension arms of 45 x 45 x 5 mm steel angle with three lines of barbed wire and droppers	m	5.00
Supply and erect palisade security fence Jacksons "Barbican" 2500 mm high with rectangular hollow section steel pales at 150 mm centres on three 50 x 50 x 6 mm rails; rails bolted to 80 x 60 mm posts set in concrete 450 x 450 x 750 mm deep at 2750 mm centres; tops of pales to points and set at 45° angle; all metalwork to be hot-dip factory galvanized for painting on site	m	130.00
Supply and erect single gate to match above complete with welded hinges and lock		
1.0 m wide	each	994.00
4.0 m wide	each	1070.00
8.0 m wide	pair	2220.00
Supply and erect Orsogril proprietary welded steel mesh panel fencing on steel posts set 750 mm deep in concrete foundations 600 x 600 mm; supply and erect proprietary single gate 2.0 m wide to match fencing		
930 mm high	100 m	11800.00
1326 mm high	100 m	15100.00
1722 mm high	100 m	19800.00
RAILINGS		
Conservation of historic railings; Eura Conservation Ltd Remove railings to workshop off site; shotblast and repair mildly damaged railings; remove rust and paint with three coats; transport back to site and re-erect		
Railings with finials 1.80 m high	m	520.00
Railings ornate cast or wrought iron	m	810.00
Supply and erect mild steel bar railing of 19 mm balusters at 115 mm centres welded to mild steel top and bottom rails 40 x 10 mm; bays 2.0 m long bolted to 51 x 51 mm ms hollow section posts set in C15P concrete; all metal work galvanized after fabrication		
900 mm high	m	75.00
1200 mm high	m	98.00
1500 mm high	m	110.00

FENCING

Item Excluding site overheads and profit	Unit	Total rate £
RAILINGS - cont'd		
Supply and erect mild steel pedestrian guard rail Class A; rails to be rectangular hollow sealed section 50 x 30 x 2.5 mm, vertical support 25 x 19 mm central between intermediate and top rail; posts to be set 300 mm into paving base; all components factory welded and factory primed for painting on site		
Panels 1000 mm high x 2000 mm wide with 150 mm toe space and 200 mm visibility gap at top	m	89.00
BALLSTOP FENCING		
Supply and erect plastic coated 30 x 30 mm netting fixed to 60.3 mm diameter 12 mm solid bar lattice galvanized dual posts; top, middle and bottom rails with 3 horizontal rails on 60.3 mm diameter nylon coated tubular steel posts at 3.0 m centres and 60.3 mm diameter straining posts with struts at 50 m centres set 750 mm into FND2 concrete footings 300 x 300 x 600 mm deep; include framed chain link gate 900 x 1800 mm high to match complete with hinges and locking latch		
4500 mm high	100 m	10260.00
5000 mm high	100 m	12950.00
6000 mm high	100 m	14430.00
STOCK GATE		
Erect stock gate, ms tubular field gate, diamond braced 1.8 m high hung on tubular steel posts set in concrete (C7P); complete with ironmongery; all galvanized		
width 3.00 m	each	319.00
width 4.20 m	each	348.00
TRIP RAILS		
Steel trip rail		
Erect trip rail of galvanized steel tube 38 mm internal diameter with sleeved joint fixed to 38 mm diameter steel posts as above 700 mm long; set in C7P concrete at 1.20 m centres; metalwork primed and painted two coats metal preservative paint	m	121.00
Birdsmouth fencing; timber		
600 mm high	m	20.00
900 mm high	m	21.00

STREET FURNITURE

Item Excluding site overheads and profit	Unit	Total rate £
BENCHES, SEATS AND BOLLARDS		
Bollards and access restriction		
Supply and install powder coated steel parking posts and bases to specified colour; fold down top locking type, complete with keys and instructions; posts and bases set in concrete footing 200 x 200 x 300 mm deep	each	178.00
Supply and install 10 nr cast iron "Doric" bollards 920 mm high above ground x 170 mm diameter bedded in concrete base 400 mm diameter x 400 mm deep	10 nr	1299.00
Benches and seating		
In grassed area excavate for base 2500 x 1575 mm and lay 100 mm hardcore, 100 mm concrete, brick pavers in stack bond bedded in 25 mm cement:lime:sand mortar; supply and fix where shown on drawing proprietary seat, hardwood slats on black powder coated steel frame, bolted down with 4 nr 24 x 90 mm recessed hex-head stainless steel anchor bolts set into concrete	set	1033.00
Cycle stand		
Supply and fix cycle stand 1250 m x 550 mm of 60.3 mm black powder coated hollow steel sections, one-piece with rounded top corners; set 250 mm into paving	each	304.00
Street planters		
Supply and locate in position precast concrete planters; fill with topsoil placed over 50 mm shingle and terram; plant with assorted 5 litre and 3 litre shrubs at to provide instant effect 970 mm diameter x 470 mm high; white exposed aggregate finish	each	40.00

DRAINAGE

Item Excluding site overheads and profit	Unit	Total rate £
SURFACE WATER DRAINAGE		
Clay gully Excavate hole; supply and set in concrete (C10P) vitrified clay trapped mud (dirt) gully with rodding eye; complete with galvanized bucket and cast iron hinged locking grate and frame, flexible joint to pipe; connect to drainage system with flexible joints	each	202.00
Concrete road gully Excavate hole and lay 100 mm concrete base (1:3:6) 150 x150 mm to suit given invert level of drain; supply and connect trapped precast concrete road gully 450 mm diameter x 1.07 m deep with 160 mm outlet; set in concrete surround; connect to vitrified clay drainage system with flexible joints; supply and fix straight bar dished top cast iron grating and frame; bedded in cement:sand mortar (1:3)	each	351.00
Gullies PVC-u Excavate hole and lay 100 mm concrete (C20P) base 150 x 150 mm to suit given invert level of drain; connect to drainage system; backfill with DoT Type 1 granular fill; install gully; complete with cast iron grate and frame		
trapped PVC-u gully	each	92.00
bottle gully 228 x 228 x 317 mm deep	each	77.00
bottle gully 228 x 228 x 642 mm deep	each	91.00
yard gully 300 mm diameter x 600 mm deep	each	258.00
Inspection chambers; brick manhole; excavate pit for inspection chamber including earthwork support and disposal of spoil to dump on site not exceeding 100 m; lay concrete (1:2:4) base 1500 mm diameter x 200 mm thick; 110 mm vitrified clay channels; benching in concrete (1:3:6) allowing one outlet and two inlets for 110 mm diameter pipe; construct inspection chamber 1 brick thick walls of engineering brick Class B; backfill with excavated material; complete with 2 nr cast iron step irons 1200 x 1200 x 1200 mm		
cover slab of precast concrete	each	878.00
access cover; Group 2; 600 x 450 mm	each	917.00
1200 x 1200 x 1500 mm		
access cover; Group 2; 600 x 450 mm	each	1081.00
recessed cover 5 tonne load; 600 x 450 mm; filled with block paviors	each	935.50
Pipe laying Excavate trench by excavator 600 mm deep; lay Type 2 bedding; backfill to 150 mm above pipe with gravel rejects; lay non woven geofabric and fill with topsoil to ground level		
160 mm PVC-u drainpipe	100 m	2546.00
110 mm PVC-u drainpipe	100 m	1556.00
150 mm vitrified clay	100 m	2266.00
100 mm vitrified clay	100 m	1586.00
Linear drainage to design sensitive areas Excavate trench by machine; lay Aco Brickslot channel drain on concrete base and surround to falls; all to manufacturers specifications		
paving surround to both sides of channel	m	166.05

DRAINAGE

Item Excluding site overheads and profit	Unit	Total rate £
Linear drainage to pedestrian area Excavate trench by machine; lay Aco Multidrain MD polymer concrete channel drain on concrete base and surround to falls; all to manufacturers specifications; paving surround to channel with brick paving PC £300.00/1000		
"Brickslot" galvanized grating; paving to both sides	m	166.05
"Brickslot" stainless steel grating; paving to both sides	m	279.71
slotted galvanized steel grating; paving surround to one side of channel	m	118.74
Excavate trench by machine; lay Aco Multidrain PPD recycled polypropylene channel drain on concrete base and surround to falls; all to manufacturers specifications; paving surround to channel with brick paving PC £300.00/1000		
"Heelguard" composite black; paving surround to both sides of channel	m	118.74
Linear drainage to light vehicular area Excavate trench by machine; lay Aco Multidrain PPD recycled polypropylene channel drain on concrete base and surround to falls; all to manufacturers specifications; paving surround to channel with brick paving PC £300.00/1000		
"Heelguard" composite black; paving surround to both sides of channel	m	135.05
ductile iron; paving surround to both sides of channel	m	173.31
Accessories for channel drain Sump unit with sediment bucket	nr	124.91
End cap; inlet/outlet	nr	17.20
AGRICULTURAL DRAINAGE		
Excavate and form ditch and bank with 45° sides in light to medium soils; all widths taken at bottom of ditch		
300 mm wide x 600 mm deep	100 m	100.00
600 mm wide x 900 mm deep	100 m	121.00
1.20 m wide x 900 mm deep	100 m	735.00
1.50 m wide x 1.20 m deep	100 m	1238.00
Clear and bottom existing ditch average 1.50 m deep, trim back vegetation and remove debris to licensed tip not exceeding 13 km, lay jointed concrete pipes; including bedding, haunching and topping with 150 mm concrete; 11.50 N/mm^2 - 40 mm aggregate; backfill with approved spoil from site		
Pipes 300 mm diameter	100 m	5906.00
Pipes 450 mm diameter	100 m	7515.00
Pipes 600 mm diameter	100 m	10020.00
Clay land drain Excavate trench by excavator to 450 mm deep; lay 100 mm vitrified clay drain with butt joints, bedding Class B; backfill with excavated material screened to remove stones over 40 mm; backfill to be laid in layers not exceeding 150 mm; top with 150 mm topsoil remove surplus material to approved dump on site not exceeding 100 m; final level of fill to allow for settlement	100 m	1396.00

DRAINAGE

Item Excluding site overheads and profit	Unit	Total rate £
DRAINAGE		
SUBSOIL DRAINAGE - BY MACHINE		
Main drain; remove 150 mm topsoil and deposit alongside trench, excavate drain trench by machine and lay flexible perforated drain; lay bed of gravel rejects 100 mm; backfill with gravel rejects or similar to within 150 mm of finished ground level; complete fill with topsoil; remove surplus spoil to approved dump on site		
Main drain 160 mm supplied in 35 m lengths		
450 mm deep	100 m	683.00
600 mm deep	100 m	801.00
900 mm deep	100 m	1372.00
Extra for couplings	each	3.00
As above but with 100 mm main drain supplied in 100 m lengths		
450 mm deep	100 m	952.00
600 mm deep	100 m	1257.00
900 mm deep	100 m	2046.00
Extra for couplings	each	2.00
Laterals to mains above; herringbone pattern; excavation and backfilling as above; inclusive of connecting lateral to main drain		
160 mm pipe to 450 mm deep trench		
laterals at 1.0 m centres	100 m^2	683.00
laterals at 2.0 m centres	100 m^2	342.00
laterals at 3.0 m centres	100 m^2	226.00
laterals at 5.0 m centres	100 m^2	137.00
laterals at 10.0 m centres	100 m^2	69.00
160 mm pipe to 600 mm deep trench		
laterals at 1.0 m centres	100 m^2	801.00
laterals at 2.0 m centres	100 m^2	401.00
laterals at 3.0 m centres	100 m^2	265.00
laterals at 5.0 m centres	100 m^2	161.00
laterals at 10.0 m centres	100 m^2	81.00
160 mm pipe to 900 mm deep trench		
laterals at 1.0 m centres	100 m^2	1372.00
laterals at 2.0 m centres	100 m^2	686.00
laterals at 3.0 m centres	100 m^2	453.00
laterals at 5.0 m centres	100 m^2	275.00
laterals at 10.0 m centres	100 m^2	138.00
Extra for 160/160 mm couplings connecting laterals to main drain		
laterals at 1.0 m centres	10 m	91.00
laterals at 2.0 m centres	10 m	46.00
laterals at 3.0 m centres	10 m	31.00
laterals at 5.0 m centres	10 m	19.00
laterals at 10.0 m centres	10 m	10.00
100 mm pipe to 450 mm deep trench		
laterals at 1.0 m centres	100 m^2	952.00
laterals at 2.0 m centres	100 m^2	476.00
laterals at 3.0 m centres	100 m^2	314.00
laterals at 5.0 m centres	100 m^2	191.00
laterals at 10.0 m centres	100 m^2	96.00
100 mm pipe to 600 mm deep trench		
laterals at 1.0 m centres	100 m^2	1257.00
laterals at 2.0 m centres	100 m^2	629.00
laterals at 3.0 m centres	100 m^2	415.00
laterals at 5.0 m centres	100 m^2	252.00
laterals at 10.0 m centres	100 m^2	126.00

DRAINAGE

Item Excluding site overheads and profit	Unit	Total rate £
100 mm pipe to 900 mm deep trench		
laterals at 1.0 m centres	100 m²	2046.00
laterals at 2.0 m centres	100 m²	1030.00
laterals at 3.0 m centres	100 m²	676.00
laterals at 5.0 m centres	100 m²	410.00
laterals at 10.0 m centres	100 m²	205.00
80 mm pipe to 450 mm deep trench		
laterals at 1.0 m centres	100 m²	909.00
laterals at 2.0 m centres	100 m²	455.00
laterals at 3.0 m centres	100 m²	300.00
laterals at 5.0 m centres	100 m²	182.00
laterals at 10.0 m centres	100 m²	91.00
80 mm pipe to 600 mm deep trench		
laterals at 1.0 m centres	100 m²	1214.00
laterals at 2.0 m centres	100 m²	607.00
laterals at 3.0 m centres	100 m²	401.00
laterals at 5.0 m centres	100 m²	243.00
laterals at 10.0 m centres	100 m²	122.00
80 mm pipe to 900 mm deep trench		
laterals at 1.0 m centres	100 m²	2003.00
laterals at 2.0 m centres	100 m²	1010.00
laterals at 3.0 m centres	100 m²	661.00
laterals at 5.0 m centres	100 m²	401.00
laterals at 10.0 m centres	100 m²	201.00
Extra for 100/80 mm junctions connecting laterals to main drain		
laterals at 1.0 m centres	10 m	42.00
laterals at 2.0 m centres	10 m	21.00
laterals at 3.0 m centres	10 m	20.00
laterals at 5.0 m centres	10 m	9.00
laterals at 10.0 m centres	10 m	5.00
SUBSOIL DRAINAGE - BY HAND		
Main drain; remove 150 mm topsoil and deposit alongside trench; excavate drain trench by machine and lay flexible perforated drain; lay bed of gravel rejects 100 mm; backfill with gravel rejects or similar to within 150 mm of finished ground level; complete fill with topsoil; remove surplus spoil to approved dump on site		
Main drain 160 mm in supplied in 35 m lengths		
450 mm deep	100 m	1902.00
600 mm deep	100 m	2394.00
900 mm deep	100 m	3217.00
Extra for couplings	each	3.00
As above but with 100 mm main drain supplied in 100 m lengths		
450 mm deep	100 m	1245.00
600 mm deep	100 m	1588.00
900 mm deep	100 m	2241.00
Extra for couplings	each	2.00
Laterals to mains above; herringbone pattern; excavation and backfilling as above; inclusive of connecting lateral to main drain		
160 mm pipe to 450 mm deep trench		
laterals at 1.0 m centres	100 m²	1902.00
laterals at 2.0 m centres	100 m²	951.00
laterals at 3.0 m centres	100 m²	628.00
laterals at 5.0 m centres	100 m²	381.00
laterals at 10.0 m centres	100 m²	191.00

DRAINAGE

Item Excluding site overheads and profit	Unit	Total rate £
SUBSOIL DRAINAGE - BY HAND - cont'd		
Laterals to mains above; herringbone pattern; excavation and backfilling as above; inclusive of connecting lateral to main drain - cont'd		
160 mm pipe to 600 mm deep trench		
laterals at 1.0 m centres	100 m²	2394.00
laterals at 2.0 m centres	100 m²	1197.00
laterals at 3.0 m centres	100 m²	790.00
laterals at 5.0 m centres	100 m²	479.00
laterals at 10.0 m centres	100 m²	240.00
160 mm pipe to 900 mm deep trench		
laterals at 1.0 m centres	100 m²	3217.00
laterals at 2.0 m centres	100 m²	1609.00
laterals at 3.0 m centres	100 m²	1070.00
laterals at 5.0 m centres	100 m²	644.00
laterals at 10.0 m centres	100 m²	322.00
Extra for 160/160 mm couplings connecting laterals to main drain		
laterals at 1.0 m centres	10 m	91.00
laterals at 2.0 m centres	10 m	46.00
laterals at 3.0 m centres	10 m	31.00
laterals at 5.0 m centres	10 m	19.00
laterals at 10.0 m centres	10 m	10.00
100 mm pipe to 450 mm deep trench		
laterals at 1.0 m centres	100 m²	1245.00
laterals at 2.0 m centres	100 m²	623.00
laterals at 3.0 m centres	100 m²	411.00
laterals at 5.0 m centres	100 m²	249.00
laterals at 10 m centres	100 m²	125.00
100 mm pipe to 600 mm deep trench		
laterals at 1.0 m centres	100 m²	1588.00
laterals at 2.0 m centres	100 m²	794.00
laterals at 3.0 m centres	100 m²	524.00
laterals at 5.0 m centres	100 m²	318.00
laterals at 10.0 m centres	100 m²	159.00
100 mm pipe to 900 mm deep trench		
laterals at 1.0 m centres	100 m²	2241.00
laterals at 2.0 m centres	100 m²	1130.00
laterals at 3.0 m centres	100 m²	740.00
laterals at 5.0 m centres	100 m²	449.00
laterals at 10.0 m centres	100 m²	225.00
80 mm pipe to 450 mm deep trench		
laterals at 1.0 m centres	100 m²	1202.00
laterals at 2.0 m centres	100 m²	601.00
laterals at 3.0 m centres	100 m²	397.00
laterals at 5.0 m centres	100 m²	241.00
laterals at 10.0 m centres	100 m²	121.00
80 mm pipe to 600 mm deep trench		
laterals at 1.0 m centres	100 m²	1545.00
laterals at 2.0 m centres	100 m²	773.00
laterals at 3.0 m centres	100 m²	510.00
laterals at 5.0 m centres	100 m²	309.00
laterals at 10.0 m centres	100 m²	155.00
80 mm pipe to 900 mm deep trench		
laterals at 1.0 m centres	100 m²	2198.00
laterals at 2.0 m centres	100 m²	1100.00
laterals at 3.0 m centres	100 m²	726.00
laterals at 5.0 m centres	100 m²	440.00
laterals at 10.0 m centres	100 m²	220.00

Approximate Estimates – Major Works

DRAINAGE

Item Excluding site overheads and profit	Unit	Total rate £
Extra for 100/80 mm couplings connecting laterals to main drain		
laterals at 1.0 m centres	10 m	**42.00**
laterals at 2.0 m centres	10 m	**21.00**
laterals at 3.0 m centres	10 m	**20.00**
laterals at 5.0 m centres	10 m	**9.00**
laterals at 10.0 m centres	10 m	**5.00**
SOAKAWAYS		
Construct soakaway from perforated concrete rings; excavation, casting in situ concrete ring beam base; filling with gravel 250 mm deep; placing perforated concrete rings; surrounding with geofabric and backfilling with 250 mm granular surround and excavated material; step irons and cover slab; inclusive of all earthwork retention and disposal offsite of surplus material		
900 mm diameter		
1.00 m deep	nr	**527.00**
2.00 m deep	nr	**858.00**
1200 mm diameter		
1.00 m deep	nr	**702.00**
2.00 m deep	nr	**1150.00**
2400 mm diameter		
1.00 m deep	nr	**1902.00**
2.00 m deep	nr	**3100.00**

IRRIGATION

Item Excluding site overheads and profit	Unit	Total rate £
LANDSCAPE IRRIGATION		
Automatic irrigation; KAR UK Ltd		
Large garden consisting of 24 stations; 7000 m^2 irrigated area		
turf only	nr	12000.00
70/30% turf/shrub beds	nr	14000.00
Medium sized garden consisting of 12 stations; 3500 m^2 irrigated area		
turf only	nr	7800.00
70/30% turf/shrub beds	nr	11000.00
Medium sized garden consisting of 6 stations of irrigated area; 1000 m^2		
turf only	nr	5200.00
70/30% turf/shrub beds	nr	5900.00
50/50% turf/shrub beds	nr	6400.00
Leaky pipe irrigation; Leaky Pipe Ltd		
Works by machine; main supply and connection to laterals; excavate trench for main or ring main 450 mm deep; supply and lay pipe; backfill and lightly compact trench		
20 mm LDPE	100 m	267.00
16 mm LDPE	100 m	230.00
Works by hand; main supply and connection to laterals; excavate trench for main or ring main 450 mm deep; supply and lay pipe; backfill and lightly compact trench		
20 mm LDPE	100 m	1329.00
16 mm LDPE	100 m	1292.00
Turf area irrigation; laterals to mains; to cultivated soil; excavate trench 150 mm deep using hoe or mattock; lay moisture leaking pipe laid 150 mm below ground at centres of 350 mm		
low leak	100 m^2	619.00
high leak	100 m^2	556.00
Landscape area irrigation; laterals to mains; moisture leaking pipe laid to the surface of irrigated areas at 600 mm centres		
low leak	100 m^2	328.00
high leak	100 m^2	291.00
Landscape area irrigation; laterals to mains; moisture leaking pipe laid to the surface of irrigated areas at 900 mm centres		
low leak	100 m^2	219.00
high leak	100 m^2	194.00
multistation controller	each	413.00
solenoid valves connected to automatic controller	each	527.00

WATER FEATURES

Item Excluding site overheads and profit	Unit	Total rate £
LAKES AND PONDS		
FAIRWATER LTD		
Excavate for small pond or lake maximum depth 1.00 m; remove arisings off site; grade and trim to shape; lay 75 mm sharp sand; line with 0.75 mm butyl liner 75 mm sharp sand and geofabric; cover over with sifted topsoil; anchor liner to anchor trench; install balancing tank and automatic top-up system		
Pond or lake of organic shape		
100 m^2; perimeter 50 m	each	4900.00
250 m^2; perimeter 90 m	each	7500.00
500 m^2; perimeter 130 m	each	16862.00
1000 m^2; perimeter 175 m	each	33222.00
Excavate for lake average depth 1.0 m, allow for bringing to specified levels; reserve topsoil; remove spoil to approved dump on site; remove all stones and debris over 75 mm; lay polythene sheet including welding all joints and seams by specialist; screen and replace topsoil 200 mm thick		
Prices are for lakes of regular shape		
500 micron sheet	1000 m	8764.80
1000 micron sheet	1000 m	11477.00
Extra for removing spoil to tip	m^3	16.00
Extra for 25 mm sand blinding to lake bed	100 m^2	102.00
Extra for screening topsoil	m^2	2.00
Extra for spreading imported topsoil	100 m^2	718.00
Plant aquatic plants in lake topsoil		
Aponogeton distachyum PC £312.00/100	100	385.00
Acorus calamus PC £198.00/100	100	271.00
Butomus umbellatus PC £198.00/100	100	271.00
Typha latifolia PC £198.00/100	100	271.00
Nymphaea PC £816.00/100	100	889.00
Formal water features		
Excavate and construct water feature of regular shape; lay 100 mm hardcore base and 150 mm concrete 1:2:4 site mixed; line base and vertical face with butyl liner 0.75 micron and construct vertical sides of reinforced blockwork; rendering two coats; anchor the liner behind blockwork; install pumps balancing tanks and all connections to mains supply		
1.00 x 1.00 x 1.00 m deep	nr	2277.00
2.00 x 1.00 x 1.00 m deep	nr	2876.00

TIMBER DECKING

Item Excluding site overheads and profit	Unit	Total rate £
AVS FENCING		
Timber decking; support structure of timber joists for decking laid on blinded base (measured separately)		
joists 50 x 150 mm	10 m²	**356.00**
joists 50 x 200 mm	10 m²	**347.00**
joists 50 x 250 mm	10 m²	**298.00**
As above but inclusive of timber decking boards in yellow cedar 142 mm wide x 42 mm thick		
joists 50 x 150 mm	10 m²	**835.00**
joists 50 x 200 mm	10 m²	**826.00**
joists 50 x 250 mm	10 m²	**777.00**
As above but boards 141 x 26 mm thick		
joists 50 x 200 mm	10 m²	**347.00**
joists 50 x 250 mm	10 m²	**298.00**
As above but decking boards in red cedar 131 mm wide x 42 mm thick		
joists 50 x 150 mm	10 m²	**730.00**
joists 50 x 200 mm	10 m²	**721.00**
joists 50 x 250 mm	10 m²	**672.00**
Add to all of the above for handrails fixed to posts 100 x 100 x 1370 mm high		
square balusters at 100 mm centres	m	**72.00**
square balusters at 300 mm centres	m	**46.00**
turned balusters at 100 mm centres	m	**92.00**
turned balusters at 300 mm centres	m	**53.00**

ESSENTIAL READING FROM TAYLOR AND FRANCIS

Design for Outdoor Recreation
Second Edition

Simon Bell

A manual for planners, designers and managers of outdoor recreation destinations, this book works through the processes of design and provides the tools to find the most appropriate balance between visitor needs and the capacity of the landscape.

A range of different aspects are covered including car parking, information signing, hiking, waterside activities, wildlife watching and camping.

This second edition incorporates new examples from overseas, including Australia, New Zealand, Japan and Eastern Europe as well as focusing on more current issues such as accessibility and the changing demands for recreational use.

July 2008: 276x219: 272pp
Pb: 978-0-415-44172-8 **£44.99**

To Order: Tel: +44 (0) 1235 400524 **Fax:** +44 (0) 1235 400525
or Post: Taylor and Francis Customer Services,
Bookpoint Ltd, Unit T1, 200 Milton Park, Abingdon, Oxon, OX14 4TA UK
Email: book.orders@tandf.co.uk

For a complete listing of all our titles visit:
www.tandf.co.uk

Taylor & Francis
Taylor & Francis Group

ESSENTIAL READING FROM TAYLOR AND FRANCIS

Rethinking Landscape
A Critical Reader

Ian H. Thompson

This unusually wide-ranging critical tool in the field of landscape architecture provides extensive excerpted materials from, and detailed critical perspectives on, standard and neglected texts from the 18th century to the present day. Considering the aesthetic, social, cultural and environmental foundations of our thinking about landscape this book explores the key writings which shaped the field in its emergence and maturity. Uniquely the book also includes original materials drawn from philosophical, ethical and political writings.

Selected Contents:

Part 1: Pluralism

Part 2: Aesthetics

Part 3: The Social Mission

Part 4: Ecology

Part 5: Some other Perspectives

Part 6: Conclusions and Suggestions

2008: 246x174: 288pp
Hb: 978-0-415-42463-9 **£85.00**
Pb: 978-0-415-42464-6 **£24.99**

To Order: Tel: +44 (0) 1235 400524 **Fax:** +44 (0) 1235 400525
or Post: Taylor and Francis Customer Services,
Bookpoint Ltd, Unit T1, 200 Milton Park, Abingdon, Oxon, OX14 4TA UK
Email: book.orders@tandf.co.uk

For a complete listing of all our titles visit:
www.tandf.co.uk

Taylor & Francis
Taylor & Francis Group

PART 3

Prices for Measured Works – Major Works

INTRODUCTION

Typical Project Profile

Contract value	£100,000.00 - £300,000.00
Labour rate (see page 7)	£18.25 per hour
Labour rate for maintenance contracts	£15.25 per hour
Number of site staff	30
Project area	6000 m^2
Project location	Outer London
Project components	50% hard landscape 50% soft landscape and planting
Access to works areas	Very good
Contract	Main contract
Delivery of materials	Full loads
Profit and site overheads	Excluded

NEW ITEMS FOR THIS EDITION

Item Excluding site overheads and profit	PC £	Labour hours	Labour £	Plant £	Material £	Unit	Total rate £
D20 LASER SURFACE GRADING							
Surface grading; Agripower Ltd; using laser controlled equipment							
Maximum variation of existing ground level							
75 mm; laser box grader and tractor	-	-	-	-	-	ha	1200.00
100 mm; laser controlled low ground pressure bulldozer	-	-	-	-	-	ha	1900.00
Fraise mowing; scarification of existing turf layer to rootzone level; Agripower Ltd							
Cart arisings to tip on site							
average thickness 50 - 75 mm	-	-	-	-	-	ha	2963.00
J50 GREEN ROOF SYSTEMS							
Bauder New Products; non-specialist installed products for laying under pavings or planting for collection or drainage							
DSE20 Drainage/Reservoir Board; 20 mm thick drainage board	-	-	-	-	-	m²	11.49
DSE40 Drainage/Reservoir Board; for intensive green roof applications	-	-	-	-	-	m²	15.71
Versicell 20 Drainage Board; specialist installed drainage layer for intensive green roof applications; predominantly hard landscape	-	-	-	-	-	m²	22.09
Bauder Extensive Substrate; 80 mm deep; for use with sedum and hardy perennials and biodiverse vegetation	-	-	-	-	-	m²	19.60
Bauder Intensive Substrate; 200 mm deep; for use in traditional green roof applications e.g. grass and bedding plants	-	-	-	-	-	m²	49.00
Q10 GRANITE SETT EDGES							
New granite sett edgings; CED Ltd; 100 x 100 x 100 mm; bedding in cement mortar (1:4); including haunching with in situ concrete; 11.50 N/mm² - 40 mm aggregate one side; concrete foundations (not included)							
Edgings 100 x 100 x 100 mm; ref 1R "better quality" with 20 mm pointing gaps							
100 mm wide	2.55	1.33	24.27	-	4.73	m	29.00
2 rows; 220 mm wide	5.09	2.22	40.52	-	7.92	m	48.43
3 rows; 340 mm wide	7.64	3.00	54.75	-	9.86	m	64.61
Edgings 100 x 100 x 200 mm; ref 1R "better quality" with 20 mm pointing gaps							
100 mm wide	2.60	0.75	13.69	-	4.78	m	18.47
2 rows; 220 mm wide	5.20	1.50	27.38	-	8.02	m	35.39
3 rows; 340 mm wide	7.80	2.25	41.06	-	10.02	m	51.08
Edgings 100 x 100 x 100 mm; ref 8R "standard quality" with 20 mm pointing gaps							
100 mm wide	2.44	1.33	24.27	-	4.62	m	28.90
2 rows; 220 mm wide	4.88	2.22	40.52	-	7.71	m	48.22
3 rows; 340 mm wide	7.32	3.00	54.75	-	9.55	m	64.30

NEW ITEMS FOR THIS EDITION

Item Excluding site overheads and profit	PC £	Labour hours	Labour £	Plant £	Material £	Unit	Total rate £
Edgings 100 x 100 x 200 mm; ref 8R "standard quality" with 20 mm pointing gaps							
100 mm wide	3.42	0.75	13.69	-	5.60	m	**19.29**
2 rows; 220 mm wide	6.84	1.50	27.38	-	9.66	m	**37.04**
3 rows; 340 mm wide	10.26	2.25	41.06	-	12.48	m	**53.54**
Reclaimed granite setts edgings; CED Ltd; 100 x 100 mm; bedding in cement mortar (1:4); including haunching with in situ concrete; 11.50 N/mm^2 - 40 mm aggregate one side; on concrete foundations (not included)							
Edgings 100 x 100 x 100 mm; with 20 mm pointing gaps							
100 mm wide	3.58	1.33	24.27	-	5.77	m	**30.04**
2 rows; 220 mm wide	7.17	2.22	40.52	-	9.99	m	**50.51**
3 rows; 340 mm wide	10.75	3.00	54.75	-	12.97	m	**67.72**
Q20 GRANULAR SUB-BASES FOR ROADS/PAVINGS							
Type 1 (recycled material) granular fill base; PC £9.95/tonne (£21.89/m^3 compacted)							
By machine							
100 mm thick	2.19	0.03	0.51	0.33	2.19	m^2	**3.03**
150 mm thick	3.28	0.03	0.46	0.49	3.28	m^2	**4.23**
250 mm thick	5.47	0.02	0.38	0.82	5.47	m^2	**6.67**
over 250 mm thick	9.95	0.20	3.65	2.88	9.95	m^3	**16.48**
By hand (mechanical compaction)							
100 mm thick	2.19	0.17	3.04	0.05	2.19	m^2	**5.28**
150 mm thick	3.28	0.25	4.56	0.08	3.28	m^2	**7.93**
250 mm thick	5.47	0.42	7.60	0.13	5.47	m^2	**13.21**
over 250 mm thick	9.95	0.47	8.53	0.13	9.95	m^3	**18.61**

NEW ITEMS FOR THIS EDITION

Item Excluding site overheads and profit	PC £	Labour hours	Labour £	Plant £	Material £	Unit	Total rate £
Q22 ADDAGRIP							
Bound aggregates; Addagrip Surface Treatments UK Ltd; natural decorative resin bonded surface dressing laid to concrete, macadam or to plywood panels priced separately							
"Angular Buff", "Terracotta", "Autumn Gold" or "Derbyshire cream"; 6 mm aggregate in Addaset							
15 mm depth	-	-	-	-	-	m²	**55.00**
18 mm depth	-	-	-	-	-	m²	**65.00**
"Cobalt Blue Glass"; 6 mm							
15 mm depth	-	-	-	-	-	m²	**80.00**
"Midnight Grey"; 6 mm							
15 mm depth	-	-	-	-	-	m²	**55.00**
"Chocolate"; 6 mm							
15 mm depth	-	-	-	-	-	m²	**55.00**
18 mm depth	-	-	-	-	-	m²	**65.00**
Q25 STEPS							
In situ concrete steps; ready mixed concrete; Gen 1; on 100 mm hardcore base; base concrete thickness 100 mm treads and risers cast simultaneously; inclusive of A142 mesh within the concrete; step riser 170 mm high; inclusive of all shuttering and concrete labours; floated finish; to graded ground (not included)							
Treads 250 mm wide							
one riser	5.29	4.00	73.00	-	7.24	m	**80.24**
two risers	8.48	6.00	109.50	-	11.44	m	**120.94**
three risers	12.28	8.00	146.00	-	20.03	m	**166.03**
four risers	16.49	12.00	219.00	-	24.78	m	**243.78**
six risers	20.88	16.00	292.00	-	30.04	m	**322.04**
Lay surface to concrete in situ steps above							
granite treads; pre-cut off site; 305 mm wide x 40 mm thick bullnose finish inclusive of 20 mm overhang; granite riser 20 mm thick all laid on 15 mm mortar bed	21.18	2.00	36.50	-	22.06	m	**58.56**
riven yorkstone tread and riser 50 mm thick; cut on site	30.02	2.50	45.63	-	30.90	m	**76.53**
riven yorkstone treads 225 mm wide x 50 mm thick cut on site; two courses brick risers laid on flat stretcher bond; bricks PC £500.00/1000	14.68	3.00	54.75	-	20.33	m	**75.08**
Marshalls Saxon paving slab 450 x 450 mm cut to 225 mm wide x 38 mm thick; cut on site; two courses brick risers laid on flat stretcher bond; bricks PC £500.00/1000	4.56	3.00	54.75	-	10.21	m	**64.96**
Treads 450 mm wide							
one riser	9.53	5.00	91.25	-	11.92	m	**103.17**
two risers	19.47	7.00	127.75	-	24.34	m	**152.09**
three risers	27.52	9.00	164.25	-	36.60	m	**200.85**
four risers	47.60	13.00	237.25	-	57.33	m	**294.58**
six risers	60.49	18.00	328.50	-	70.84	m	**399.34**

NEW ITEMS FOR THIS EDITION

Item Excluding site overheads and profit	PC £	Labour hours	Labour £	Plant £	Material £	Unit	Total rate £
Lay surface to concrete in situ steps							
granite treads; pre-cut off site; 470 mm wide x 40 mm thick bullnose finish inclusive of 20 mm overhang; granite riser 20 mm thick all laid on 15 mm mortar bed	30.28	4.00	73.00	-	31.39	m	104.39
riven yorkstone tread and riser 50 mm thick; cut on site	40.79	4.50	82.13	-	41.89	m	124.02
riven yorkstone treads 460 mm wide 50 mm thick cut on site; two courses brick risers laid on flat stretcher bond; bricks PC £500.00/1000	30.02	5.00	91.25	-	35.84	m	127.09
Marshalls Saxon paving slab 450 x 450 x 38 mm thick cut on site; two courses brick risers laid on flat stretcher bond; bricks PC £500.00/1000	9.12	2.75	50.19	-	14.77	m	64.96
Q25 SUDS - SUSTAINABLE PAVINGS							
SUDS paving; Concrete Block Permeable Paving (CBPP) Note: Sustainable Drainage Systems (SUDS) aim to reduce flood risk by controlling the rate and volume of surface water run off from developments. Permeable Paving is a SUDS technique. In addition to managing the quantity of surface water run off more effectively, one of the primary benefits of CBPP is the potential enhancement of water quality. Permeable pavements improve water quality by mirroring nature in providing filtration and allowing for natural biodegradation of hydrocarbons and the dilution of other contaminants as water passes through the system.							
Bases for Sustainable Urban Drainage System (SUDS) sub-base for use below impervious paving; Aggregate Industries Ltd; Bardon DrainAgg 20 4/20 mm open graded material laid loose on prepared sub-grade; within edge restraints to paving area (not included); graded to levels and falls as per paving manufacturer's instructions							
150 mm thick	5.10	0.03	0.46	0.49	5.10	m²	6.05
200 mm thick	6.80	0.02	0.30	0.66	6.80	m²	7.76
250 mm thick	8.50	0.02	0.38	0.82	8.50	m²	9.70
350 mm thick	11.90	0.03	0.52	1.13	11.90	m²	13.55
450 mm thick	15.30	0.04	0.73	1.58	15.30	m²	17.61

NEW ITEMS FOR THIS EDITION

Item Excluding site overheads and profit	PC £	Labour hours	Labour £	Plant £	Material £	Unit	Total rate £
Q25 SUDS - SUSTAINABLE PAVINGS - cont'd							
Sub-base replacement system for SUDS paving "Charcon Permavoid"; Aggregate Industries Ltd; geocellular, interlocking, high strength; high void capacity (92%); typically a 4:1 ratio when compared with alternative granular systems; allows for water storage to be confined at shallow depth within the subbase layer; inclusive of connection, ties, pins; does not include for encapsulation in geotextile/geomembranes							
708 x 354 x 150 mm							
150 mm thick	51.64	0.17	3.04	-	51.64	m²	**54.68**
300 mm thick	107.12	0.33	6.08	-	107.12	m²	**113.20**
Encapsulation of Permavoid; 150 mm thick systems; geomembrane geofabric or combinatiion of both for filtration or attenuation of water; please see the manufacturers design guide							
Charcon geomembrane, heat welded	10.58	0.05	0.91	-	12.70	m²	**13.61**
Charcon non reinforced geotextile	6.55	0.03	0.61	-	7.87	m²	**8.47**
Charcon reinforced geotextile	3.52	0.03	0.61	-	4.22	m²	**4.83**
Encapsulation of Permavoid; 300 mm thick systems; geomembrane geofabric or combinatiion of both for filtration or attenuation of water; please see the manufacturers design guide							
Charcon geomembrane, heat welded	11.96	0.06	1.01	-	14.35	m²	**15.37**
Charcon non reinforced geotextile	7.41	0.04	0.78	-	8.89	m²	**9.67**
Charcon reinforced geotextile	3.98	0.04	0.65	-	3.98	m²	**4.63**
Pervious surfacing materials; Aggregate Industries Ltd; Charcon Infilta CBPP Rectangular permeable block paving system; incorporating a 5 mm spacer design that provides a 5 mm void allowing ingress of water through to the sub base storage system (priced separately); various colours; bedded on 50 mm thick 2 - 6.3 mm clean, angular, free draining uncompacted aggregate; joints infilled with 3 mm clean grit							
"Infilta" 200 x 100 x 80 mm thick; natural grey	14.25	0.44	8.11	0.10	16.75	m²	**24.97**
"Infilta" 200 x 100 x 80 mm thick; coloured	14.75	0.44	8.11	0.10	17.25	m²	**25.47**
Q30 TURF							
Industrially grown turf							
Inturf							
"RTF" Rhizomatous Tall Fescue; deep rooted self repairing for difficult site or drought conditions or poor soils	2.42	0.08	1.46	-	2.42	m²	**3.88**

NEW ITEMS FOR THIS EDITION

Item Excluding site overheads and profit	PC £	Labour hours	Labour £	Plant £	Material £	Unit	**Total rate £**
Q31 PLANTING							
MARKET PRICES OF NATIVE SPECIES; Oakover Nurseries Ltd; The following prices are typically some of the more popular species used in native plantations; prices vary between species; readers should check the catalogue of the supplier							
1+1; 1 year seedling transplanted and grown for a year							
30 - 40 cm	-	-	-	-	0.45	nr	-
40 - 60 cm	-	-	-	-	0.55	nr	-
1+2; 1 year seedling transplanted and grown for 2 years							
60 - 80 cm	-	-	-	-	0.65	nr	-
Whips							
80 -100 cm	-	-	-	-	0.90	nr	-
100 -125 cm	-	-	-	-	1.30	nr	-
Feathered trees							
125 -150 cm	-	-	-	-	2.10	nr	-
150 -175 cm	-	-	-	-	2.50	nr	-
175 - 200 cm	-	-	-	-	4.00	nr	-
200 -250 cm	-	-	-	-	6.50	nr	-
250 -300 cm	-	-	-	-	8.00	nr	-
Native species planting Forming planting holes; in cultivated ground (cultivating not included); by mechanical auger; trimming holes by hand; depositing excavated material alongside holes							
250 mm diameter	-	0.03	0.61	0.04	-	nr	**0.65**
250 x 250 mm	-	0.04	0.73	0.08	-	nr	**0.81**
300 x 300 mm	-	0.08	1.37	0.10	-	nr	**1.46**
Hand excavation; forming planting holes; in cultivated ground (cultivating not included); depositing excavated material alongside holes							
100 x 100 x 100 mm deep; with mattock or hoe	-	0.01	0.12	-	-	nr	**0.12**
250 x 250 x 300 mm deep	-	0.04	0.73	-	-	nr	**0.73**
300 x 300 x 300 mm deep	-	0.06	1.01	-	-	nr	**1.01**

NEW ITEMS FOR THIS EDITION

Item Excluding site overheads and profit	PC £	Labour hours	Labour £	Plant £	Material £	Unit	Total rate £
Q35 LANDSCAPE CHEMICALS							
Chemical applications labours only							
Controlled droplet application (CDA)							
plants at 1.50 m centres; 4444 nr/ha	-	18.52	337.99	-	-	ha	337.99
plants at 1.50 m centres; 44.44 nr/100 m^2	-	1.85	33.80	-	-	100 m^2	33.80
plants at 1.75 m centres; 3265 nr/ha	-	13.60	248.20	-	-	ha	248.20
plants at 1.75 m centres; 3265 nr/ha	-	1.36	24.82	-	-	100 m^2	24.82
plants at 2.00 m centres; 2500 nr/ha	-	10.42	190.16	-	-	ha	190.16
plants at 2.00 m centres; 2500 nr/ha	-	1.04	19.02	-	-	100 m^2	19.02
spot spraying to shrub beds	-	0.40	7.30	-	-	100 m^2	7.30
spot spraying to hard surfaces	-	0.25	4.56	-	-	100 m^2	4.56
mass spraying	-	5.00	91.25	-	-	ha	91.25
mass spraying	-	0.50	9.13	-	-	100 m^2	9.13
Power spraying							
mass spraying to areas containing vegetation	-	4.00	73.00	60.02	-	ha	133.02
mass spraying to hard surfaces	-	0.01	0.15	0.12	-	100 m^2	0.28
spraying to kerbs and channels along roadways 300 mm wide	-	2.50	45.63	37.51	-	1 km	83.14
General herbicides; in accordance with manufacturer's instructions; "Knapsack" spray application; see "Market rates of chemicals" for costs of specific applications							
knapsack sprayer; selective spraying around bases of plants	-	0.35	6.40	-	-	100 m^2	6.40
knapsack sprayer; spot spraying	-	0.20	3.65	-	-	100 m^2	3.65
knapsack sprayer; mass spraying	-	0.20	3.65	-	-	100 m^2	3.65
granular distribution by hand or hand applicator	-	0.40	7.30	-	-	100 m^2	7.30
Backpack spraying; keeping weed free at bases of plants							
plants at 1.50 m centres; 4444 nr/ha	-	27.78	506.99	-	-	ha	506.99
plants at 1.75 m centres; 3265 nr/ha	-	20.40	372.30	-	-	ha	372.30
plants at 2.00 m centres; 2500 nr/ha	-	15.63	285.25	-	-	ha	285.25
mass spraying	-	15.00	273.75	-	-	ha	273.75
The following table provides the areas of spraying in each planting scenario for plants which require 1.00 m diameter weed free circles							
Area of weed free circles required in each planting density							
plants at 500 mm centres; 400 nr/100 m^2	-	-	-	-	400.00	m^2	-
plants at 600 m centres; 278 nr/100 m^2	-	-	-	-	278.00	m^2	-
plants at 750 mm centres; 178 nr/100 m^2	-	-	-	-	178.00	m^2	-
plants at 1.00 m centres; 100 nr/100 m^2	-	-	-	-	100.00	m^2	-
plants at 1.50 m centres; 4444 nr/ha	-	-	-	-	3490.00	m^2	-
plants at 1.75 m centres; 3265 nr/ha	-	-	-	-	2564.00	m^2	-
plants at 2.00 m centres; 2500 nr/ha	-	-	-	-	1963.00	m^2	-

NEW ITEMS FOR THIS EDITION

Item Excluding site overheads and profit	PC £	Labour hours	Labour £	Plant £	Material £	Unit	Total rate £
MARKET PRICES OF LANDSCAPE CHEMICALS AT SUGGESTED APPLICATION RATES							
Note: All chemicals are standard knapsack or backpack applied unless specifically stated as CDA or TDA.							
Total herbicides							
Herbicide applications; CDA; chemical application via low pressure specialized wands to landscape planting; application to maintain 1.00 m diameter clear circles (0.79 m²) around new planting Vanquish Biactive; ALS Ltd; Bayer; enhanced movement glyphosate; application rate 15 litre/ha							
plants at 1.50 m centres; 4444 nr/ha	-	-	-	-	46.02	ha	-
plants at 1.75 m centres; 3265 nr/ha	-	-	-	-	33.97	ha	-
plants at 2.00 m centres; 2500 nr/ha	-	-	-	-	26.05	ha	-
mass spraying	-	-	-	-	132.00	ha	-
spot spraying	-	-	-	-	1.32	ha	-
Total herbicides; CDA; low pressure sustainable applications "Hilite"; Nomix Enviro Ltd; glyphosate oil-emulsion; Total Droplet Control (TDC formerly CDA); low volume sustainable chemical applications; formulation controls annual and perennial grasses and non-woody broad-leaved weeds in amenity and industrial areas, parks, highways, paved areas and factory sites; also effective for clearing vegetation prior to planting of ornamental species, around the bases of trees, shrubs and roses, as well as in forestry areas							
10 l/ha	-	-	-	-	1.58	100 m²	-
7.5 l/ha	-	-	-	-	1.18	100 m²	-
"Tribute Plus"; Nomix Enviro Ltd ; TDA; MCPA/dicamba/mecaprop-p; TDC; for selective control of broad-leaved weeds in turf							
6 l/ha; white clover, plantains, creeping buttercup, thistle	-	-	-	-	0.26	100 m²	-
8 l/ha; yarrow, ragwort, parsley, piert, cinquefoil, mouse eared chickweed, selfheal, pearlwort, sorrel, black meddick, dandelion and related weeds, bulbous buttercup	-	-	-	-	0.35	100 m²	-
"Stirrup"; Rigby Taylor Ltd; glyphosate; CDA; 192 g							
10 l/ha	-	-	-	-	1.58	100 m²	-
7.5 l/ha	-	-	-	-	1.58	100 m²	-

NEW ITEMS FOR THIS EDITION

Item Excluding site overheads and profit	PC £	Labour hours	Labour £	Plant £	Material £	Unit	Total rate £
MARKET PRICES OF LANDSCAPE CHEMICALS AT SUGGESTED APPLICATION RATES - cont'd							
Total herbicides - cont'd							
"Gallup Biograde Amenity"; Rigby Taylor Ltd; glyphosate 360 g/l formulation							
general use; woody weeds; ash, beech, bracken, bramble; 3 l/ha	-	-	-	-	0.38	100 m^2	-
annual and perennial grasses; heather (peat soils); 4 l/ha	-	-	-	-	0.51	100 m^2	-
pre-planting; general clearance; 5 l/ha	-	-	-	-	0.63	100 m^2	-
heather; mineral soils; 6 l/ha	-	-	-	-	0.76	100 m^2	-
rhododendron; 10 l/ha	-	-	-	-	1.26	100 m^2	-
"Gallup Hi-aktiv Amenity"; Rigby Taylor Ltd; 490 g/l formulation							
general use; woody weeds; ash, beech, bracken, bramble; 2.2 l/ha	-	-	-	-	0.34	100 m^2	-
annual and perennial grasses; heather (peat soils); 2.9 l/ha	-	-	-	-	0.45	100 m^2	-
pre-planting; general clearance; 3.7 l/ha	-	-	-	-	0.57	100 m^2	-
heather; mineral soils; 4.4 l/ha	-	-	-	-	0.68	100 m^2	-
rhododendron; 7.3 l/ha	-	-	-	-	1.12	100 m^2	-
"Discman" CDA Biograde; Rigby Taylor Ltd							
application rate; 7 l/ha; light vegetation and annual weeds	-	-	-	-	1.01	100 m^2	-
application rate; 8.5 l/ha; forestry/woodlands; before planting, to control broad-leaved and grass weeds	-	-	-	-	1.23	100 m^2	-
application rate; 10 l/ha; established deep-rooted annuals, perennial grasses and broad-leaved weeds	-	-	-	-	1.45	100 m^2	-
Total herbicides; standard backpack or power sprayer applications							
"Roundup Pro Biactive 360"; Scotts UK							
application rate; 5 l/ha; land not intended for cropping	-	-	-	-	0.61	100 m^2	-
application rate; 10 l/ha; forestry; weed control	-	-	-	-	1.21	100 m^2	-
"Roundup Pro Biactive 450"; Scotts UK; 450 g/litre glyphosate							
application rate 4 l/ha; natural surfaces not intended to bear vegetation; amenity vegetation control; permeable surfaces overlaying soil; hard surfaces	-	-	-	-	0.58	100 m^2	-
application rate 8 l/ha; forest; forest nursery weed control	-	-	-	-	1.15	100 m^2	-
application rate 16 mm/l; stump application	-	-	-	-	0.23	litre	-
"Roundup Pro Biactive 450" Scotts UK; enhanced movement glyphosate; application rate 4 litre/ha							
plants at 1.50 m centres; 4444 nr/ha	75.74	-	-	-	83.32	ha	-
plants at 1.75 m centres; 3265 nr/ha	55.71	-	-	-	61.28	ha	-
plants at 2.00 m centres; 2500 nr/ha	42.62	-	-	-	46.89	ha	-
mass spraying	57.60	-	-	-	63.36	ha	-
spot spraying	2.16	-	-	-	2.38	100 m^2	-

Prices for Measured Works – Major Works

NEW ITEMS FOR THIS EDITION

Item Excluding site overheads and profit	PC £	Labour hours	Labour £	Plant £	Material £	Unit	Total rate £
Kerb							
"Kerb Flowable"; propyzamide	-	-	-	-	2.16	100 m²	-
"Kerb Granules"; 2 x 2000 tree pack	-	-	-	-	0.06	tree	-
"Proliance Quattro"; Scotts UK; 360 g/l glyphosate; advanced glyphosate formulation which incorporates a fully optimized surfactant system and water conditioner; to hazard classification; offers a high standard of operator and environmental safety							
application rate 6 l/ha; perennial broad-leaved weeds	-	-	-	-	1.05	100 m²	-
application rate 4 l/ha; perennial grasses	-	-	-	-	0.70	100 m²	-
application rate 1.5 l/ha; annual weeds	-	-	-	-	0.26	100 m²	-
"Dual"; Nomix Enviro Ltd; glyphosate/ sulfosulfuron; TDC; low volume sustainable chemical applications; residual effect lasting up to 6 months; shrub beds, gravelled areas, tree bases, parks, industrial areas and around obstacles							
application rate 9 l/ha	-	-	-	-	1.98	100 m²	-
"Glyfos Dakar Pro"; Headland Amenity Ltd; 68% glyphosate as water dispersible granules; broad-leaved weeds and grasses							
application rate 2.5 kg/ha	-	-	-	-	0.40	100 m²	-
Total herbicides with residual pre-emergent action							
"Pistol"; ALS; Bayer; barrier prevention for emergence control and glyphosate for control of emerged weeds at a single application per season; application rate 4.5 litre/ha; 40 g/l diflufenican and 250 g/l glyphosate							
plants at 1.50 m centres; 4444 nr/ha	-	-	-	-	32.81	ha	-
plants at 1.75 m centres; 3265 nr/ha	-	-	-	-	24.04	ha	-
plants at 2.00 m centres; 2500 nr/ha	-	-	-	-	23.51	ha	-
Residual herbicides							
"Chikara"; Rigby Taylor Ltd; "Belchim Crop Protection"; flazasulfuron; systemic pre-emergent and early post-emergent herbicide for the control of annual and perennial weeds on natural surfaces not intended to bear vegetation and permeable surfaces over-lying soil							
150 g/ha	-	-	-	-	2.49	100 m²	-
Contact herbicides							
"Jewel"; Scotts UK; carfentrazone and mecoprop-p; selective weed and mosskiller							
application rate; 1.5 kg/ha	-	-	-	-	1.25	100 m²	-
"Natural Weed and Moss spray"; Headland Amenity Ltd; acetic acid							
application rate; 100 l/ha	-	-	-	-	4.50	100 m²	-

NEW ITEMS FOR THIS EDITION

Item Excluding site overheads and profit	PC £	Labour hours	Labour £	Plant £	Material £	Unit	Total rate £
MARKET PRICES OF LANDSCAPE CHEMICALS AT SUGGESTED APPLICATION RATES - cont'd							
Selective herbicides CDA "Hilite"; Nomix Enviro; glyphosate oil-emulsion; TD ; low volume sustainable chemical applications; formulation controls annual and perennial grasses and non-woody broad-leaved weeds in amenity and industrial areas, parks, highways, paved areas and factory sites; also effective for clearing vegetation prior to planting of ornamental species, around the bases of trees, shrubs and roses, as well as in forestry areas							
10 l/ha	–	–	–	–	1.58	100 m²	–
7.5 l/ha	–	–	–	–	1.18	100 m²	–
"Tribute Plus"; Nomix Enviro; MCPA/ dicamba/mecaprop-p; TDC; for selective control of broad-leaved weeds in turf							
6 l/ha; white clover, plantains, creeping buttercup, thistle	–	–	–	–	0.26	100 m²	–
8 l/ha; yarrow, ragwort, parsley, piert, cinquefoil, mouse eared chickweed, selfheal, pearlwort, sorrel, black meddick, dandelion and related weeds, bulbous buttercup	–	–	–	–	0.35	100 m²	–
Contact and systemic combined herbicides "Jewel"; Scotts UK; carfentrazone and mecoprop-p; selective weed and mosskiller							
application rate; 1.5 kg/ha	–	–	–	–	1.25	100 m²	–
"Finale"; ALS; Bayer; glufosinate-ammonium; controls grasses and broad-leaved weeds; it can also be used for weed control around non-edible ornamental crops including trees and shrubs and in tree nurseries; may be used to "burn off" weeds in shrub beds prior to treatment with a residual herbicide; is also ideal for preparing sports turf for linemarking							
application rate; 5 l/ha; seedlings of all species; established annual weeds and grasses as specified	–	–	–	–	0.79	100 m²	–
application rate; 8 l/ha; established annual and perennial weeds and grasses as specified in the product guide	–	–	–	–	1.26	100 m²	–

NEW ITEMS FOR THIS EDITION

Item Excluding site overheads and profit	PC £	Labour hours	Labour £	Plant £	Material £	Unit	Total rate £
Selective herbicides; power or knapsack sprayer							
"Junction"; Rigby Taylor Ltd; selective post-emergence turf herbicide containing 6.25 g/litre florasulam plus 452 g/litre 2,4-D; specially designed for use by contractors for maintenance treatments of a wide range of broad-leaved weeds whilst giving excellent safety to turfgrass							
application rate; 1.2 l/ha	-	-	-	-	0.50	100 m²	-
"Greenor"; Rigby Taylor Ltd; fluroxypyr, clopyralid and MCPA; systemic selective treatment of weeds in turfgrass							
application rate; 4 l/ha	-	-	-	-	1.00	100 m²	-
"Super Selective Plus"; Rigby Taylor Ltd; combination of three active ingredients (mecoprop-p, MCPA and dicamba) enables this formulation to give good control of most commonly occurring broad-leaved weeds found in sports and amenity turf							
application rate; 3.5 l/ha	-	-	-	-	0.21	100 m²	-
"Crossbar"; Rigby Taylor Ltd; fluroxypyr, 2, 4-D and dicamba for control of weeds in amenity turf including fine turf							
application rate; 2 l/ha	-	-	-	-	0.76	100 m²	-
"Barclay Greengard"; Rigby Taylor Ltd; 2, 4-D, triclopyr, and dicamba; selective herbicide for the control of certain broad-leaved weeds in amenity grassland and scrub control in forestry and uncropped areas including Japanese Knotweed and weeds listed under the Weeds Act of 1959							
application rate; 5 l/ha	-	-	-	-	2.48	100 m²	-
application rate; 2 l/ha	-	-	-	-	0.99	100 m²	-
"Intrepid 2"; Scotts UK; for broad-leaved weed control in grass							
application rate; 7700 m²/5 l	-	-	-	-	0.51	100 m²	-
Re-Act; Scotts UK; MCPA, mecoprop-p and dicamba; selective herbicide; established managed amenity turf and newly seeded grass; controls many annual and perennial weeds; will not vaporise in hot conditions							
application rate; 9090 - 14,285 m²/5 l	-	-	-	-	0.46	100 m²	-
"Praxys"; Scotts UK; fluroxypyr-meptyl, clopyralid and florasulam; selective, systemic post-emergence herbicide for managed amenity turf, lawns and amenity grassland with high selectivity to established and young turf							
application rate; 9090 - 14,285 m²/5 l	-	-	-	-	1.39	100 m²	-
"Asulox"; ALS; Bayer; post-emergence translocated herbicide for the control of bracken and docks							
application rate; docks; 17857 m²/5 l	-	-	-	-	0.34	100 m²	-
application rate; bracken; 4545 m²/5 l	-	-	-	-	1.32	100 m²	-
"Spearhead"; ALS; Bayer; selective herbicide for the control of broad-leaved weeds in established turf							
application rate; 4.5 l/ha	-	-	-	-	0.85	100 m²	-

NEW ITEMS FOR THIS EDITION

Item Excluding site overheads and profit	PC £	Labour hours	Labour £	Plant £	Material £	Unit	Total rate £
MARKET PRICES OF LANDSCAPE CHEMICALS AT SUGGESTED APPLICATION RATES - cont'd							
Selective herbicides - cont'd							
"Cabadex"; Headland Amenity Ltd; fluroxypyr and florasulam; newly sown and established turf (particularly speedwell, yarrow and yellow suckling clover)							
application rate 2 l/ha	-	-	-	-	0.65	100 m²	-
"Relay Turf"; Headland Amenity Ltd; mecoprop-p, MCPA and dicamba; many broad-leaved weeds in managed amenity turf							
application rate; 5 l/ha	-	-	-	-	0.45	100 m²	-
"Blaster"; Headland Amenity Ltd; triclopyr clopyralid; nettles, docks, thistles, brambles, woody weeds in amenity grassland							
application rate; 4 l/ha	-	-	-	-	1.40	100 m²	-
Woody weed herbicides; backpack or power sprayers							
"Timbrel"; Rigby Taylor Ltd; Bayer; summer applied; water based; selective scrub and brushwood herbicide							
2.0 l/ha; bramble, briar, broom, gorse, nettle	-	-	-	-	1.24	100 m²	-
4.0 l/ha; alder, birch, blackthorn, dogwood, elder, poplar, rosebay willowherb, sycamore	-	-	-	-	2.49	100 m²	-
6.0 l/ha; beech, box, buckthorn, elm, hazel, hornbeam, horse chestnut, lime, maple, privet, rowan, Spanish chestnut, willow, wild pear	-	-	-	-	3.73	100 m²	-
8.0 l/ha; ash, oak, rhododendron	-	-	-	-	4.98	100 m²	-
"Timbrel"; Rigby Taylor Ltd; Bayer; winter applied; paraffin or diesel based; selective scrub and brushwood herbicide							
3.0 l/ha; bramble, briar, broom, gorse, nettle	-	-	-	-	1.87	100 m²	-
6.0 l/ha; alder, ash, beech, birch, blackthorn, box, buckthorn, dogwood, elder, elm, hazel, hornbeam, horse chestnut, lime, maple, oak, poplar, privet, rowan, Spanish chestnut, sycamore, willow, wild pear	-	-	-	-	3.73	100 m²	-
10.0 l/ha; hawthorn, laurel, rhododendron	-	-	-	-	6.22	100 m²	-
"Garlon 4"; Nomix Enviro Ltd; 667 g/litre triclopyr; control of brushwood, scrub, woody and perennial herbaceous weeds in amenity, forestry, rail and industrial areas; also suitable for forestry production							
2 l/ha; bramble, briar, broom, gorse, nettle	-	-	-	-	0.98	100 m²	-
4 l/ha; alder, birch, blackthorn, dogwood, elder, poplar, rosebay, willow-herb, sycamore	-	-	-	-	1.96	100 m²	-

NEW ITEMS FOR THIS EDITION

Item Excluding site overheads and profit	PC £	Labour hours	Labour £	Plant £	Material £	Unit	Total rate £
6 l/ha; beech, box, elm, buckthorn, hazel, hornbeam, horse chestnut, lime, maple, privet, rowan, Spanish chestnut, willow, wild pear	-	-	-	-	2.94	100 m²	-
8 l/ha; rhododendron, ash, oak	-	-	-	-	3.92	100 m²	-
"Tordon 22K"; Nomix Enviro; picloram; selective action; control of tough woody weeds, invasive species, scrub and deep-rooted perennials; residual action grass cover is maintained							
2.8 l/ha; white clover, creeping thistle, hogweed, plantains	-	-	-	-	1.19	100 m²	-
4.3 - 5.6 l/ha; common bird s-foot trefoil, field bindweed, creeping buttercup, burdock, cat s ear, common chickweed, colt s foot, creeping cinquefoil, dandelion, docks, hawkweed, Japanese knotweed, mugwort, common nettle, silverweed, tansy, cotton thistle, rosebay willowherb, yarrow	-	-	-	-	2.12	100 m²	-
8.4 l/ha; wild mignonettte, common ragwort	-	-	-	-	3.56	100 m²	-
11.2 l/ha; bracken, brambles	-	-	-	-	7.72	100 m²	-
Japanese knotweed and ragwort control "Loram 24"; ALS Ltd; Bayer; picloram							
4.2 l/ha; Japanese Knotweed and other weeds as listed in the suppliers literature	-	-	-	-	1.60	100 m²	-
5.6 l/ha; common ragwort, wild mignonette.	-	-	-	-	2.13	100 m²	-
Aquatic herbicides; backpack or power sprayers "Roundup Pro Biactive 360"; Scotts UK							
application rate; 6 l/ha; aquatic weed control	-	-	-	-	0.73	100 m²	-
"Roundup Pro Biactive 450"; Scotts UK; 450 g/litre glyphosate							
application rate 4.8 l/ha; enclosed waters, open waters, land immediately adjacent to aquatic areas	-	-	-	-	0.69	100 m²	-
Aquatic herbicides; TDC "Conqueror"; Nomix Enviro; low volume sustainable chemical applications glyphosate with aquatic approval							
10 l/ha; established annuals, deep-rooted perennials and broad-leaved weeds	-	-	-	-	0.16	100 m²	-
7.5 l/ha; light vegetation and small annual weeds aquatic situations	-	-	-	-	1.18	100 m²	-
12.5 l/ha; aquatic situations emergent weeds e.g. reed, grasses and watercress	-	-	-	-	0.20	100 m²	-
15 l/ha; aquatic situations floating weeds such as water lilies	-	-	-	-	0.24	100 m²	-

NEW ITEMS FOR THIS EDITION

Item Excluding site overheads and profit	PC £	Labour hours	Labour £	Plant £	Material £	Unit	Total rate £
MARKET PRICES OF LANDSCAPE CHEMICALS AT SUGGESTED APPLICATION RATES - cont'd							
Fungicides							
"Masalon"; Rigby Taylor Ltd; systemic							
application rate; 8 l/ha	-	-	-	-	7.54	100 m^2	-
"Defender"; Rigby Taylor Ltd; broad spectrum fungicide displaying protective and curative activity against fusarium patch and red thread							
application rate; 700 g/ha	-	-	-	-	6.23	100 m^2	-
"Fusion"; Rigby Taylor Ltd; contact and systemic fungicide with curative and preventative activity against major turf diseases such as fusarium, red thread, dollar spot, anthracnose and rust							
application rate; 700 g/ha	-	-	-	-	6.09	100 m^2	-
"Rayzor"; Rigby Taylor Ltd; broad spectrum contact fungicide with both protectant and eradicant activity for the control of fusarium patch and red thread in managed amenity turf							
application rate; 5 l/2000 m^2	-	-	-	-	8.69	100 m^2	-
"Mascot Contact"; multi-site protectant fungicide for the control of a range of diseases on turf							
application rate; 5 l/1667 m^2	-	-	-	-	6.59	100 m^2	-
"Daconil Weatherstik"; Scotts UK; turf							
application rate; 333 m^2/l	-	-	-	-	5.07	100 m^2	-
Moss control							
"Jewel"; Scotts UK; carfentrazone and mecoprop-p; selective weed and moss killer							
application rate; 1.5 kg/ha	-	-	-	-	1.25	100 m^2	-
Insect control							
"Crossfire 480"; Rigby Taylor Ltd; control of leatherjackets and fruit fly in sports and amenity turf							
application rate; 1.5 l/ha	-	-	-	-	0.56	100 m^2	-
"Merit Turf"; Bayer; insecticide for the control of chafer grubs and leatherjackets							
application rate; 10 kg/3333 m^2	-	-	-	-	5.95	100 m^2	-

Prices for Measured Works – Major Works

NEW ITEMS FOR THIS EDITION

Item Excluding site overheads and profit	PC £	Labour hours	Labour £	Plant £	Material £	Unit	Total rate £
Q35 TREATMENT OF JAPANESE KNOTWEED							
Japanese Knotweed Management Plan; PBA Solutions Ltd Formulate Japanese Knotweed Management Plan (KMP) as recommended by the Environment Agency Code of Practice 2006 (EA CoP).	-	-	-	-	-	nr	1200.00
Treatment of Japanese Knotweed; PBA Solutions Ltd; removal and/or treatment in accordance with 'The Knotweed Code of Practice' published by the Environment Agency 2006 (EACoP); Japanese knotweed must be treated by qualified practitioners Commercial chemical treatments based on areas less than 50 m² knotweed area; allows for the provision of contractual arrangements including method statements and spraying records; also allows for the provision of risk, environment and COSHH assessments							
rate per visit	-	-	-	-	-	nr	225.00
rate per year	-	-	-	-	-	nr	900.00
rate for complete treatment programme lasting 3 years	-	-	-	-	-	nr	2700.00
Commercial chemical treatments based on areas between 50-300 m² knotweed area; allows for the provision of contractual arrangements including method statements and spraying recordsl also allows for the provision of risk, environment and COSHH assessments							
rate per visit	-	-	-	-	-	nr	300.00
rate per year	-	-	-	-	-	nr	1200.00
rate for complete treatment programme lasting 3 years	-	-	-	-	-	nr	3600.00
Extra over minimum rate above; for areas 300 - 900 m²							
rate per visit	-	-	-	-	-	m²	0.48
rate per year	-	-	-	-	-	m²	1.92
rate for complete treatment programme lasting 3 years	-	-	-	-	-	m²	5.76
Extra over minimum rate above; for areas over 900 m²							
rate per visit	-	-	-	-	-	m²	0.40
rate per year	-	-	-	-	-	m²	1.60
rate for complete treatment programme lasting 3 years	-	-	-	-	-	m²	4.80

NEW ITEMS FOR THIS EDITION

Item Excluding site overheads and profit	PC £	Labour hours	Labour £	Plant £	Material £	Unit	Total rate £
Q35 TREATMENT OF JAPANESE KNOTWEED - cont'd							
Treatment of Japanese Knotweed - cont'd Note: Taking waste off site will require pre-treatment. Landfill operators are required by the Environment Agency to obtain evidence of pre-treatment to show that the waste has been treated to either reduce its volume or hazardousness nature, or facilitate its handling, or enhance its recovery by the waste produce (site of origin) prior being taken to landfill. Generally pre-treatment of Japanese Knotweed can include chemical treatment, cutting or burning. The cost to implement this operation would be similar to item 4/5 with a minimum charge of £300.00. To enable waste to go to landfill soil testing would need to be completed and consultation with a landfill operator or invasive weed specialist is recommended in the first instance.							
Japanese knotweed contaminated waste removal; PBA Solutions Ltd (see note above) Removal offsite to a licensed landfill site using 20 tonne gross vehicle weight lorries; price is for haulage and tipping only; the mileage stated below is the distance from point of collection to licensed landfill site; price excludes landfill tax and loading of material into lorries; rates will fluctuate between landfill operators and the figures given are for initial guidance only							
less than 20 miles	-	-	-	-	-	load	575.00
20 - 40 miles	-	-	-	-	-	load	637.00
40 - 60 miles	-	-	-	-	-	load	835.00
Rootbarriers; PBA Solutions Ltd; used in conjunctiont with on site control methods such as bund treatment, cell burial, vertical boundary protection and capping. The following barriers have been reliably used in connection with Japanese knotweed control and as such are effective when correctly installed as part of a control programme which would usually include for chemical control							
Linear root deflection barriers; installed to trench measured separately							
"Flexiroot" non permeable; reinforced LDPE coated polypropylene geomembrane	5.36	0.05	0.91	-	5.89	m	**6.80**
"bioroot-barrier x" permeable root barrier; mechanically bonded geocomposite consisting of a cooper-foil bonded between 2 layers of geotextile	10.76	0.05	0.91	-	10.76	m	**11.67**

NEW ITEMS FOR THIS EDITION

Item Excluding site overheads and profit	PC £	Labour hours	Labour £	Plant £	Material £	Unit	**Total rate £**
Landfill tax rates; based on waste being within landfill operator thresholds for controlled waste and with Japanese knotweed being the only form of contamination; the rates below are based on visual inspections of the load at the gate of the tip							
Rates to April 2010							
higher rate (where knotweed is visible at more than 5% on inspection at landfill)	-	-	-	-	40.00	tonne	-
lower rate (where knotweed is visible at less than 5% on inspection at landfill)	-	-	-	-	2.50	tonne	-
Rates from April 2010 - April 2011							
higher rate (where knotweed is visible at more than 5% on inspection at landfill)	-	-	-	-	48.00	tonne	-
lower rate (where knotweed is visible at less than 5% on inspection at landfill)	-	-	-	-	2.50	tonne	-
Q50 STRUCTURES							
Pergolas; AVS Fencing Supplies Ltd; construct timber pergola; posts 150 x 150 mm x 2.40 m finished height in 600 mm deep minimum concrete pits; beams of 200 x 50 mm x 2.00 m wide; fixed to posts with dowels; rafters 200 x 38 mm notched to beams; inclusive of all mechanical excavation and disposal off site							
Pergola 2.00 m wide in green oak; posts at 2.00 m centres; dowel fixed							
rafters at 600 mm centres	-	8.40	172.20	3.59	240.04	m	**415.83**
rafters at 400 mm centres	-	10.80	221.40	3.59	290.47	m	**515.46**
Pergola 3.00 m wide in green oak; posts at 1.50 m centres							
rafters at 600 mm centres	-	7.20	155.70	4.31	318.79	m	**478.80**
rafters at 400 mm centres	-	8.00	173.00	4.31	394.29	m	**571.60**
Pergola 2.00 wide in prepared softwood; beams fixed with bolts; posts at 2.00 m centres							
rafters at 600 mm centres	-	5.00	108.13	3.59	141.38	m	**253.10**
rafters at 400 mm centres	-	6.00	129.75	3.59	162.84	m	**296.18**
Pergola 3.00 m wide in prepared softwood; posts at 1.50 m centres							
rafters at 600 mm centres	-	7.20	155.70	4.31	140.92	m	**300.94**
rafters at 400 mm centres	-	8.00	173.00	4.31	161.35	m	**338.67**

A PRELIMINARIES

Item Excluding site overheads and profit	PC £	Labour hours	Labour £	Plant £	Material £	Unit	Total rate £
A11 TENDER AND CONTRACT DOCUMENTS							
Health and safety							
Produce health and safety file including preliminary meeting and subsequent progress meetings with external planning officer in connection with health and safety; project value							
£35,000	-	8.00	244.00	-	-	nr	**244.00**
£75,000	-	12.00	366.00	-	-	nr	**366.00**
£100,000	-	16.00	488.00	-	-	nr	**488.00**
£200,000 to £500,000	-	40.00	1220.00	-	-	nr	**1220.00**
Maintain health and safety file for project duration; project value							
£35,000	-	4.00	122.00	-	-	week	**122.00**
£75,000	-	4.00	122.00	-	-	week	**122.00**
£100,000	-	8.00	244.00	-	-	week	**244.00**
£200,000 to £500,000	-	8.00	244.00	-	-	week	**244.00**
Produce written risk assessments on all areas of operations within the scope of works of the contract; project value							
£35,000	-	2.00	61.00	-	-	week	**61.00**
£75,000	-	3.00	91.50	-	-	week	**91.50**
£100,000	-	5.00	152.50	-	-	week	**152.50**
£200,000 to £500,000	-	8.00	244.00	-	-	week	**244.00**
Produce COSHH assessments on all substances to be used in connection with the contract; project value							
£35,000	-	1.50	45.75	-	-	nr	**45.75**
£75,000	-	2.00	61.00	-	-	nr	**61.00**
£100,000	-	3.00	91.50	-	-	nr	**91.50**
£200,000 to £500,000	-	3.00	91.50	-	-	nr	**91.50**
Method statements							
Provide detailed method statements on all aspects of the works; project value							
£30,000	-	2.50	76.25	-	-	nr	**76.25**
£50,000	-	3.50	106.75	-	-	nr	**106.75**
£75,000	-	4.00	122.00	-	-	nr	**122.00**
£100,000	-	5.00	152.50	-	-	nr	**152.50**
£200,000	-	6.00	183.00	-	-	nr	**183.00**

A PRELIMINARIES

Item Excluding site overheads and profit	PC £	Labour hours	Labour £	Plant £	Material £	Unit	Total rate £
A30 EMPLOYERS REQUIREMENTS: TENDERING/SUBLETTING/SUPPLY							
Tendering costs for an employed estimator for the acquisition of a landscape main or subcontract; inclusive of measurement, sourcing of materials, and suppliers, cost and area calculations, pre and post tender meetings, bid compliance method statements and submission documents							
Remeasureable contract from prepared bills; contract value							
£25,000		-	-	-	-	nr	180.00
£50,000		-	-	-	-	nr	300.00
£100,000		-	-	-	-	nr	720.00
£200,000		-	-	-	-	nr	1200.00
£500,000		-	-	-	-	nr	1560.00
£1,000,000		-	-	-	-	nr	2400.00
Lump sum contract from specifications and drawings only; contract value							
£25,000		-	-	-	-	nr	480.00
£50,000		-	-	-	-	nr	600.00
£100,000		-	-	-	-	nr	1200.00
£200,000		-	-	-	-	nr	2160.00
£500,000		-	-	-	-	nr	3000.00
£1,000,000		-	-	-	-	nr	4800.00
Material costs for tendering							
Drawing and plan printing and distribution costs; plans issued by employer on CD; project value							
£30,000		-	-	-	19.00	nr	-
£30,000 to £80,000		-	-	-	47.50	nr	-
£80,000 to £150,000		-	-	-	95.00	nr	-
£150,000 to £300,000		-	-	-	133.00	nr	-
£300,000 to £1,000,000		-	-	-	228.00	nr	-
A32 EMPLOYERS REQUIREMENTS: MANAGEMENT OF THE WORKS							
Programmes							
Allow for production of works programmes prior to the start of the works; project value							
£30,000		3.00	91.50	-	-	nr	91.50
£50,000		6.00	183.00	-	-	nr	183.00
£75,000		8.00	244.00	-	-	nr	244.00
£100,000		10.00	305.00	-	-	nr	305.00
£200,000		14.00	427.00	-	-	nr	427.00
Allow for updating the works programme during the course of the works; project value							
£30,000		1.00	30.50	-	-	nr	30.50
£50,000		1.50	45.75	-	-	nr	45.75
£75,000		2.00	61.00	-	-	nr	61.00
£100,000		3.00	91.50	-	-	nr	91.50
£200,000		5.00	152.50	-	-	nr	152.50

A PRELIMINARIES

Item Excluding site overheads and profit	PC £	Labour hours	Labour £	Plant £	Material £	Unit	Total rate £
A32 EMPLOYERS REQUIREMENTS: MANAGEMENT OF THE WORKS - cont'd							
Setting out							
Setting out for external works operations comprising hard and soft works elements; placing of pegs and string lines to Landscape Architect's drawings; surveying levels and placing level pegs; obtaining approval from the Landscape Architect to commence works; areas of entire site							
1,000 m²	-	5.00	91.25	-	6.00	nr	**97.25**
2,500 m²	-	8.00	146.00	-	12.00	nr	**158.00**
5,000 m²	-	8.00	146.00	-	12.00	nr	**158.00**
10,000 m²	-	32.00	584.00	-	30.00	nr	**614.00**
A34 SECURITY/SAFETY/PROTECTION							
Temorary security fence; HSS Hire; mesh framed unclimbable fencing; including precast concrete supports and couplings Weekly hire; 2.85 x 2.00 m high							
weekly hire rate	-	-	-	1.28	-	m	**1.28**
erection of fencing; labour only	-	0.10	1.82	-	-	m	**1.82**
removal of fencing loading to collection vehicle	-	0.07	1.22	-	-	m	**1.22**
delivery charge	-	-	-	0.80	-	m	**0.80**
return haulage charge	-	-	-	0.60	-	m	**0.60**
A41 CONTRACTORS GENERAL COST ITEMS: SITE ACCOMODATION							
General The following items are instances of the commonly found preliminary costs associated with external works contracts. The assumption is made that the external works contractor is subcontracted to a main contractor.							
Elliot Hire; erect temporary accommodation and storage on concrete base measured separately Prefabricated office hire; jackleg; open plan							
3.6 x 2.4 m	-	-	-	-	-	week	**36.75**
4.8 x 2.4 m	-	-	-	-	-	week	**36.75**
Armoured store; erect temporary secure storage container for tools and equipment							
3.0 x 2.4 m	-	-	-	-	-	week	**12.00**
3.6 x 2.4 m	-	-	-	-	-	week	**12.00**
Delivery and collection charges on site offices							
delivery charge	-	-	-	-	-	load	**126.00**
collection charge	-	-	-	-	-	load	**126.00**

A PRELIMINARIES

Item Excluding site overheads and profit	PC £	Labour hours	Labour £	Plant £	Material £	Unit	Total rate £
Toilet facilities John Anderson Ltd; serviced self-contained toilet delivered to and collected from site; maintained by toilet supply company							
single chemical toilet including wash hand basin and water	-	-	-	35.00	-	week	**35.00**
delivery and collection; each way	-	-	-	25.00	-	nr	**25.00**
A43 CONTRACTOR'S GENERAL COST ITEMS: MECHANICAL PLANT							
Southern Conveyors; moving only of granular material by belt conveyor; conveyors fitted with troughed belts, receiving hopper, front and rear undercarriages and driven by electrical and air motors; support work for conveyor installation (scaffolding), delivery, collection and installation all excluded							
Conveyor belt width 400 mm; mechanically loaded and removed at offload point; conveyor length							
up to 5 m	-	1.50	27.38	6.55	-	m^3	**33.92**
10 m	-	1.50	27.38	7.31	-	m^3	**34.69**
12.5 m	-	1.50	27.38	7.67	-	m^3	**35.04**
15 m	-	1.50	27.38	8.09	-	m^3	**35.46**
20 m	-	1.50	27.38	8.87	-	m^3	**36.25**
25 m	-	1.50	27.38	9.56	-	m^3	**36.93**
30 m	-	1.50	27.38	10.32	-	m^3	**37.70**
Conveyor belt width 600 mm; mechanically loaded and removed at offload point; conveyor length							
up to 5 m	-	1.00	18.25	4.77	-	m^3	**23.03**
10 m	-	1.00	18.25	5.38	-	m^3	**23.64**
12.5 m	-	1.00	18.25	5.74	-	m^3	**23.99**
15 m	-	1.00	18.25	6.07	-	m^3	**24.32**
20 m	-	1.00	18.25	6.78	-	m^3	**25.03**
25 m	-	1.00	18.25	7.44	-	m^3	**25.69**
30 m	-	1.00	18.25	8.10	-	m^3	**26.36**
Conveyor belt width 400 mm; mechanically loaded and removed by hand at offload point; conveyor length							
up to 5 m	-	3.19	58.22	5.09	-	m^3	**63.31**
10 m	-	3.19	58.22	6.10	-	m^3	**64.33**
12.5 m	-	3.19	58.22	6.58	-	m^3	**64.80**
15 m	-	3.19	58.22	7.14	-	m^3	**65.36**
20 m	-	3.19	58.22	8.18	-	m^3	**66.41**
25 m	-	3.19	58.22	9.10	-	m^3	**67.32**
30 m	-	3.19	58.22	10.11	-	m^3	**68.34**

A PRELIMINARIES

Item Excluding site overheads and profit	PC £	Labour hours	Labour £	Plant £	Material £	Unit	Total rate £
A43 CONTRACTOR'S GENERAL COST ITEMS: MECHANICAL PLANT - cont'd							
Terranova Cranes Ltd; crane hire; materials handling and lifting; telescopic cranes supply and management; exclusive of roadway management or planning applications; prices below illustrate lifts of 1 tonne at maximum crane reach; lift cycle of 0.25 hours; mechanical filling of material skip if appropriate							
35 tonne mobile crane; lifting capacity of 1 tonne at 26 m; C.P.A hire (all management by hirer)							
granular materials including concrete	-	0.75	13.69	15.86	-	tonne	**29.55**
palletized or packed materials	-	0.50	9.13	13.13	-	tonne	**22.25**
35 tonne mobile crane; lifting capacity of 1 tonne at 26 m; contract lift							
granular materials or concrete	-	0.75	13.69	43.36	-	tonne	**57.05**
palletized or packed materials	-	0.50	9.13	40.63	-	tonne	**49.75**
50 tonne mobile crane; lifting capacity of 1 tonne at 34 m; C.P.A hire (all management by hirer)							
granular materials including concrete	-	0.75	13.69	19.92	-	tonne	**33.61**
palletized or packed materials	-	1.00	18.25	17.19	-	tonne	**35.44**
50 tonne mobile crane; lifting capacity of 1 tonne at 34 m; contract lift							
granular materials or concrete	-	0.75	13.69	49.55	-	tonne	**63.24**
palletized or packed materials	-	0.50	9.13	46.82	-	tonne	**55.94**
80 tonne mobile crane; lifting capacity of 1 tonne at 44 m; C.P.A. hire (all management by hirer)							
granular materials including concrete	-	0.75	13.69	27.73	-	tonne	**41.42**
palletized or packed materials	-	0.50	9.13	25.00	-	tonne	**34.13**
80 tonne mobile crane; lifting capacity of 1 tonne at 44 m; contract lift							
granular materials or concrete	-	0.75	13.69	55.86	-	tonne	**69.55**
palletized or packed materials	-	0.50	9.13	53.13	-	tonne	**62.25**
A44 CONTR'S GEN'L COSTS/ TEMPORARY ROADS							
Eve Trakway; portable roadway systems Temporary roadway system laid directly onto existing surface or onto PVC matting to protect existing surface; most systems are based on a weekly hire charge with transportation, installation and recovery charges included							
Heavy Duty Trakpanel; per panel (3.05 x 2.59 m) per week	-	-	-	-	-	m^2	**5.32**
Outrigger Mats for use in conjunction with Heavy Duty Trakpanels; per set of 4 mats per week	-	-	-	-	-	set	**85.00**
Medium Duty Trakpanel; per panel (2.44 x 3.00 m) per week	-	-	-	-	-	m^2	**5.74**

A PRELIMINARIES

Item Excluding site overheads and profit	PC £	Labour hours	Labour £	Plant £	Material £	Unit	Total rate £
LD20 Eveolution; Light Duty Trakway; roll out system minimum delivery 50 m	-	-	-	-	-	m²	**5.00**
Terraplas Walkways; turf protection system; per section (1 x 1 m); per week	-	-	-	-	-	m²	**5.25**

B COMPLETE BUILDINGS/STRUCTURES/UNITS

Item Excluding site overheads and profit	PC £	Labour hours	Labour £	Plant £	Material £	Unit	Total rate £
B10 PREFABRICATED BUILDINGS/ STRUCTURES							
Cast stone buildings; Haddonstone Ltd; ornamental garden buildings in Portland Bath or Terracotta finished cast stone; prices for stonework and facades only; excavations, foundations, reinforcement, concrete infill, roofing and floors all priced separately							
Pavilion Venetian Folly L9400; Tuscan columns, pedimented arch, quoins and optional balustrading							
4184 mm high x 4728 mm wide x 3147 mm deep	10500.00	175.00	3193.75	222.60	10616.70	nr	**14033.05**
Pavilion L9300; Tuscan columns							
3496 mm high x 3634 mm wide	6500.00	144.00	2628.00	190.80	6616.70	nr	**9435.50**
Small Classical Temple L9250; 6 column with fibreglass lead effect finish dome roof overall height 3610 mm; diameter							
2540 mm	5600.00	130.00	2372.50	159.00	5716.70	nr	**8248.20**
Large Classical Temple L9100; 8 column with fibreglass lead effect finish dome roof overall height 4664 mm; diameter							
3190 mm	10500.00	165.00	3011.25	159.00	10616.70	nr	**13786.95**
Stepped floors to temples							
single step; Small Classical Temple	1025.00	24.00	438.00	-	1072.04	nr	**1510.04**
single step; Large Classical Temple	1450.00	26.00	474.50	-	1506.45	nr	**1980.95**
Stone structures; Architectural Heritage Ltd; hand carved from solid natural limestone with wrought iron domed roof, decorated frieze and base and integral seats; supply and erect only; excavations and concrete bases priced separately							
"The Park Temple"; 5 columns; 3500 mm high x 1650 mm diameter	9800.00	96.00	1752.00	60.00	9800.00	nr	**11612.00**
"The Estate Temple"; 6 columns; 4000 mm high x 2700 mm diameter	17600.00	120.00	2190.00	120.00	17600.00	nr	**19910.00**
Stone structures; Architectural Heritage Ltd; hand carved from solid natural limestone with solid oak trelliage; prices for stonework and facades only; supply and erect only; excavations and concrete bases priced separately							
"The Pergola", 2240 mm high x 2640 mm wide x 6990 mm long	14000.00	96.00	1752.00	240.00	14030.79	nr	**16022.79**
Ornamental timber buildings; Architectural Heritage Ltd; hand built in English oak with fire retardant wheat straw thatch; hand painted, fired lead glass windows							
"The Thatched Edwardian Summer House"; 3710 mm high x 3540 mm wide x 2910 mm deep overall; 2540 x 1910 mm internal	22000.00	32.00	584.00	-	22000.00	nr	**22584.00**

B COMPLETE BUILDINGS/STRUCTURES/UNITS

Item Excluding site overheads and profit	PC £	Labour hours	Labour £	Plant £	Material £	Unit	Total rate £
Ornamental stone structures; **Architectural Heritage Ltd; setting to** **bases or plinths (not included)** "The Obelisk"; classic natural stone obelisk; tapering square form on panelled square base; 1860 mm high x 360 mm square	1200.00	2.00	36.50	-	1200.00	nr	**1236.50**
"The Narcissus Column"; natural limestone column on pedestal surmounted by bronze sculpture; overall height 3440 mm with 440 mm square base	3200.00	2.00	36.50	-	3200.00	nr	**3236.50**
"The Armillary Sundial"; artificial stone baluster pedestal base surmounted by verdigris copper armillary sphere calibrated sundial	2200.00	1.00	18.25	-	2200.00	nr	**2218.25**

C EXISTING SITE/BUILDING/SERVICES

Item Excluding site overheads and profit	PC £	Labour hours	Labour £	Plant £	Material £	Unit	Total rate £
C12 UNDERGROUND SERVICES SURVEY							
Ground penetrating radar surveys for artefacts, services and cable avoidance							
Areas not exceeding 1000 m²	-	-	-	-	-	nr	4800.00
C13 BUILDING FABRIC SURVEY							
Surveys							
Asbestos surveys							
type 3 asbestos survey (destructive testing) to single storey free standing buildings such as park buildings, storage sheds and the like; total floor area less than 100 m²	-	-	-	-	-	nr	950.00
extra over for the removal only of asbestos off site to tip (tipping charges only)	-	-	-	-	-	tonne	68.00
C20 DEMOLITION							
Demolish existing structures; disposal offsite; mechanical demolition; with 3 tonne excavator and dumper							
Brick wall							
112.5 mm thick	-	0.13	2.43	0.69	3.05	m²	6.17
225 mm thick	-	0.17	3.04	0.86	6.10	m²	10.00
337.5 mm thick	-	0.20	3.65	1.03	10.23	m²	14.91
450 mm thick	-	0.27	4.87	1.37	12.20	m²	18.44
Demolish existing structures; disposal offsite mechanically loaded; all other works by hand							
Brick wall							
112.5 mm thick	-	0.33	6.08	4.16	3.05	m²	13.29
225 mm thick	-	0.50	9.13	-	6.10	m²	15.22
337.5 mm thick	-	0.67	12.17	-	10.23	m²	22.40
450 mm thick	-	1.00	18.25	-	12.20	m²	30.45
Demolish existing structures; disposal offsite mechanically loaded; by diesel or electric breaker; all other works by hand							
Brick wall							
112.5 mm thick	-	0.17	3.04	0.41	3.05	m²	6.50
225 mm thick	-	0.20	3.65	0.50	6.10	m²	10.24
337.5 mm thick	-	0.25	4.56	0.62	9.04	m²	14.23
450 mm thick	-	0.33	6.08	0.83	12.20	m²	19.10
Break out concrete footings associated with free standing walls; inclusive of all excavation and backfilling with excavated material; disposal offsite mechanically loaded							
By mechanical breaker; diesel or electric							
plain concrete	-	1.50	27.37	8.90	15.11	m³	51.38
reinforced concrete	-	2.50	45.63	12.58	27.10	m³	85.31

C EXISTING SITE/BUILDING/SERVICES

Item Excluding site overheads and profit	PC £	Labour hours	Labour £	Plant £	Material £	Unit	Total rate £
Remove existing free standing buildings; demolition by hand							
Timber buildings with suspended timber floor; hardstanding or concrete base not included; disposal off site							
shed 6.0 m²	-	2.00	36.50	-	28.23	nr	**64.73**
shed 10.0 m²	-	3.00	54.76	-	50.81	nr	**105.57**
shed 15.0 m²	-	3.50	63.88	-	79.05	nr	**142.92**
Timber building; insulated; with timber or concrete posts set in concrete, felt covered timber or tiled roof; internal walls cladding with timber or plasterboard; load arisings to skip							
timber structure 6.0 m²	-	2.50	45.63	-	121.96	nr	**167.58**
timber structure 12.0 m²	-	3.50	63.88	-	170.51	nr	**234.39**
timber structure 20.0 m²	-	8.00	146.00	377.81	180.67	nr	**704.48**
Demolition of free standing brick buildings with tiled or sheet roof; concrete foundations measured separately; mechanical demolition; maximum distance to stockpile 25 m; inclusive for all access scaffolding and the like; maximum height of roof 4.0 m; inclusive of all doors, windows, guttering and down pipes; including disposal by grab							
Half brick thick							
10 m²	-	8.00	146.00	142.43	204.08	nr	**492.51**
20 m²	-	16.00	292.00	253.03	262.47	nr	**807.49**
1 brick thick							
10 m²	-	10.00	182.50	221.22	408.16	nr	**811.88**
20 m²	-	16.00	292.00	331.90	524.93	nr	**1148.84**
Cavity wall with blockwork inner skin and brick outer skin; insulated							
10 m²	-	12.00	219.09	300.10	566.57	nr	**1085.75**
20 m²	-	20.00	365.00	410.83	729.93	nr	**1505.76**
Extra over to the above for disconnection of services							
Electrical							
disconnection	-	2.00	36.50	-	-	nr	**36.50**
grub out cables and dispose; backfilling; by machine	-	-	-	0.48	0.45	m	**0.93**
grub out cables and dispose; backfilling; by hand	-	0.50	9.13	-	0.45	m	**9.58**
Water supply, foul or surface water drainage							
disconnection; capping off	-	1.00	18.25	-	36.50	nr	**54.75**
grub out pipes and dispose; backfilling; by machine	-	-	-	0.48	0.45	m	**0.93**
grub out pipes and dispose; backfilling; by hand	-	0.50	9.13	-	0.45	m	**9.58**

C EXISTING SITE/BUILDING/SERVICES

Item Excluding site overheads and profit	PC £	Labour hours	Labour £	Plant £	Material £	Unit	Total rate £
C50 REPAIRING/RENOVATING/ CONSERVING METAL							
Eura Conservation Ltd; conservation of metal railings; works to heritage conservation standards							
Taking down and transporting existing metalwork to an off site workshop for conservation							
pair of gates; maximum overall width 4.00 m	-	-	-	-	-	pair	522.00
side screens to gates	-	-	-	-	-	pair	423.00
railings; 1.20 m high; plain	-	-	-	-	-	m	45.00
railings; 1.80 m high; with finials	-	-	-	-	-	m	85.00
ornate cast or wrought iron railings	-	-	-	-	-	m	95.00
Conserving metalwork off site; inclusive of repairs to metal work; rust removal rubbing down and preparing for re-erection on site							
gates; 2.50 m high; width not exceeding 4.00 m	-	-	-	-	-	m	1046.00
railings; 1.20 m high; plain	-	-	-	-	-	m	135.00
railings; 1.80 m high; with finials	-	-	-	-	-	m	205.00
ornate cast or wrought iron railings	-	-	-	-	-	m	410.00
Transporting from store nd re-erection of existing gates into locations recorded with position orientation and features of the original installation							
gates; 2.50 m high; width not exceeding 4.00 m	-	-	-	-	-	nr	892.00
Refixing existing railings previously taken down and repaired on site							
railings; 1.20 m high; plain	-	-	-	-	-	m	89.00
railings; 1.80 m high; with finials	-	-	-	-	-	m	173.00
ornate cast or wrought iron railings	-	-	-	-	-	m	216.00
Eura Conservation Ltd; supply of conservation grade railings to match existing materials on site; inclusive of all surveys, measurements and analysis of materials, installation methods and the like							
Wavy bar railings							
1.20 m high; plain	-	-	-	-	-	m	92.00
1.70 m high; straight	-	-	-	-	-	m	285.00
1.70 m high; curved	-	-	-	-	-	m	346.00
1.80 m high; with finials	-	-	-	-	-	m	287.00
ornate cast or wrought iron railings	-	-	-	-	-	m	787.00
Painting of railings: see section M60							

D GROUNDWORK

Item Excluding site overheads and profit	PC £	Labour hours	Labour £	Plant £	Material £	Unit	Total rate £
D11 SOIL STABILIZATION							
Soil stabilization - General Preamble: Earth-retaining and stabilizing materials are often specified as part of the earth-forming work in landscape contracts, and therefore this section lists a number of products specially designed for large-scale earth control. There are two types: rigid units for structural retention of earth on steep slopes; and flexible meshes and sheets for control of soil erosion where structural strength is not required. Prices for these items depend on quantity, difficulty of access to the site and availability of suitable filling material; estimates should be obtained from the manufacturer when the site conditions have been determined.							
Crib walls; Keller Ground Engineering "Timbercrib" timber crib walling system; machine filled with crushed rock inclusive of reinforced concrete footing cribfill stone, rear wall land drain and rear wall drainage/ separation membrane; excluding excavation							
ref 600/36; for retaining walls up to 1.50 m high	-	-	-	-	-	m²	160.00
ref 750/36; for retaining walls up to 2.40 m high	-	-	-	-	-	m²	170.00
ref 900/36; for retaining walls up to 2.90 m high	-	-	-	-	-	m²	185.00
ref 1050/46; for retaining walls up to 3.50 m high	-	-	-	-	-	m²	195.00
ref 1200/46; for retaining walls up to 4.20 m high	-	-	-	-	-	m²	205.00
ref 1350/46; for retaining walls up to 4.80 m high	-	-	-	-	-	m²	215.00
ref 1500/60; for retaining walls up to 5.30 m high	-	-	-	-	-	m²	230.00
ref 11650/60; for retaining walls up to 6.00 m high	-	-	-	-	-	m²	250.00
Retaining walls; Keller Ground Engineering "Textomur" reinforced soil system embankments							
reinforced soil slopes at angles of 60 - 70° to the horizontal; as an alternative to reinforced concrete or gabion solutions; finishing with excavated material/grass or shrubs	-	-	-	-	-	m²	110.00

D GROUNDWORK

Item Excluding site overheads and profit	PC £	Labour hours	Labour £	Plant £	Material £	Unit	Total rate £
D11 SOIL STABILIZATION - cont'd							
Retaining walls; RCC; Tarmac Precast Concrete Retaining walls of units with plain concrete finish; prices based on 24 tonne loads but other quantities available (excavation, temporary shoring, foundations and backfilling not included)							
1000 mm wide x 1000 mm high	86.00	1.50	27.38	6.53	118.75	m	**152.66**
1000 mm wide x 1250 mm high	128.00	1.27	23.18	6.53	168.25	m	**197.97**
1000 mm wide x 1750 mm high	146.00	1.27	23.18	6.53	201.87	m	**231.59**
1000 mm wide x 2400 mm high	216.00	1.27	23.18	6.53	292.98	m	**322.70**
1000 mm wide x 2690 mm high	268.00	1.27	23.18	43.31	355.04	m	**421.53**
1000 mm wide x 3000 mm high	277.00	1.47	26.83	54.14	375.17	m	**456.14**
1000 mm wide x 3750 mm high	396.00	1.57	28.74	64.97	522.04	m	**615.76**
Retaining walls; Maccaferri Ltd Wire mesh gabions; galvanized mesh 80 x 100 mm; filling with broken stones 125 - 200 mm size; wire down securely to manufacturer's instructions; filling front face by hand							
2 x 1 x 0.50 m	21.05	2.00	36.50	4.35	88.25	nr	**129.10**
2 x 1 x 1.0 m	29.49	4.00	73.00	8.71	163.89	nr	**245.60**
PVC coated gabions							
2 x 1 x 0.5m	27.24	2.00	36.50	4.35	94.44	nr	**135.29**
2 x 1 x 1.0 m	38.34	4.00	73.00	8.71	172.74	nr	**254.45**
"Reno" mattress gabions							
6 x 2 x 0.17 m	80.06	3.00	54.75	6.53	217.15	nr	**278.43**
6 x 2 x 0.23 m	86.66	4.50	82.13	8.71	272.13	nr	**362.97**
6 x 2 x 0.30 m	101.86	6.00	109.50	9.80	343.78	nr	**463.08**
Retaining walls; Tensar International "Tensartech TW1" retaining wall system; modular dry laid concrete blocks; 220 x 400 mm long x 150 mm high connected to "Tensar RE" geogrid with proprietary connectors; geogrid laid horizontally within the fill at 300 mm centres; on 150 x 450 mm concrete foundation; filling with imported granular material							
1.00 m high	75.00	3.00	54.75	8.25	83.47	m²	**146.47**
2.00 m high	75.00	4.00	73.00	8.25	83.47	m²	**164.72**
3.00 m high	75.00	4.50	82.13	8.25	83.47	m²	**173.84**

D GROUNDWORK

Item Excluding site overheads and profit	PC £	Labour hours	Labour £	Plant £	Material £	Unit	Total rate £
Retaining walls; Grass Concrete Ltd "Betoflor" precast concrete landscape retaining walls including soil filling to pockets (excavation, concrete foundations, backfilling stones to rear of walls and planting not included)							
"Betoflor" interlocking units; 250 mm long x 250 mm x 200 mm modular deep; in walls 250 mm wide	-	-	-	-	-	m²	81.54
extra over "Betoflor" interlocking units for colours	-	-	-	-	-	m²	6.68
"Betoatlas" earth retaining walls; 250 mm long x 500 mm wide x 200 mm modular deep; in walls 500 mm wide	-	-	-	-	-	m²	113.63
extra over "Betoatlas" interlocking units for colours	-	-	-	-	-	m²	13.86
ABG Ltd; "Webwall"; honeycomb structured HDPE polymer strips with interlocking joints; laid to prepared terraces and backfilled with excavated material to form raked wall for planting 500 mm high units laid and backfilled horizontally within the face of the slope to produce stepped gradient; measured as bank face area; wall height							
up to 1.00 m high	-	-	-	-	-	m²	80.00
up to 2.00 m high	-	-	-	-	-	m²	100.00
over 2.00 m high	-	-	-	-	-	m²	110.00
250 mm high units laid and backfilled horizontally within the face of the slope to produce stepped gradient; measured as bank face area; wall height							
up to 1.00 m high	-	-	-	-	-	m²	90.00
up to 2.00 m high	-	-	-	-	-	m²	110.00
over 2.00 m high	-	-	-	-	-	m²	120.00
Setting out; grading and levelling; compacting bottoms of excavations	-	0.13	2.43	0.43	-	m	2.86
Extra over "Betoflor" retaining walls for C7P concrete foundations							
700 x 300 mm deep	17.79	0.27	4.87	-	17.79	m	22.66
Retaining walls; Forticrete Ltd "Keystone" pc concrete block retaining wall; geogrid included for walls over 1.00 m high; excavation, concrete foundation, stone backfill to rear of wall all measured separately							
1.0 m high	86.00	2.40	43.80	-	88.15	m²	131.95
2.0 m high	86.00	2.40	43.80	7.95	99.95	m²	151.70
3.0 m high	86.00	2.40	43.80	10.60	108.15	m²	162.55
4.0 m high	86.00	2.40	43.80	12.72	114.50	m²	171.02

D GROUNDWORK

Item Excluding site overheads and profit	PC £	Labour hours	Labour £	Plant £	Material £	Unit	Total rate £
D11 SOIL STABILIZATION - cont'd							
Retaining walls; Forticrete Ltd Stepoc Blocks; interlocking blocks; 10 mm reinforcing laid loose horizontally to preformed notches and vertical reinforcing nominal size 10 mm fixed to starter bars; infilling with concrete; foundations and starter bars measured separately							
type 325; 400 x 225 x 325 mm	47.08	1.40	25.55	-	66.47	m^2	**92.02**
type 256; 400 x 225 x 256 mm	42.49	1.20	21.90	-	62.27	m^2	**84.17**
type 190; 400 x 225 x 190 mm	32.58	1.00	18.25	-	48.41	m^2	**66.66**
Embankments; Tensar International Embankments; reinforced with "Tensar Uniaxial Geogrid"; ref 40 RE; 40 kN/m width; geogrid laid horizontally within fill to 100% of vertical height of slope at 1.00 m centres; filling with excavated material							
slopes less than 45°; "Tensartech Natural Green System"	2.00	0.19	3.39	1.74	2.10	m^3	**7.22**
slopes exceeding 45°; "Tensartech Greenslope System"	4.30	0.19	3.39	2.57	4.51	m^3	**10.47**
extra over to slopes exceeding 45° for temporary shuttering; shuttering measured per m^2 of vertical elevation	-	-	-	-	10.00	m^2	**-**
Embankments; reinforced with "Tensar Uniaxial Geogrid"; ref 55 RE; 55 kN/m width; geogrid laid horizontally within fill to 100% of vertical height of slope at 1.00 m centres; filling with excavated material							
slopes less than 45°; "Tensartech Natural Green System"	2.50	0.19	3.39	1.74	2.63	m^3	**7.75**
slopes exceeding 45°; "Tensartech Greenslope System"	5.38	0.19	3.39	2.57	5.64	m^3	**11.60**
extra over to slopes exceeding 45° for temporary shuttering; shuttering measured per m^2 of vertical elevation	-	-	-	-	10.00	m^2	**-**
Extra for "Tensar Mat"; erosion control mats; to faces of slopes of 45° or less; filling with 20 mm fine topsoil; seeding	3.50	0.02	0.36	0.17	4.81	m^2	**5.34**
Embankments; reinforced with "Tensar Uniaxial Geogrid" ref 55 RE; 55 kN/m width; geogrid laid horizontally within fill to 50% of horizontal length of slopes at specified centres; filling with imported fill PC £8.00/m^3							
300 mm centres; slopes up to 45°; "Tensartech Natural Green System"	4.15	0.23	4.26	1.98	13.96	m^3	**20.20**
600 mm centres; slopes up to 45°; "Tensartech Natural Green System"	2.08	0.19	3.39	2.57	11.78	m^3	**17.73**
extra over to slopes exceeding 45° for temporary shuttering; shuttering measured per m^2 of vertical elevation	-	-	-	-	10.00	m^2	**-**
Extra for "Tensar Mat"; erosion control mats; to faces of slopes of 45° or less; filling with 20 mm fine topsoil; seeding	3.50	0.02	0.36	0.17	4.81	m^2	**5.34**

D GROUNDWORK

Item Excluding site overheads and profit	PC £	Labour hours	Labour £	Plant £	Material £	Unit	Total rate £
Embankments; reinforced with "Tensar Uniaxial Geogrid" ref 55 RE; 55 kN/m width; geogrid laid horizontally within fill; filling with excavated material; wrapping around at faces							
300 mm centres; slopes exceeding 45°; "Tensartech Natural Green System"	8.32	0.19	3.39	1.74	8.74	m³	**13.87**
600 mm centres; slopes exceeding 45°; "Tensartech Greenslope System"	4.15	0.19	3.39	1.74	4.36	m³	**9.48**
extra over to slopes exceeding 45° for temporary shuttering; shuttering measured per m² of vertical elevation	-	-	-	-	10.00	m²	**-**
Extra over embankments for bagwork face supports for slopes exceeding 45°	-	0.50	9.13	-	13.00	m²	**22.13**
Extra over embankments for seeding of bags	-	0.10	1.82	-	0.27	m²	**2.09**
Extra over embankments for "Bodkin" joints	-	-	-	-	1.00	m	**-**
Anchoring systems; surface stabilization; Platipus Anchors Ltd; centres and depths shown should be verified with a design engineer; they may vary either way dependent on circumstances							
Concrete revetments; hard anodized stealth anchors with stainless steel accessories installed to a depth of 1 m							
"SO4" anchors to a depth of 1.00 - 1.50 m at 1.00 m centres	25.36	0.33	6.08	1.53	25.36	m²	**32.97**
Loadlocking and crimping tool; for stressing and crimping the anchors							
purchase price	385.88	-	-	-	385.88	nr	**-**
hire rate per week	66.15	-	-	-	66.15	nr	**-**
Surface erosion; geotextile anchoring on slopes; aluminium alloy Stealth anchors							
"SO4", "SO6" to a depth of 1.0 - 2.0 m at 2.00 m centres	12.12	0.20	3.65	1.53	12.12	m²	**17.30**
Brick block or in situ concrete distressed retaining walls; anchors installed in two rows at varying tensions between 18 kN - 50 kN along the length of the retaining wall; core drilling of wall not included							
"SO8", "BO6", "BO8" bronze Stealth and Bat anchors installed in combination with stainless steel accessories to a depth of 6 m; average cost base on 1.50 m centres; long term solution	363.83	0.50	9.13	1.53	363.83	m²	**374.48**
"SO8", "BO6", "BO8" cast SG iron Stealth and Bat anchors installed in combination with galvanized steel accessories to a depth of 6 m; average cost base on 1.50 m centres; short term solution	155.45	0.50	9.13	1.53	155.45	m²	**166.11**
Timber retaining wall support; anchors installed 19 kN along the length of the retaining wall							
"SO6", "SO8", bronze Stealth anchors installed in combination with stainless steel accessories to a depth of 3.0 m; 1.50 centres; average cost	54.58	0.50	9.13	1.22	54.58	m²	**64.93**

D GROUNDWORK

Item Excluding site overheads and profit	PC £	Labour hours	Labour £	Plant £	Material £	Unit	Total rate £
D11 SOIL STABILIZATION - cont'd							
Anchoring systems - cont'd							
Gabions; anchors for support and stability to moving, overturning or rotating gabion retaining walls							
"SO8", "BO6" anchors including base plate at 1.00 m centres to a depth of 4.00 m	275.63	0.33	6.08	1.53	275.63	m²	**283.24**
Soil nailing of geofabrics; anchors fixed through surface of fabric or erosion control surface treatment (not included)							
"SO8", "BO6" anchors including base plate at 1.50 m centres to a depth of 3.00 m	110.25	0.17	3.04	1.02	110.25	m²	**114.31**
Embankments; Grass Concrete Ltd							
"Grasscrete"; in situ reinforced concrete surfacing; to 20 mm thick sand blinding layer (not included); including soiling and seeding							
ref GC1; 100 mm thick	-	-	-	-	-	m²	**31.33**
ref GC2; 150 mm thick	-	-	-	-	-	m²	**38.75**
"Grassblock 103"; solid matrix precast concrete blocks; to 20 mm thick sand blinding layer; excluding edge restraint; including soiling and seeding							
406 x 406 x 103 mm; fully interlocking	-	-	-	-	-	m²	**31.95**
Grass reinforcement; Farmura Environmental Ltd							
"Matrix" grass paver; recycled polyethylene and polypropylene mixed interlocking erosion control and grass reinforcement system laid to rootzone prepared separately and filled with screened topsoil and seeded with grass seed; green							
640 x 330 x 38 mm	13.50	0.13	2.28	-	16.53	m²	**18.81**
Embankments; Cooper Clarke Civils and Lintels							
"Ecoblock" polyethylene; 925 x 310 x 50 mm; heavy duty for car parking and fire paths							
to firm subsoil (not included)	19.04	0.04	0.73	-	20.14	m²	**20.87**
to 100 mm granular fill and "Geotextile"	19.04	0.08	1.46	0.13	24.42	m²	**26.01**
to 250 mm granular fill and "Geotextile"	19.04	0.20	3.65	0.21	29.77	m²	**33.64**
Extra for filling "Ecoblock" with topsoil; seeding with rye grass at 50 g/m²	1.32	0.02	0.37	0.17	1.58	m²	**2.12**
"Neoweb" polyethylene soil-stabilizing panels; to soil surfaces brought to grade (not included); filling with excavated material							
panels; 2.5 x 8.0 m x 100 mm deep	7.70	0.10	1.82	2.47	7.86	m²	**12.15**
panels; 2.5 x 8.0 m x 200 mm deep	15.39	0.13	2.43	2.95	15.70	m²	**21.08**
"Neoweb" polyethylene soil-stabilizing panels; to soil surfaces brought to grade (not included); filling with ballast							
panels; 2.5 x 8.0 m x 100 mm deep	7.70	0.11	2.03	2.47	9.76	m²	**14.26**
panels; 2.5 x 8.0 m x 200 mm deep	15.39	0.15	2.81	2.95	19.51	m²	**25.26**

D GROUNDWORK

Item Excluding site overheads and profit	PC £	Labour hours	Labour £	Plant £	Material £	Unit	Total rate £
"Neoweb" polyethylene soil-stabilizing panels; to soil surfaces brought to grade (not included); filling with ST2 10 N/mm² concrete							
panels; 2.5 x 8.0 m x 100 mm deep	7.70	0.16	2.92	-	16.27	m²	19.19
panels; 2.5 x 8.0 m x 200 mm deep	15.39	0.20	3.65	-	32.53	m²	36.18
Extra over for filling "Neoweb" with imported topsoil; seeding with rye grass at 50 g/m²	2.76	0.07	1.22	0.29	3.29	m²	4.79
Timber log retaining walls; AVS Fencing Supplies Ltd; all timber posts are "kiln dried redwood" with 15 year guarantee; posts which have been cut would not be subject to the supplier's guarantee							
Machine rounded softwood logs to trenches priced separately; disposal of excavated material priced separately; inclusive of 75 mm hardcore blinding to trench and backfilling trench with site mixed concrete 1:3:6; geofabric pinned to rear of logs; heights of logs above ground							
500 mm (constructed from 1.80 m lengths)	28.05	1.50	27.38	-	37.06	m	64.43
1.20 m (constructed from 1.80 m lengths)	56.10	1.30	23.73	-	75.12	m	98.84
1.60 m (constructed from 2.40 m lengths)	74.90	2.50	45.63	-	99.25	m	144.88
2.00 m (constructed from 3.00 m lengths)	93.60	3.50	63.88	-	123.29	m	187.17
As above but with 150 mm machine rounded timbers							
500 mm	38.44	2.50	45.63	-	47.45	m	93.07
1.20 m	128.01	1.75	31.94	-	147.02	m	178.96
1.60 m	128.13	3.00	54.75	-	152.49	m	207.24
As above but with 200 mm machine rounded timbers							
1.80 m (constructed from 2.40 m lengths)	170.80	4.00	73.00	-	200.49	m	273.49
Railway sleeper walls; AVS Fencing Supplies Ltd; retaining wall from railway sleepers; fixed with steel galvanized pins 12 mm driven into the ground; sleepers laid flat							
Grade 1 softwood; 2590 x 250 x 125 mm							
150 mm; 1 sleeper high	6.81	0.50	9.13	-	7.17	m	16.29
300 mm; 2 sleepers high	13.62	1.00	18.25	-	14.34	m	32.59
450 mm; 3 sleepers high	20.27	1.50	27.38	-	21.21	m	48.59
600 mm; 4 sleepers high	27.30	2.00	36.50	-	28.25	m	64.75
Grade 1 softwood as above but with 2 nr galvanized angle iron stakes set into concrete internally and screwed to the inside face of the sleepers							
750 mm; 5 sleepers high	33.77	2.50	45.63	-	59.08	m²	104.71
900 mm; 6 sleepers high	40.42	2.75	50.19	-	68.45	m²	118.64
Grade 1 hardwood; 2590 x 250 x 150 mm							
150 mm; 1 sleeper high	7.00	0.50	9.13	-	7.36	m	16.49
300 mm; 2 sleepers high	14.01	1.00	18.25	-	14.73	m	32.98
450 mm; 3 sleepers high	20.84	1.50	27.38	-	21.79	m	49.16
600 mm; 4 sleepers high	28.08	2.00	36.50	-	29.02	m	65.53

D GROUNDWORK

Item Excluding site overheads and profit	PC £	Labour hours	Labour £	Plant £	Material £	Unit	Total rate £
D11 SOIL STABILIZATION - cont'd							
Railway sleeper walls - cont'd							
Grade 1 hardwood as above but with 2 nr galvanized angle iron stakes set into concrete internally and screwed to the inside face of the sleepers							
750 mm; 5 sleepers high	34.74	2.50	45.63	-	60.05	m²	**105.67**
900 mm; 6 sleepers high	41.58	2.75	50.19	-	69.60	m²	**119.79**
New pine softwood; 2400 x 250 x 125 mm							
120 mm; 1 sleeper high	7.40	0.50	9.13	-	7.76	m	**16.88**
240 mm; 2 sleepers high	14.80	1.00	18.25	-	15.52	m	**33.77**
360 mm; 3 sleepers high	22.20	1.50	27.38	-	23.14	m	**50.52**
480 mm; 4 sleepers high	29.60	2.00	36.50	-	30.55	m	**67.05**
New pine softwood as above but with 2 nr galvanized angle iron stakes set into concrete internally and screwed to the inside face of the sleepers							
600 mm; 5 sleepers high	37.00	2.50	45.63	-	62.31	m²	**107.93**
720 mm; 6 sleepers high	44.40	2.75	50.19	-	72.42	m²	**122.61**
New oak hardwood 2600 x 220 x 130 mm							
130 mm; 1 sleeper high	9.80	0.50	9.13	-	10.16	m	**19.29**
260 mm; 2 sleepers high	19.60	1.00	18.25	-	20.32	m	**38.57**
390 mm; 3 sleepers high	29.40	1.50	27.38	-	30.34	m	**57.72**
520 mm; 4 sleepers high	39.20	2.00	36.50	-	40.15	m	**76.65**
New oak hardwood as above but with 2 nr galvanized angle iron stakes set into concrete internally and screwed to the inside face of the sleepers							
640 mm; 5 sleepers high	49.00	2.50	45.63	-	74.31	m²	**119.93**
760 mm; 6 sleepers high	58.80	2.75	50.19	-	86.82	m²	**137.01**
Excavate foundation trench; set railway sleepers vertically on end in concrete 1:3:6 continuous foundation to 33.3% of their length to form retaining wall							
Grade 1 hardwood; finished height above ground level							
300 mm	12.42	3.00	54.75	1.80	22.96	m	**79.50**
500 mm	18.00	3.00	54.75	1.80	31.33	m	**87.87**
600 mm	25.02	3.00	54.75	1.80	45.87	m	**102.41**
750 mm	27.00	3.50	63.88	1.80	48.84	m	**114.51**
1.00 m	36.90	3.75	68.44	1.80	55.27	m	**125.50**
Excavate and place vertical steel universal beams 165 wide in concrete base at 2.590 centres; fix railway sleepers set horizontally between beams to form horizontal fence or retaining wall							
Grade 1 hardwood; bay length 2.590 m							
500 mm high; 2 sleepers	15.18	2.50	45.63	-	54.10	bay	**99.72**
750 mm high; 3 sleepers	22.31	2.60	47.45	-	80.36	bay	**127.81**
1.00 m high; 4 sleepers	29.90	3.00	54.75	-	104.09	bay	**158.84**
1.25 m high; 5 sleepers	38.18	3.50	63.88	-	133.85	bay	**197.73**
1.50 m high; 6 sleepers	44.85	3.00	54.75	-	159.33	bay	**214.08**
1.75 m high; 7 sleepers	52.21	3.00	54.75	-	166.69	bay	**221.44**

D GROUNDWORK

Item Excluding site overheads and profit	PC £	Labour hours	Labour £	Plant £	Material £	Unit	Total rate £
Willow walling to riverbanks; LDC Ltd Woven willow walling as retention to riverbanks; driving or concreting posts in at 2 m centres; intermediate posts at 500 mm centres							
1.20 m high	-	-	-	-	-	m	**114.86**
1.50 m high	-	-	-	-	-	m	**146.44**
Flexible sheet materials; Tensar International "Tensar Mat"; erosion mats; 3 - 4.50 m wide; securing with "Tensar" pegs; lap rolls 100 mm; anchors at top and bottom of slopes; in trenches	3.65	0.02	0.30	-	4.01	m^2	**4.31**
Topsoil filling to "Tensar Mat"; including brushing and raking	0.65	0.02	0.30	0.10	0.81	m^2	**1.22**
"Tensar Bi-axial Geogrid"; to graded compacted base; filling with 200 mm granular fill; compacting (turf or paving to surfaces not included); 400 mm laps							
ref SS20; 39 x 39 mm mesh	1.40	0.01	0.24	0.13	5.24	m^2	**5.62**
ref SS30; 39 x 39 mm mesh	1.40	0.01	0.24	0.13	5.24	m^2	**5.62**
Flexible sheet materials; Terram Ltd "Terram" synthetic fibre filter fabric; to graded base (not included)							
"Terram 1000"; 0.70 mm thick; mean water flow 50 $l/m^2/s$	0.42	0.02	0.30	-	0.42	m^2	**0.73**
"Terram 2000"; 1.00 mm thick; mean water flow 33 $l/m^2/s$	0.91	0.02	0.30	-	1.09	m^2	**1.40**
"Terram Minipack"	0.55	-	0.06	-	0.55	m^2	**0.61**
Flexible sheet materials; Greenfix Ltd "Greenfix"; erosion control mats; 10 -15 mm thick; fixing with 4 nr crimped pins in accordance with manufacturer's instructions; to graded surface (not included)							
unseeded "Eromat 1"; 2.40 m wide	89.11	2.00	36.50	-	158.02	100 m^2	**194.52**
unseeded "Eromat 2"; 2.40 m wide	113.86	2.00	36.50	-	185.25	100 m^2	**221.75**
unseeded "Eromat 3"; 2.40 m wide	128.71	2.00	36.50	-	201.58	100 m^2	**238.08**
seeded "Covamat 1"; 2.40 m wide	150.00	2.00	36.50	-	217.50	100 m^2	**254.00**
seeded "Covamat 2"; 2.40 m wide	170.00	2.00	36.50	-	238.50	100 m^2	**275.00**
seeded "Covamat 3"; 2.40 m wide	190.00	2.00	36.50	-	259.50	100 m^2	**296.00**
"Bioroll" 300 mm diameter to river banks and revetments	19.00	0.13	2.28	-	20.76	m	**23.05**
Extra over "Greenfix" erosion control mats for fertilizer applied at 70 g/m^2	7.91	0.20	3.65	-	7.91	100 m^2	**11.56**
Extra over "Greenfix" erosion control mats for Geojute; fixing with steel pins	1.42	0.02	0.41	-	1.54	m^2	**1.95**
Extra over "Greenfix" erosion control mats for laying to slopes exceeding 30°	-	-	-	-	-	25%	**-**

D GROUNDWORK

Item Excluding site overheads and profit	PC £	Labour hours	Labour £	Plant £	Material £	Unit	Total rate £
D11 SOIL STABILIZATION - cont'd							
Extra for the following operations							
Spreading 25 mm approved topsoil							
by machine	0.65	0.01	0.11	0.10	0.78	m²	1.00
by hand	0.65	0.04	0.64	-	0.78	m²	1.42
Grass seed; PC £4.50/kg; spreading in two operations; by hand							
35 g/m²	15.75	0.17	3.04	-	15.75	100 m²	18.79
50 g/m²	22.50	0.17	3.04	-	22.50	100 m²	25.54
70 g/m²	31.50	0.17	3.04	-	31.50	100 m²	34.54
100 g/m²	45.00	0.20	3.65	-	45.00	100 m²	48.65
125 g/m²	56.25	0.20	3.65	-	56.25	100 m²	59.90
Extra over seeding by hand for slopes over 30° (allowing for the actual area but measured in plan)							
35 g/m²	2.34	-	0.07	-	2.34	100 m²	2.41
50 g/m²	3.38	-	0.07	-	3.38	100 m²	3.44
70 g/m²	4.72	-	0.07	-	4.72	100 m²	4.79
100 g/m²	6.75	-	0.08	-	6.75	100 m²	6.83
125 g/m²	8.41	-	0.08	-	8.41	100 m²	8.49
Grass seed; PC £4.50/kg; spreading in two operations; by machine							
35 g/m²	15.75	-	-	0.51	15.75	100 m²	16.26
50 g/m²	22.50	-	-	0.51	22.50	100 m²	23.01
70 g/m²	31.50	-	-	0.51	31.50	100 m²	32.01
100 g/m²	45.00	-	-	0.51	45.00	100 m²	45.51
125 kg/ha	562.50	-	-	50.95	562.50	ha	613.45
150 kg/ha	675.00	-	-	50.95	675.00	ha	725.95
200 kg/ha	900.00	-	-	50.95	900.00	ha	950.95
250 kg/ha	1125.00	-	-	50.95	1125.00	ha	1175.95
300 kg/ha	1350.00	-	-	50.95	1350.00	ha	1400.95
350 kg/ha	1575.00	-	-	50.95	1575.00	ha	1625.95
400 kg/ha	1800.00	-	-	50.95	1800.00	ha	1850.95
500 kg/ha	2250.00	-	-	50.95	2250.00	ha	2300.95
700 kg/ha	3150.00	-	-	50.95	3150.00	ha	3200.95
1400 kg/ha	6300.00	-	-	50.95	6300.00	ha	6350.95
Extra over seeding by machine for slopes over 30° (allowing for the actual area but measured in plan)							
35 g/m²	15.75	-	-	0.08	2.36	100 m²	2.44
50 g/m²	22.50	-	-	0.08	3.38	100 m²	3.45
70 g/m²	31.50	-	-	0.08	4.72	100 m²	4.80
100 g/m²	45.00	-	-	0.08	6.75	100 m²	6.83
125 kg/ha	562.50	-	-	7.64	84.38	ha	92.02
150 kg/ha	675.00	-	-	7.64	101.25	ha	108.89
200 kg/ha	900.00	-	-	7.64	135.00	ha	142.64
250 kg/ha	1125.00	-	-	7.64	168.75	ha	176.39
300 kg/ha	1350.00	-	-	7.64	202.50	ha	210.14
350 kg/ha	1575.00	-	-	7.64	236.25	ha	243.89
400 kg/ha	1800.00	-	-	7.64	270.00	ha	277.64
500 kg/ha	2250.00	-	-	7.64	337.50	ha	345.14
700 kg/ha	3150.00	-	-	7.64	472.50	ha	480.14
1400 kg/ha	6300.00	-	-	7.64	945.00	ha	952.64

D GROUNDWORK

Item Excluding site overheads and profit	PC £	Labour hours	Labour £	Plant £	Material £	Unit	Total rate £
D20 EXCAVATING AND FILLING							
MACHINE SELECTION TABLE							
Road Equipment Ltd; machine volumes for excavating/filling only and placing excavated material alongside or to a dumper; no bulkages are allowed for in the material volumes; these rates should be increased by user-preferred percentages to suit prevailing site conditions; the figures in the next section for "Excavation mechanical" and filling allow for the use of banksmen within the rates shown below							
1.5 tonne excavators: digging volume							
1 cycle/minute; 0.04 m³	-	0.42	7.60	2.42	-	m³	10.03
2 cycles/minute; 0.08 m³	-	0.21	3.80	1.54	-	m³	5.34
3 cycles/minute; 0.12 m³	-	0.14	2.53	1.26	-	m³	3.79
3 tonne excavators; digging volume							
1 cycle/minute; 0.13 m³	-	0.13	2.34	1.39	-	m³	3.73
2 cycles/minute; 0.26 m³	-	0.06	1.17	2.21	-	m³	3.38
3 cycles/minute; 0.39 m³	-	0.04	0.78	0.96	-	m³	1.74
5 tonne excavators; digging volume							
1 cycle/minute; 0.28 m³	-	0.06	1.09	1.94	-	m³	3.02
2 cycles/minute; 0.56 m³	-	0.03	0.54	1.63	-	m³	2.17
3 cycles/minute; 0.84 m³	-	0.02	0.36	1.70	-	m³	2.06
7 tonne excavators; supplied with operator; digging volume							
1 cycle/minute; 0.28 m³	-	0.06	1.09	2.99	-	m³	4.07
2 cycles/minute; 0.56 m³	-	0.03	0.54	1.46	-	m³	2.01
3 cycles/minute; 0.84 m³	-	0.02	0.36	1.70	-	m³	2.06
21 tonne excavators; supplied with operator; digging volume							
1 cycle/minute; 1.21 m³	-	-	-	0.68	-	m³	0.68
2 cycles/minute; 2.42 m³	-	-	-	0.28	-	m³	0.28
3 cycles/minute; 3.63 m³	-	-	-	0.19	-	m³	0.19
Backhoe loader; excavating; JCB 3 CX rear bucket capacity 0.28 m³							
1 cycle/minute; 0.28 m³	-	-	-	1.78	-	m³	1.78
2 cycles/minute; 0.56 m³	-	-	-	0.89	-	m³	0.89
3 cycles/minute; 0.84 m³	-	-	-	0.59	-	m³	0.59
Backhoe loader; loading from stockpile; JCB 3 CX front bucket capacity 1.00 m³							
1 cycle/minute; 1.00 m³	-	-	-	0.50	-	m³	0.50
2 cycles/minute; 2.00 m³	-	-	-	0.25	-	m³	0.25
Note: All volumes below are based on excavated "earth" moist at 1,997 kg/m³ solid or 1,598 kg/m³ loose; a 25% bulkage factor has been used; the weight capacities below exceed the volume capacities of the machine in most cases; see the memorandum section at the back of this book for further weights of materials.							

D GROUNDWORK

Item Excluding site overheads and profit	PC £	Labour hours	Labour £	Plant £	Material £	Unit	Total rate £
D20 EXCAVATING AND FILLING - cont'd							
Dumpers; Road Equipment Ltd							
1 tonne high tip skip loader; volume 0.485 m³ (775 kg)							
5 loads per hour	-	0.41	7.53	1.29	-	m³	8.81
7 loads per hour	-	0.29	5.38	0.97	-	m³	6.34
10 loads per hour	-	0.21	3.76	0.73	-	m³	4.49
3 tonne dumper; maximum volume 2.40 m³ (3.38 t); available volume 1.9 m³							
4 loads per hour	-	0.14	2.53	0.47	-	m³	3.00
5 loads per hour	-	0.11	2.03	0.38	-	m³	2.41
7 loads per hour	-	0.08	1.45	0.29	-	m³	1.74
10 loads per hour	-	0.06	1.01	0.22	-	m³	1.23
6 tonne dumper; maximum volume 3.40 m³ (5.4 t); available volume 3.77 m³							
4 loads per hour	-	0.07	1.21	0.30	-	m³	1.51
5 loads per hour	-	0.05	0.97	0.25	-	m³	1.21
7 loads per hour	-	0.04	0.69	0.19	-	m³	0.88
10 loads per hour	-	0.03	0.48	0.15	-	m³	0.64
Market prices of topsoil; prices shown include for 20% settlement;							
Recycled topsoil	-	-	-	-	21.60	m³	-
Multiple source screened topsoil	-	-	-	-	31.20	m³	-
Single source topsoil; British Sugar PLC	-	-	-	-	38.40	m³	-
High grade topsoil for planting	-	-	-	-	43.20	m³	-
Site preparation							
Felling and removing trees off site							
girth 600 mm - 1.50 m (95 - 240 mm trunk diameter)	-	5.00	91.25	19.04	-	nr	110.29
girth 1.50 - 3.00 m (240 - 475 mm trunk diameter)	-	17.00	310.25	76.15	-	nr	386.40
girth 3.00 - 4.00 m (475 - 630 mm trunk diameter)	-	48.53	885.64	121.83	-	nr	1007.47
Removing tree stumps							
girth 600 mm - 1.50 m	-	2.00	36.50	66.00	-	nr	102.50
girth 1.50 - 3.00 m	-	7.00	127.75	91.44	-	nr	219.19
girth over 3.00 m	-	12.00	219.00	156.75	-	nr	375.75
Stump grinding; disposing to spoil heaps							
girth 600 mm - 1.50 m	-	2.00	36.50	32.20	-	nr	68.70
girth 1.50 - 3.00 m	-	2.50	45.63	64.41	-	nr	110.03
girth over 3.00 m	-	4.00	73.00	56.35	-	nr	129.35
Clearing site vegetation							
mechanical clearance	-	0.25	4.56	11.33	-	100 m²	15.89
hand clearance	-	2.00	36.50	-	-	100 m²	36.50
Lifting turf for preservation							
machine lift and stack	-	0.75	13.69	9.96	-	100 m²	23.64
hand lift and stack	-	8.33	152.08	-	-	100 m²	152.08

D GROUNDWORK

Item Excluding site overheads and profit	PC £	Labour hours	Labour £	Plant £	Material £	Unit	Total rate £
Site clearance; by machine; clear site of mature shrubs from existing cultivated beds; dig out roots by machine							
Mixed shrubs in beds; planting centres 500 mm average							
height less than 1 m	-	0.03	0.46	0.71	-	m²	1.17
1.00 - 1.50 m	-	0.04	0.73	1.14	-	m²	1.87
1.50 - 2.00 m; pruning to ground level by hand	-	0.10	1.82	2.85	-	m²	4.67
2.00 - 3.00 m; pruning to ground level by hand	-	0.10	1.82	5.70	-	m²	7.52
3.00 - 4.00 m; pruning to ground level by hand	-	0.20	3.65	9.49	-	m²	13.14
Site clearance; by hand; clear site of mature shrubs from existing cultivated beds; dig out roots							
Mixed shrubs in beds; planting centres 500 mm average							
height less than 1 m	-	0.33	6.08	-	-	m²	6.08
1.00 - 1.50 m	-	0.50	9.13	-	-	m²	9.13
1.50 - 2.00 m	-	1.00	18.25	-	-	m²	18.25
2.00 - 3.00 m	-	2.00	36.50	-	-	m²	36.50
3.00 - 4.00 m	-	3.00	54.76	-	-	m²	54.76
Disposal of material from site clearance operations							
Shrubs and groundcovers less than 1.00 m height; disposal by 15 m³ self loaded truck							
deciduous shrubs; not chipped; winter	-	0.05	0.91	0.50	8.00	m²	9.41
deciduous shrubs; chipped; winter	-	0.05	0.91	1.28	1.20	m²	3.40
evergreen or deciduous shrubs; chipped; summer	-	0.08	1.37	0.79	4.80	m²	6.96
Shrubs 1.00 - 2.00 m height; disposal by 15 m³ self loaded truck							
deciduous shrubs; not chipped; winter	-	0.17	3.04	0.50	24.00	m²	27.54
deciduous shrubs; chipped; winter	-	0.25	4.56	0.79	4.80	m²	10.15
evergreen or deciduous shrubs; chipped; summer	-	0.30	5.47	0.88	9.60	m²	15.96
Shrubs or hedges 2.00 - 3.00 m height; disposal by 15 m³ self loaded truck							
deciduous plants non woody growth; not chipped; winter	-	0.25	4.56	0.50	24.00	m²	29.06
deciduous shrubs; chipped; winter	-	0.50	9.13	0.79	12.00	m²	21.91
evergreen or deciduous shrubs non woody growth; chipped; summer	-	0.25	4.56	0.79	4.80	m²	10.15
evergreen or deciduous shrubs woody growth; chipped; summer	-	0.67	12.17	1.07	14.40	m²	27.64
Shrubs and groundcovers less than 1.00 m height; disposal to spoil heap							
deciduous shrubs; not chipped; winter	-	0.05	0.91	-	-	m²	0.91
deciduous shrubs; chipped; winter	-	0.05	0.91	0.29	-	m²	1.20
evergreen or deciduous shrubs; chipped; summer	-	0.08	1.37	0.29	-	m²	1.65

D GROUNDWORK

Item Excluding site overheads and profit	PC £	Labour hours	Labour £	Plant £	Material £	Unit	Total rate £
D20 EXCAVATING AND FILLING - cont'd							
Disposal of material from site clearance operations - cont'd							
Shrubs 1.00 - 2.00 m height; disposal to spoil heap							
deciduous shrubs; not chipped; winter	-	0.17	3.04	-	-	m²	3.04
deciduous shrubs; chipped; winter	-	0.25	4.56	0.29	-	m²	4.85
evergreen or deciduous shrubs; chipped; summer	-	0.30	5.47	0.38	-	m²	5.86
Shrubs or hedges 2.00 - 3.00 m height; disposal to spoil heaps							
deciduous plants non woody growth; not chipped; winter	-	0.25	4.56	-	-	m²	4.56
deciduous shrubs; chipped; winter	-	0.50	9.13	0.29	-	m²	9.41
evergreen or deciduous shrubs non woody growth; chipped; summer	-	0.67	12.17	0.57	-	m²	12.74
evergreen or deciduous shrubs woody growth; chipped; summer	-	1.50	27.37	1.14	-	m²	28.52
Note: The figures in this section relate to the machine capacities shown earlier in this section. The figures below however allow for dig efficiency based on depth. The figures below also allow for a banksman. The figures below allow for bulkages of 25% on loamy soils; adjustments should be made for different soil types.							
Excavating; mechanical; topsoil for preservation							
3 tonne tracked excavator (bucket volume 0.13 m³)							
average depth 100 mm	-	2.40	43.80	68.35	-	100 m²	112.15
average depth 150 mm	-	3.36	61.32	95.68	-	100 m²	157.00
average depth 200 mm	-	4.00	73.00	113.91	-	100 m²	186.91
average depth 250 mm	-	4.40	80.30	125.30	-	100 m²	205.60
average depth 300 mm	-	4.80	87.60	136.69	-	100 m²	224.29
JCB Sitemaster 3CX (bucket volume 0.28 m³)							
average depth 100 mm	-	1.00	18.25	33.00	-	100 m²	51.25
average depth 150 mm	-	1.25	22.81	41.25	-	100 m²	64.06
average depth 200 mm	-	1.78	32.48	58.74	-	100 m²	91.22
average depth 250 mm	-	1.90	34.67	62.70	-	100 m²	97.38
average depth 300 mm	-	2.00	36.50	66.00	-	100 m²	102.50
Excavating; mechanical; to reduce levels							
5 tonne excavator (bucket volume 0.28 m³)							
maximum depth not exceeding 0.25 m	-	0.14	2.56	0.77	-	m³	3.32
maximum depth not exceeding 1.00 m	-	0.10	1.74	0.52	-	m³	2.26
JCB Sitemaster 3CX (bucket volume 0.28 m³)							
maximum depth not exceeding 0.25 m	-	0.07	1.28	2.31	-	m³	3.59
maximum depth not exceeding 1.00 m	-	0.06	1.09	1.96	-	m³	3.05
21 tonne 360 tracked excavator (bucket volume 1.21 m³)							
maximum depth not exceeding 1.00 m	-	0.01	0.20	0.45	-	m³	0.65
maximum depth not exceeding 2.00 m	-	0.02	0.30	0.68	-	m³	0.98

D GROUNDWORK

Item Excluding site overheads and profit	PC £	Labour hours	Labour £	Plant £	Material £	Unit	Total rate £
Pits; 3 tonne tracked excavator							
maximum depth not exceeding 0.25 m	-	0.33	6.08	2.40	-	m³	8.48
maximum depth not exceeding 1.00 m	-	0.50	9.13	3.59	-	m³	12.72
maximum depth not exceeding 2.00 m	-	0.60	10.95	4.31	-	m³	15.26
Trenches; width not exceeding 0.30 m; 3 tonne excavator							
maximum depth not exceeding 0.25 m	-	1.33	24.33	5.75	-	m³	30.08
maximum depth not exceeding 1.00 m	-	0.69	12.50	2.95	-	m³	15.46
maximum depth not exceeding 2.00 m	-	0.60	10.95	2.59	-	m³	13.54
Trenches; width exceeding 0.30 m; 3 tonne excavator							
maximum depth not exceeding 0.25 m	-	0.60	10.95	2.59	-	m³	13.54
maximum depth not exceeding 1.00 m	-	0.50	9.13	2.16	-	m³	11.28
maximum depth not exceeding 2.00 m	-	0.38	7.02	1.66	-	m³	8.68
Extra over any types of excavating irrespective of depth for breaking out existing materials; JCB with breaker attachment							
hard rock	-	0.50	9.13	62.50	-	m³	71.63
concrete	-	0.50	9.13	23.44	-	m³	32.56
reinforced concrete	-	1.00	18.25	39.41	-	m³	57.66
brickwork, blockwork or stonework	-	0.25	4.56	23.44	-	m³	28.00
Extra over any types of excavating irrespective of depth for breaking out existing hard pavings; JCB with breaker attachment							
concrete; 100 mm thick	-	-	-	1.72	-	m²	1.72
concrete; 150 mm thick	-	-	-	2.86	-	m²	2.86
concrete; 200 mm thick	-	-	-	3.44	-	m²	3.44
concrete; 300 mm thick	-	-	-	5.16	-	m²	5.16
reinforced concrete; 100 mm thick	-	0.08	1.52	2.07	-	m²	3.59
reinforced concrete; 150 mm thick	-	0.08	1.37	2.80	-	m²	4.17
reinforced concrete; 200 mm thick	-	0.10	1.82	3.74	-	m²	5.56
reinforced concrete; 300 mm thick	-	0.15	2.74	5.61	-	m²	8.34
tarmacadam; 75 mm thick	-	-	-	1.72	-	m²	1.72
tarmacadam and hardcore; 150 mm thick	-	-	-	2.75	-	m²	2.75
Extra over any types of excavating irrespective of depth for taking up							
precast concrete paving slabs	-	0.07	1.22	0.61	-	m²	1.83
natural stone paving	-	0.10	1.82	0.92	-	m²	2.74
cobbles	-	0.13	2.28	1.15	-	m²	3.43
brick paviors	-	0.13	2.28	1.15	-	m²	3.43
Excavating; hand							
Topsoil for preservation; loading to barrows							
average depth 100 mm	-	0.24	4.38	-	-	m²	4.38
average depth 150 mm	-	0.36	6.57	-	-	m²	6.57
average depth 200 mm	-	0.58	10.51	-	-	m²	10.51
average depth 250 mm	-	0.72	13.14	-	-	m²	13.14
average depth 300 mm	-	0.86	15.77	-	-	m²	15.77

D GROUNDWORK

Item Excluding site overheads and profit	PC £	Labour hours	Labour £	Plant £	Material £	Unit	Total rate £
D20 EXCAVATING AND FILLING - cont'd							
Excavating; hand							
Topsoil to reduce levels							
maximum depth not exceeding 0.25 m	-	2.40	43.80	-	-	m³	43.80
maximum depth not exceeding 1.00 m	-	3.12	56.94	-	-	m³	56.94
Pits							
maximum depth not exceeding 0.25 m	-	2.67	48.67	-	-	m³	48.67
maximum depth not exceeding 1.00 m	-	3.47	63.27	-	-	m³	63.27
maximum depth not exceeding 2.00 m							
(includes earthwork support)	-	6.93	126.53	-	-	m³	181.63
Trenches; width not exceeding 0.30 m							
maximum depth not exceeding 0.25 m	-	2.86	52.14	-	-	m³	52.14
maximum depth not exceeding 1.00 m	-	3.72	67.91	-	-	m³	67.91
maximum depth not exceeding 2.00 m							
(includes earthwork support)	-	3.72	67.91	-	-	m³	95.45
Trenches; width exceeding 0.30 m wide							
maximum depth not exceeding 0.25 m	-	2.86	52.14	-	-	m³	52.14
maximum depth not exceeding 1.00 m	-	4.00	73.00	-	-	m³	73.00
maximum depth not exceeding 2.00 m							
(includes earthwork support)	-	6.00	109.50	-	-	m³	164.59
Extra over any types of excavating irrespective of depth for breaking out existing materials; hand held pneumatic breaker							
rock	-	5.00	91.25	30.60	-	m³	121.85
concrete	-	2.50	45.63	15.30	-	m³	60.92
reinforced concrete	-	4.00	73.00	28.56	-	m³	101.56
brickwork, blockwork or stonework	-	1.50	27.38	9.18	-	m³	36.55
Filling to make up levels; mechanical (JCB rear bucket)							
Arising from the excavations							
average thickness not exceeding 0.25 m	-	0.07	1.34	2.21	-	m³	3.55
average thickness less than 500 mm	-	0.06	1.14	1.88	-	m³	3.02
average thickness 1.00 m	-	0.05	0.84	1.39	-	m³	2.23
Obtained from on site spoil heaps; average 25 m distance; multiple handling							
average thickness less than 250 mm	-	0.07	1.30	2.65	-	m³	3.95
average thickness less than 500 mm	-	0.06	1.07	2.18	-	m³	3.25
average thickness 1.00 m	-	0.05	0.91	1.85	-	m³	2.77
Obtained off site; planting quality topsoil PC £26.00/m³							
average thickness less than 250 mm	-	0.07	1.30	2.65	31.20	m³	35.15
average thickness less than 500 mm	-	0.06	1.07	2.18	31.20	m³	34.45
average thickness 1.00 m	-	0.05	0.91	1.85	31.20	m³	33.97
Obtained off site; hardcore; PC £10.80/m³							
average thickness less than 250 mm	-	0.08	1.52	3.09	12.96	m³	17.57
average thickness less than 500 mm	-	0.06	1.14	2.32	12.96	m³	16.42
average thickness 1.00 m	-	0.05	0.96	1.95	12.96	m³	15.87

D GROUNDWORK

Item Excluding site overheads and profit	PC £	Labour hours	Labour £	Plant £	Material £	Unit	Total rate £
Filling to make up levels; hand							
Arising from the excavations							
average thickness exceeding 0.25 m; depositing in layers 150 mm maximum thickness	-	0.60	10.95	-	-	m³	**10.95**
Obtained from on site spoil heaps; average 25 m distance; multiple handling							
average thickness exceeding 0.25 m thick; depositing in layers 150 mm maximum thickness	-	1.00	18.25	-	-	m³	**18.25**
Disposal							
Note: Most commercial site disposal is carried out by 20 tonne, 8 wheeled vehicles. It has been customary to calculate disposal from construction sites in terms of full 15 m³ loads. Spon's research has found that based on weights of common materials such as clean hardcore and topsoil, vehicles could not load more than 12 m³ at a time. Most hauliers do not make it apparent that their loads are calculated by weight and not by volume. The rates below reflect these lesser volumes which are limited by the 20 tonne limit. The volumes shown below are based on volumes "in the solid". Weight and bulking factors have been applied. For further information please see the weights of typical materials in the Earthworks section of the Memoranda at the back of this book.							
Disposal; mechanical							
Light soils and loams (bulking factor of 1.25); 40 tonnes (2 loads per hour) Excavated material; off site; to tip; mechanically loaded (JCB)							
inert	-	0.04	0.76	1.25	13.86	m³	**15.87**
hardcore	-	0.04	0.76	1.25	12.96	m³	**14.97**
macadam	-	0.04	0.76	1.25	17.28	m³	**19.29**
clean concrete	-	0.04	0.76	1.25	10.08	m³	**12.09**
soil (sandy and loam) dry	-	0.04	0.76	1.25	13.86	m³	**15.87**
soil (sandy and loam) wet	-	0.04	0.76	1.25	14.70	m³	**16.71**
broken out compacted materials such as road bases and the like	-	0.04	0.76	1.25	15.75	m³	**17.76**
soil (clay) dry	-	0.04	0.76	1.25	17.99	m³	**20.00**
soil (clay) wet	-	0.04	0.76	1.25	18.69	m³	**20.70**
rubbish (mixed loads)	-	0.04	0.76	1.25	21.33	m³	**23.35**
green waste	-	0.04	0.76	1.25	24.00	m³	**26.01**
As above but allowing for 3 loads (36 m³, 60 tonne) per hour removed							
inert material	-	0.03	0.51	0.83	13.86	m³	**15.20**

D GROUNDWORK

Item Excluding site overheads and profit	PC £	Labour hours	Labour £	Plant £	Material £	Unit	Total rate £
D20 EXCAVATING AND FILLING - cont'd							
Disposal - cont'd							
Excavated material; off site; to tip; mechanically loaded by grab; capacity of load 7.5 m³ (12 tonne)							
inert material	-	-	-	-	-	m³	39.60
hardcore	-	-	-	-	-	m³	53.40
macadam	-	-	-	-	-	m³	34.80
clean concrete	-	-	-	-	-	m³	36.00
soil (sandy and loam) dry	-	-	-	-	-	m³	39.60
soil (sandy and loam) wet	-	-	-	-	-	m³	42.00
broken out compacted materials such as road bases and the like	-	-	-	-	-	m³	44.00
soil (clay) dry	-	-	-	-	-	m³	51.40
soil (clay) wet	-	-	-	-	-	m³	53.40
rubbish (mixed loads)	-	-	-	-	-	m³	36.67
green waste	-	-	-	-	-	m³	33.33
Disposal by skip; 6 yd3 (4.6 m³)							
Excavated material loaded to skip							
by machine	-	-	-	48.40	-	m³	48.40
by hand	-	3.00	54.75	46.19	-	m³	100.94
Excavated material; on site							
In spoil heaps							
average 25 m distance	-	0.02	0.41	0.85	-	m³	1.26
average 50 m distance	-	0.04	0.71	1.48	-	m³	2.19
average 100 m distance	-	0.06	1.13	2.35	-	m³	3.48
average 200 m distance	-	0.12	2.19	4.54	-	m³	6.74
Spreading on site							
average 25 m distance	-	0.02	0.41	4.04	-	m³	4.45
average 50 m distance	-	0.04	0.71	4.53	-	m³	5.25
average 100 m distance	-	0.06	1.13	5.54	-	m³	6.67
average 200 m distance	-	0.12	2.19	5.21	-	m³	7.40
Disposal; hand							
Excavated material; on site; in spoil heaps							
average 25 m distance	-	2.40	43.80	-	-	m³	43.80
average 50 m distance	-	2.64	48.18	-	-	m³	48.18
average 100 m distance	-	3.00	54.75	-	-	m³	54.75
average 200 m distance	-	3.60	65.70	-	-	m³	65.70
Excavated material; spreading on site							
average 25 m distance	-	2.64	48.18	-	-	m³	48.18
average 50 m distance	-	3.00	54.75	-	-	m³	54.75
average 100 m distance	-	3.60	65.70	-	-	m³	65.70
average 200 m distance	-	4.20	76.65	-	-	m³	76.65

D GROUNDWORK

Item Excluding site overheads and profit	PC £	Labour hours	Labour £	Plant £	Material £	Unit	Total rate £
Grading operations; surface previously excavated to reduce levels to prepare to receive subsequent treatments; grading to accurate levels and falls 20 mm tolerances							
Clay or heavy soils or hardcore							
5 tonne excavator	-	0.04	0.73	0.22	-	m^2	0.95
JCB Sitemaster 3CX	-	0.02	0.36	0.66	-	m^2	1.02
21 tonne 360 tracked excavator	-	0.01	0.18	0.41	-	m^2	0.59
by hand	-	0.10	1.82	-	-	m^2	1.82
Loamy topsoils							
5 tonne excavator	-	0.03	0.49	0.15	-	m^2	0.63
JCB Sitemaster 3CX	-	0.01	0.24	0.44	-	m^2	0.68
21 tonne 360 tracked excavator	-	0.01	0.14	0.31	-	m^2	0.44
by hand	-	0.05	0.91	-	-	m^2	0.91
Sand or graded granular materials							
5 tonne excavator	-	0.02	0.36	0.11	-	m^2	0.47
JCB Sitemaster 3CX	-	0.01	0.18	0.33	-	m^2	0.51
21 tonne 360 tracked excavator	-	0.01	0.10	0.24	-	m^2	0.34
by hand	-	0.03	0.61	-	-	m^2	0.61
Grading operations; surface recently filled to raise levels to prepare to receive subsequent treatments; grading to accurate levels and falls 20 mm tolerances							
Clay or heavy soils or hardcore							
5 tonne excavator	-	0.03	0.55	0.16	-	m^2	0.71
JCB Sitemaster 3CX	-	0.02	0.27	0.50	-	m^2	0.77
21 tonne 360 tracked excavator	-	0.01	0.12	0.28	-	m^2	0.40
by hand	-	0.08	1.52	-	-	m^2	1.52
Loamy topsoils							
5 tonne excavator	-	0.02	0.38	0.11	-	m^2	0.49
JCB Sitemaster 3CX	-	0.01	0.19	0.34	-	m^2	0.53
21 tonne 360 tracked excavator	-	0.01	0.12	0.28	-	m^2	0.40
by hand	-	0.04	0.73	-	-	m^2	0.73
Sand or graded granular materials							
5 tonne excavator	-	0.02	0.27	0.08	-	m^2	0.36
JCB Sitemaster 3CX	-	0.01	0.14	0.25	-	m^2	0.39
21 tonne 360 tracked excavator	-	0.01	0.09	0.21	-	m^2	0.30
by hand	-	0.03	0.52	-	-	m^2	0.52
Surface grading; Agripower Ltd; using laser controlled equipment							
Maximum variation of existing ground level							
75 mm; laser box grader and tractor	-	-	-	-	-	ha	1200.00
100 mm; laser controlled low ground pressure bulldozer	-	-	-	-	-	ha	1900.00
Fraise mowing; scarification of existing turf layer to rootzone level; Agripower Ltd							
Cart arisings to tip on site							
average thickness 50 - 75 mm	-	-	-	-	-	ha	2963.00

D GROUNDWORK

Item Excluding site overheads and profit	PC £	Labour hours	Labour £	Plant £	Material £	Unit	Total rate £
D20 EXCAVATING AND FILLING - cont'd							
Cultivating							
Ripping up subsoil; using approved subsoiling machine; minimum depth 250 mm below topsoil; at 1.20 m centres; in							
gravel or sandy clay	-	-	-	2.40	-	100 m^2	2.40
soil compacted by machines	-	-	-	2.80	-	100 m^2	2.80
clay	-	-	-	3.00	-	100 m^2	3.00
chalk or other soft rock	-	-	-	6.01	-	100 m^2	6.01
Extra for subsoiling at 1 m centres	-	-	-	0.60	-	100 m^2	0.60
Breaking up existing ground; using pedestrian operated tine cultivator or rotavator; loam or sandy soil							
100 mm deep	-	0.22	4.01	2.12	-	100 m^2	6.14
150 mm deep	-	0.28	5.02	2.65	-	100 m^2	7.67
200 mm deep	-	0.37	6.69	3.54	-	100 m^2	10.23
As above but in heavy clay or wet soils							
100 mm deep	-	0.44	8.03	4.25	-	100 m^2	12.28
150 mm deep	-	0.66	12.04	6.37	-	100 m^2	18.41
200 mm deep	-	0.82	15.06	7.96	-	100 m^2	23.02
Breaking up existing ground; using tractor drawn tine cultivator or rotavator							
100 mm deep	-	-	-	0.28	-	100 m^2	0.28
150 mm deep	-	-	-	0.36	-	100 m^2	0.36
200 mm deep	-	-	-	0.47	-	100 m^2	0.47
400 mm deep	-	-	-	1.42	-	100 m^2	1.42
Cultivating ploughed ground; using disc, drag, or chain harrow							
4 passes	-	-	-	1.71	-	100 m^2	1.71
Rolling cultivated ground lightly; using self-propelled agricultural roller	-	0.06	1.01	0.60	-	100 m^2	1.62
Importing and storing selected and approved topsoil; 20 tonne load = 11.88 m^3 average							
20 tonne loads	26.00	-	-	-	31.20	m^3	-
Surface treatments							
Compacting							
bottoms of excavations	-	0.01	0.09	0.03	-	m^2	0.12
Surface preparation; trimming surfaces of cultivated ground to final levels; removing roots, stones and debris exceeding 50 mm in any direction to tip offsite; slopes less than 15°							
clean ground with minimal stone content	-	0.25	4.56	-	-	100 m^2	4.56
slightly stony; 0.5 kg stones per m^2	-	0.33	6.08	-	-	100 m^2	6.08
very stony; 1.0 - 3.0 kg stones per m^2	-	0.50	9.13	-	0.01	100 m^2	9.13
clearing mixed, slightly contaminated rubble; inclusive of roots and vegetation	-	0.50	9.13	-	0.06	100 m^2	9.19
clearing brick-bats, stones and clean rubble	-	0.60	10.95	-	0.02	100 m^2	10.97

E IN SITU CONCRETE/LARGE PRECAST CONRETE

Item Excluding site overheads and profit	PC £	Labour hours	Labour £	Plant £	Material £	Unit	Total rate £
E10 MIXING/CASTING/CURING IN SITU CONCRETE							
General The concrete mixes used here are referred to as "Designed", "Standard" and "Designated" mixes.							
Designed mix User specifies the performance of the concrete. Producer is responsible for selecting the appropriate mix. Strength testing is essential.							
Prescribed mix User specifies the mix constituents and is responsible for ensuring that the concrete meets performance requirements. Accurate mix proportions are essential.							
Standard mix Made with a restricted range of materials. Specification to include the proposed use of the material as well as the standard mix reference, type of cement, type and size of aggregate, and slump (workability). Quality assurance is required.							
Designated mix Producer to hold current product conformity certification and quality approval. Quality assurance is essential. The mix must not be modified. Note: The mixes below are all mixed in 113 litre mixers. Please see the Minor Works section of this book for mixes containing aggregates delivered in bulk bags.							
Concrete mixes; mixed on site; costs for producing concrete; prices for commonly used mixes for various types of work; based on bulk load 20 tonne rates for aggregates							
Roughest type mass concrete such as footings and road haunchings; 300 mm thick							
1:3:6	59.18	1.20	21.85	-	59.18	m³	81.02
1:3:6 sulphate resisting	74.90	1.20	21.85	-	74.90	m³	96.75
As above but aggregates delivered in 10 tonne loads							
1:3:6	63.88	1.20	21.85	-	63.88	m³	85.72
1:3:6 sulphate resisting	79.60	1.20	21.85	-	79.60	m³	101.45
As above but aggregates delivered in 850 kg bulk bags							
1:3:6	89.18	1.20	21.85	-	89.18	m³	111.02
1:3:6 sulphate resisting	104.90	1.20	21.85	-	104.90	m³	126.75
Most ordinary use of concrete such as mass walls above ground, road slabs, etc. and general reinforced concrete work							
1:2:4	71.12	1.20	21.85	-	71.12	m³	92.96
1:2:4 sulphate resisting	94.10	1.20	21.85	-	94.10	m³	115.95

E IN SITU CONCRETE/LARGE PRECAST CONRETE

Item Excluding site overheads and profit	PC £	Labour hours	Labour £	Plant £	Material £	Unit	Total rate £
E10 MIXING/CASTING/CURING IN SITU CONCRETE - cont'd							
Concrete mixes - cont'd							
As above but aggregates delivered in 10 tonne loads							
1:2:4	75.82	1.20	21.85	-	75.82	m^3	**97.66**
1:2:4 sulphate resisting	98.80	1.20	21.85	-	98.80	m^3	**120.65**
As above but aggregates delivered in 850 kg bulk bags							
1:2:4	101.12	1.20	21.85	-	101.12	m^3	**122.96**
1:2:4 sulphate resisting	124.10	1.20	21.85	-	124.10	m^3	**145.95**
Watertight floors, pavements, walls, tanks, pits, steps, paths, surface of two course roads; reinforced concrete where extra strength is required							
1:1.5:3	81.69	1.20	21.85	-	81.69	m^3	**103.54**
As above but aggregates delivered in 10 tonne loads							
1:1.5:3	86.39	1.20	21.85	-	86.39	m^3	**108.24**
As above but aggregates delivered in 850 kg bulk bags							
1:1.5:3	111.69	1.20	21.85	-	111.69	m^3	**133.54**
Plain in situ concrete; site mixed; 10 N/mm^2 - 40 mm aggregate (1:3:6) (aggregate delivery indicated)							
Foundations							
ordinary portland cement; 20 tonne ballast loads	81.02	1.00	18.25	-	83.05	m^3	**101.30**
ordinary portland cement; 10 tonne ballast loads	85.72	1.00	18.25	-	87.86	m^3	**106.11**
ordinary portland cement; 850 kg bulk bags	111.02	1.00	18.25	-	113.80	m^3	**132.05**
sulphate resistant cement; 10 tonne ballast loads	120.65	1.00	18.25	-	123.67	m^3	**141.92**
sulphate resistant cement; 850 kg bulk bags	126.75	1.00	18.25	-	129.92	m^3	**148.17**
Foundations; poured on or against earth or unblinded hardcore							
ordinary portland cement; 20 tonne ballast loads	81.02	1.00	18.25	-	85.07	m^3	**103.32**
ordinary portland cement; 10 tonne ballast loads	85.72	1.00	18.25	-	90.01	m^3	**108.26**
ordinary portland cement; 850 kg bulk bags	111.02	1.00	18.25	-	116.57	m^3	**134.82**
sulphate resistant cement; 20 tonne ballast loads	115.95	1.00	18.25	-	121.75	m^3	**140.00**
sulphate resistant cement; 850 kg bulk bags	126.75	1.00	18.25	-	133.09	m^3	**151.34**
Isolated foundations							
ordinary portland cement; 20 tonne ballast loads	81.02	1.10	20.07	-	83.05	m^3	**103.12**
ordinary portland cement; 850 kg bulk bags	111.02	1.10	20.07	-	113.80	m^3	**133.87**
sulphate resistant cement; 20 tonne ballast loads	115.95	1.10	20.07	-	115.95	m^3	**136.03**
sulphate resistant cement; 850 kg bulk bags	126.75	1.10	20.07	-	129.92	m^3	**149.99**

E IN SITU CONCRETE/LARGE PRECAST CONRETE

Item Excluding site overheads and profit	PC £	Labour hours	Labour £	Plant £	Material £	Unit	Total rate £
Plain in situ concrete; site mixed; 21 N/mm² - 20 mm aggregate (1:2:4) Foundations							
ordinary portland cement; 20 tonne ballast loads	92.96	1.00	18.25	-	95.28	m³	**113.53**
ordinary portland cement; 850 kg bulk bags	122.96	1.00	18.25	-	126.03	m³	**144.28**
sulphate resistant cement; 20 tonne ballast loads	115.95	1.00	18.25	-	118.85	m³	**137.10**
sulphate resistant cement; 850 kg bulk bags	126.75	1.00	18.25	-	129.92	m³	**148.17**
Foundations; poured on or against earth or unblinded hardcore							
ordinary portland cement; 20 tonne ballast loads	92.96	1.00	18.25	-	97.61	m³	**115.86**
ordinary portland cement; 850 kg bulk bags	122.96	1.00	18.25	-	129.11	m³	**147.36**
sulphate resistant cement; 20 tonne ballast loads	115.95	1.00	18.25	-	121.75	m³	**140.00**
sulphate resistant cement; 850 kg bulk bags	126.75	1.00	18.25	-	133.09	m³	**151.34**
Isolated foundations							
ordinary portland cement; 20 tonne ballast loads	92.96	1.10	20.07	-	95.28	m³	**115.36**
ordinary portland cement; 850 kg bulk bags	122.96	1.10	20.07	-	126.03	m³	**146.11**
sulphate resistant cement; 20 tonne ballast loads	115.95	1.10	20.07	-	118.85	m³	**138.92**
sulphate resistant cement; 850 kg bulk bags	126.75	1.10	20.07	-	129.92	m³	**149.99**
Ready mix concrete; concrete mixed on site Euromix Ready mix concrete mixed on site; placed by barrow not more than 25 m distance (16 barrows/m³)							
concrete 1:3:6	95.00	1.00	18.25	-	95.00	m³	**113.25**
concrete 1:2:4	95.00	1.00	18.25	-	95.00	m³	**113.25**
concrete C30	105.00	1.00	18.25	-	105.00	m³	**123.25**
Reinforced in situ concrete; site mixed; 21 N/mm² - 20 mm aggregate (1:2:4); aggregates delivered in 10 tonne loads Foundations							
ordinary portland cement	92.96	2.20	40.15	-	95.28	m³	**135.43**
sulphate resistant cement	115.95	2.20	40.15	-	118.85	m³	**159.00**
Foundations; poured on or against earth or unblinded hardcore							
ordinary portland cement	92.96	2.20	40.15	-	95.28	m³	**135.43**
sulphate resistant cement	120.65	2.20	40.15	-	126.68	m³	**166.83**
Isolated foundations							
ordinary portland cement	92.96	2.75	50.19	-	95.28	m³	**145.47**
sulphate resistant cement	120.65	2.75	50.19	-	126.68	m³	**176.87**

E IN SITU CONCRETE/LARGE PRECAST CONRETE

Item Excluding site overheads and profit	PC £	Labour hours	Labour £	Plant £	Material £	Unit	Total rate £
E10 MIXING/CASTING/CURING IN SITU CONCRETE - cont'd							
Plain in situ concrete; ready mixed; Tarmac Southern; 10 N/mm mixes; suitable for mass concrete fill and blinding							
Foundations							
GEN 1; Designated mix	78.44	1.50	27.38	-	78.44	m^3	105.81
ST 2; Standard mix	84.15	1.50	27.38	-	88.36	m^3	115.73
Foundations; poured on or against earth or unblinded hardcore							
GEN 1; Designated mix	78.44	1.57	28.74	-	78.44	m^3	107.18
ST 2; Standard mix	84.15	1.57	28.74	-	90.46	m^3	119.20
Isolated foundations							
GEN 1; Designated mix	78.44	2.00	36.50	-	78.44	m^3	114.94
ST 2; Standard mix	84.15	2.00	36.50	-	88.36	m^3	124.86
Plain in situ concrete; ready mixed; Tarmac Southern; 15 N/mm mixes; suitable for oversite below suspended slabs and strip footings in non aggressive soils							
Foundations							
GEN 2; Designated mix	80.89	1.50	27.38	-	84.93	m^3	112.31
ST 3; Standard mix	84.49	1.50	27.38	-	88.71	m^3	116.09
Foundations; poured on or against earth or unblinded hardcore							
GEN 2; Designated mix	80.89	1.57	28.74	-	80.89	m^3	109.63
ST 3; Standard mix	84.49	1.57	28.74	-	84.49	m^3	113.23
Isolated foundations							
GEN 2; Designated mix	80.89	2.00	36.50	-	80.89	m^3	117.39
ST 3; Standard mix	84.49	2.00	36.50	-	84.49	m^3	120.99
Plain in situ concrete; ready mixed; Tarmac Southern; air entrained mixes suitable for paving							
Beds or slabs; house drives, parking and external paving							
PAV 1; 35 N/mm^2; Designated mix	83.28	1.50	27.38	-	87.44	m^3	114.82
Beds or slabs; heavy duty external paving							
PAV 2; 40 N/mm^2; Designated mix	84.05	1.50	27.38	-	88.25	m^3	115.63
Reinforced in situ concrete; ready mixed; Tarmac Southern; 35 N/mm^2 mix; suitable for foundations in class 2 sulphate conditions							
Foundations							
RC 35; Designated mix	82.40	2.00	36.50	-	86.52	m^3	123.02
Foundations; poured on or against earth or unblinded hardcore							
RC 35; Designated mix	82.40	2.10	38.33	-	88.58	m^3	126.91
Isolated foundations							
RC 35; Designated mix	82.40	2.00	36.50	-	82.40	m^3	118.90

E IN SITU CONCRETE/LARGE PRECAST CONRETE

Item Excluding site overheads and profit	PC £	Labour hours	Labour £	Plant £	Material £	Unit	Total rate £
E20 FORMWORK FOR IN SITU CONCRETE							
Plain vertical formwork; basic finish							
Sides of foundations							
height not exceeding 250 mm	-	1.00	18.25	-	1.31	m	19.56
height 250 - 500 mm	-	1.00	18.25	-	2.27	m	20.52
height 500 mm - 1.00 m	-	1.50	27.38	-	4.54	m	31.91
height exceeding 1.00 m	-	2.00	36.50	-	4.54	m²	41.04
Sides of foundations; left in							
height not exceeding 250 mm	-	1.00	18.25	-	4.92	m	23.17
height 250 - 500 mm	-	1.00	18.25	-	9.48	m	27.73
height 500 mm - 1.00 m	-	1.50	27.38	-	18.96	m	46.33
height over 1.00 m	-	2.00	36.50	-	9.41	m²	45.91
E30 REINFORCEMENT FOR IN SITU CONCRETE							
The rates for reinforcement shown for steel bars below are based on prices which would be supplied on a typical landscape contract. The steel prices shown have been priced on a selection of steel delivered to site where the total order quantity is in the region of 2 tonnes. The assumption is that should larger quantities be required, the work would fall outside the scope of the typical landscape contract defined in the front of this book. Keener rates can be obtained for larger orders.							
Reinforcement bars; hot rolled plain round mild steel; straight or bent							
Bars							
8 mm nominal size	450.00	27.00	492.75	-	450.00	tonne	942.75
10 mm nominal size	450.00	26.00	474.50	-	450.00	tonne	924.50
12 mm nominal size	450.00	25.00	456.25	-	450.00	tonne	906.25
16 mm nominal size	450.00	24.00	438.00	-	450.00	tonne	888.00
20 mm nominal size	450.00	23.00	419.75	-	450.00	tonne	869.75
Reinforcement bar to concrete formwork							
8 mm bar							
100 ccs	1.75	0.33	6.02	-	1.75	m²	7.78
200 ccs	0.85	0.25	4.56	-	0.85	m²	5.42
300 ccs	0.58	0.15	2.74	-	0.58	m²	3.32
10 mm bar							
100 ccs	2.75	0.33	6.02	-	2.75	m²	8.77
200 ccs	1.35	0.25	4.56	-	1.35	m²	5.91
300 ccs	0.90	0.15	2.74	-	0.90	m²	3.64
12 mm bar							
100 ccs	3.96	0.33	6.02	-	3.96	m²	9.98
200 ccs	1.98	0.25	4.56	-	1.98	m²	6.54
300 ccs	1.30	0.15	2.74	-	1.30	m²	4.04
16 mm bar							
100 ccs	7.07	0.40	7.30	-	7.07	m²	14.37
200 ccs	3.56	0.33	6.02	-	3.56	m²	9.58
300 ccs	2.38	0.25	4.56	-	2.38	m²	6.95

E IN SITU CONCRETE/LARGE PRECAST CONRETE

Item Excluding site overheads and profit	PC £	Labour hours	Labour £	Plant £	Material £	Unit	Total rate £
E30 REINFORCEMENT FOR IN SITU CONCRETE - cont'd							
Reinforcement bar to concrete formwork - cont'd							
25 mm bar							
100 ccs	17.32	0.40	7.30	-	17.32	m^2	**24.63**
200 ccs	8.64	0.33	6.02	-	8.64	m^2	**14.66**
300 ccs	5.76	0.25	4.56	-	5.76	m^2	**10.32**
32 mm bar							
100 ccs	28.39	0.40	7.30	-	28.39	m^2	**35.70**
200 ccs	14.18	0.33	6.02	-	14.18	m^2	**20.20**
300 ccs	9.45	0.25	4.56	-	9.45	m^2	**14.01**
Reinforcement fabric; lapped; in beds or suspended slabs							
Fabric							
ref A98 (1.54 kg/m^2)	0.69	0.22	4.01	-	0.76	m^2	**4.77**
ref A142 (2.22 kg/m^2)	1.00	0.22	4.01	-	1.10	m^2	**5.12**
ref A193 (3.02 kg/m^2)	1.40	0.22	4.01	-	1.54	m^2	**5.55**
ref A252 (3.95 kg/m^2)	1.78	0.24	4.38	-	1.96	m^2	**6.34**
ref A393 (6.16 kg/m^2)	2.80	0.28	5.11	-	3.08	m^2	**8.19**

F MASONRY

Item Excluding site overheads and profit	PC £	Labour hours	Labour £	Plant £	Material £	Unit	Total rate £
F MASONRY							
Cement; Builder Centre							
Portland cement	-	-	-	-	3.11	25 kg	-
Sulphate resistant cement	-	-	-	-	5.00	25 kg	-
White cement	-	-	-	-	7.30	25 kg	-
Sand; Builder Centre							
Building sand							
loose	-	-	-	-	35.00	m³	-
850 kg bulk bags	-	-	-	-	30.00	nr	-
Sharp sand							
loose	-	-	-	-	28.72	m³	-
850 kg bulk bags	-	-	-	-	30.00	nr	-
Sand; Yeoman Aggregates Ltd							
Sharp sand	-	-	-	-	17.95	tonne	-
Bricks; E.T. Clay Products							
Reclaimed (second hand) bricks							
Yellows	-	-	-	-	980.00	1000	-
Yellow Multi	-	-	-	-	980.00	1000	-
Mixed London Stock	-	-	-	-	700.00	1000	-
Red Multi	-	-	-	-	650.00	1000	-
Gaults	-	-	-	-	540.00	1000	-
Red Rubbers	-	-	-	-	780.00	1000	-
Common bricks	-	-	-	-	192.40	1000	-
Engineering bricks	-	-	-	-	291.20	1000	-
F10 BRICK/BLOCK WALLING Note: Batching quantities for these mortar mixes may be found in the Tables and Memoranda section of this book.							
Mortar mixes; common mixes for various types of work; mortar mixed on site; prices based on builders merchant rates for cement Aggregates delivered in 850 kg bulk bags; mechanically mixed							
1:3	-	0.75	13.69	-	133.18	m³	**146.86**
1:4	-	0.75	13.69	-	106.34	m³	**120.03**
1:1:6	-	0.75	13.69	-	127.58	m³	**141.27**
1:1:6 sulphate resisting	-	0.75	13.69	-	147.99	m³	**161.68**
Mortar mixes; common mixes for various types of work; mortar mixed on site; prices based on builders merchant rates for cement Aggregates delivered in 10 tonne loads; mechanically mixed							
1:3	-	0.75	13.69	-	103.81	m³	**117.49**
1:4	-	0.75	13.69	-	80.40	m³	**94.08**
1:1:6	-	0.75	13.69	-	103.02	m³	**116.70**
1:1:6 sulphate resisting	-	0.75	13.69	-	123.43	m³	**137.12**

F MASONRY

Item Excluding site overheads and profit	PC £	Labour hours	Labour £	Plant £	Material £	Unit	Total rate £
F10 BRICK/BLOCK WALLING - cont'd							
Variation in brick prices Add or subtract the following amounts for every £1.00/1000 difference in the PC price of the measured items below							
half brick thick	-	-	-	-	0.06	m²	-
one brick thick	-	-	-	-	0.13	m²	-
one and a half brick thick	-	-	-	-	0.19	m²	-
two brick thick	-	-	-	-	0.25	m²	-
Mortar (1:3) required per m² of brickwork; brick size 215 x 102.5 x 65 mm Half brick wall (103 mm)							
no frog	-	-	-	-	2.71	m²	-
single frog	-	-	-	-	3.33	m²	-
double frog	-	-	-	-	3.94	m²	-
2 x half brick cavity wall (270 mm)							
no frog	-	-	-	-	4.68	m²	-
single frog	-	-	-	-	5.55	m²	-
double frog	-	-	-	-	6.79	m²	-
One brick wall (215 mm)							
no frog	-	-	-	-	5.67	m²	-
single frog	-	-	-	-	6.79	m²	-
double frog	-	-	-	-	7.89	m²	-
One and a half brick wall (328 mm)							
no frog	-	-	-	-	7.77	m²	-
single frog	-	-	-	-	9.12	m²	-
double frog	-	-	-	-	10.85	m²	-
Mortar (1:3) required per m² of blockwork; blocks 440 x 215 mm Block thickness							
100 mm	-	-	-	-	0.87	m²	-
140 mm	-	-	-	-	1.12	m²	-
hollow blocks 440 x 215 mm	-	-	-	-	0.28	m²	-
Movement of materials Loading to wheelbarrows and transporting to location; per 215 mm thick walls; maximum distance 25 m	-	0.42	7.60	-	-	m²	7.60
Class B engineering bricks; PC £239.00/ 1000; double Flemish bond in cement mortar (1:3) Offloading mechanically; loading to wheelbarrows; transporting to location maximum 25 m distance per 215 mm thick walls	-	0.42	7.60	-	-	m²	7.60
Construct walls							
half brick thick	-	1.80	32.85	-	16.82	m²	49.67
one brick thick	-	3.60	65.69	-	33.64	m²	99.33
one and a half brick thick	-	5.40	98.54	-	63.30	m²	161.84
two brick thick	-	7.20	131.39	-	67.27	m²	198.66
Construct walls; curved; mean radius 6 m							
half brick thick	-	2.70	49.32	-	17.37	m²	66.70
one brick thick	-	5.41	98.65	-	33.64	m²	132.28

F MASONRY

Item Excluding site overheads and profit	PC £	Labour hours	Labour £	Plant £	Material £	Unit	Total rate £
Construct walls; curved; mean radius 1.50 m							
half brick thick	-	3.60	65.69	-	17.41	m²	83.10
one brick thick	-	7.20	131.39	-	33.64	m²	165.03
Construct walls; tapering; one face battering; average							
one and a half brick thick	-	6.65	121.34	-	50.45	m²	171.80
two brick thick	-	8.87	161.79	-	67.27	m²	229.06
Construct walls; battering (retaining)							
one and a half brick thick	-	6.65	121.34	-	50.45	m²	171.80
two brick thick	-	8.87	161.79	-	67.27	m²	229.06
Isolated piers							
one brick thick	-	7.00	127.75	-	37.16	m²	164.91
one and a half brick thick	-	9.00	164.25	-	55.74	m²	219.99
two brick thick	-	10.00	182.50	-	75.49	m²	257.99
three brick thick	-	12.40	226.30	-	111.47	m²	337.77
Projections; vertical							
one brick x half brick	-	0.70	12.78	-	3.76	m	16.53
one brick x one brick	-	1.40	25.55	-	7.51	m	33.06
one and a half brick x one brick	-	2.10	38.33	-	11.27	m	49.60
two brick by one brick	-	2.30	41.98	-	15.03	m	57.00
Walls; half brick thick							
in honeycomb bond	-	1.80	32.85	-	12.15	m²	45.00
in quarter bond	-	1.67	30.42	-	16.46	m²	46.87
Facing bricks; PC £300.00/1000; English garden wall bond; in gauged mortar (1:1:6); facework one side							
Mechanically offloading; maximum 25 m distance; loading to wheelbarrows and transporting to location; walls	-	0.42	7.60	-	-	m²	7.60
Construct walls							
half brick thick	-	1.80	32.85	-	20.65	m²	53.50
half brick thick (using site cut snap headers to form bond)	-	2.41	44.07	-	20.65	m²	64.72
one brick thick	-	3.60	65.69	-	41.30	m²	107.00
one and a half brick thick	-	5.40	98.54	-	61.95	m²	160.49
two brick thick	-	7.20	131.39	-	82.60	m²	213.99
Walls; curved; mean radius 6 m							
half brick thick	-	2.70	49.32	-	22.13	m²	71.46
one brick thick	-	5.41	98.65	-	43.10	m²	141.75
Walls; curved; mean radius 1.50 m							
half brick thick	-	3.60	65.69	-	21.68	m²	87.38
one brick thick	-	7.20	131.39	-	42.20	m²	173.59
Walls; tapering; one face battering; average							
one and a half brick thick	-	6.65	121.34	-	64.65	m²	185.99
two brick thick	-	8.87	161.79	-	86.20	m²	247.99
Walls; battering (retaining)							
one and a half brick thick	-	5.98	109.21	-	64.65	m²	173.86
two brick thick	-	7.98	145.61	-	86.20	m²	231.81
Isolated piers; English bond; facework all round							
one brick thick	-	7.00	127.75	-	44.84	m²	172.59
one and a half brick thick	-	9.00	164.25	-	67.27	m²	231.52
two brick thick	-	10.00	182.50	-	90.86	m²	273.36
three brick thick	-	12.40	226.30	-	134.53	m²	360.83

F MASONRY

Item Excluding site overheads and profit	PC £	Labour hours	Labour £	Plant £	Material £	Unit	Total rate £
F10 BRICK/BLOCK WALLING - cont'd							
Facing bricks - cont'd							
Projections; vertical							
one brick x half brick	-	0.70	12.78	-	4.61	m	**17.38**
one brick x one brick	-	1.40	25.55	-	9.21	m	**34.76**
one and a half brick x one brick	-	2.10	38.33	-	13.83	m	**52.15**
two brick x one brick	-	2.30	41.98	-	18.43	m	**60.41**
Brickwork fair faced both sides; facing bricks in gauged mortar (1:1:6)							
extra for fair face both sides; flush, struck, weathered, or bucket-handle pointing	-	0.67	12.17	-	-	m²	**12.17**
Extra for cement mortar (1:3) in lieu of gauged mortar							
half brick thick	-	-	-	-	0.01	m²	-
one brick thick	-	-	-	-	0.02	m²	-
one and a half brick thick	-	-	-	-	0.03	m²	-
two brick thick	-	-	-	-	0.04	m²	-
Reclaimed bricks; PC £800.00/1000; English garden wall bond; in gauged mortar (1:1:6)							
Walls							
half brick thick (stretcher bond)	-	1.80	32.85	-	52.15	m²	**85.00**
half brick thick (using site cut snap headers to form bond)	-	2.41	44.07	-	52.15	m²	**96.22**
one brick thick	-	3.60	65.69	-	102.38	m²	**168.08**
one and a half brick thick	-	5.40	98.54	-	153.57	m²	**252.11**
two brick thick	-	7.20	131.39	-	204.76	m²	**336.15**
Walls; curved; mean radius 6 m							
half brick thick	-	2.70	49.32	-	55.13	m²	**104.46**
one brick thick	-	7.20	131.39	-	109.10	m²	**240.49**
Walls; curved; mean radius 1.50 m							
half brick thick	-	1.60	29.20	-	57.53	m²	**86.73**
one brick thick	-	3.20	58.40	-	113.90	m²	**172.30**
Extra for cement mortar (1:3) in lieu of gauged mortar							
half brick thick	-	-	-	-	0.01	m²	-
one brick thick	-	-	-	-	0.02	m²	-
one and a half brick thick	-	-	-	-	0.03	m²	-
two brick thick	-	-	-	-	0.04	m²	-
Walls; stretcher bond; wall ties at 450 mm centres vertically and horizontally							
one brick thick	1.33	3.27	59.74	-	103.75	m²	**163.49**
two brick thick	4.00	4.91	89.59	-	208.86	m²	**298.45**
Brickwork fair faced both sides; facing bricks in gauged mortar (1:1:6)							
extra for fair face both sides; flush, struck, weathered or bucket-handle pointing	-	0.67	12.17	-	-	m²	**12.17**

F MASONRY

Item Excluding site overheads and profit	PC £	Labour hours	Labour £	Plant £	Material £	Unit	Total rate £
Brick copings							
Copings; all brick headers-on-edge; two angles rounded 53 mm radius; flush pointing top and both sides as work proceeds; one brick wide; horizontal							
machine-made specials	25.65	0.49	8.99	-	27.12	m	**36.11**
hand-made specials	25.65	0.49	8.99	-	27.13	m	**36.12**
Extra over copings for two courses machine-made tile creasings; projecting 25 mm each side; 260 mm wide copings; horizontal	5.54	0.50	9.13	-	6.17	m	**15.29**
Copings; all brick headers-on-edge; flush pointing top and both sides as work proceeds; one brick wide; horizontal							
facing bricks PC £300.00/1000	4.00	0.49	8.99	-	4.30	m	**13.29**
engineering bricks PC £239.00/1000	3.19	0.49	8.99	-	3.35	m	**12.34**
Dense aggregate concrete blocks; "Tarmac Topblock" or other equal and approved; in gauged mortar (1:2:9)							
Walls							
Solid blocks 7 N/mm²							
440 x 215 x 100 mm thick	6.95	1.20	21.90	-	7.89	m²	**29.79**
440 x 215 x 140 mm thick	11.18	1.30	23.73	-	12.24	m²	**35.96**
Solid blocks 7 N/mm² laid flat							
440 x 100 x 215 mm thick	14.18	3.22	58.77	-	16.70	m²	**75.47**
Hollow concrete blocks							
440 x 215 x 215 mm thick	17.57	1.30	23.73	-	18.63	m²	**42.35**
Filling of hollow concrete blocks with concrete as work proceeds; tamping and compacting							
440 x 215 x 215 mm thick	16.99	0.20	3.65	-	16.99	m²	**20.64**
F20 NATURAL STONE RUBBLE WALLING							
Granite walls							
Granite random rubble walls; laid dry							
200 mm thick; single faced	45.80	6.66	121.55	-	45.80	m²	**167.34**
Granite walls; one face battering to 50°; pointing faces							
450 mm (average) thick	103.05	5.00	91.25	-	111.52	m²	**202.77**

F MASONRY

Item Excluding site overheads and profit	PC £	Labour hours	Labour £	Plant £	Material £	Unit	Total rate £
F20 NATURAL STONE RUBBLE WALLING - cont'd							
Dry stone walling - General Preamble: In rural areas where natural stone is a traditional material, it may be possible to use dry stone walling or dyking as an alternative to fences or brick walls. Many local authorities are willing to meet the extra cost of stone walling in areas of high landscape value, and they may hold lists of available craftsmen. DSWA Office, Westmorland County Showground, Lane Farm, Crooklands, Milnthorpe, Cumbria, LA7 7NH; Tel: 01539 567753; E-mail: information@dswa.org.uk Note: Traditional walls are not built on concrete foundations. Dry stone wall; wall on concrete foundation (not included); dry stone coursed wall inclusive of locking stones and filling to wall with broken stone or rubble; walls up to 1.20 m high; battered; 2 sides fair faced							
Yorkstone	-	6.50	118.63	-	60.67	m²	**179.29**
Cotswold stone	-	6.50	118.63	-	74.15	m²	**192.78**
Purbeck	-	6.50	118.63	-	80.89	m²	**199.52**
Rockery stone - General Preamble: Rockery stone prices vary considerably with source, carriage, distance and load. Typical PC prices are in the range of £60 - £80 per tonne collected.							
Rockery stone; CED Ltd Boulders; maximum distance 25 m; by machine							
750 mm diameter	74.42	0.90	16.43	6.53	74.42	nr	**97.38**
1 m diameter	230.05	2.00	36.50	13.06	230.05	nr	**279.61**
1.5 m diameter	765.05	2.00	36.50	47.50	765.05	nr	**849.05**
2 m diameter	1856.45	2.00	36.50	47.50	1856.45	nr	**1940.45**
Boulders; maximum distance 25 m; by hand							
750 mm diameter	74.42	0.75	13.69	-	74.42	nr	**88.11**
1 m diameter	230.05	1.89	34.41	-	230.05	nr	**264.46**

F MASONRY

Item Excluding site overheads and profit	PC £	Labour hours	Labour £	Plant £	Material £	Unit	Total rate £
F22 CAST STONE WALLING ASHLAR WALLING/DRESSINGS							
Haddonstone Ltd; cast stone piers; ornamental masonry in Portland Bath or Terracotta finished cast stone							
Gate Pier S120; to foundations and underground work measured separately; concrete infill							
S120G base unit to pier; 699 x 699 x 172 mm	157.45	0.75	13.69	-	161.99	nr	**175.67**
S120F/F shaft base unit; 533 x 533 x 280 mm	125.73	1.00	18.25	-	130.41	nr	**148.66**
S120E/E main shaft unit; 533 x 533 x 280 mm; nr of units required dependent on height of pier	125.73	1.00	18.25	-	130.41	nr	**148.66**
S120D/D top shaft unit; 33 x 533 x 280 mm	125.73	1.00	18.25	-	130.41	nr	**148.66**
S120C pier cap unit; 737 x 737 x 114 mm	149.23	0.50	9.13	-	153.91	nr	**163.03**
S120B pier block unit; base for finial; 533 x 533 x 64 mm	63.45	0.33	6.02	-	63.59	nr	**69.61**
Pier blocks; flat to receive gate finial							
S100B; 440 x 440 x 63 mm	41.13	0.33	6.08	-	41.46	nr	**47.54**
S120B; 546 x 546 x 64 mm	63.45	0.33	6.08	-	63.78	nr	**69.86**
S150B; 330 x 330 x 51 mm	22.33	0.33	6.08	-	22.66	nr	**28.74**
Pier caps; part weathered							
S100C; 915 x 915 x 150 mm	354.85	0.50	9.13	-	355.50	nr	**364.63**
S120C; 737 x 737 x 114 mm	149.23	0.50	9.13	-	149.56	nr	**158.68**
S150C; 584 x 584 x 120 mm	108.10	0.50	9.13	-	108.43	nr	**117.55**
Pier caps; weathered							
S230C; 1029 x 1029 x 175 mm	465.30	0.50	9.13	-	465.63	nr	**474.75**
S215C; 687 x 687 x 175 mm	203.28	0.50	9.13	-	203.61	nr	**212.73**
S210C; 584 x 584 x 175 mm	128.08	0.50	9.13	-	128.41	nr	**137.53**
Pier strings							
S100S; 800 x 800 x 55 mm	121.03	0.50	9.13	-	121.23	nr	**130.35**
S120S; 555 x 555 x 44 mm	55.23	0.50	9.13	-	55.43	nr	**64.55**
S150S; 457 x 457 x 48 mm	219.73	0.50	9.13	-	219.93	nr	**229.05**
Balls and bases							
E150A Ball 535 mm and E150C collared base	328.00	0.50	9.13	-	328.20	nr	**337.32**
E120A Ball 330 mm and E120C collared base	122.00	0.50	9.13	-	122.20	nr	**131.32**
E110A Ball 230 mm and E110C collared base	85.00	0.50	9.13	-	85.20	nr	**94.32**
E100A Ball 170 mm and E100B plain base	53.00	0.50	9.13	-	53.20	nr	**62.32**

F MASONRY

Item Excluding site overheads and profit	PC £	Labour hours	Labour £	Plant £	Material £	Unit	Total rate £
F22 CAST STONE WALLING ASHLAR WALLING/DRESSINGS - cont'd							
Haddonstone Ltd; cast stone copings; ornamental masonry in Portland Bath or Terracotta finished cast stone							
Copings for walls; bedded, jointed and pointed in approved coloured cement-lime mortar 1:2:9							
T100 weathered coping; 102 mm high x 178 mm wide x 914 mm	38.42	0.33	6.08	-	38.70	m	**44.78**
T140 weathered coping; 102 mm high x 337 mm wide x 914 mm	62.76	0.33	6.08	-	63.04	m	**69.13**
T200 weathered coping; 127 mm high x 508 mm wide x 750 mm	97.34	0.33	6.08	-	97.62	m	**103.70**
T170 weathered coping; 108 mm high x 483 mm wide x 914 mm	105.02	0.33	6.08	-	105.30	m	**111.38**
T340 raked coping; 75 - 100 mm high x 290 mm wide x 900 mm	60.20	0.33	6.08	-	60.48	m	**66.56**
T310 raked coping; 76 - 89 mm high x 381 mm wide x 914 mm	69.16	0.33	6.08	-	69.44	m	**75.52**
Bordeaux walling; Forticrete Ltd							
Dry stacked random sized units 150 - 400 mm long x 150 mm high cast stone wall mechanically interlocked with fibreglass pins; constructed to levelling pad of coarse compacted granular material back filled behind the elevation with 300 mm wide granular drainage material; walls to 5.00 m high (retaining walls over heights shown below require individual design)							
gravity wall; near vertical; 250 mm thick	81.00	1.00	18.25	-	81.00	m²	**99.25**
gravity wall; 9.5°; battered	81.00	2.00	36.50	-	81.00	m²	**117.50**
copings to Bordeaux wall; 70 mm thick; random lengths	13.00	0.25	4.56	-	13.00	m	**17.56**
Retaining wall; as above but reinforced with Tensar geogrid 40 RE laid between every two courses horizontally into the face of the excavation (excavation not included)							
Near vertical; 250 mm thick; 1.50 m of geogrid length							
up to 1.20 m high; 2 layers of geogrid	81.00	1.50	27.38	-	87.00	m²	**114.38**
1.20 - 1.50 m high; 3 layers of geogrid	81.00	2.00	36.50	-	90.00	m²	**126.50**
1.50 - 1.8 m high; 4 layers of geogrid	81.00	2.25	41.06	-	93.00	m²	**134.06**
Battered; maximum 1:3 slope							
up to 1.20 m high; 3 layers of geogrid	81.00	2.50	45.63	-	90.00	m²	**135.63**
1.20 - 1.50 m high; 4 layers of geogrid	81.00	2.50	45.63	-	93.00	m²	**138.63**
1.50 - 1.8 m high; 5 layers of geogrid	81.00	3.00	54.75	-	95.50	m²	**150.25**

F MASONRY

Item Excluding site overheads and profit	PC £	Labour hours	Labour £	Plant £	Material £	Unit	Total rate £
F30 ACCESSORIES/SUNDRY ITEMS FOR BRICK/BLOCK/STONE WALLING							
Damp proof courses; pitch polymer; 150 mm laps							
Horizontal							
width not exceeding 225 mm	4.30	1.14	20.80	-	5.91	m^2	26.71
width exceeding 225 mm	4.30	0.58	10.59	-	6.44	m^2	17.03
Vertical							
width not exceeding 225 mm	4.30	1.72	31.39	-	5.91	m^2	37.30
Two courses slates in cement mortar (1:3)							
Horizontal							
width exceeding 225 mm	8.86	3.46	63.15	-	12.38	m^2	75.53
Vertical							
width exceeding 225 mm	8.86	5.18	94.53	-	12.38	m^2	106.92
F31 PRECAST CONCRETE SILLS/ LINTELS/COPINGS/FEATURES							
Mix 21.00 N/mm^2 - 20 mm aggregate (1:2:4)							
Copings; once weathered; twice grooved							
152 x 75 mm	5.15	0.40	7.30	-	5.46	m	12.76
178 x 65 mm	7.46	0.40	7.30	-	7.75	m	15.05
305 x 75 mm	10.39	0.50	9.13	-	10.86	m	19.98
Pier caps; four sides weathered							
305 x 305 mm	6.20	1.00	18.25	-	6.34	nr	24.59
381 x 381 mm	6.51	1.00	18.25	-	6.65	nr	24.90
533 x 533 mm	6.56	1.20	21.90	-	6.70	nr	28.60

G STRUCTURAL/CARCASSING METAL/TIMBER

Item Excluding site overheads and profit	PC £	Labour hours	Labour £	Plant £	Material £	Unit	Total rate £
G31 PREFABRICATED TIMBER UNIT DECKING							
Timber decking							
Supports for timber decking; softwood joists to receive decking boards; joists at 400 mm centres; Southern Yellow pine							
38 x 88 mm	14.20	1.00	18.25	-	17.33	m^2	**35.58**
47 x 150 mm	13.35	1.00	18.25	-	16.39	m^2	**34.65**
47 x 100 mm	8.90	1.00	18.25	-	11.50	m^2	**29.75**
Hardwood decking; Yellow Balau; grooved or smooth; 6 mm joints							
deck boards; 90 mm wide x 19 mm thick	17.18	1.00	18.25	-	19.12	m^2	**37.37**
deck boards; 145 mm wide x 21 mm thick	25.16	1.00	18.25	-	29.62	m^2	**47.87**
deck boards; 145 mm wide x 28 mm thick	25.82	1.00	18.25	-	30.35	m^2	**48.60**
Hardwood decking; Ipe; smooth; 6 mm joints							
deck boards; 90 mm wide x 19 mm thick	61.65	1.00	18.25	-	63.60	m^2	**81.85**
deck boards; 145 mm wide x 19 mm thick	44.95	1.00	18.25	-	46.90	m^2	**65.15**
AVS "Profile" Western Red Cedar; 6 mm joints smooth or reeded							
ex 125 wide x 38 mm thick	18.15	1.00	18.25	-	20.10	m^2	**38.35**
ex 150 wide x 32 mm thick	14.90	1.00	18.25	-	18.34	m^2	**36.59**
Handrails and base rail; fixed to posts at 2.00 m centres							
posts; 100 x 100 x 1370 mm high	8.66	1.00	18.25	-	9.74	m	**27.99**
posts; turned; 1220 mm high	14.78	1.00	18.25	-	15.85	m	**34.10**
Handrails; balusters							
square balusters at 100 mm centres	34.20	0.50	9.13	-	34.65	m	**43.77**
square balusters at 300 mm centres	11.39	0.33	6.08	-	11.75	m	**17.82**
turned balusters at 100 mm centres	54.00	0.50	9.13	-	54.45	m	**63.57**
turned balusters at 300 mm centres	17.98	0.33	6.02	-	18.34	m	**24.36**

H CLADDING/COVERING

Item Excluding site overheads and profit	PC £	Labour hours	Labour £	Plant £	Material £	Unit	Total rate £
H51 NATURAL STONE SLAB CLADDING/ FEATURES							
Sawn yorkstone cladding; Johnsons Wellfield Quarries Ltd Six sides sawn stone; rubbed face; sawn and jointed edges; fixed to blockwork (not included) with stainless steel fixings "Ancon Ltd" grade 304 stainless steel frame cramp and dowel 7 mm; cladding units drilled 4 x to receive dowels							
440 x 200 x 50 mm thick	88.95	1.87	34.13	-	92.35	m^2	**126.48**
H52 CAST STONE SLAB CLADDING/ FEATURES							
Cast stone cladding; Haddonstone Ltd Reconstituted stone in Portland Bath or Terracotta; fixed to blockwork or concrete (not included) with stainless steel fixings; stainless steel frame cramp and dowel M6 mm; cladding units drilled 4 x to receive dowels							
1000 x 300 x 50 mm thick	40.00	1.87	34.13	-	43.40	m^2	**77.53**

J WATERPROOFING

Item Excluding site overheads and profit	PC £	Labour hours	Labour £	Plant £	Material £	Unit	Total rate £
J10 SPECIALIST WATERPROOF RENDERING							
"Sika" waterproof rendering; steel trowelled							
Walls; 20 mm thick; three coats; to concrete base							
width exceeding 300 mm	-	-	-	-	-	m²	48.84
width not exceeding 300 mm	-	-	-	-	-	m²	76.53
Walls; 25 mm thick; three coats; to concrete base							
width exceeding 300 mm	-	-	-	-	-	m²	55.37
width not exceeding 300 mm	-	-	-	-	-	m²	87.93
J20 MASTIC ASPHALT TANKING/DAMP PROOFING							
Tanking and damp proofing; mastic asphalt; type T 1097; Bituchem							
13 mm thick; one coat covering; to concrete base; flat; work subsequently covered							
width exceeding 300 mm	-	-	-	-	-	m²	11.24
20 mm thick; two coat coverings; to concrete base; flat; work subsequently covered							
width exceeding 300 mm	-	-	-	-	-	m²	14.09
30 mm thick; three coat coverings; to concrete base; flat; work subsequently covered							
width exceeding 300 mm	-	-	-	-	-	m²	19.13
13 mm thick; two coat coverings; to brickwork base; vertical; work subsequently covered							
width exceeding 300 mm	-	-	-	-	-	m²	37.34
20 mm thick; three coat coverings; to brickwork base; vertical; work subsequently covered							
width exceeding 300 mm	-	-	-	-	-	m²	50.98
Internal angle fillets; work subsequently covered	-	-	-	-	-	m	3.97
Turning asphalt nibs into grooves; 20 mm deep	-	-	-	-	-	m	2.52
J30 LIQUID APPLIED TANKING/DAMP PROOFING							
Tanking and damp proofing; Ruberoid Building Products; "Synthaprufe" cold applied bituminous emulsion waterproof coating							
"Synthaprufe"; to smooth finished concrete or screeded slabs; flat; blinding with sand							
two coats	2.01	0.22	4.06	-	2.35	m²	6.41
three coats	3.01	0.31	5.66	-	3.45	m²	9.11
"Synthaprufe"; to fair faced brickwork with flush joints, rendered brickwork or smooth finished concrete walls; vertical							
two coats	2.26	0.29	5.21	-	2.63	m²	7.84
three coats	3.31	0.40	7.30	-	3.79	m²	11.09

J WATERPROOFING

Item Excluding site overheads and profit	PC £	Labour hours	Labour £	Plant £	Material £	Unit	Total rate £
Tanking and damp proofing; RIW Ltd Liquid asphaltic composition; to smooth finished concrete screeded slabs or screeded slabs; flat							
two coats	7.92	0.33	6.08	-	7.92	m^2	**14.00**
Liquid asphaltic composition; fair faced brickwork with flush joints, rendered brickwork or smooth finished concrete walls; vertical							
two coats	7.92	0.50	9.13	-	7.92	m^2	**17.05**
"Heviseal"; to smooth finished concrete or screeded slabs; to surfaces of ponds, tanks or planters; flat							
two coats	6.00	0.33	6.08	-	6.60	m^2	**12.69**
"Heviseal"; to fair faced brickwork with flush joints, rendered brickwork or smooth finished concrete walls; to surfaces of retaining walls, ponds, tanks or planters; vertical							
two coats	6.00	0.50	9.13	-	6.60	m^2	**15.73**
J40 FLEXIBLE SHEET TANKING/DAMP PROOFING							
Tanking and damp proofing; Grace Construction Products "Bitu-thene 2000"; 1.00 mm thick; overlapping and bonding; including sealing all edges							
to concrete slabs; flat	6.60	0.25	4.56	-	7.26	m^2	**11.82**
to brick/concrete walls; vertical	6.60	0.40	7.30	-	7.59	m^2	**14.89**

J WATERPROOFING

Item Excluding site overheads and profit	PC £	Labour hours	Labour £	Plant £	Material £	Unit	Total rate £
J50 GREEN ROOF SYSTEMS Preamble: A variety of systems are available which address all the varied requirements for a successful Green Roof. For installation by approved contractors only. The prices shown are for budgeting purposes only, as each installation is site specific and may incorporate some or all of the resources shown. Specifiers should verify that the systems specified include for design liability and inspections by the suppliers. The systems below assume commercial insulation levels are required to the space below the proposed Green Roof. Extensive Green roofs are those of generally lightweight construction with low maintenance planting and shallow soil designed for aesthetics only; Intensive Green roofs are designed to allow use for recreation and trafficking. They require more maintenance and allow a greater variety of surfaces and plant types.							
Intensive Green Roof; Bauder Ltd; soil based systems able to provide a variety of hard and soft landscaping; laid to the surface of an unprepared roof deck Vapour barrier laid to prevent intersticial condensation from spaces below the roof applied by torching to the roof deck							
VB4-Expal aluminium lined	-	-	-	-	-	m²	15.13
Insulation laid and hot bitumen bonded to vapour barrier							
PIR Insulation 100 mm	-	-	-	-	-	m²	32.78
Underlayer to receive root barrier partially bonded to insulation by torching							
G4E	-	-	-	-	-	m²	14.93
Root barrier "Plant E" chemically treated root resistant capping sheet fully bonded to G4E underlayer by torching	-	-	-	-	-	m²	20.37
Slip layers to absorb differential movement							
PE Foil; 2 layers laid to root barriers	-	-	-	-	-	m²	3.75
Optional protection layer to prevent mechanical damage "Protection mat"; 6 mm thick rubber matting; loose laid	-	-	-	-	-	m²	10.41
Drainage medium laid to root barrier "Drainage board" free draining EPS 50 mm thick	-	-	-	-	-	m²	13.72
"Reservoir board" up to 21.5 litre water storage capacity EPS 75 mm thick	-	-	-	-	-	m²	17.09
Filtration to prevent soil migration to drainage system "Filter fleece"; 3 mm thick polyester geotextile; loose laid over drainage/ reservoir layer	-	-	-	-	-	m²	3.77

J WATERPROOFING

Item Excluding site overheads and profit	PC £	Labour hours	Labour £	Plant £	Material £	Unit	Total rate £
For hard landscaped areas incorporate rigid drainage board laid to the protection mat							
PLT 60 drainage board	-	-	-	-	-	m²	22.62
Extensive Green Roof System; Bauder Ltd; low maintenance soil free system incorporating single layer growing and planting medium							
Vapour barrier laid to prevent intersticial condensation applied by torching to the roof deck							
VB4-Expal aluminium lined	-	-	-	-	-	m²	15.13
Insulation laid and hot bitumen bonded to vapour barrier							
PIR Insulation 100 mm	-	-	-	-	-	m²	32.78
Underlayer to receive rootbarrier partially bonded to insulation by torching							
G4E	-	-	-	-	-	m²	14.93
Root barrier							
"Plant E" chemically treated root resistant capping sheet fully bonded to G4E underlayer by torching	-	-	-	-	-	m²	20.37
Landscape options to the above systems; Bauder Ltd							
Hydroplanting system; to Extensive Green Roof as detailed above							
SDF mat 20 mm thick drainage layer; loose laid	-	-	-	-	-	m²	42.70
Xeroflor vegetation blanket; Bauder Ltd; to Extensive Green Roof as detailed above							
Xeroflor Xf 301 pre-cultivated sedum blanket incorporating 800 gram recycled fibre water retention layer; laid loose	-	-	-	-	-	m²	46.20
Waterproofing to upstands; Bauder Ltd							
Bauder Vapour Barrier; Bauder G4E & Bauder Plant E							
up to 200 mm high	-	-	-	-	-	m	23.78
up to 400 mm high	-	-	-	-	-	m	33.46
up to 600 mm high	-	-	-	-	-	m	44.62
Reservoir and drainage boards; Bauder; non-specialist installed products for laying under pavings or planting for collection or drainage							
DSE20 Drainage/Reservoir Board; 20 mm thick drainage board	-	-	-	-	-	m²	11.49
DSE40 Drainage/Reservoir Board; for intensive green roof applications	-	-	-	-	-	m²	15.71
Versicell 20 Drainage Board; specialist installed drainage layer for intensive green roof applications; predominantly hard landscape	-	-	-	-	-	m²	22.09
Bauder Extensive Substrate; 80 mm deep; for use with sedum and hardy perennials and biodiverse vegetation	-	-	-	-	-	m²	19.60

J WATERPROOFING

Item Excluding site overheads and profit	PC £	Labour hours	Labour £	Plant £	Material £	Unit	Total rate £
J50 GREEN ROOF SYSTEMS - cont'd							
Reservoir and drainage boards - cont'd Bauder Intensive Substrate; 200 mm deep; for use in traditional green roof applications e.g. grass and bedding plants	-	-	-	-	-	m²	49.00
Inverted waterproofing systems; Alumasc Exterior Building Products Ltd; to roof surfaces to receive Green Roof systems "Hydrotech 6125"; monolithic hot melt rubberized bitumen; applied in two 3 mm layers incorporating a polyester reinforcing sheet with 4 mm thick protection sheet and chemically impregnated root barrier; fully bonded into the Hydrotech; applied to plywood or suitably prepared wood float finish and primed concrete deck or screeds							
10 mm thick	-	-	-	-	-	m²	31.52
"Alumasc Roofmate"; extruded polystyrene insulation; optional system; thickness to suit required U value; calculated at design stage; indicative thicknesses; laid to Hydrotech 6125							
0.25 U value; average requirement 120 mm	-	-	-	-	-	m²	21.85
Warm Roof waterproofing systems; Alumasc Exterior Building Products Ltd (Euroroof); to roof surfaces to receive Green Roof systems "Derbigum" system							
"Nilperm" aluminium lined vapour barrier; 2 mm thick; bonded in hot bitumen to the roof deck	-	-	-	-	-	m²	9.88
"Korklite" insulation bonded to the vapour barrier in hot bitumen; U value dependent; 80 mm thick	-	-	-	-	-	m²	18.70
"Hi-Ten Universal" 2 mm thick underlayer; fully bonded to the insulation	-	-	-	-	-	m²	8.04
"Derbigum Anti-Root" cap sheet impregnated with root resisting chemical bonded to the underlayer	-	-	-	-	-	m²	16.55

J WATERPROOFING

Item Excluding site overheads and profit	PC £	Labour hours	Labour £	Plant £	Material £	Unit	Total rate £
Intensive Green Roof Systems; Alumasc Exterior Building Products Ltd; components laid to the insulation over the Hydrotech or Derbigum waterproofing above							
Optional inclusion; moisture retention layer							
SSM-45; moisture mat	-	-	-	-	-	m²	5.73
Drainage layer; "Floradrain" recycled polypropylene; providing water reservoir, multi-directional drainage and mechanical damage protection							
FD25; 25 mm deep; rolls inclusive of filter sheet SF	-	-	-	-	-	m²	19.44
FD40; 40 mm deep; rolls inclusive of filter sheet SF	-	-	-	-	-	m²	25.32
FD25; 25 mm deep; sheets excluding filter sheet	-	-	-	-	-	m²	15.76
FD40; 40 mm deep; sheets excluding filter sheet	-	-	-	-	-	m²	19.44
FD60; 60 mm deep; sheets excluding filter sheet	-	-	-	-	-	m²	30.68
Drainage layer; "Elastodrain" recycled rubber mat; providing multi-directional drainage and mechanical damage protection							
EL200; 20 mm deep	-	-	-	-	-	m²	25.22
"Zincolit" recycled crushed brick; optional drainage infil to Floradrain layers							
FD40; 17 litres per m²	-	0.03	0.62	-	6.30	m²	6.92
FD60; 27 litres per m²	-	0.05	0.99	-	7.46	m²	8.45
Filter sheet; rolled out onto drainage layer							
"Filter sheet TG" for Elastodrain range	-	-	-	-	-	m²	4.94
"Filter sheet SF" for Floradrain range	-	-	-	-	-	m²	4.20
Intensive substrate; lightweight growing medium laid to filter sheet	-	-	-	-	-	m²	22.06
Extensive substrate; sedum carpet laid to filter sheet	-	-	-	-	-	m²	15.65
Extensive substrate; rockery type plants laid to filter sheet	-	-	-	-	-	m²	20.49
Semi-intensive substrate; heather with lavender laid to filter sheet	-	-	-	-	-	m²	28.16
Extensive Green Roof Systems; Alumasc Exterior Building Products Ltd; components laid to the Hydrotech or insulation layers above							
Moisture retention layer							
SSM-45; moisture mat	-	-	-	-	-	m²	5.73

J WATERPROOFING

Item Excluding site overheads and profit	PC £	Labour hours	Labour £	Plant £	Material £	Unit	Total rate £
J50 GREEN ROOF SYSTEMS - cont'd							
Extensive Green Roof Systems - cont'd							
Drainage layer for flat roofs; "Floradrain"; recycled polypropylene; providing water reservoir, multi-directional drainage and mechanical damage protection; supplied in sheet or roll form							
FD25; 25 mm deep; rolls inclusive of filter sheet SF	-	-	-	-	-	m²	19.44
FD40; 40 mm deep; rolls inclusive of filter sheet SF	-	-	-	-	-	m²	25.32
FD25; 25 mm deep; sheets excluding filter sheet	-	-	-	-	-	m²	15.76
FD40; 40 mm deep; sheets excluding filter sheet	-	-	-	-	-	m²	19.44
FD60; 60 mm deep; sheets excluding filter sheet	-	-	-	-	-	m²	30.68
Drainage layer for pitched roofs; "Floratec"; recycled polystyrene; providing water reservoir and multi-directional drainage; laid to moisture mat or to the waterproofing							
FS50	-	-	-	-	-	m²	15.02
FS75	-	-	-	-	-	m²	17.65
Landscape options							
Sedum Mat vegetation layer; Alumasc Exterior Building Products Ltd; to Extensive Green Roof as detailed above	-	-	-	-	-	m²	36.77
Green Roof components; Alumasc Exterior Building Products Ltd							
Outlet inspection chambers							
"KS 15"; 150 mm deep	-	-	-	-	-	nr	123.97
height extension piece 100 mm	-	-	-	-	-	nr	44.86
height extension piece 200 mm	-	-	-	-	-	nr	48.54
Outlet and irrigation control chambers							
"B32"; 300 x 300 x 300 mm high	-	-	-	-	-	nr	372.97
"B52"; 400 x 500 x 500 mm high	-	-	-	-	-	nr	472.78
Outlet damming piece for water retention	-	-	-	-	-	nr	98.76
Linear drainage channel; collects surface water from adjacent hard surfaces or down pipes for distribution to the drainage layer	-	-	-	-	-	m	97.71

M SURFACE FINISHES

Item Excluding site overheads and profit	PC £	Labour hours	Labour £	Plant £	Material £	Unit	Total rate £
M20 PLASTERED/RENDERED/ROUGH CAST COATING							
Cement:lime:sand (1:1:6); 19 mm thick; two coats; wood floated finish							
Walls							
width exceeding 300 mm; to brickwork or blockwork base	-	-	-	-	-	m²	41.80
Extra over cement:sand:lime (1:1:6) coatings for decorative texture finish with water repellent cement							
combed or floated finish	-	-	-	-	-	m²	3.80
M40 STONE/CONCRETE/QUARRY/ CERAMIC TILING							
Ceramic tiles; unglazed slip resistant; various colours and textures; jointing							
Floors							
level or to falls only not exceeding 15° from horizontal; 150 x 150 x 8 mm thick	18.57	1.00	18.25	-	21.85	m²	40.10
level or to falls only not exceeding 15° from horizontal; 150 x 150 x 12 mm thick	22.63	1.00	18.25	-	26.11	m²	44.36
Clay tiles - General							
Preamble: Typical specification - Clay tiles shall be reasonably true to shape, flat, free from flaws, frost resistant and true to sample approved by the Landscape Architect prior to laying. Quarry tiles (or semi-vitrified tiles) shall be of external quality, either heather brown or blue, to size specified, laid on 1:2:4 concrete, with 20 mm maximum aggregate 100 mm thick, on 100 mm hardcore. The hardened concrete should be well wetted and the surplus water taken off. Clay tiles shall be thoroughly wetted immediately before laying and then drained and shall be bedded to 19 mm thick cement:sand (1:3) screed. Joints should be approximately 4 mm (or 3 mm for vitrified tiles) grouted in cement: sand (1:2) and cleaned off immediately.							
Quarry tiles; external quality; including bedding; jointing							
Floors							
level or to falls only not exceeding 15° from horizontal; 150 x 150 x 12.5 mm thick; heather brown	23.22	0.80	14.60	-	26.72	m²	41.32
level or to falls only not exceeding 15° from horizontal; 225 x 225 x 29 mm thick; heather brown	23.22	0.67	12.17	-	26.72	m²	38.89
level or to falls only not exceeding 15° from horizontal; 150 x 150 x 12.5 mm thick; blue/black	14.74	1.00	18.25	-	17.82	m²	36.07
level or to falls only not exceeding 15° from horizontal; 194 x 194 x 12.5 mm thick; heather brown	13.58	0.80	14.60	-	16.61	m²	31.21

M SURFACE FINISHES

Item Excluding site overheads and profit	PC £	Labour hours	Labour £	Plant £	Material £	Unit	Total rate £
M60 PAINTING/CLEAR FINISHING							
Eura Conservation Ltd; restoration of railings; works carried out off site; excludes removal of railings from site							
Shotblast railings; remove all traces of paint and corrosion; apply temporary protective primer; measured overall 2 sides							
to plain railings; 2 sides	-	-	-	-	-	m²	**39.00**
to decorative railings; 2 sides	-	-	-	-	-	m²	**77.00**
Paint new galvanized railings in situ with 3 coats system							
vertical bar railings or gates	-	-	-	-	-	m²	**11.00**
vertical bar railings or gates with dog mesh panels	-	-	-	-	-	m²	**12.00**
gate posts not exceeding 300 mm girth	-	-	-	-	-	m²	**6.00**
Paint previously painted railings; shotblasted offsite							
vertical bar railings or gates	-	-	-	-	-	m²	**12.00**
gate posts not exceeding 300 mm girth	-	-	-	-	-	m²	**6.00**
Prepare; touch up primer; two undercoats and one finishing coat of gloss oil paint; on metal surfaces							
General surfaces							
girth exceeding 300 mm	0.85	0.33	6.08	-	0.85	m²	**6.93**
isolated surfaces; girth not exceeding 300 mm	0.30	0.13	2.43	-	0.30	m	**2.73**
isolated areas not exceeding 0.50 m² irrespective of girth	0.42	0.13	2.43	-	0.42	nr	**2.85**
Ornamental railings; each side measured separately							
girth exceeding 300 mm	0.85	0.75	13.69	-	0.85	m²	**14.54**
Prepare; one coat primer; two undercoats and one finishing coat of gloss oil paint; on wood surfaces							
General surfaces							
girth exceeding 300 mm	1.20	0.40	7.30	-	1.20	m²	**8.50**
isolated surfaces; girth not exceeding 300 mm	0.30	0.20	3.65	-	0.30	m	**3.95**
isolated areas not exceeding 0.50 m² irrespective of girth	0.30	0.36	6.64	-	0.30	nr	**6.94**
Prepare, proprietary solution primer; two coats of dark stain; on wood surfaces							
General surfaces							
girth exceeding 300 mm	2.05	0.10	1.82	-	2.11	m²	**3.94**
isolated surfaces; girth not exceeding 300 mm	0.26	0.05	0.91	-	0.27	m	**1.18**
Three coats "Dimex Shield"; to clean, dry surfaces; in accordance with manufacturer's instructions							
Brick or block walls							
girth exceeding 300 mm	0.96	0.28	5.11	-	0.96	m²	**6.07**
Cement render or concrete walls							
girth exceeding 300 mm	0.76	0.25	4.56	-	0.76	m²	**5.32**

Prices for Measured Works – Major Works

M SURFACE FINISHES

Item Excluding site overheads and profit	PC £	Labour hours	Labour £	Plant £	Material £	Unit	Total rate £
Two coats resin based paint; "Sandtex Matt"; in accordance with manufacturer's instructions							
Brick or block walls							
girth exceeding 300 mm	6.47	0.20	3.65	-	6.47	m²	**10.12**
Cement render or concrete walls							
girth exceeding 300 mm	4.40	0.17	3.04	-	4.40	m²	**7.44**

P BUILDING FABRIC SUNDRIES

Item Excluding site overheads and profit	PC £	Labour hours	Labour £	Plant £	Material £	Unit	Total rate £
P30 TRENCHES/PIPEWAYS/PITS FOR BURIED ENGINEERING SERVICES							
Excavating trenches; using 3 tonne tracked excavator; to receive pipes; grading bottoms; earthwork support; filling with excavated material to within 150 mm of finished surfaces and compacting; completing fill with topsoil; disposal of surplus soil							
Services not exceeding 200 mm nominal size							
average depth of run not exceeding 0.50 m	1.17	0.12	2.19	0.99	1.40	m	**4.59**
average depth of run not exceeding 0.75 m	1.17	0.16	2.97	1.36	1.40	m	**5.74**
average depth of run not exceeding 1.00 m	1.17	0.28	5.17	2.37	1.40	m	**8.94**
average depth of run not exceeding 1.25 m	1.17	0.38	7.00	3.20	1.17	m	**11.36**
Excavating trenches; using 3 tonne tracked excavator; to receive pipes; grading bottoms; earthwork support; filling with imported granular material and compacting; disposal of surplus soil							
Services not exceeding 200 mm nominal size							
average depth of run not exceeding 0.50 m	1.44	0.09	1.58	0.70	2.70	m	**4.99**
average depth of run not exceeding 0.75 m	2.16	0.11	1.98	0.88	4.05	m	**6.91**
average depth of run not exceeding 1.00 m	2.88	0.14	2.55	1.14	5.40	m	**9.09**
average depth of run not exceeding 1.25 m	3.60	0.23	4.17	1.91	6.75	m	**12.83**
Excavating trenches, using 3 tonne tracked excavator, to receive pipes; grading bottoms; earthwork support; filling with lean mix concrete; disposal of surplus soil							
Services not exceeding 200 mm nominal size							
average depth of run not exceeding 0.50 m	12.62	0.11	1.95	0.34	13.88	m	**16.17**
average depth of run not exceeding 0.75 m	18.93	0.13	2.37	0.43	20.82	m	**23.63**
average depth of run not exceeding 1.00 m	25.25	0.17	3.04	0.57	27.77	m	**31.38**
average depth of run not exceeding 1.25 m	31.56	0.23	4.11	0.86	34.71	m	**39.68**
Earthwork support; providing support to opposing faces of excavation; moving along as work proceeds							
Maximum depth not exceeding 2.00 m							
distance between opposing faces not exceeding 2.00 m	-	0.80	14.60	17.70	-	m	**32.30**

Q PAVING/PLANTING/FENCING/SITE FURNITURE

Item Excluding site overheads and profit	PC £	Labour hours	Labour £	Plant £	Material £	Unit	Total rate £
Q10 KERBS/EDGINGS/CHANNELS/ PAVING ACCESSORIES							
Foundations to kerbs							
Excavating trenches; width 300 mm; 3 tonne excavator; disposal off site							
depth 300 mm	-	0.10	1.82	0.43	1.04	m	3.29
depth 400 mm	-	0.11	2.03	0.48	1.39	m	3.89
By hand	-	0.60	10.99	-	1.73	m	12.72
Excavating trenches; width 450 mm; 3 tonne excavator; disposal off site							
depth 300 mm	-	0.13	2.28	0.54	1.56	m	4.38
depth 400 mm	-	0.14	2.61	0.62	2.08	m	5.30
Foundations; to kerbs, edgings or channels; in situ concrete; 21 N/mm^2 - 20 mm aggregate ((1:2:4) site mixed); one side against earth face, other against formwork (not included); site mixed concrete							
Site mixed concrete							
150 mm wide x 100 mm deep	-	0.13	2.43	-	1.39	m	3.83
150 mm wide x 150 mm deep	-	0.17	3.04	-	2.09	m	5.13
200 mm wide x 150 mm deep	-	0.20	3.65	-	2.79	m	6.44
300 mm wide x 150 mm deep	-	0.23	4.26	-	4.18	m	8.44
600 mm wide x 200 mm deep	-	0.29	5.21	-	11.16	m	16.37
Ready mixed concrete							
150 mm wide x 100 mm deep	-	0.13	2.43	-	1.23	m	3.67
150 mm wide x 150 mm deep	-	0.17	3.04	-	1.85	m	4.89
200 mm wide x 150 mm deep	-	0.20	3.65	-	2.47	m	6.12
300 mm wide x 150 mm deep	-	0.23	4.26	-	3.70	m	7.96
600 mm wide x 200 mm deep	-	0.29	5.21	-	9.87	m	15.08
Formwork; sides of foundations (this will usually be required to one side of each kerb foundation adjacent to road sub-bases)							
100 mm deep	-	0.04	0.76	-	0.15	m	0.91
150 mm deep	-	0.04	0.76	-	0.22	m	0.98
Precast concrete kerbs, channels, edgings etc.; Marshalls Mono; bedding, jointing and pointing in cement mortar (1:3); including haunching with in situ concrete; 11.50 N/mm^2 - 40 mm aggregate one side							
Kerbs; straight							
150 x 305 mm; ref HB1	7.84	0.50	9.13	-	10.27	m	19.39
125 x 255 mm; ref HB2; SP	3.75	0.44	8.11	-	6.18	m	14.29
125 x 150 mm; ref BN	2.54	0.40	7.30	-	4.97	m	12.27
Dropper kerbs; left and right handed							
125 x 255 - 150 mm; ref DL1 or DR1	5.67	0.50	9.13	-	8.10	m	17.23
125 x 255 - 150 mm; ref DL2 or DR2	5.67	0.50	9.13	-	8.10	m	17.23
Quadrant kerbs							
305 mm radius	9.53	0.50	9.13	-	11.15	nr	20.28
455 mm radius	10.25	0.50	9.13	-	12.68	nr	21.81

Q PAVING/PLANTING/FENCING/SITE FURNITURE

Item Excluding site overheads and profit	PC £	Labour hours	Labour £	Plant £	Material £	Unit	Total rate £
Q10 KERBS/EDGINGS/CHANNELS/ PAVING ACCESSORIES - cont'd							
Precast concrete kerbs, channels, edgings etc. - cont'd							
Straight kerbs or channels; to radius; 125 x 255 mm							
0.90 m radius (2 units per quarter circle)	7.17	0.80	14.60	-	9.67	m	**24.27**
1.80 m radius (4 units per quarter circle)	7.17	0.73	13.27	-	9.67	m	**22.94**
2.40 m radius (5 units per quarter circle)	7.17	0.68	12.37	-	9.67	m	**22.04**
3.00 m radius (5 units per quarter circle)	7.17	0.67	12.17	-	9.67	m	**21.84**
4.50 m radius (2 units per quarter circle)	7.17	0.60	10.96	-	9.67	m	**20.63**
6.10 m radius (11 units per quarter circle)	7.17	0.58	10.58	-	9.67	m	**20.25**
7.60 m radius (14 units per quarter circle)	7.17	0.57	10.43	-	9.67	m	**20.10**
9.15 m radius (17 units per quarter circle)	7.17	0.56	10.14	-	9.67	m	**19.81**
10.70 m radius (20 units per quarter circle)	7.17	0.56	10.14	-	9.67	m	**19.81**
12.20 m radius (22 units per quarter circle)	7.17	0.53	9.61	-	9.67	m	**19.28**
Kerbs; "Conservation Kerb" units; to simulate natural granite kerbs							
255 x 150 x 914 mm; laid flat	21.91	0.57	10.43	-	26.27	m	**36.70**
150 x 255 x 914 mm; laid vertical	21.91	0.57	10.43	-	25.56	m	**35.99**
145 x 255 mm; radius internal 3.25 m	26.68	0.83	15.21	-	31.00	m	**46.21**
150 x 255 mm; radius external 3.40 m	25.48	0.83	15.21	-	29.77	m	**44.98**
145 x 255 mm; radius internal 6.50 m	24.44	0.67	12.17	-	28.08	m	**40.25**
145 x 255 mm; radius external 6.70 m	23.71	0.67	12.17	-	27.94	m	**40.11**
150 x 255 mm; radius internal 9.80 m	23.57	0.67	12.17	-	27.81	m	**39.97**
150 x 255 mm; radius external 10.00 m	22.91	0.67	12.17	-	26.55	m	**38.72**
305 x 305 x 255 mm; solid quadrants	30.01	0.57	10.43	-	33.66	nr	**44.08**
Channels; square							
125 x 225 x 915 mm long; ref CS1	3.57	0.40	7.30	-	11.55	m	**18.85**
125 x 150 x 915 mm long; ref CS2	2.90	0.40	7.30	-	9.40	m	**16.70**
Channels; dished							
305 x 150 x 915 mm long; ref CD	10.11	0.40	7.30	-	18.78	m	**26.08**
150 x 100 x 915 mm	5.93	0.40	7.30	-	10.82	m	**18.12**
Precast concrete edging units; including haunching with in situ concrete; 11.50 N/ mm^2 - 40 mm aggregate both sides							
Edgings; rectangular, bullnosed or chamfered							
50 x 150 mm	1.55	0.33	6.08	-	5.60	m	**11.68**
125 x 150 mm bullnosed	1.60	0.33	6.08	-	5.65	m	**11.73**
50 x 200 mm	2.32	0.33	6.08	-	6.37	m	**12.46**
50 x 250 mm	2.69	0.33	6.08	-	6.74	m	**12.83**
50 x 250 mm flat top	3.15	0.33	6.08	-	7.20	m	**13.29**

Q PAVING/PLANTING/FENCING/SITE FURNITURE

Item Excluding site overheads and profit	PC £	Labour hours	Labour £	Plant £	Material £	Unit	Total rate £
Marshalls Mono; small element precast concrete kerb system; Keykerb Large (KL) upstand of 100 - 125 mm; on 150 mm concrete foundation including haunching with in situ concrete (1:3:6) 1 side							
Bullnosed or half battered; 100 x 127 x 200 mm							
laid straight	11.41	0.80	14.60	-	12.26	m	**26.86**
radial blocks laid to curve; 8 blocks/1/4 circle - 500 mm radius	8.78	1.00	18.25	-	9.63	m	**27.88**
radial blocks laid to curve; 8 radial blocks, alternating 8 standard blocks/1/4 circle - 1000 mm radius	14.48	1.25	22.81	-	15.34	m	**38.15**
radial blocks, alternating 16 standard blocks/1/4 circle - 1500 mm radius	13.33	1.50	27.38	-	14.18	m	**41.55**
internal angle 90°	4.82	0.20	3.65	-	4.82	nr	**8.47**
external angle	4.82	0.20	3.65	-	4.82	nr	**8.47**
drop crossing kerbs; KL half battered to KL Splay; LH and RH	20.16	1.00	18.25	-	21.01	pair	**39.26**
drop crossing kerbs; KL half battered to KS bullnosed	19.60	1.00	18.25	-	20.45	pair	**38.70**
Marshalls Mono; small element precast concrete kerb system; Keykerb Small (KS) upstand of 25 - 50 mm; on 150 mm concrete foundation including haunching with in situ concrete (1:3:6) 1 side							
Half battered							
laid straight	8.01	0.80	14.60	-	8.86	m	**23.46**
radial blocks laid to curve; 8 blocks/1/4 circle - 500 mm radius	6.58	1.00	18.25	-	7.43	m	**25.68**
radial blocks laid to curve; 8 radial blocks, alternating 8 standard blocks/1/4 circle - 1000 mm radius	10.59	1.25	22.81	-	11.44	m	**34.25**
radial blocks, alternating 16 standard blocks/1/4 circle - 1500 mm radius	11.33	1.50	27.38	-	12.18	m	**39.55**
internal angle 90°	4.82	0.20	3.65	-	4.82	nr	**8.47**
external angle	4.82	0.20	3.65	-	4.82	nr	**8.47**

Dressed natural stone kerbs - General
Preamble: The kerbs are to be good, sound and uniform in texture and free from defects; worked straight or to radius, square and out of wind, with the top front and back edges parallel or concentric to the dimensions specified. All drill and pick holes shall be removed from dressed faces. Standard dressings shall be designated as either fine picked, single axed or nidged or rough punched.

Q PAVING/PLANTING/FENCING/SITE FURNITURE

Item Excluding site overheads and profit	PC £	Labour hours	Labour £	Plant £	Material £	Unit	Total rate £
Q10 KERBS/EDGINGS/CHANNELS/ PAVING ACCESSORIES - cont'd							
Dressed natural stone kerbs; CED Ltd; on concrete foundations (not included); including haunching with in situ concrete; 11.50 N/mm^2 - 40 mm aggregate one side							
Granite kerbs; 125 x 250 mm							
special quality; straight; random lengths	16.30	0.80	14.60	-	19.21	m	33.81
Granite kerbs; 125 x 250 mm; curved to mean radius 3 m							
special quality; random lengths	18.20	0.91	16.61	-	21.11	m	37.72
New granite sett edgings; CED Ltd; 100 x 100 x 100 mm; bedding in cement mortar (1:4); including haunching with in situ concrete; 11.50 N/mm^2 - 40 mm aggregate one side; concrete foundations (not included)							
Edgings 100 x 100 x 100 mm; ref 1R "better quality" with 20 mm pointing gaps							
100 mm wide	2.55	1.33	24.27	-	4.73	m	29.00
2 rows; 220 mm wide	5.09	2.22	40.52	-	7.92	m	48.43
3 rows; 340 mm wide	7.64	3.00	54.75	-	9.86	m	64.61
Edgings 100 x 100 x 200 mm; ref 1R "better quality" with 20 mm pointing gaps							
100 mm wide	2.60	0.75	13.69	-	4.78	m	18.47
2 rows; 220 mm wide	5.20	1.50	27.38	-	8.02	m	35.39
3 rows; 340 mm wide	7.80	2.25	41.06	-	10.02	m	51.08
Edgings 100 x 100 x 100 mm; ref 8R "standard quality" with 20 mm pointing gaps							
100 mm wide	2.44	1.33	24.27	-	4.62	m	28.90
2 rows; 220 mm wide	4.88	2.22	40.52	-	7.71	m	48.22
3 rows; 340 mm wide	7.32	3.00	54.75	-	9.55	m	64.30
Edgings 100 x 100 x 200 mm; ref 8R "standard quality" with 20 mm pointing gaps							
100 mm wide	3.42	0.75	13.69	-	5.60	m	19.29
2 rows; 220 mm wide	6.84	1.50	27.38	-	9.66	m	37.04
3 rows; 340 mm wide	10.26	2.25	41.06	-	12.48	m	53.54
Reclaimed granite setts edgings; CED Ltd; 100 x 100 mm; bedding in cement mortar (1:4); including haunching with in situ concrete; 11.50 N/mm^2 - 40 mm aggregate one side; on concrete foundations (not included)							
Edgings; with 20 mm pointing gaps							
100 mm wide	3.58	1.33	24.27	-	5.77	m	30.04
2 rows; 220 mm wide	7.17	2.22	40.52	-	9.99	m	50.51
3 rows; 340 mm wide	10.75	3.00	54.75	-	12.97	m	67.72

Prices for Measured Works – Major Works

Q PAVING/PLANTING/FENCING/SITE FURNITURE

Item Excluding site overheads and profit	PC £	Labour hours	Labour £	Plant £	Material £	Unit	**Total rate £**
Brick or block stretchers; bedding in cement mortar (1:4); on 150 mm deep concrete foundations, including haunching with in situ concrete; 11.50 N/mm² - 40 mm aggregate one side							
Single course							
concrete paving blocks; PC £8.29/m²; 200 x 100 x 60 mm	1.66	0.31	5.61	-	4.04	m	**9.66**
engineering bricks; PC £239.00/1000; 215 x 102.5 x 65 mm	1.06	0.40	7.30	-	3.50	m	**10.80**
paving bricks; PC £450.00/1000; 215 x 102.5 x 65 mm	2.00	0.40	7.30	-	4.39	m	**11.69**
Two courses							
concrete paving blocks; PC £8.29/m²; 200 x 100 x 60 mm	3.32	0.40	7.30	-	6.13	m	**13.43**
engineering bricks; PC £239.00/1000; 215 x 102.5 x 65 mm	2.12	0.57	10.43	-	5.04	m	**15.47**
paving bricks; PC £450.00/1000; 215 x 102.5 x 65 mm	4.00	0.57	10.43	-	6.81	m	**17.24**
Three courses							
concrete paving blocks; PC £8.29/m²; 200 x 100 x 60 mm	4.97	0.44	8.11	-	7.78	m	**15.89**
engineering bricks; PC £239.00/1000; 215 x 102.5 x 65 mm	3.19	0.67	12.17	-	6.16	m	**18.32**
paving bricks; PC £450.00/1000; 215 x 102.5 x 65 mm	6.00	0.67	12.17	-	8.81	m	**20.97**
Bricks on edge; bedding in cement mortar (1:4); on 150 mm deep concrete foundations; including haunching with in situ concrete; 11.50 N/mm² - 40 mm aggregate one side							
One brick wide							
engineering bricks; 215 x 102.5 x 65 mm	3.19	0.57	10.43	-	6.26	m	**16.69**
paving bricks; 215 x 102.5 x 65 mm	6.00	0.57	10.43	-	8.91	m	**19.34**
Two courses; stretchers laid on edge; 225 mm wide							
engineering bricks; 215 x 102.5 x 65 mm	6.37	1.20	21.90	-	10.55	m	**32.45**
paving bricks; 215 x 102.5 x 65 mm	12.00	1.20	21.90	-	15.85	m	**37.75**
Extra over bricks on edge for standard kerbs to one side; haunching in concrete							
125 x 255 mm; ref HB2; SP	3.75	0.44	8.11	-	6.18	m	**14.29**
Channels; bedding in cement mortar (1:3); joints pointed flush; on concrete foundations (not included)							
Three courses stretchers; 350 mm wide; quarter bond to form dished channels							
engineering bricks; PC £239.00/1000; 215 x 102.5 x 65 mm	3.19	1.00	18.25	-	4.30	m	**22.55**
paving bricks; PC £450.00/1000; 215 x 102.5 x 65 mm	6.00	1.00	18.25	-	6.95	m	**25.20**
Three courses granite setts; 340 mm wide; to form dished channels							
340 mm wide	12.03	2.00	36.50	-	14.21	m	**50.71**

Q PAVING/PLANTING/FENCING/SITE FURNITURE

Item Excluding site overheads and profit	PC £	Labour hours	Labour £	Plant £	Material £	Unit	Total rate £
Q10 KERBS/EDGINGS/CHANNELS/ PAVING ACCESSORIES - cont'd							
Timber edging boards; fixed with 50 x 50 x 750 mm timber pegs at 1000 mm centres (excavations and hardcore under edgings not included)							
Straight							
38 x 150 mm treated softwood edge boards	2.23	0.10	1.82	-	2.23	m	**4.06**
50 x 150 mm treated softwood edge boards	2.42	0.10	1.82	-	2.42	m	**4.25**
38 x 150 mm hardwood (iroko) edge boards	7.17	0.10	1.82	-	7.17	m	**9.00**
50 x 150 mm hardwood (iroko) edge boards	9.03	0.10	1.82	-	9.03	m	**10.85**
Curved							
38 x 150 mm treated softwood edge boards	2.23	0.20	3.65	-	2.23	m	**5.88**
50 x 150 mm treated softwood edge boards	2.42	0.25	4.56	-	2.42	m	**6.99**
38 x 150 mm hardwood (iroko) edge boards	7.17	0.20	3.65	-	7.17	m	**10.82**
50 x 150 mm hardwood (iroko) edge boards	9.03	0.25	4.56	-	9.03	m	**13.59**
Permaloc "AshphaltEdge"; RTS Ltd; extruded aluminium alloy L shaped edging with 5.33 mm exposed upper lip; edging fixed to roadway base and edge profile with 250 mm steel fixing spike; laid to straight or curvilinear road edge; subsequently filled with macadam (not included)							
Depth of macadam							
38 mm	7.34	0.02	0.30	-	7.52	m	**7.83**
51 mm	8.27	0.02	0.31	-	8.48	m	**8.79**
64 mm	9.14	0.02	0.33	-	9.37	m	**9.70**
76 mm	11.05	0.02	0.36	-	11.05	m	**11.42**
102 mm	12.86	0.02	0.38	-	13.18	m	**13.56**
Permaloc "Cleanline"; RTS Ltd; heavy duty straight profile edging; for edgings to soft landscape beds or turf areas; 3.2 mm x 102 mm high; 3.2 mm thick with 4.75 mm exposed upper lip; fixed to form straight or curvilinear edge with 305 mm fixing spike							
Milled aluminium							
100 mm deep	8.41	0.02	0.33	-	8.62	m	**8.95**
Black							
100 mm deep	8.41	0.02	0.33	-	8.62	m	**8.95**

Q PAVING/PLANTING/FENCING/SITE FURNITURE

Item Excluding site overheads and profit	PC £	Labour hours	Labour £	Plant £	Material £	Unit	**Total rate £**
Permaloc "Permastrip"; RTS Ltd; heavy duty L shaped profile maintenance strip; 3.2 mm x 89 mm high with 5.2 mm exposed top lip; for straight or gentle curves on paths or bed turf interfaces; fixed to form straight or curvilinear edge with standard 305 mm stake; other stake lengths available							
Milled aluminium							
89 mm deep	8.41	0.02	0.33	-	8.62	m	**8.95**
Black							
89 mm deep	8.41	0.02	0.33	-	8.62	m	**8.95**
Permaloc "Proline"; RTS Ltd; medium duty straight profiled maintenance strip; 3.2 mm x 102 mm high with 3.18 mm exposed top lip; for straight or gentle curves on paths or bed turf interfaces; fixed to form straight or curvilinear edge with standard 305 mm stake; other stake lengths available							
Milled aluminium							
89 mm deep	7.32	0.02	0.33	-	7.50	m	**7.84**
Black							
89 mm deep	7.32	0.02	0.33	-	7.50	m	**7.84**
Q20 GRANULAR SUB-BASES TO ROADS/PAVINGS							
Hardcore bases; obtained off site; PC £8.00/m³							
By machine							
100 mm thick	0.80	0.05	0.91	0.99	0.80	m²	**2.70**
150 mm thick	1.20	0.07	1.22	1.26	1.20	m²	**3.68**
200 mm thick	1.60	0.08	1.46	1.36	1.60	m²	**4.42**
300 mm thick	2.40	0.07	1.22	1.40	2.40	m²	**5.02**
exceeding 300 mm thick	8.00	0.17	3.04	3.28	9.60	m³	**15.92**
By hand							
100 mm thick	0.80	0.20	3.65	0.16	0.80	m²	**4.61**
150 mm thick	1.20	0.30	5.48	0.16	1.44	m²	**7.08**
200 mm thick	1.60	0.40	7.30	0.26	1.92	m²	**9.48**
300 mm thick	2.40	0.60	10.95	0.16	2.88	m²	**13.99**
exceeding 300 mm thick	8.00	2.00	36.50	0.54	9.60	m³	**46.64**
Hardcore; difference for each £1.00 increase/decrease in PC price per m³; price will vary with type and source of hardcore							
average 75 mm thick	-	-	-	-	0.08	m²	-
average 100 mm thick	-	-	-	-	0.10	m²	-
average 150 mm thick	-	-	-	-	0.15	m²	-
average 200 mm thick	-	-	-	-	0.20	m²	-
average 250 mm thick	-	-	-	-	0.25	m²	-
average 300 mm thick	-	-	-	-	0.30	m²	-
exceeding 300 mm thick	-	-	-	-	1.10	m³	-

Q PAVING/PLANTING/FENCING/SITE FURNITURE

Item Excluding site overheads and profit	PC £	Labour hours	Labour £	Plant £	Material £	Unit	Total rate £
Q20 GRANULAR SUB-BASES TO ROADS/PAVINGS - cont'd							
Type 1 granular fill base; PC £15.950/ tonne (£35.09/m^3 compacted)							
By machine							
100 mm thick	3.51	0.03	0.51	0.33	3.51	m^2	4.35
150 mm thick	5.26	0.03	0.46	0.49	5.26	m^2	6.21
250 mm thick	8.77	0.02	0.38	0.82	8.77	m^2	9.97
over 250 mm thick	35.09	0.20	3.65	2.88	35.09	m^3	41.62
By hand (mechanical compaction)							
100 mm thick	3.51	0.17	3.04	0.05	3.51	m^2	6.60
150 mm thick	5.26	0.25	4.56	0.08	5.26	m^2	9.91
250 mm thick	8.77	0.42	7.60	0.13	8.77	m^2	16.51
over 250 mm thick	35.09	0.47	8.53	0.13	35.09	m^3	43.75
Type 1 (recycled material) granular fill base; PC £9.95/tonne (£21.89/m^3 compacted)							
By machine							
100 mm thick	2.19	0.03	0.51	0.33	2.19	m^2	3.03
150 mm thick	3.28	0.03	0.46	0.49	3.28	m^2	4.23
250 mm thick	5.47	0.02	0.38	0.82	5.47	m^2	6.67
over 250 mm thick	9.95	0.20	3.65	2.88	9.95	m^3	16.48
By hand (mechanical compaction)							
100 mm thick	2.19	0.17	3.04	0.05	2.19	m^2	5.28
150 mm thick	3.28	0.25	4.56	0.08	3.28	m^2	7.93
250 mm thick	5.47	0.42	7.60	0.13	5.47	m^2	13.21
over 250 mm thick	9.95	0.47	8.53	0.13	9.95	m^3	18.61
Surface treatments							
Sand blinding; to hardcore base (not included); 25 mm thick	0.72	0.03	0.61	-	0.72	m^2	1.33
Sand blinding; to hardcore base (not included); 50 mm thick	1.44	0.05	0.91	-	1.44	m^2	2.35
Filter fabrics; to hardcore base (not included)	0.42	0.01	0.18	-	0.44	m^2	0.63
Q21 IN SITU CONCRETE ROADS/ PAVINGS							
Unreinforced concrete; on prepared sub-base (not included)							
Roads; 21.00 N/mm^2 - 20 mm aggregate (1:2:4) mechanically mixed on site							
100 mm thick	9.30	0.13	2.28	-	9.76	m^2	12.04
150 mm thick	13.94	0.17	3.04	-	14.64	m^2	17.68
Reinforced in situ concrete; mechanically mixed on site; normal Portland cement; on hardcore base (not included); reinforcement (not included)							
Roads; 11.50 N/mm^2 - 40 mm aggregate (1:3:6)							
100 mm thick	8.10	0.40	7.30	-	8.30	m^2	15.60
150 mm thick	12.15	0.60	10.95	-	12.46	m^2	23.41
200 mm thick	16.20	0.80	14.60	-	17.01	m^2	31.61
250 mm thick	20.25	1.00	18.25	0.22	20.76	m^2	39.23
300 mm thick	24.31	1.20	21.90	0.22	24.91	m^2	47.03

Q PAVING/PLANTING/FENCING/SITE FURNITURE

Item Excluding site overheads and profit	PC £	Labour hours	Labour £	Plant £	Material £	Unit	Total rate £
Roads; 21.00 N/mm² - 20 mm aggregate (1:2:4)							
100 mm thick	9.30	0.40	7.30	-	9.53	m²	**16.83**
150 mm thick	13.94	0.60	10.95	-	14.30	m²	**25.25**
200 mm thick	18.59	0.80	14.60	-	19.06	m²	**33.66**
250 mm thick	23.24	1.00	18.25	0.22	23.82	m²	**42.28**
300 mm thick	27.89	1.20	21.90	0.22	28.59	m²	**50.70**
Roads; 25.00 N/mm² - 20 mm aggregate GEN 4 ready mixed							
100 mm thick	8.24	0.40	7.30	-	8.45	m²	**15.75**
150 mm thick	12.37	0.60	10.95	-	12.68	m²	**23.63**
200 mm thick	16.49	0.80	14.60	-	16.90	m²	**31.50**
250 mm thick	20.61	1.00	18.25	0.22	21.12	m²	**39.59**
300 mm thick	24.73	1.20	21.90	0.22	25.35	m²	**47.47**
Reinforced in situ concrete; ready mixed; discharged directly into location from supply lorry; normal Portland cement; on hardcore base (not included); reinforcement (not included)							
Roads; 11.50 N/mm² - 40 mm aggregate (1:3:6)							
100 mm thick	8.41	0.16	2.92	-	8.41	m²	**11.33**
150 mm thick	12.62	0.24	4.38	-	12.62	m²	**17.00**
200 mm thick	16.83	0.36	6.57	-	16.83	m²	**23.40**
250 mm thick	21.04	0.54	9.86	0.22	21.04	m²	**31.11**
300 mm thick	25.25	0.66	12.04	0.22	25.25	m²	**37.51**
Roads; 21.00 N/mm² - 20 mm aggregate (1:2:4)							
100 mm thick	8.47	0.16	2.92	-	8.47	m²	**11.39**
150 mm thick	12.71	0.24	4.38	-	12.71	m²	**17.09**
200 mm thick	16.94	0.36	6.57	-	16.94	m²	**23.51**
250 mm thick	21.18	0.54	9.86	0.22	21.18	m²	**31.25**
300 mm thick	25.41	0.66	12.04	0.22	25.41	m²	**37.68**
Roads; 26.00 N/mm² - 20 mm aggregate (1:1.5:3)							
100 mm thick	8.49	0.16	2.92	-	8.49	m²	**11.41**
150 mm thick	12.73	0.24	4.38	-	12.73	m²	**17.11**
200 mm thick	16.94	0.36	6.57	-	16.94	m²	**23.51**
250 mm thick	21.21	0.54	9.86	0.22	21.21	m²	**31.29**
300 mm thick	25.46	0.66	12.04	0.22	25.46	m²	**37.72**
Roads; PAV1 concrete - 35 N/mm²; Designated mix							
100 mm thick	8.33	0.16	2.92	-	8.54	m²	**11.46**
150 mm thick	12.49	0.24	4.38	-	12.81	m²	**17.19**
200 mm thick	16.66	0.36	6.57	-	17.07	m²	**23.64**
250 mm thick	20.82	0.54	9.86	0.22	21.34	m²	**31.41**
300 mm thick	24.98	0.66	12.04	0.22	25.61	m²	**37.87**
Concrete sundries							
Treating surfaces of unset concrete; grading to cambers; tamping with 75 mm thick steel shod tamper or similar	-	0.13	2.43	-	-	m²	**2.43**
Expansion joints							
13 mm thick joint filler; formwork							
width or depth not exceeding 150 mm	1.89	0.20	3.65	-	2.50	m	**6.15**
width or depth 150 - 300 mm	3.78	0.25	4.56	-	5.00	m	**9.57**
width or depth 300 - 450 mm	5.66	0.30	5.47	-	7.51	m	**12.98**

Q PAVING/PLANTING/FENCING/SITE FURNITURE

Item Excluding site overheads and profit	PC £	Labour hours	Labour £	Plant £	Material £	Unit	Total rate £
Q21 IN SITU CONCRETE ROADS/ PAVINGS - cont'd							
Expansion joints - cont'd							
25 mm thick joint filler; formwork							
width or depth not exceeding 150 mm	5.06	0.20	3.65	-	5.67	m	9.32
width or depth 150 - 300 mm	10.12	0.25	4.56	-	11.35	m	15.91
width or depth 300 - 450 mm	15.18	0.30	5.47	-	17.02	m	22.50
Sealants; sealing top 25 mm of joint with rubberized bituminous compound	1.19	0.25	4.56	-	1.19	m	5.76
Formwork for in situ concrete							
Sides of foundations							
height not exceeding 250 mm	0.37	0.03	0.61	-	0.49	m	1.10
height 250 - 500 mm	0.50	0.04	0.73	-	0.76	m	1.49
height 500 mm - 1.00 m	0.50	0.05	0.91	-	0.84	m	1.75
height exceeding 1.00 m	1.50	2.00	36.50	-	4.54	m^2	41.04
Extra over formwork for curved work 6 m radius	-	0.25	4.56	-	-	m	4.56
Steel road forms; to edges of beds or faces of foundations							
150 mm wide	-	0.20	3.65	1.10	-	m	4.75
Reinforcement; fabric; side laps 150 mm; head laps 300 mm; mesh 200 x 200 mm; in roads, footpaths or pavings							
Fabric							
ref A142 (2.22 kg/m^2)	1.00	0.08	1.51	-	1.10	m^2	2.61
ref A193 (3.02 kg/m^2)	1.40	0.08	1.51	-	1.54	m^2	3.05
Q22 COATED MACADAM/ASPHALT ROADS/PAVINGS							
Coated macadam/asphalt roads/pavings - General							
Preamble: The prices for all in situ finishings to roads and footpaths include for work to falls, crossfalls or slopes not exceeding 15° from horizontal; for laying on prepared bases (not included) and for rolling with an appropriate roller. Users should note the new terminology for the surfaces described below which is to European standard descriptions. The now redundant descriptions for each course are shown in brackets.							
Macadam surfacing; Spadeoak Construction Co Ltd; surface (wearing) course; 20 mm of 6 mm dense bitumen macadam							
Machine lay; areas 1000 m^2 and over							
limestone aggregate	-	-	-	-	-	m^2	5.26
granite aggregate	-	-	-	-	-	m^2	5.32
red	-	-	-	-	-	m^2	9.45
Hand lay; areas 400 m^2 and over							
limestone aggregate	-	-	-	-	-	m^2	6.69
granite aggregate	-	-	-	-	-	m^2	6.75
red	-	-	-	-	-	m^2	11.12

Prices for Measured Works – Major Works

Q PAVING/PLANTING/FENCING/SITE FURNITURE

Item Excluding site overheads and profit	PC £	Labour hours	Labour £	Plant £	Material £	Unit	Total rate £
Macadam surfacing; Spadeoak Construction Co Ltd; surface (wearing) course; 30 mm of 10 mm dense bitumen macadam							
Machine lay; areas 1000 m² and over							
limestone aggregate	-	-	-	-	-	m²	6.35
granite aggregate	-	-	-	-	-	m²	6.43
red	-	-	-	-	-	m²	13.31
Hand lay; areas 400 m² and over							
limestone aggregate	-	-	-	-	-	m²	7.84
granite aggregate	-	-	-	-	-	m²	7.93
red	-	-	-	-	-	m²	15.24
Macadam surfacing; Spadeoak Construction Co Ltd; surface (wearing) course; 40 mm of 10 mm dense bitumen macadam							
Machine lay; areas 1000 m² and over							
limestone aggregate	-	-	-	-	-	m²	8.08
granite aggregate	-	-	-	-	-	m²	8.20
red	-	-	-	-	-	m²	15.09
Hand lay; areas 400 m² and over							
limestone aggregate	-	-	-	-	-	m²	9.69
granite aggregate	-	-	-	-	-	m²	9.81
red	-	-	-	-	-	m²	17.13
Macadam surfacing; Spadeoak Construction Co Ltd; binder (base) course; 50 mm of 20 mm dense bitumen macadam							
Machine lay; areas 1000 m² and over							
limestone aggregate	-	-	-	-	-	m²	8.10
granite aggregate	-	-	-	-	-	m²	8.23
Hand lay; areas 400 m² and over							
limestone aggregate	-	-	-	-	-	m²	9.71
granite aggregate	-	-	-	-	-	m²	9.85
Macadam surfacing; Spadeoak Construction Co Ltd; binder (base) course; 60 mm of 20 mm dense bitumen macadam							
Machine lay; areas 1000 m² and over							
limestone aggregate	-	-	-	-	-	m²	8.91
granite aggregate	-	-	-	-	-	m²	9.06
Hand lay; areas 400 m² and over							
limestone aggregate	-	-	-	-	-	m²	10.57
granite aggregate	-	-	-	-	-	m²	10.73
Macadam surfacing; Spadeoak Construction Co Ltd; base (roadbase) course; 75 mm of 28 mm dense bitumen macadam							
Machine lay; areas 1000 m² and over							
limestone aggregate	-	-	-	-	-	m²	10.93
granite aggregate	-	-	-	-	-	m²	11.12
Hand lay; areas 400 m² and over							
limestone aggregate	-	-	-	-	-	m²	12.73
granite aggregate	-	-	-	-	-	m²	12.94

Q PAVING/PLANTING/FENCING/SITE FURNITURE

Item Excluding site overheads and profit	PC £	Labour hours	Labour £	Plant £	Material £	Unit	Total rate £
Q22 COATED MACADAM/ASPHALT ROADS/PAVINGS - cont'd							
Macadam surfacing; Spadeoak Construction Co Ltd; base (roadbase) course; 100 mm of 28 mm dense bitumen macadam							
Machine lay; areas 1000 m^2 and over							
limestone aggregate	-	-	-	-	-	m^2	13.54
granite aggregate	-	-	-	-	-	m^2	13.80
Hand lay; areas 400 m^2 and over							
limestone aggregate	-	-	-	-	-	m^2	15.51
granite aggregate	-	-	-	-	-	m^2	15.79
Macadam surfacing; Spadeoak Construction Co Ltd; base (roadbase) course; 150 mm of 28 mm dense bitumen macadam in two layers							
Machine lay; areas 1000 m^2 and over							
limestone aggregate	-	-	-	-	-	m^2	21.35
granite aggregate	-	-	-	-	-	m^2	21.74
Hand lay; areas 400 m^2 and over							
limestone aggregate	-	-	-	-	-	m^2	24.92
granite aggregate	-	-	-	-	-	m^2	25.33
Base (roadbase) course; 200 mm of 28 mm dense bitumen macadam in two layers							
Machine lay; areas 1000 m^2 and over							
limestone aggregate	-	-	-	-	-	m^2	27.13
granite aggregate	-	-	-	-	-	m^2	27.65
Hand lay; areas 400 m^2 and over							
limestone aggregate	-	-	-	-	-	m^2	21.08
granite aggregate	-	-	-	-	-	m^2	31.63
Resin bound macadam pavings; machine laid; Bituchem							
"Natratex" wearing course; clear resin bound macadam to pedestrian or vehicular hard landscape areas; laid to base course (not included)							
25 mm thick to pedestrian areas	15.50	-	-	-	15.50	m^2	-
30 mm thick to vehicular areas	18.60	-	-	-	18.60	m^2	-
"Colourtex" coloured resin bound macadam to pedestrian or vehicular hard landscape areas							
25 mm thick	15.50	-	-	-	15.50	m^2	-
Marking car parks							
Car parking space division strips; laid hot at 115° C; on bitumen macadam surfacing							
minimum daily rate	-	-	-	-	-	item	500.00
Stainless metal road studs							
100 x 100 mm	5.50	0.25	4.56	-	5.50	nr	10.06

Q PAVING/PLANTING/FENCING/SITE FURNITURE

Item Excluding site overheads and profit	PC £	Labour hours	Labour £	Plant £	Material £	Unit	Total rate £
Bound aggregates; Addagrip Surface Treatments UK Ltd; natural decorative resin bonded surface dressing laid to concrete, macadam or to plywood panels priced separately							
Primer coat to macadam or concrete base	-	-	-	-	-	m²	4.60
Golden pea gravel; 1 - 3 mm							
buff adhesive	-	-	-	-	-	m²	22.00
red adhesive	-	-	-	-	-	m²	22.00
green adhesive	-	-	-	-	-	m²	22.00
Golden pea gravel; 2 - 5 mm							
buff adhesive	-	-	-	-	-	m²	25.00
Chinese bauxite; 1 - 3 mm							
buff adhesive	-	-	-	-	-	m²	22.00
"Angular Buff", "Terracotta", "Autumn Gold" or "Derbyshire cream"; 6 mm aggregate in Addaset							
15 mm depth	-	-	-	-	-	m²	55.00
18 mm depth	-	-	-	-	-	m²	65.00
"Cobalt Blue Glass"; 6 mm							
15 mm depth	-	-	-	-	-	m²	80.00
"Midnight Grey"; 6 mm							
15 mm depth	-	-	-	-	-	m²	55.00
"Chocolate"; 6 mm							
15 mm depth	-	-	-	-	-	m²	55.00
18 mm depth	-	-	-	-	-	m²	65.00
Q23 GRAVEL/HOGGIN/WOODCHIP ROADS/PAVINGS							
Excavation and path preparation							
Excavating; 300 mm deep; to width of path; depositing excavated material at sides of excavation							
width 1.00 m	-	-	-	2.20	-	m²	2.20
width 1.50 m	-	-	-	1.84	-	m²	1.84
width 2.00 m	-	-	-	1.57	-	m²	1.57
width 3.00 m	-	-	-	1.32	-	m²	1.32
Excavating trenches; in centre of pathways; 100 mm flexible drain pipes; filling with clean broken stone or gravel rejects							
300 x 450 mm deep	5.65	0.10	1.82	1.10	5.77	m	8.69
Hand trimming and compacting reduced surface of pathway; by machine							
width 1.00 m	-	0.05	0.91	0.16	-	m	1.07
width 1.50 m	-	0.04	0.81	0.14	-	m	0.95
width 2.00 m	-	0.04	0.73	0.13	-	m	0.86
width 3.00 m	-	0.04	0.73	0.13	-	m	0.86
Permeable membranes; to trimmed and compacted surface of pathway							
"Terram 1000"	0.42	0.02	0.36	-	0.44	m²	0.81

Q PAVING/PLANTING/FENCING/SITE FURNITURE

Item Excluding site overheads and profit	PC £	Labour hours	Labour £	Plant £	Material £	Unit	Total rate £
Q23 GRAVEL/HOGGIN/WOODCHIP ROADS/PAVINGS - cont'd							
Permaloc "AsphaltEdge"; RTS Ltd; extruded aluminium alloy L shaped edging with 5.33 mm exposed upper lip; edging fixed to roadway base and edge profile with 250 mm steel fixing spike; laid to straight or curvilinear road edge; subsequently filled with macadam (not included)							
Depth of macadam							
38 mm	7.34	0.02	0.30	-	7.52	m	**7.83**
51 mm	8.27	0.02	0.31	-	8.48	m	**8.79**
64 mm	9.14	0.02	0.33	-	9.37	m	**9.70**
76 mm	11.05	0.02	0.36	-	11.05	m	**11.42**
102 mm	12.86	0.02	0.38	-	13.18	m	**13.56**
Permaloc "Cleanline"; RTS Ltd; heavy duty straight profile edging; for edgings to soft landscape beds or turf areas; 3.2 mm x 102 mm high; 3.2 mm thick with 4.75 mm exposed upper lip; fixed to form straight or curvilinear edge with 305 mm fixing spike							
Milled aluminium							
100 mm deep	8.41	0.02	0.33	-	8.62	m	**8.95**
Black							
100 mm deep	8.41	0.02	0.33	-	8.62	m	**8.95**
Permaloc "Permastrip"; RTS Ltd; heavy duty L shaped profile maintenance strip; 3.2 mm x 89 mm high with 5.2 mm exposed top lip; for straight or gentle curves on paths or bed turf interfaces; fixed to form straight or curvilinear edge with standard 305 mm stake; other stake lengths available							
Milled aluminium							
89 mm deep	8.41	0.02	0.33	-	8.62	m	**8.95**
Black							
89 mm deep	8.41	0.02	0.33	-	8.62	m	**8.95**
Permaloc "Proline"; RTS Ltd; medium duty straight profiled maintenance strip; 3.2 mm x 102 mm high with 3.18 mm exposed top lip; for straight or gentle curves on paths or bed turf interfaces; fixed to form straight or curvilinear edge with standard 305 mm stake; other stake lengths available							
Milled aluminium							
89 mm deep	7.32	0.02	0.33	-	7.50	m	**7.84**
Black							
89 mm deep	7.32	0.02	0.33	-	7.50	m	**7.84**

Q PAVING/PLANTING/SITE FURNITURE

Item Excluding site overheads and profit	PC £	Labour hours	Labour £	Plant £	Material £	Unit	Total rate £
Filling to make up levels							
Obtained off site; hardcore; PC £8.00/m³							
150 mm thick	1.28	0.04	0.73	0.46	1.28	m²	**2.47**
Obtained off site; granular fill type 1; PC £15.95/tonne (£35.09/m³ compacted)							
100 mm thick	3.51	0.03	0.51	0.33	3.51	m²	**4.35**
150 mm thick	5.26	0.03	0.46	0.49	5.26	m²	**6.21**
Surface treatments							
Sand blinding; to hardcore (not included)							
50 mm thick	1.44	0.04	0.73	-	1.58	m²	**2.31**
Filter fabric; to hardcore (not included)	0.42	0.01	0.18	-	0.44	m²	**0.63**
Granular pavings							
Footpath gravels; porous self binding gravel							
CED Ltd; Cedec gravel; self-binding; laid to inert (non-limestone) base measured separately; compacting							
red, silver or gold; 50 mm thick	10.99	0.03	0.46	0.38	10.99	m²	**11.82**
Grundon Ltd; Coxwell self-binding path gravels laid and compacted to excavation or base measured separately							
50 mm thick	3.27	0.03	0.46	0.38	3.27	m²	**4.10**
Breedon Special Aggregates; "Golden Gravel" or equivalent; rolling wet; on hardcore base (not included); for pavements; to falls and crossfalls and to slopes not exceeding 15° from horizontal; over 300 mm wide							
50 mm thick	10.37	0.03	0.46	0.31	10.37	m²	**11.13**
75 mm thick	15.55	0.10	1.82	0.39	15.55	m²	**17.76**
Breedon Special Aggregates; "Wayfarer" specially formulated fine gravel for use on golf course pathways							
50 mm thick	9.36	0.03	0.46	0.31	9.36	m²	**10.13**
75 mm thick	14.03	0.05	0.91	0.55	14.03	m²	**15.49**
Hoggin (stabilized); PC £54.58/m³ on hardcore base (not included); to falls and crossfalls and to slopes not exceeding 15° from horizontal; over 300 mm wide							
100 mm thick	5.46	0.03	0.61	0.50	8.19	m²	**9.30**
150 mm thick	8.19	0.05	0.91	0.75	12.28	m²	**13.95**
Ballast; as dug; watering; rolling; on hardcore base (not included)							
100 mm thick	3.33	0.03	0.61	0.50	4.16	m²	**5.27**
150 mm thick	5.00	0.05	0.91	0.75	6.24	m²	**7.91**
Footpath gravels; porous loose gravels							
Breedon Special Aggregates; Breedon Buff decorative limestone chippings							
50 mm thick	4.57	0.01	0.18	0.09	4.80	m²	**5.07**
75 mm thick	12.96	0.01	0.23	0.11	13.61	m²	**13.95**
Breedon Special Aggregates; Breedon Buff decorative limestone chippings							
50 mm thick	4.57	0.01	0.18	0.09	4.80	m²	**5.07**
75 mm thick	6.86	0.01	0.23	0.11	7.20	m²	**7.54**

Q PAVING/PLANTING/FENCING/SITE FURNITURE

Item Excluding site overheads and profit	PC £	Labour hours	Labour £	Plant £	Material £	Unit	Total rate £
Q23 GRAVEL/HOGGIN/WOODCHIP ROADS/PAVINGS - cont'd							
Footpath gravels - cont'd							
Breedon Special Aggregates; Brindle or Moorland Black chippings							
50 mm thick	6.01	0.01	0.18	0.09	6.31	m^2	**6.58**
75 mm thick	9.01	0.01	0.23	0.11	9.47	m^2	**9.80**
Breedon Special Aggregates; Slate Chippings; plum, blue or green							
50 mm thick	8.65	0.01	0.18	0.09	9.09	m^2	**9.36**
75 mm thick	12.98	0.01	0.23	0.11	13.64	m^2	**13.98**
Washed shingle; on prepared base (not included)							
25 - 50 mm size; 25 mm thick	0.91	0.02	0.32	0.08	0.95	m^2	**1.35**
25 - 50 mm size; 75 mm thick	2.72	0.05	0.96	0.23	2.86	m^2	**4.05**
50 - 75 mm size; 25 mm thick	0.91	0.02	0.36	0.09	0.95	m^2	**1.41**
50 - 75 mm size; 75 mm thick	2.72	0.07	1.22	0.29	2.86	m^2	**4.36**
Pea shingle; on prepared base (not included)							
10 - 15 mm size; 25 mm thick	0.91	0.02	0.32	0.08	0.95	m^2	**1.35**
5 - 10 mm size; 75 mm thick	2.72	0.05	0.96	0.23	2.86	m^2	**4.05**
Wood chip surfaces; Melcourt Industries Ltd (items labelled FSC are Forest Stewardship Council certified)							
Wood chips; to surface of pathways by machine; material delivered in 80 m^3 loads; levelling and spreading by hand (excavation and preparation not included)							
Walk Chips; 100 mm thick; FSC (25 m^3 loads)	3.42	0.03	0.61	0.27	3.59	m^2	**4.47**
Walk Chips; 100 mm thick; FSC (80 m^3 loads)	2.17	0.03	0.61	0.27	2.28	m^2	**3.16**
Woodfibre; 100 mm thick; FSC (80 m^3 loads)	2.00	0.03	0.61	0.27	2.09	m^2	**2.98**
Q24 INTERLOCKING BRICK/BLOCK ROADS/PAVINGS							
Precast concrete block edgings; PC £7.63/m^2; 200 x 100 x 60 mm; on prepared base (not included); haunching one side							
Edgings; butt joints							
stretcher course	0.76	0.17	3.04	-	3.01	m	**6.05**
header course	1.53	0.27	4.87	-	3.92	m	**8.79**
Precast concrete vehicular paving blocks; Marshalls Plc; on prepared base (not included); on 50 mm compacted sharp sand bed; blocks laid in 7 mm loose sand and vibrated; joints filled with sharp sand and vibrated; level and to falls only							
"Trafica" paving blocks; 450 x 450 x 70 mm							
"Perfecta" finish; colour natural	27.26	0.50	9.13	0.10	29.52	m^2	**38.75**
"Perfecta" finish; colour buff	31.55	0.50	9.13	0.10	33.92	m^2	**43.15**
"Saxon" finish; colour natural	23.96	0.50	9.13	0.10	26.14	m^2	**35.37**
"Saxon" finish; colour buff	27.56	0.50	9.13	0.10	29.83	m^2	**39.06**

Q PAVING/PLANTING/FENCING/SITE FURNITURE

Item Excluding site overheads and profit	PC £	Labour hours	Labour £	Plant £	Material £	Unit	Total rate £
Precast concrete vehicular paving blocks; "Keyblok" Marshalls Plc; on prepared base (not included); on 50 mm compacted sharp sand bed; blocks laid in 7 mm loose sand and vibrated; joints filled with sharp sand and vibrated; level and to falls only							
Herringbone bond							
200 x 100 x 60 mm; natural grey	7.63	1.50	27.37	0.10	9.66	m²	37.13
200 x 100 x 60 mm; colours	8.29	1.50	27.37	0.10	10.35	m²	37.83
200 x 100 x 80 mm; natural grey	8.49	1.50	27.37	0.10	10.56	m²	38.04
200 x 100 x 80 mm; colours	9.59	1.50	27.37	0.10	11.72	m²	39.19
Basketweave bond							
200 x 100 x 60 mm; natural grey	7.63	1.20	21.90	0.10	9.66	m²	31.66
200 x 100 x 60 mm; colours	8.29	1.20	21.90	0.10	10.35	m²	32.35
200 x 100 x 80 mm; natural grey	8.49	1.20	21.90	0.10	10.56	m²	32.56
200 x 100 x 80 mm; colours	9.59	1.20	21.90	0.10	11.72	m²	33.72
Precast concrete vehicular paving blocks; Charcon Hard Landscaping; on prepared base (not included); on 50 mm compacted sharp sand bed; blocks laid in 7 mm loose sand and vibrated; joints filled with sharp sand and vibrated; level and to falls only							
"Europa" concrete blocks							
200 x 100 x 60 mm; natural grey	9.39	1.50	27.37	0.10	11.51	m²	38.98
200 x 100 x 60 mm; colours	10.24	1.50	27.37	0.10	12.40	m²	39.87
"Parliament" concrete blocks							
200 x 100 x 65 mm; natural grey	25.91	1.50	27.37	0.10	28.20	m²	55.68
200 x 100 x 65 mm; colours	25.91	1.50	27.37	0.10	28.20	m²	55.68
Recycled polyethylene grassblocks; Boddingtons Grass Reinforcement Solutions; Boddingtons Ltd; interlocking units laid to prepared base or rootzone (not included)							
"Bodpave"; load bearing >300 tonnes per m²; 500 x 500 x 50 mm deep							
1 - 50 m²	16.50	0.20	3.65	0.17	19.40	m²	23.22
51 - 500 m²	13.75	0.20	3.65	0.17	16.65	m²	20.47
501 - 1000 m²	13.20	0.20	3.65	0.17	16.10	m²	19.92
1001 - 1559 m²	12.50	0.20	3.65	0.17	15.40	m²	19.22
1560 m² or over	12.00	0.20	3.65	0.17	14.90	m²	18.72
"Boddingtons GrassProtecta"; extruded expanded polyethylene flexible mesh laid to existing grass surface or newly seeded areas to provide heavy surface protection from traffic and pedestrians							
"Standard"; 1.2 kg/m²; 20 x 2 m; up to 320 m²	7.50	0.01	0.15	-	7.50	m²	7.65
"Standard"; 1.2 kg/m²; 20 x 2 m; 321 m² to 3000 m²	6.50	0.01	0.15	-	6.50	m²	6.65
"Standard"; 1.2 kg/m²; 20 x 2 m; over 3000 m²	6.00	0.01	0.15	-	6.00	m²	6.15
"Heavy"; 2 kg/m²; 20 x 2 m; up to 320 m²	8.50	0.01	0.15	-	8.50	m²	8.65
"Heavy"; 2 kg/m²; 20 x 2 m; 321 m² to 3000 m²	8.00	0.01	0.15	-	8.00	m²	8.15
"Heavy"; 2 kg/m²; 20 x 2 m; over 3000 m²	7.50	0.01	0.15	-	7.50	m²	7.65

Q PAVING/PLANTING/FENCING/SITE FURNITURE

Item Excluding site overheads and profit	PC £	Labour hours	Labour £	Plant £	Material £	Unit	Total rate £
Q24 INTERLOCKING BRICK/BLOCK ROADS/PAVINGS - cont'd							
Recycled polyethylene grassblocks - cont'd "Boddingtons TurfProtecta"; extruded polyethylene flexible mesh laid to existing grass surface or newly seeded areas to provide surface protection from traffic including vehicle or animal wear and tear							
"Standard" 30 x 2 m; up to 300 m^2	2.90	0.01	0.15	-	2.90	m^2	3.05
"Standard" 30 x 2 m; 301 m^2 to 600 m^2	2.49	0.01	0.15	-	2.49	m^2	2.64
"Standard" 30 x 2 m; 601 m^2 to 1440 m^2	2.41	0.01	0.15	-	2.41	m^2	2.56
"Standard" 30 x 2 m; over 1440 m^2	2.24	0.01	0.15	-	2.24	m^2	2.39
"Premium" 30 x 2 m; up to 300 m^2	2.99	0.01	0.15	-	2.99	m^2	3.14
"Premium" 30 x 2 m; 301 m^2 to 600 m^2	2.66	0.01	0.15	-	2.66	m^2	2.81
"Premium" 30 x 2 m; 601 m^2 to 1440 m^2	2.57	0.01	0.15	-	2.57	m^2	2.72
"Premium" 30 x 2 m; over 1440 m^2	2.41	0.01	0.15	-	2.41	m^2	2.56
"Grassroad"; Cooper Clarke Civils and Lintels; heavy duty for car parking and fire paths verge hardening and shallow embankments Honeycomb cellular polyproylene interconnecting paviors with integral downstead anti-shear cleats including topsoil but excluding edge restraints; to granular sub-base (not included)							
635 x 330 x 42 mm overall laid to a module of 622 x 311 x 32 mm	-	-	-	-	-	m^2	29.35
extra over for green colour	-	-	-	-	-	m^2	0.54
Q25 SLAB/BRICK/SETT/COBBLE PAVINGS							
Bricks - General Preamble: Bricks shall be hard, well burnt, non-dusting, resistant to frost and sulphate attack and true to shape, size and sample.							
Movement of materials Mechanically offloading bricks; loading wheelbarrows; transporting maximum 25 m distance	-	0.20	3.65	-	-	m^2	3.65
Edge restraints; to brick paving; on prepared base (not included); 65 mm thick bricks; PC £300.00/1000; haunching one side Header course							
200 x 100 mm; butt joints	3.00	0.27	4.87	-	5.40	m	10.26
210 x 105 mm; mortar joints	2.67	0.50	9.13	-	5.28	m	14.41
Stretcher course							
200 x 100 mm; butt joints	1.50	0.17	3.04	-	3.75	m	6.79
210 x 105 mm; mortar joints	1.36	0.33	6.08	-	3.71	m	9.80

Q PAVING/PLANTING/FENCING/SITE FURNITURE

Item Excluding site overheads and profit	PC £	Labour hours	Labour £	Plant £	Material £	Unit	Total rate £
Variation in brick prices; add or subtract the following amounts for every £1.00/ 1000 difference in the PC price							
Edgings							
100 mm wide	-	-	-	-	0.05	10 m	-
200 mm wide	-	-	-	-	0.10	10 m	-
102.5 mm wide	-	-	-	-	0.04	10 m	-
215 mm wide	-	-	-	-	0.09	10 m	-
Clay brick pavings; on prepared base (not included); bedding on 50 mm sharp sand; kiln dried sand joints							
Pavings; 200 x 100 x 65 mm wirecut chamfered paviors							
brick; PC £450.00/1000	22.50	1.44	26.35	0.22	24.87	m²	51.43
Clay brick pavings; 200 x 100 x 50 mm; laid to running stretcher or stack bond only; on prepared base (not included); bedding on cement:sand (1:4) pointing mortar as work proceeds							
PC £600.00/1000							
laid on edge	47.62	4.76	86.90	-	55.11	m²	142.02
laid on edge but pavior 65 mm thick	40.00	3.81	69.52	-	47.30	m²	116.82
laid flat	25.97	2.20	40.15	-	30.38	m²	70.53
PC £500.00/1000							
laid on edge	39.68	4.76	86.90	-	46.98	m²	133.88
laid on edge but pavior 65 mm thick	33.33	3.81	69.52	-	40.47	m²	109.99
laid flat	21.64	2.20	40.15	-	25.94	m²	66.09
PC £400.00/1000							
laid flat	17.32	2.20	40.15	-	21.51	m²	61.66
laid on edge	31.74	4.76	86.90	-	38.84	m²	125.75
laid on edge but pavior 65 mm thick	26.66	3.81	69.52	-	33.63	m²	103.16
PC £300.00/1000							
laid on edge	23.81	4.76	86.90	-	30.71	m²	117.61
laid on edge but pavior 65 mm thick	20.00	3.81	69.52	-	26.80	m²	96.33
laid flat	12.99	2.20	40.15	-	17.07	m²	57.22
Clay brick pavings; 200 x 100 x 50 mm; butt jointed laid herringbone or basketweave pattern only; on prepared base (not included); bedding on 50 mm sharp sand							
PC £600.00/1000							
laid flat	30.00	1.44	26.35	0.28	32.56	m²	59.19
PC £500.00/1000							
laid flat	25.00	1.44	26.35	0.28	27.43	m²	54.06
PC £400.00/1000							
laid flat	20.00	1.44	26.35	0.28	22.31	m²	48.94
PC £300.00/1000							
laid flat	15.00	1.44	26.35	0.28	17.18	m²	43.81

Q PAVING/PLANTING/FENCING/SITE FURNITURE

Item Excluding site overheads and profit	PC £	Labour hours	Labour £	Plant £	Material £	Unit	Total rate £
Q25 SLAB/BRICK/SETT/COBBLE PAVINGS - cont'd							
Clay brick pavings; 215 x 102.5 x 65 mm; on prepared base (not included); bedding on cement:sand (1:4) pointing mortar as work proceeds							
Paving bricks; PC £600.00/1000; herringbone bond							
laid on edge	35.55	3.55	64.88	-	41.45	m^2	**106.33**
laid flat	23.70	2.37	43.26	-	28.71	m^2	**71.97**
Paving bricks; PC £600.00/1000; basketweave bond							
laid on edge	35.55	2.37	43.26	-	41.45	m^2	**84.71**
laid flat	23.70	1.58	28.84	-	28.71	m^2	**57.55**
Paving bricks; PC £600.00/1000; running or stack bond							
laid on edge	35.55	1.90	34.61	-	41.45	m^2	**76.06**
laid flat	23.70	1.26	23.07	-	28.71	m^2	**51.79**
Paving bricks; PC £500.00/1000; Herringbone bond							
laid on edge	29.63	3.55	64.88	-	33.24	m^2	**98.12**
laid flat	19.75	2.37	43.26	-	24.76	m^2	**68.02**
Paving bricks; PC £500.00/1000; basketweave bond							
laid on edge	29.63	2.37	43.26	-	33.24	m^2	**76.50**
laid flat	19.75	1.58	28.84	-	24.76	m^2	**53.60**
Paving bricks; PC £500.00/1000; running or stack bond							
laid on edge	29.63	1.90	34.61	-	33.24	m^2	**67.86**
laid flat	19.75	1.26	23.07	-	24.76	m^2	**47.84**
Paving bricks; PC £400.00/1000; herringbone bond							
laid on edge	23.70	3.55	64.88	-	27.17	m^2	**92.05**
laid flat	15.80	2.37	43.26	-	20.72	m^2	**63.98**
Paving bricks; PC £400.00/1000; basketweave bond							
laid on edge	23.70	2.37	43.26	-	27.17	m^2	**70.43**
laid flat	15.80	1.58	28.84	-	20.72	m^2	**49.56**
Paving bricks; PC £400.00/1000; running or stack bond							
laid on edge	23.70	1.90	34.61	-	27.17	m^2	**61.78**
laid flat	15.80	1.26	23.07	-	20.72	m^2	**43.79**
Paving bricks; PC £300.00/1000; herringbone bond							
laid on edge	17.77	3.55	64.88	-	20.65	m^2	**85.53**
laid flat	11.85	2.37	43.26	-	16.86	m^2	**60.12**
Paving bricks; PC £300.00/1000; basketweave bond							
laid on edge	17.77	2.37	43.26	-	20.65	m^2	**63.91**
laid flat	11.85	1.58	28.84	-	16.86	m^2	**45.70**
Paving bricks; PC £300.00/1000; running or stack bond							
laid on edge	17.77	1.90	34.60	-	20.65	m^2	**55.26**
laid flat	11.85	1.26	23.07	-	16.86	m^2	**39.94**
Cutting							
curved cutting	-	0.44	8.11	6.25	-	m	**14.36**
raking cutting	-	0.33	6.08	5.15	-	m	**11.23**

Q PAVING/PLANTING/FENCING/SITE FURNITURE

Item Excluding site overheads and profit	PC £	Labour hours	Labour £	Plant £	Material £	Unit	Total rate £
Add or subtract the following amounts for every £10.00/1000 difference in the prime cost of bricks							
Butt joints							
200 x 100 mm	-	-	-	-	0.50	m²	-
215 x 102.5 mm	-	-	-	-	0.45	m²	-
10 mm mortar joints							
200 x 100 mm	-	-	-	-	0.43	m²	-
215 x 102.5 mm	-	-	-	-	0.40	m²	-
SUDS paving; Concrete Block Permeable Paving (CBPP)							
Note: Sustainable Drainage Systems (SUDS) aim to reduce flood risk by controlling the rate and volume of surface water run off from developments. Permeable Paving is a SUDS technique. In addition to managing the quantity of surface water run off more effectively, one of the primary benefits of CBPP is the potential enhancement of water quality. Permeable pavements improve water quality by mirroring nature in providing filtration and allowing for natural biodegradation of hydrocarbons and the dilution of other contaminants as water passes through the system.							
Bases for Sustainable Urban Drainage System (SUDS) sub-base for use below impervious paving; Aggregate Industries Ltd; Bardon DrainAgg 20							
4/20 mm open graded material laid loose on prepared sub-grade; within edge restraints to paving area (not included); graded to levels and falls as per paving manufacturer's instructions							
150 mm thick	5.10	0.03	0.46	0.49	5.10	m²	6.05
200 mm thick	6.80	0.02	0.30	0.66	6.80	m²	7.76
250 mm thick	8.50	0.02	0.38	0.82	8.50	m²	9.70
350 mm thick	11.90	0.03	0.52	1.13	11.90	m²	13.55
450 mm thick	15.30	0.04	0.73	1.58	15.30	m²	17.61
Sub-base replacement system for SUDS paving "Charcon Permavoid"; Aggregate Industries Ltd; geocellular, interlocking, high strength; high void capacity (92%); typically a 4:1 ratio when compared with alternative granular systems; allows for water storage to be confined at shallow depth within the sub-base layer; inclusive of connection, ties, pins; does not include for encapsulation in geotextile/ geomembranes							
708 x 354 x 150 mm							
150 mm thick	51.64	0.17	3.04	-	51.64	m²	54.68
300 mm thick	107.12	0.33	6.08	-	107.12	m²	113.20

Q PAVING/PLANTING/FENCING/SITE FURNITURE

Item Excluding site overheads and profit	PC £	Labour hours	Labour £	Plant £	Material £	Unit	Total rate £
Q25 SLAB/BRICK/SETT/COBBLE PAVINGS - cont'd							
Sub-base replacement system for SUDS paving "Charcon Permavoid" - cont'd							
Encapsulation of Permavoid; 150 mm thick systems; geomembrane geofabric or combinatiion of both for filtration or attenuation of water; please see the manufacturers design guide							
Charcon geomembrane, heat welded	10.58	0.05	0.91	-	12.70	m^2	13.61
Charcon non reinforced geotextile	6.55	0.03	0.61	-	7.87	m^2	8.47
Charcon reinforced geotextile	3.52	0.03	0.61	-	4.22	m^2	4.83
Encapsulation of Permavoid; 300 mm thick systems; geomembrane geofabric or combinatiion of both for filtration or attenuation of water; please see the manufacturers design guide							
Charcon geomembrane, heat welded	11.96	0.06	1.01	-	14.35	m^2	15.37
Charcon non reinforced geotextile	7.41	0.04	0.78	-	8.89	m^2	9.67
Charcon reinforced geotextile	3.98	0.04	0.65	-	3.98	m^2	4.63
Pervious surfacing materials; Aggregate Industries Ltd; Charcon Infilta CBPP							
Rectangular permeable block paving system; incorporating a 5 mm spacer design that provides a 5 mm void allowing ingress of water through to the sub base storage system (priced separately); various colours; bedded on 50 mm thick 2 - 6.3 mm clean, angular, free draining uncompacted aggregate; joints infilled with 3 mm clean grit							
"Infilta" 200 x 100 x 80 mm thick; natural grey	14.25	0.44	8.11	0.10	16.75	m^2	24.97
"Infilta" 200 x 100 x 80 mm thick; coloured	14.75	0.44	8.11	0.10	17.25	m^2	25.47
Slab pavings; precast concrete; Charcon Hard Landscaping; on prepared sub-base (not included); bedding on 25 mm thick cement:sand mortar (1:4); butt joints; straight both ways; on 50 mm thick sharp sand base							
Pavings; natural grey							
450 x 450 x 70 mm chamfered	16.05	0.44	8.11	-	19.84	m^2	27.95
450 x 450 x 50 mm chamfered	12.54	0.44	8.11	-	16.33	m^2	24.44
600 x 300 x 50 mm	10.89	0.44	8.11	-	14.68	m^2	22.79
400 x 400 x 65 mm chamfered	21.81	0.40	7.30	-	25.60	m^2	32.90
450 x 600 x 50 mm	10.44	0.44	8.11	-	14.23	m^2	22.34
600 x 600 x 50 mm	8.56	0.40	7.30	-	12.34	m^2	19.64
750 x 600 x 50 mm	8.18	0.40	7.30	-	11.97	m^2	19.27
900 x 600 x 50 mm	7.00	0.40	7.30	-	10.79	m^2	18.09
Pavings; coloured							
450 x 450 x 70 mm chamfered	21.38	0.44	8.11	-	25.17	m^2	33.28
450 x 600 x 50 mm	12.93	0.44	8.11	-	16.71	m^2	24.82
400 x 400 x 65 mm chamfered	21.81	0.40	7.30	-	25.60	m^2	32.90
600 x 600 x 50 mm	10.39	0.40	7.30	-	14.18	m^2	21.48
750 x 600 x 50 mm	9.62	0.40	7.30	-	13.41	m^2	20.71
900 x 600 x 50 mm	8.27	0.40	7.30	-	12.06	m^2	19.36

Q PAVING/PLANTING/FENCING/SITE FURNITURE

Item Excluding site overheads and profit	PC £	Labour hours	Labour £	Plant £	Material £	Unit	Total rate £
Precast concrete pavings; Charcon Hard Landscaping; on prepared sub-base (not included); bedding on 25 mm thick cement:sand mortar (1:4); butt joints; straight both ways; jointing in cement: sand (1:3) brushed in; on 50 mm thick sharp sand base							
"Appalacian" rough textured exposed aggregate pebble paving							
600 x 600 x 65 mm	26.70	0.50	9.13	-	29.72	m²	**38.84**
Pavings; Marshalls Plc; spot bedding on 5 nr pads of cement:sand mortar (1:4); on sharp sand							
"Blister Tactile" pavings; specially textured slabs for guidance of blind pedestrians; red or buff							
400 x 400 x 50 mm	26.63	0.50	9.13	-	28.77	m²	**37.90**
450 x 450 x 50 mm	23.51	0.50	9.13	-	25.58	m²	**34.70**
"Metric Four Square" pavings							
496 x 496 x 50 mm; exposed river gravel aggregate	75.11	0.50	9.13	-	76.59	m²	**85.72**
"Metric Four Square" cycle blocks							
496 x 496 x 50 mm; exposed aggregate	75.11	0.25	4.56	-	77.14	m²	**81.70**
Precast concrete pavings; Marshalls Plc; "Heritage" imitation riven yorkstone paving; on prepared sub-base measured separately; bedding on 25 mm thick cement:sand mortar (1:4); pointed straight both ways cement:sand (1:3)							
Square and rectangular paving							
450 x 300 x 38 mm	35.88	1.00	18.25	-	39.21	m²	**57.46**
450 x 450 x 38 mm	22.05	0.75	13.69	-	25.38	m²	**39.07**
600 x 300 x 38 mm	26.50	0.80	14.60	-	29.83	m²	**44.43**
600 x 450 x 38 mm	24.42	0.75	13.69	-	27.75	m²	**41.44**
600 x 600 x 38 mm	23.13	0.50	9.13	-	26.52	m²	**35.65**
Extra labours for laying the a selection of the above sizes to random rectangular pattern	-	0.33	6.08	-	-	m²	**6.08**
Radial paving for circles							
circle with centre stone and first ring (8 slabs); 450 x 230/560 x 38 mm; diameter 1.54 m (total area 1.86 m²)	62.46	1.50	27.38	-	66.81	nr	**94.18**
circle with second ring (16 slabs); 450 x 300/460 x 38 mm; diameter 2.48 m (total area 4.83 m²)	62.46	4.00	73.00	-	172.19	nr	**245.19**
circle with third ring (16 slabs); 450 x 470/625 x 38 mm; diameter 3.42 m (total area 9.18 m²)	62.46	8.00	146.00	-	313.90	nr	**459.90**
Stepping stones							
380 mm diameter x 38 mm	4.50	0.20	3.65	-	9.20	nr	**12.85**
asymmetrical; 560 x 420 x 38 mm	6.00	0.20	3.65	-	10.70	nr	**14.35**

Q PAVING/PLANTING/FENCING/SITE FURNITURE

Item Excluding site overheads and profit	PC £	Labour hours	Labour £	Plant £	Material £	Unit	Total rate £
Q25 SLAB/BRICK/SETT/COBBLE PAVINGS - cont'd							
Precast concrete pavings; Marshalls Plc; "Chancery" imitation reclaimed riven yorkstone paving; on prepared sub-base measured separately; bedding on 25 mm thick cement:sand mortar (1:4); pointed straight both ways cement:sand (1:3)							
Square and rectangular paving							
300 x 300 x 45 mm	26.42	1.00	18.25	-	29.75	m^2	48.00
450 x 300 x 45 mm	24.91	0.90	16.43	-	28.25	m^2	44.67
600 x 300 x 45 mm	24.17	0.80	14.60	-	27.50	m^2	42.10
600 x 450 x 45 mm	24.27	0.75	13.69	-	27.61	m^2	41.29
450 x 450 x 45 mm	21.90	0.75	13.69	-	25.24	m^2	38.92
600 x 600 x 45 mm	23.46	0.50	9.13	-	26.80	m^2	35.92
Extra labours for laying the a selection of the above sizes to random rectangular pattern	-	0.33	6.08	-	-	m^2	6.08
Radial paving for circles							
circle with centre stone and first ring (8 slabs); 450 x 230/560 x 38 mm; diameter 1.54 m (total area 1.86 m^2)	79.95	1.50	27.38	-	84.30	nr	111.67
circle with second ring (16 slabs); 450 x 300/460 x 38 mm; diameter 2.48 m (total area 4.83 m^2)	206.35	4.00	73.00	-	217.52	nr	290.52
circle with third ring (16 slabs); 450 x 470/625 x 38 mm; diameter 3.42 m (total area 9.18 m^2)	375.63	8.00	146.00	-	396.67	nr	542.67
Squaring off set for 2 ring circle							
16 slabs; 2.72 m^2	219.69	1.00	18.25	-	219.69	nr	237.94
Marshalls Plc; "La Linia Pavings"; fine textured exposed aggregate 80 mm thick pavings in various sizes to designed laying patterns; laid to 50 mm sharp sand bed on Type 1 base all priced separately							
Bonded laying patterns; 300 x 300 mm							
light granite/anthracite basalt	30.97	0.75	13.69	-	31.74	m^2	45.43
Indian granite/yellow	30.97	0.75	13.69	-	31.74	m^2	45.43
Random scatter pattern incorporating 100 x 200 mm, 200 x 200 mm and 300 x 200 mm units							
light granite/anthracite basalt	30.97	1.00	18.25	-	30.97	m^2	49.22
Indian granite/yellow	30.97	1.00	18.25	-	30.97	m^2	49.22
Marshalls Plc; "La Linia Grande Paving"; as above but 140 mm thick							
Bonded laying patterns; 300 x 300 mm							
all colours	60.27	0.75	13.69	-	60.27	m^2	73.96
Random scatter pattern incorporating 100 x 200 mm, 200 x 200 mm and 300 x 200 mm units							
light granite/anthracite basalt	60.27	1.00	18.25	-	60.27	m^2	78.52
Extra over to the above for incorporating inlay stones to patterns; blue, light granite or anthracite basalt							
to prescribed patterns	380.55	1.50	27.38	-	380.55	m^2	407.93
as individual units; triangular 200 x 200 x 282 mm	2.15	0.05	0.91	-	2.15	nr	3.06

Q PAVING/PLANTING/FENCING/SITE FURNITURE

Item Excluding site overheads and profit	PC £	Labour hours	Labour £	Plant £	Material £	Unit	Total rate £
Pedestrian deterrent pavings; Marshalls Plc; on prepared base (not included); bedding on 25 mm cement:sand (1:3); cement:sand (1:3) joints							
"Lambeth" pyramidal pavings							
600 x 600 x 75 mm	17.80	0.25	4.56	-	21.13	m²	**25.70**
"Thaxted" pavings; granite sett appearance							
600 x 600 x 75 mm	21.00	0.33	6.08	-	24.33	m²	**30.42**
Pedestrian deterrent pavings; Townscape Products Ltd; on prepared base (not included); bedding on 25 mm cement:sand (1:3); cement:sand (1:3) joints							
"Strata" striated textured slab pavings; giving bonded appearance; grey							
600 x 600 x 60 mm	27.92	0.50	9.13	-	31.25	m²	**40.38**
"Geoset" raised chamfered studs pavings; grey							
600 x 600 x 60 mm	27.92	0.50	9.13	-	31.25	m²	**40.38**
"Abbey" square cobble pattern pavings; reinforced							
600 x 600 x 65 mm	26.39	0.80	14.60	-	29.72	m²	**44.32**
Edge restraints; to block paving; on prepared base (not included); 200 x 100 x 80 mm; PC £8.49/m²; haunching one side							
Header course							
200 x 100 mm; butt joints	1.70	0.27	4.87	-	4.09	m	**8.96**
Stretcher course							
200 x 100 mm; butt joints	0.85	0.17	3.04	-	3.10	m	**6.14**
Concrete paviors; Marshalls Plc; on prepared base (not included); bedding on 50 mm sand; kiln dried sand joints swept in							
"Keyblok" paviors							
200 x 100 x 60 mm; grey	7.63	0.40	7.30	0.10	9.28	m²	**16.68**
200 x 100 x 60 mm; colours	8.29	0.40	7.30	0.10	9.94	m²	**17.34**
200 x 100 x 80 mm; grey	8.49	0.44	8.11	0.10	10.14	m²	**18.35**
200 x 100 x 80 mm; colours	9.59	0.44	8.11	0.10	11.24	m²	**19.45**
Concrete cobble paviors; Charcon Hard Landscaping; Concrete Products; on prepared base (not included); bedding on 50 mm sand; kiln dried sand joints swept in							
Paviors							
"Woburn" blocks; 100 - 201 x 134 x 80 mm; random sizes	27.23	0.67	12.17	0.10	29.56	m²	**41.83**
"Woburn" blocks; 100 - 201 x 134 x 80 mm; single size	27.23	0.50	9.13	0.10	29.56	m²	**38.79**
"Woburn" blocks; 100 - 201 x 134 x 60 mm; random sizes	22.44	0.67	12.17	0.10	24.65	m²	**36.92**
"Woburn" blocks; 100 - 201 x 134 x 60 mm; single size	22.44	0.50	9.13	0.10	24.62	m²	**33.85**

Q PAVING/PLANTING/FENCING/SITE FURNITURE

Item Excluding site overheads and profit	PC £	Labour hours	Labour £	Plant £	Material £	Unit	Total rate £
Q25 SLAB/BRICK/SETT/COBBLE PAVINGS - cont'd							
Concrete setts; on 25 mm sand; compacted; vibrated; joints filled with sand; natural or coloured; well rammed hardcore base (not included)							
Marshalls Plc; Tegula cobble paving							
60 mm thick; random sizes	20.53	0.57	10.43	0.10	23.35	m^2	**33.88**
60 mm thick; single size	20.53	0.45	8.29	0.10	23.35	m^2	**31.74**
80 mm thick; random sizes	23.65	0.57	10.43	0.10	26.62	m^2	**37.15**
80 mm thick; single size	23.65	0.45	8.29	0.10	26.62	m^2	**35.02**
cobbles; 80 x 80 x 60 mm thick; traditional	35.84	0.56	10.14	0.10	38.53	m^2	**48.77**
Cobbles							
Charcon Hard Landscaping; Country setts							
100 mm thick; random sizes	37.19	1.00	18.25	0.10	38.98	m^2	**57.33**
100 mm thick; single size	37.19	0.67	12.17	0.10	39.91	m^2	**52.18**
Natural stone, slab or granite paving - General							
Preamble: Provide paving slabs of the specified thickness in random sizes but not less than 25 slabs per 10 m^2 of surface area, to be laid in parallel courses with joints alternately broken and laid to falls.							
Reconstituted yorkstone aggregate pavings; Marshalls Plc; "Saxon" on prepared sub-base measured separately; bedding on 25 mm thick cement:sand mortar (1:4) ;on 50 mm thick sharp sand base							
Square and rectangular paving in buff; butt joints straight both ways							
300 x 300 x 35 mm	32.74	0.88	16.06	-	37.06	m^2	**53.12**
600 x 300 x 35 mm	21.62	0.71	13.05	-	25.93	m^2	**38.98**
450 x 450 x 50 mm	23.65	0.82	15.06	-	27.96	m^2	**43.02**
600 x 600 x 35 mm	16.21	0.55	10.04	-	20.52	m^2	**30.56**
600 x 600 x 50 mm	20.38	0.60	11.04	-	24.69	m^2	**35.74**
Square and rectangular paving in natural; butt joints straight both ways							
300 x 300 x 35 mm	27.20	0.88	16.06	-	31.51	m^2	**47.57**
450 x 450 x 50 mm	20.14	0.77	14.05	-	24.45	m^2	**38.50**
600 x 300 x 35 mm	19.04	0.82	15.06	-	23.35	m^2	**38.41**
600 x 600 x 35 mm	13.97	0.55	10.04	-	18.28	m^2	**28.32**
600 x 600 x 50 mm	16.97	0.66	12.04	-	21.28	m^2	**33.32**
Radial paving for circles; 20 mm joints							
circle with centre stone and first ring (8 slabs); 450 x 230/560 x 35 mm; diameter 1.54 m (total area 1.86 m^2)	64.48	1.50	27.38	-	69.07	nr	**96.45**
circle with second ring (16 slabs); 450 x 300/460 x 35 mm; diameter 2.48 m (total area 4.83 m^2)	157.12	4.00	73.00	-	168.97	nr	**241.97**
circle with third ring (24 slabs); 450 x 310/430 x 35 mm; diameter 3.42 m (total area 9.18 m^2)	296.08	8.00	146.00	-	318.75	nr	**464.75**

Q PAVING/PLANTING/FENCING/SITE FURNITURE

Item Excluding site overheads and profit	PC £	Labour hours	Labour £	Plant £	Material £	Unit	Total rate £
Granite setts; bedding on 25 mm cement: sand (1:3) Natural granite setts; 100 x 100 mm to 125 x 150 mm; x 150 to 250 mm length; riven surface; silver grey							
new; standard grade	33.93	2.00	36.50	-	40.97	m²	77.47
new; high grade	25.78	2.00	36.50	-	32.82	m²	69.32
reclaimed; cleaned	35.55	2.00	36.50	-	43.30	m²	79.80
Natural stone, slate or granite flag pavings; CED Ltd; on prepared base (not included); bedding on 25 mm cement: sand (1:3); cement:sand (1:3) joints Yorkstone; riven laid random rectangular							
new slabs; 40 - 60 mm thick	65.26	1.71	31.28	-	74.87	m²	106.16
reclaimed slabs; Cathedral grade; 50 - 75 mm thick	79.58	2.80	51.10	-	90.63	m²	141.73
Donegal quartzite slabs; standard tiles							
200 mm x random lengths x 15-25 mm	64.69	3.50	63.88	-	67.78	m²	131.65
250 mm x random lengths x 15-25 mm	64.69	3.30	60.23	-	67.78	m²	128.00
300 mm x random lengths x 15-25 mm	64.69	3.10	56.58	-	67.78	m²	124.35
350 mm x random lengths x 15-25 mm	64.69	2.80	51.10	-	67.78	m²	118.88
400 mm x random lengths x 15-25 mm	64.69	2.50	45.63	-	67.78	m²	113.40
450 mm x random lengths x 15-25 mm	64.69	2.33	42.58	-	67.78	m²	110.36
Natural yorkstone, pavings or edgings; Johnsons Wellfield Quarries; sawn 6 sides; 50 mm thick; on prepared base measured separately; bedding on 25 mm cement:sand (1:3); cement:sand (1:3) joints Paving							
laid to random rectangular pattern	56.25	1.71	31.28	-	62.56	m²	93.84
laid to coursed laying pattern; 3 sizes	61.25	1.72	31.46	-	67.25	m²	98.71
Paving; single size							
600 x 600 mm	66.25	0.85	15.51	-	72.50	m²	88.01
600 x 400 mm	66.25	1.00	18.25	-	72.50	m²	90.75
300 x 200 mm	75.00	2.00	36.50	-	81.69	m²	118.19
215 x 102.5 mm	77.50	2.50	45.63	-	84.31	m²	129.94
Paving; cut to template off site; 600 x 600 mm; radius							
1.00 m	170.00	3.33	60.83	-	172.94	m²	233.77
2.50 m	170.00	2.00	36.50	-	172.94	m²	209.44
5.00 m	170.00	2.00	36.50	-	172.94	m²	209.44
Edgings							
100 mm wide x random lengths	7.00	0.50	9.13	-	7.64	m	16.77
100 x 100 mm	7.75	0.50	9.13	-	11.07	m	20.20
100 x 200 mm	15.50	0.50	9.13	-	19.21	m	28.34
250 mm wide x random lengths	15.31	0.40	7.30	-	19.01	m	26.31
500 mm wide x random lengths	30.63	0.33	6.08	-	35.10	m	41.18
Yorkstone edgings; 600 mm long x 250 mm wide; cut to radius							
1.00 m to 3.00	48.75	0.50	9.13	-	51.48	m	60.61
3.00 m to 5.00 m	48.75	0.44	8.11	-	51.48	m	59.59
exceeding 5.00 m	48.75	0.40	7.30	-	51.48	m	58.78

Q PAVING/PLANTING/FENCING/SITE FURNITURE

Item Excluding site overheads and profit	PC £	Labour hours	Labour £	Plant £	Material £	Unit	Total rate £
Q25 SLAB/BRICK/SETT/COBBLE PAVINGS - cont'd							
Natural yorkstone pavings or edgings; Johnsons Wellfield Quarries; sawn 6 sides; 75 mm thick; on prepared base measured separately; bedding on 25 mm cement:sand (1:3); cement:sand (1:3) joints							
Paving							
laid to random rectangular pattern	68.15	0.95	17.34	-	74.49	m²	**91.83**
laid to coursed laying pattern; 3 sizes	73.15	0.95	17.34	-	79.74	m²	**97.08**
Paving; single size							
600 x 600 mm	78.15	0.95	17.34	-	84.99	m²	**102.33**
600 x 400 mm	78.15	0.95	17.34	-	84.99	m²	**102.33**
300 x 200 mm	88.00	0.75	13.69	-	95.34	m²	**109.02**
215 x 102.5 mm	90.00	2.50	45.63	-	97.44	m²	**143.06**
Paving; cut to template off site; 600 x 600 mm; radius							
1.00 m	200.00	4.00	73.00	-	202.94	m²	**275.94**
2.50 m	200.00	2.50	45.63	-	202.94	m²	**248.56**
5.00 m	200.00	2.50	45.63	-	202.94	m²	**248.56**
Edgings							
100 mm wide x random lengths	8.75	0.60	10.95	-	9.48	m	**20.43**
100 x 100 mm	0.90	0.60	10.95	-	3.88	m	**14.83**
100 x 200 mm	1.80	0.50	9.13	-	4.83	m	**13.95**
250 mm wide x random lengths	18.29	0.50	9.13	-	22.14	m	**31.26**
500 mm wide x random lengths	36.58	0.40	7.30	-	41.34	m	**48.64**
Edgings; 600 mm long x 250 mm wide; cut to radius							
1.00 m to 3.00	56.25	0.60	10.95	-	59.36	m	**70.31**
3.00 m to 5.00 m	56.25	0.50	9.13	-	59.36	m	**68.48**
exceeding 5.00 m	56.25	0.44	8.11	-	59.36	m	**67.47**
CED Ltd; Indian sandstone, riven pavings or edgings; 25-35 mm thick; on prepared base measured separately; bedding on 25 mm cement:sand (1:3); cement:sand (1:3) joints							
Paving							
laid to random rectangular pattern	19.50	2.40	43.81	-	23.41	m²	**67.22**
laid to coursed laying pattern; 3 sizes	19.50	2.00	36.50	-	23.41	m²	**59.91**
Paving; single size							
600 x 600 mm	19.50	1.00	18.25	-	23.41	m²	**41.66**
600 x 400 mm	19.50	1.25	22.81	-	23.41	m²	**46.22**
400 x 400 mm	19.50	1.67	30.42	-	23.41	m²	**53.83**
Natural stone, slate or granite flag pavings; CED Ltd; on prepared base (not included); bedding on 25 mm cement:sand (1:3); cement:sand (1:3) joints							
Granite paving; sawn 6 sides; textured top							
new slabs; silver grey; 50 mm thick	33.20	1.71	31.28	-	37.95	m²	**69.23**
new slabs; blue grey; 50 mm thick	38.93	1.71	31.28	-	43.96	m²	**75.25**
new slabs; yellow grey; 50 mm thick	41.22	1.71	31.28	-	46.37	m²	**77.65**
new slabs; black; 50 mm thick	60.68	1.71	31.28	-	66.80	m²	**98.09**

Q PAVING/PLANTING/FENCING/SITE FURNITURE

Item Excluding site overheads and profit	PC £	Labour hours	Labour £	Plant £	Material £	Unit	Total rate £
Edgings; silver grey							
100 mm wide x random lengths	4.61	1.00	18.25	-	6.74	m	24.99
100 x 100mm	4.61	1.50	27.38	-	7.66	m	35.04
100 mm long x 200 mm wide	9.21	0.60	10.95	-	12.27	m	23.22
250 mm wide x random lengths	11.52	0.75	13.69	-	15.60	m	29.29
300 mm wide x random lengths	13.82	0.75	13.69	-	16.75	m	30.44
Cobble pavings - General							
Cobbles should be embedded by hand, tight-butted, endwise to a depth of 60% of their length. A dry grout of rapid-hardening cement:sand (1:2) shall be brushed over the cobbles until the interstices are filled to the level of the adjoining paving. Surplus grout shall then be brushed off and a light, fine spray of water applied over the area.							
Cobble pavings							
Cobbles; to present a uniform colour in panels; or varied in colour as required							
Scottish Beach Cobbles; 50 - 75 mm	21.50	3.33	60.83	-	27.32	m²	88.15
Scottish Beach Cobbles; 75 - 100 mm	36.92	2.50	45.63	-	42.75	m²	88.37
Scottish Beach Cobbles; 100 - 200 mm	52.20	2.00	36.50	-	58.02	m²	94.52
Concrete cycle blocks; Marshalls Plc; bedding in cement:sand (1:4)							
"Metric 4 Square" cycle stand blocks; smooth grey concrete							
496 x 496 x 100 mm	38.25	0.25	4.56	-	38.69	nr	43.25
Concrete cycle blocks; Townscape Products Ltd; on 100 mm concrete (1:2:4); on 150 mm hardcore; bedding in cement:sand (1:4)							
Cycle blocks							
"Cycle Bloc"; in white concrete	20.40	0.25	4.56	-	20.99	nr	25.56
"Mountain Cycle Bloc"; in white concrete	28.23	0.25	4.56	-	28.82	nr	33.39
Grass concrete - General							
Preamble: Grass seed should be a perennial ryegrass mixture, with the proportion depending on expected traffic. Hardwearing winter sportsground mixtures are suitable for public areas. Loose gravel, shingle or sand is liable to be kicked out of the blocks; rammed hoggin or other stabilized material should be specified.							

Q PAVING/PLANTING/FENCING/SITE FURNITURE

Item Excluding site overheads and profit	PC £	Labour hours	Labour £	Plant £	Material £	Unit	Total rate £
Q25 SLAB/BRICK/SETT/COBBLE PAVINGS - cont'd							
Grass concrete; Grass Concrete Ltd; on blinded granular Type 1 sub-base (not included)							
"Grasscrete" in situ concrete continuously reinforced surfacing; including expansion joints at 10 m centres; soiling; seeding							
ref GC2; 150 mm thick; traffic up to 40.00 tonnes	-	-	-	-	-	m^2	38.75
ref GC1; 100 mm thick; traffic up to 13.30 tonnes	-	-	-	-	-	m^2	31.33
ref GC3; 76 mm thick; traffic up to 4.30 tonnes	-	-	-	-	-	m^2	27.73
Grass concrete; Grass Concrete Ltd; 406 x 406 mm blocks; on 20 mm sharp sand; on blinded MOT type 1 sub-base (not included); level and to falls only; including filling with topsoil; seeding with dwarf rye grass at PC £4.50/kg							
Pavings							
ref GB103; 103 mm thick	14.23	0.40	7.30	0.17	16.63	m^2	24.10
ref GB83; 83 mm thick	12.21	0.36	6.64	0.17	14.61	m^2	21.41
Grass concrete; Charcon Hard Landscaping; on 25 mm sharp sand; including filling with topsoil; seeding with dwarf rye grass at PC £4.50/kg							
"Grassgrid" grass/concrete paving blocks 366 x 274 x 100 mm thick	16.73	0.38	6.84	0.17	19.13	m^2	26.14
Grass concrete; Marshalls Plc; on 25 mm sharp sand; including filling with topsoil; seeding with dwarf rye grass at PC £4.50/kg							
Concrete grass pavings							
"Grassguard 130"; for light duty applications (80 mm prepared base not included)	17.74	0.38	6.84	0.17	20.74	m^2	27.75
"Grassguard 160"; for medium duty applications (80 - 150 mm prepared base not included)	20.91	0.46	8.42	0.17	23.99	m^2	32.58
"Grassguard 180"; for heavy duty applications (150 mm prepared base not included)	23.41	0.60	10.95	0.17	26.55	m^2	37.67
Full mortar bedding							
Extra over pavings for bedding on 25 mm cement:sand (1:4); in lieu of spot bedding on sharp sand	-	0.03	0.46	-	1.01	m^2	1.47

Q PAVING/PLANTING/FENCING/SITE FURNITURE

Item Excluding site overheads and profit	PC £	Labour hours	Labour £	Plant £	Material £	Unit	Total rate £
STEPS							
In situ concrete steps; ready mixed concrete; Gen 1; on 100 mm hardcore base; base concrete thickness 100 mm treads and risers cast simultaneously; inclusive of A142 mesh within the concrete; step riser 170 mm high; inclusive of all shuttering and concrete labours; floated finish; to graded ground (not included)							
Treads 250 mm wide							
one riser	5.29	4.00	73.00	-	7.24	m	80.24
two risers	8.48	6.00	109.50	-	11.44	m	120.94
three risers	12.28	8.00	146.00	-	20.03	m	166.03
four risers	16.49	12.00	219.00	-	24.78	m	243.78
six risers	20.88	16.00	292.00	-	30.04	m	322.04
Lay surface to concrete in situ steps above							
granite treads; pre-cut off site; 305 mm wide x 40 mm thick bullnose finish inclusive of 20 mm overhang; granite riser 20 mm thick all laid on 15 mm mortar bed	21.18	2.00	36.50	-	22.06	m	58.56
riven yorkstone tread and riser 50 mm thick; cut on site	30.02	2.50	45.63	-	30.90	m	76.53
riven yorkstone treads 225 mm wide x 50 mm thick cut on site; two courses brick risers laid on flat stretcher bond; bricks PC £500.00/1000	14.68	3.00	54.75	-	20.33	m	75.08
Marshalls Saxon paving slab 450 x 450 mm cut to 225 mm wide x 38 mm thick; cut on site; two courses brick risers laid on flat stretcher bond; bricks PC £500.00/1000	4.56	3.00	54.75	-	10.21	m	64.96
Treads 450 mm wide							
one riser	9.53	5.00	91.25	-	11.92	m	103.17
two risers	19.47	7.00	127.75	-	24.34	m	152.09
three risers	27.52	9.00	164.25	-	36.60	m	200.85
four risers	47.60	13.00	237.25	-	57.33	m	294.58
six risers	60.49	18.00	328.50	-	70.84	m	399.34
Lay surface to concrete in situ steps							
granite treads; pre-cut off site; 470 mm wide x 40 mm thick bullnose finish inclusive of 20 mm overhang; granite riser 20 mm thick all laid on 15 mm mortar bed	30.28	4.00	73.00	-	31.39	m	104.39
riven yorkstone tread and riser 50 mm thick; cut on site	40.79	4.50	82.13	-	41.89	m	124.02
riven yorkstone treads 460 mm wide 50 mm thick cut on site; two courses brick risers laid on flat stretcher bond; bricks PC £500.00/1000	30.02	5.00	91.25	-	35.84	m	127.09
Marshalls Saxon paving slab 450 x 450 x 38 mm thick cut on site; two courses brick risers laid on flat stretcher bond; bricks PC £500.00/1000	9.12	2.75	50.19	-	14.77	m	64.96

Q PAVING/PLANTING/FENCING/SITE FURNITURE

Item Excluding site overheads and profit	PC £	Labour hours	Labour £	Plant £	Material £	Unit	Total rate £
Q25 SLAB/BRICK/SETT/COBBLE PAVINGS - cont'd							
Specialized mortars; SteinTec Ltd							
Primer for specialized mortar; "SteinTec Tuffbond" priming mortar immediately prior to paving material being placed on bedding mortar							
maximum thickness 1.5 mm (1.5 kg/m^2)	-	0.02	0.36	-	1.02	m^2	**1.39**
"Tuffbed"; 2 pack hydraulic mortar for paving applications							
30 mm thick	6.90	0.17	3.04	-	6.90	m^2	**9.94**
40 mm thick	9.20	0.20	3.65	-	9.20	m^2	**12.85**
50 mm thick	11.50	0.25	4.56	-	11.50	m^2	**16.06**
SteinTec jointing mortar; joint size							
100 x 100 x 100 mm; 10 mm joints; 31.24 kg/m^2	12.50	0.50	9.13	-	12.50	m^2	**21.62**
100 x 100 x 100 mm; 20 mm joints; 31.24 kg/m^2	22.00	1.00	18.25	-	22.00	m^2	**40.25**
200 x 100 x 65 mm; 10 mm joints; 15.7 kg/m^2	6.28	0.10	1.82	-	6.28	m^2	**8.11**
215 x 112.5 x 65 mm; 10 mm joints; 14.3 kg/m^2	5.72	0.08	1.52	-	5.72	m^2	**7.24**
300 x 450 x 65 mm; 10 mm joints; 6.24 kg/m^2	2.50	0.04	0.73	-	2.50	m^2	**3.23**
300 x 450 x 65 mm; 20 mm joints; 11.98 kg/m^2	4.79	0.10	1.82	-	4.79	m^2	**6.62**
450 x 450 x 65 mm; 20 mm joints; 9.75 kg/m^2	3.90	0.11	2.03	-	3.90	m^2	**5.93**
450 x 600 x 65 mm; 20 mm joints; 8.6 kg/m^2	3.44	0.09	1.66	-	3.44	m^2	**5.10**
600 x 600 x 65 mm; 20 mm joints; 7.43 kg/m^2	2.97	0.07	1.30	-	2.97	m^2	**4.28**
Q26 SPECIAL SURFACINGS/PAVINGS FOR SPORT/GENERAL AMENITY							
Market prices of surfacing materials							
Surfacings; Melcourt Industries Ltd (items labelled FSC are Forest Stewardship Council certified)							
Playbark 10/50®; per 25 m^3 load	-	-	-	-	57.10	m^3	-
Playbark 10/50®; per 80 m^3 load	-	-	-	-	44.10	m^3	-
Playbark 8/25®; per 25 m^3 load	-	-	-	-	55.60	m^3	-
Playbark 8/25®; per 80 m^3 load	-	-	-	-	42.60	m^3	-
Playchips®; per 25 m^3 load; FSC	-	-	-	-	39.75	m^3	-
Playchips®; per 80 m^3 load; FSC	-	-	-	-	25.75	m^3	-
Kushyfall; per 25 m^3 load; FSC	-	-	-	-	35.20	m^3	-
Kushyfall; per 80 m^3 load; FSC	-	-	-	-	22.20	m^3	-
Softfall; per 25 m^3 load	-	-	-	-	28.90	m^3	-
Softfall; per 80 m^3 load	-	-	-	-	15.90	m^3	-
Playsand; per 10 t load	-	-	-	-	94.86	m^3	-
Playsand; per 20 t load	-	-	-	-	82.62	m^3	-
Walk Chips; per 25 m^3 load; FSC	-	-	-	-	34.20	m^3	-
Walk Chips; per 80 m^3 load; FSC	-	-	-	-	21.70	m^3	-
Woodfibre; per 25 m^3 load; FSC	-	-	-	-	32.95	m^3	-
Woodfibre; per 80 m^3 load; FSC	-	-	-	-	19.95	m^3	-

Prices for Measured Works – Major Works

Q PAVING/PLANTING/FENCING/SITE FURNITURE

Item Excluding site overheads and profit	PC £	Labour hours	Labour £	Plant £	Material £	Unit	Total rate £
Natural sports pitches; Agripower Ltd; sports pitches; gravel raft construction							
Professional standard							
football pitch; 6500 m²	-	-	-	-	-	nr	150000.00
cricket wicket; county standard	-	-	-	-	-	nr	40000.00
Playing fields; Sport England Compliant							
football pitch; 6500 m²	-	-	-	-	-	nr	80000.00
cricket wicket; playing field standard	-	-	-	-	-	nr	25000.00
track and infield	-	-	-	-	-	nr	125000.00
Artificial surfaces and finishes - General							
Preamble: Advice should also be sought from the Technical Unit for Sport, the appropriate regional office of the Sports Council or the National Playing Fields Association. Some of the following prices include base work whereas others are for a specialist surface only on to a base prepared and costed separately.							
Artificial sports pitches; Agripower Ltd; third generation rubber crumb							
Sports pitches to respective National governing body standards; inclusive of all excavation, drainage, lighting, fencing, etc.							
football pitch; 6500 m²	-	-	-	-	-	nr	490000.00
rugby union multi-use pitch; 120 x 75 m	-	-	-	-	-	nr	590000.00
hockey pitch; water based; 101.4 x 63 m	-	-	-	-	-	nr	550000.00
athletic track; 8 lane; International amateur athletic federation	-	-	-	-	-	nr	750000.00
Multi-sport pitches; inclusive of all excavation, drainage, lighting, fencing, etc.							
sand dressed; 101.4 x 63 m	-	-	-	-	-	nr	390000.00
sand filled; 101.4 x 63 m	-	-	-	-	-	nr	340000.00
polymeric (synthetic bound rubber)	-	-	-	-	-	m²	130.00
macadam	-	-	-	-	-	m²	80.00
Sports areas; Baylis Landscape Contractors							
Sports tracks; polyurethane rubber surfacing; on bitumen-macadam (not included); prices for 5500 m² minimum							
"International"	-	-	-	-	-	m²	40.00
"Club Grade"	-	-	-	-	-	m²	31.00
Sports areas; multi-component polyurethane rubber surfacing; on bitumen-macadam; finished with polyurethane or acrylic coat; green or red							
"Permaprene"	-	-	-	-	-	m²	41.00
Sports areas; Exclusive Leisure Ltd							
Cricketweave artificial wicket system; including proprietary sub-base (all by specialist subcontractor)							
28 x 2.74 m; hard porous base	-	-	-	-	-	nr	6702.00
28 x 2.74 m; tarmac base	-	-	-	-	-	nr	8039.00

Q PAVING/PLANTING/FENCING/SITE FURNITURE

Item Excluding site overheads and profit	PC £	Labour hours	Labour £	Plant £	Material £	Unit	Total rate £
Q26 SPECIAL SURFACINGS/PAVINGS FOR SPORT/GENERAL AMENITY - cont'd							
Sports areas - cont'd							
Cricketweave practice batting end							
11 x 2 .74 m; hard porous base	-	-	-	-	-	each	3515.00
11 x 2 .74 m; tarmac base	-	-	-	-	-	each	4378.00
11 x 3.65 m; hard porous base	-	-	-	-	-	each	4584.00
11 x 3.65 m; tarmac base	-	-	-	-	-	each	5732.00
Cricketweave practice bowling end							
8 x 2.74 m; hard porous base	-	-	-	-	-	each	2639.00
8 x 2.74 m; tarmac base	-	-	-	-	-	each	3265.00
Supply and erection of single bay cricket cage							
18.3 x 3.65 x 3.6 m high	-	-	-	-	-	each	2599.00
Tennis courts; Duracourt (Spadeoak) Ltd							
Hard playing surfaces to SAPCA Code of Practice minimum requirements; laid on 65 mm thick macadam base on 150 mm thick stone foundation; to include lines, nets, posts, brick edging, 2.75 m high fence and perimeter drainage (excluding excavation, levelling or additional foundation); based on court size 36.6 x 18.3 m (670 m^2)							
"Durapore"; all weather porous macadam and acrylic colour coating	-	-	-	-	-	m^2	55.56
"Tiger Turf TT"; sand filled artificial grass	-	-	-	-	-	m^2	72.39
"Tiger Turf Grand Prix"; short pile sand filled artificial grass	-	-	-	-	-	m^2	75.76
"DecoColour"; impervious acrylic hardcourt (200 mm foundation)	-	-	-	-	-	m^2	70.71
"DecoTurf"; cushioned impervious acrylic tournament surface (200 mm foundation)	-	-	-	-	-	m^2	80.81
"Porous Kushion Kourt"; porous cushioned acrylic surface	-	-	-	-	-	m^2	79.12
"EasiClay"; synthetic clay system	-	-	-	-	-	m^2	82.49
"Canada Tenn"; American green clay; fast dry surface (no macadam but including irrigation)	-	-	-	-	-	m^2	84.17
Playgrounds; Baylis Landscape Contractors; "Rubaflex" in situ playground surfacing, porous; on prepared stone/granular base Type 1 (not included) and macadam base course (not included)							
Black							
15 mm thick (0.50 m critical fall height)	-	-	-	-	-	m^2	27.15
35 mm thick (1.00 m critical fall height)	-	-	-	-	-	m^2	41.85
60 mm thick (1.50 m critical fall height)	-	-	-	-	-	m^2	55.45

Q PAVING/PLANTING/FENCING/SITE FURNITURE

Item Excluding site overheads and profit	PC £	Labour hours	Labour £	Plant £	Material £	Unit	Total rate £
Playgrounds; Baylis Landscape Contractors; "Rubaflex" in situ playground surfacing, porous; on prepared stone/granular base Type 1 (not included) and macadam base course (not included)							
Coloured							
15 mm thick (0.50 m critical fall height)	-	-	-	-	-	m²	64.48
35 mm thick (1.00 m critical fall height)	-	-	-	-	-	m²	70.15
60 mm thick (1.50 m critical fall height)	-	-	-	-	-	m²	79.19
Playgrounds; Wicksteed Leisure Ltd							
Safety tiles; on prepared base (not included)							
1000 x 1000 x 60 mm; red or green	57.00	0.13	2.28	-	57.00	m²	59.28
1000 x 1000 x 60 mm; black	54.00	0.13	2.28	-	54.00	m²	56.28
1000 x 1000 x 43 mm; red or green	54.00	0.13	2.28	-	54.00	m²	56.28
1000 x 1000 x 43 mm; black	49.00	0.13	2.28	-	49.00	m²	51.28
Playgrounds; SMP Playgrounds Ltd							
Tiles; on prepared base (not included)							
Premier 25; 1000 x 1000 x 25 mm; black; for general use	46.00	0.20	3.65	-	54.25	m²	57.90
Premier 70; 1000 x 1000 x 70 mm; black; for higher equipment	80.00	0.20	3.65	-	88.25	m²	91.90
Playgrounds; Melcourt Industries Ltd; specifiers and users should contact the supplier for performance specifications of the materials below							
"Play surfaces"; on drainage layer (not included); minimum 300 mm settled depth							
Playbark® 8/25; 8 - 25 mm particles; red/brown	16.68	0.35	6.40	-	16.68	m²	23.08
Playbark 8/25; 8 - 25 mm particles; red/brown	-	0.35	6.40	-	14.06	m²	20.46
Playbark 10/50; 10 - 50 mm particles; red/brown	14.70	0.35	6.40	-	16.17	m²	22.57
Playchips; FSC graded woodchips	8.58	0.35	6.40	-	8.58	m²	14.98
Kushyfall; "fiberised" woodchips	7.40	0.35	6.40	-	7.40	m²	13.80
Softfall; conifer shavings	5.30	0.35	6.40	-	5.30	m²	11.70
Playgrounds; timber edgings							
Timber edging boards; 50 x 50 x 750 mm timber pegs at 1000 mm centres; excavations and hardcore under edgings (not included)							
50 x 150 mm; hardwood (iroko) edge boards	9.03	0.10	1.82	-	9.03	m	10.85
38 x 150 mm; hardwood (iroko) edge boards	7.17	0.10	1.82	-	7.17	m	9.00
50 x 150 mm; treated softwood edge boards	2.42	0.10	1.82	-	2.42	m	4.25
38 x 150 mm; treated softwood edge boards	2.23	0.10	1.82	-	2.23	m	4.06

Q PAVING/PLANTING/FENCING/SITE FURNITURE

Item Excluding site overheads and profit	PC £	Labour hours	Labour £	Plant £	Material £	Unit	Total rate £
Q30 SEEDING/TURFING							
Seeding/turfing - General							
Preamble: The following market prices generally reflect the manufacturer's recommended retail prices. Trade and bulk discounts are often available on the prices shown. The manufacturers of these products generally recommend application rates. Note: the following rates reflect the average rate for each product.							
Market prices of pre-seeding materials							
Rigby Taylor Ltd							
turf fertilizer; "Mascot Sports Amenity 9+7+7"	-	-	-	-	3.07	100 m²	-
Boughton Loam							
screened topsoil; 100 mm	-	-	-	-	63.00	m³	-
screened Kettering loam; 3 mm	-	-	-	-	99.00	m³	-
screened Kettering loam; sterilized; 3 mm	-	-	-	-	108.00	m³	-
top dressing; sand soil mixtures; 90/10 to 50/50	-	-	-	-	90.00	m³	-
Market prices of turf fertilizers; recommended "average" application rates							
Scotts UK; turf fertilizers							
grass fertilizer; slow release; "Sierraform GT Preseeder"; 18+22+05	-	-	-	-	5.01	100 m²	-
grass fertilizer; slow release; "Sierraform GT All Season"; 18+06+18	-	-	-	-	4.17	100 m²	-
grass fertilizer; slow release; "Sierraform GT Anti-Stress"; 15+00+26	-	-	-	-	5.01	100 m²	-
grass fertilizer; slow release; "Sierraform GT Momentum"; 22+05+11	-	-	-	-	3.34	100 m²	-
grass fertilizer; slow release; "Sierraform GT Spring Start"; 16+00+16	-	-	-	-	3.68	100 m²	-
grass fertilizer; water soluble; "Sierrasol"; 28+05+18+TE	-	-	-	-	11.68	100 m²	-
grass fertilizer; water soluble; "Sierrasol"; 20+05+30+TE	-	-	-	-	9.08	100 m²	-
grass fertilizer; controlled release; "Sierrablen"; 27+05+05 (8-9 months)	-	-	-	-	5.41	100 m²	-
grass fertilizer; controlled release; "Sierrablen"; 28+05+05 (5-6 months)	-	-	-	-	5.16	100 m²	-
grass fertilizer; controlled release; "Sierrablen"; 15+00+22 (5-6 months)	-	-	-	-	4.55	100 m²	-
grass fertilizer; "Sierrablen Fine"; 38+00+00 (2-3 months)	-	-	-	-	1.85	100 m²	-
grass fertilizer; "Sierrablen Mini"; 00+00+37 (5-6 months)	-	-	-	-	3.52	100 m²	-
grass fertilizer; "Greenmaster Pro-Lite Turf Tonic"; 08+00+00	-	-	-	-	2.50	100 m²	-
grass fertilizer; "Greenmaster Pro-Lite Spring & Summer"; 14+05+10	-	-	-	-	3.57	100 m²	-
grass fertilizer; "Greenmaster Pro-Lite Zero Phosphate"; 14+00+10	-	-	-	-	3.57	100 m²	-
grass fertilizer; "Greenmaster Pro-Lite Mosskiller"; 14+00+00	-	-	-	-	3.98	100 m²	-

Q PAVING/PLANTING/FENCING/SITE FURNITURE

Item Excluding site overheads and profit	PC £	Labour hours	Labour £	Plant £	Material £	Unit	**Total rate £**
grass fertilizer; "Greenmaster Pro-Lite Invigorator"; 04+00+08	-	-	-	-	2.70	100 m²	-
grass fertilizer; "Greenmaster Pro-Lite Extra"; 14+02+04	-	-	-	-	4.52	100 m²	-
grass fertilizer; "Greenmaster Pro-Lite Autumn"; 06+05+10	-	-	-	-	3.85	100 m²	-
grass fertilizer; "Greenmaster Pro-Lite Double K"; 07+00+14	-	-	-	-	3.85	100 m²	-
grass fertilizer; "Greenmaster Pro-Lite NK"; 12+00+12	-	-	-	-	3.85	100 m²	-
outfield turf fertilizer; "Sportsmaster Standard Spring & Summer"; 09+07+07	-	-	-	-	3.19	100 m²	-
outfield turf fertilizer; "Sportsmaster Standard Autumn"; 04+12+12	-	-	-	-	3.40	100 m²	-
outfield turf fertilizer; "Sportsmaster Standard Zero Phosphate"; 12+00+09	-	-	-	-	3.57	100 m²	-
outfield turf fertilizer; "Sportsmaster Specialities Municipal"; 15+05+15	-	-	-	-	5.59	100 m²	-
"TPMC"; tree planting and mulching compost	-	-	-	-	4.86	75 l	-
Rigby Taylor Ltd; 35 g/m²							
grass fertilizer; "Mascot Microfine"; 14+4+2 + 2% Mg	-	-	-	-	7.82	100 m²	-
grass fertilizer; "Mascot Microfine"; 12+0+10 + 2% Mg + 2% Fe	-	-	-	-	6.82	100 m²	-
grass fertilizer; "Mascot Microfine"; 8+0+6 + 2% Mg + 4% Fe	-	-	-	-	6.25	100 m²	-
grass fertilizer; "Mascot Microfine Organic OC1"; 8+0+0 + 2% Fe	-	-	-	-	7.01	100 m²	-
grass fertilizer; "Mascot Microfine Organic OC2"; 5+2+10	-	-	-	-	8.07	100 m²	-
grass fertilizer; "Mascot Guardian"; 6+1+12 + 2% Mg + 2% Fe + seaweed meal, amino acids and ERD (enhanced root developer)	-	-	-	-	9.56	100 m²	-
grass fertilizer; "Mascot Premier-HG"; 12+3+24	-	-	-	-	7.79	100 m²	-
grass fertilizer; "Mascot Fine Turf"; 11+5+5	-	-	-	-	5.62	100 m²	-
grass fertilizer; "Mascot Fine Turf"; 12+0+9 + 1% Mg + 1% Fe + seaweed	-	-	-	-	6.20	100 m²	-
outfield fertilizer; "Mascot Sports Amenity"; 9+7+7	-	-	-	-	4.30	100 m²	-
outfield fertilizer; "Mascot Sports Amenity"; 7+7+7	-	-	-	-	3.95	100 m²	-
outfield fertilizer; "Mascot Sports Amenity"; 15+5+10	-	-	-	-	4.72	100 m²	-
outfield fertilizer; "Mascot Sports Amenity"; 3+12+12	-	-	-	-	4.50	100 m²	-
liquid fertilizer; "Mascot Microflow-CX"; 4+2+18 + trace elements; 200-1400 ml /100 m²	-	-	-	-	6.03	100 m²	-
liquid fertilizer; "Mascot Microflow-CX"; 12+0+8 + trace elements; 200-1400 ml /100 m²	-	-	-	-	5.47	100 m²	-
liquid fertilizer; "Mascot Microflow-C"; 25+0+0 + trace elements; 200-1400 ml /100 m²	-	-	-	-	5.70	100 m²	-

Q PAVING/PLANTING/FENCING/SITE FURNITURE

Item Excluding site overheads and profit	PC £	Labour hours	Labour £	Plant £	Material £	Unit	Total rate £
Q30 SEEDING/TURFING - cont'd							
Market prices of grass seed Preamble: The prices shown are for supply only at one number 20 kg or 25 kg bag purchase price unless otherwise stated. Rates shown are based on the manufacturer's maximum recommendation for each seed type. Trade and bulk discounts are often available on the prices shown for quantities of more than one bag. Bowling greens; fine lawns; ornamental turf; croquet lawns							
"British Seed Houses"; ref A1 Greens; 35 g/m²	-	-	-	-	21.85	100 m²	-
"DLF Trifolium"; ref J Green; 34 - 50 g/m²	-	-	-	-	38.75	100 m²	-
"DLF Trifolium"; ref J Premier Green; 34 - 50 g/m²	-	-	-	-	61.85	100 m²	-
"DLF Trifolium"; ref Promaster 20; 35 - 50 g/m²	-	-	-	-	20.20	100 m²	-
"DLF Trifolium"; ref Promaster 10; 35 - 50 g/m²	-	-	-	-	34.15	100 m²	-
Tennis courts; cricket squares							
"British Seed Houses"; ref A2 Lawns & Tennis; 35 g/m²	-	-	-	-	16.10	100 m²	-
"British Seed Houses"; ref A5 Cricket Square; 35 g/m²	-	-	-	-	25.11	100 m²	-
"DLF Trifolium"; ref J Premier Green; 34 - 50 g/m²	-	-	-	-	61.85	100 m²	-
"DLF Trifolium"; ref J Court; 18 - 25 g/m²	-	-	-	-	12.68	100 m²	-
"DLF Trifolium"; ref Promaster 35; 35 g/m²	-	-	-	-	14.35	100 m²	-
Amenity grassed areas; general purpose lawns							
"British Seed Houses"; ref A3 Landscape; 25 - 50 g/m²	-	-	-	-	23.18	100 m²	-
"DLF Trifolium"; ref J Rye Fairway; 18 - 25 g/m²	-	-	-	-	14.68	100 m²	-
"DLF Trifolium"; ref J Court; 18 - 25 g/m²	-	-	-	-	12.68	100 m²	-
"DLF Trifolium"; ref Promaster 50; 25 - 35 g/m²	-	-	-	-	13.51	100 m²	-
"DLF Trifolium"; ref Promaster 120; 25 - 35 g/m²	-	-	-	-	14.21	100 m²	-
"Rigby Taylor"; ref Mascot R15 General Landscape; 35 g/m²	-	-	-	-	21.30	100 m²	-
Conservation; country parks; slopes and banks							
"British Seed Houses"; ref A4 Low-maintenance; 17 - 35 g/m²	-	-	-	-	16.59	100 m²	-
"British Seed Houses"; ref A16 Country Park; 8 - 19 g/m²	-	-	-	-	9.74	100 m²	-
"British Seed Houses"; ref A17 Legume and Clover; 2 g/m²	-	-	-	-	24.16	100 m²	-

Q PAVING/PLANTING/FENCING/SITE FURNITURE

Item Excluding site overheads and profit	PC £	Labour hours	Labour £	Plant £	Material £	Unit	Total rate £
Shaded areas							
"British Seed Houses"; ref A6 Supra Shade; 50 g/m^2	-	-	-	-	32.54	100 m^2	-
"DLF Trifolium"; ref J Green; 34 - 50 g/m^2	-	-	-	-	38.75	100 m^2	-
"DLF Trifolium"; ref Promaster 60; 35 - 50 g/m^2	-	-	-	-	24.15	100 m^2	-
"Rigby Taylor"; ref Mascot R18 Shade/ Drought Tolerance; 35 g/m^2	-	-	-	-	34.52	100 m^2	-
Sports pitches; rugby; soccer pitches							
"British Seed Houses"; ref A7 Sportsground; 20 g/m^2	-	-	-	-	12.41	100 m^2	-
"DLF Trifolium"; ref J Pitch; 18 - 30 g/m^2	-	-	-	-	12.45	100 m^2	-
"DLF Trifolium"; ref Promaster 70; 15 - 35 g/m^2	-	-	-	-	12.25	100 m^2	-
"DLF Trifolium"; ref Promaster 75; 15 - 35 g/m^2	-	-	-	-	16.70	100 m^2	-
"DLF Trifolium"; ref Promaster 80; 17 - 35 g/m^2	-	-	-	-	12.14	100 m^2	-
"Rigby Taylor"; ref Mascot R11 Football & Rugby; 35 g/m^2	-	-	-	-	18.41	100 m^2	-
"Rigby Taylor"; ref Mascot R12 General Playing Fields; 35 g/m^2	-	-	-	-	19.19	100 m^2	-
Outfields							
"British Seed Houses"; ref A7 Sportsground; 20 g/m^2	-	-	-	-	12.41	100 m^2	-
"British Seed Houses"; ref A9 Outfield; 17 - 35 g/m^2	-	-	-	-	17.10	100 m^2	-
"DLF Trifolium"; ref J Fairway; 12 - 25 g/m^2	-	-	-	-	12.70	100 m^2	-
"DLF Trifolium"; ref Promaster 40; 35 g/m^2	-	-	-	-	15.15	100 m^2	-
"DLF Trifolium"; ref Promaster 70; 15 - 35 g/m^2	-	-	-	-	12.25	100 m^2	-
"Rigby Taylor"; ref Mascot R4 Cricket Outfields; 35 g/m^2	-	-	-	-	25.72	100 m^2	-
Hockey pitches							
"DLF Trifolium"; ref J Fairway; 12 - 25 g/m^2	-	-	-	-	12.70	100 m^2	-
"DLF Trifolium"; ref Promaster 70; 15 - 35 g/m^2	-	-	-	-	12.25	100 m^2	-
"Rigby Taylor"; ref Mascot R10 Cricket & Hockey Outfield; 35 g/m^2	-	-	-	-	20.39	100 m^2	-
Parks							
"British Seed Houses"; ref A7 Sportsground 20 g/m^2	-	-	-	-	12.41	100 m^2	-
"British Seed Houses"; ref A9 Outfield; 17 - 35 g/m^2	-	-	-	-	17.10	100 m^2	-
"DLF Trifolium"; ref Promaster 120; 25 - 35 g/m^2	-	-	-	-	14.21	100 m^2	-
"Rigby Taylor"; ref Mascot R16 Landscape & Ornamental; 35 g/m^2	-	-	-	-	21.26	100 m^2	-
Informal playing fields							
"DLF Trifolium"; ref J Rye Fairway; 18 - 25 g/m^2	-	-	-	-	14.68	100 m^2	-
"DLF Trifolium"; ref Promaster 45; 35 g/m^2	-	-	-	-	14.60	100 m^2	-
Caravan sites							
"British Seed Houses"; ref A9 Outfield; 17 - 35 g/m^2	-	-	-	-	17.10	100 m^2	-

Q PAVING/PLANTING/FENCING/SITE FURNITURE

Item Excluding site overheads and profit	PC £	Labour hours	Labour £	Plant £	Material £	Unit	Total rate £
Q30 SEEDING/TURFING - cont'd							
Market prices of grass seed - cont'd							
Sports pitch re-seeding and repair							
"British Seed Houses"; ref A8 Pitch Renovator; 20 - 35 g/m^2	-	-	-	-	15.93	100 m^2	-
"British Seed Houses"; ref A20 Ryesport; 20 - 35 g/m^2	-	-	-	-	16.11	100 m^2	-
"DLF Trifolium"; ref J Pitch; 18 - 30 g/m^2	-	-	-	-	12.45	100 m^2	-
"DLF Trifolium"; ref Promaster 80; 17 - 35 g/m^2	-	-	-	-	12.14	100 m^2	-
"DLF Trifolium"; ref Promaster 81; 17 - 35 g/m^2	-	-	-	-	12.85	100 m^2	-
"Rigby Taylor"; ref Mascot R14 Premier Winter Games Renovation; 35 g/m^2	-	-	-	-	18.88	100 m^2	-
Racecourses; gallops; polo grounds; horse rides							
"British Seed Houses"; ref A14 Racecourse; 25 - 30 g/m^2	-	-	-	-	15.51	100 m^2	-
"DLF Trifolium"; ref J Court; 18 - 25 g/m^2	-	-	-	-	12.68	100 m^2	-
"DLF Trifolium"; ref Promaster 65; 17 - 35 g/m^2	-	-	-	-	20.30	100 m^2	-
Motorway and road verges							
"British Seed Houses"; ref A18 Road Verge; 6 - 15 g/m^2	-	-	-	-	7.22	100 m^2	-
"DLF Trifolium"; ref Promaster 85; 10 g/m^2	-	-	-	-	4.20	100 m^2	-
"DLF Trifolium"; ref Promaster 120; 25 - 35 g/m^2	-	-	-	-	14.21	100 m^2	-
Golf courses; tees							
"British Seed Houses"; ref A10 Golf Tee, 35 - 50 g/m^2	-	-	-	-	29.85	100 m^2	-
"DLF Trifolium"; ref J Premier Fairway; 18 - 30 g/m^2	-	-	-	-	22.29	100 m^2	-
"DLF Trifolium"; ref J Court; 18 - 25 g/m^2	-	-	-	-	12.68	100 m^2	-
"DLF Trifolium"; ref Promaster 40; 35 g/m^2	-	-	-	-	15.15	100 m^2	-
"DLF Trifolium"; ref Promaster 45; 35 g/m^2	-	-	-	-	14.60	100 m^2	-
"Rigby Taylor"; ref Mascot R4 Golf Tees & Fairways; 35 g/m^2	-	-	-	-	51.45	100 m^2	-
Golf courses; greens							
"British Seed Houses"; ref A11 Golf Green; 35 g/m^2	-	-	-	-	22.30	100 m^2	-
"British Seed Houses"; ref A13 Golf Roughs; 8 g/m^2	-	-	-	-	3.94	100 m^2	-
"DLF Trifolium"; ref J All Bent; 8 g/m^2	-	-	-	-	13.10	100 m^2	-
"DLF Trifolium"; ref J Green; 34 - 50 g/m^2	-	-	-	-	38.75	100 m^2	-
"DLF Trifolium"; ref Promaster 5; 35 - 50 g/m^2	-	-	-	-	21.70	100 m^2	-
"Rigby Taylor"; ref Mascot R1 Greenkeeper; 35 g/m^2	-	-	-	-	65.20	100 m^2	-

Q PAVING/PLANTING/SITE FURNITURE

Item Excluding site overheads and profit	PC £	Labour hours	Labour £	Plant £	Material £	Unit	Total rate £
Golf courses; fairways							
"British Seed Houses"; ref A12 Golf Fairway; 15 - 25 g/m²	-	-	-	-	13.54	100 m²	-
"DLF Trifolium"; ref J Rye Fairway; 18 - 30 g/m²	-	-	-	-	17.61	100 m²	-
"DLF Trifolium"; ref J Premier Fairway; 18 - 30 g/m²	-	-	-	-	22.29	100 m²	-
"DLF Trifolium"; ref Promaster 40; 35 g/m²	-	-	-	-	15.15	100 m²	-
"DLF Trifolium"; ref Promaster 45; 35 g/m²	-	-	-	-	14.60	100 m²	-
"Rigby Taylor"; ref Mascot R6 Fescue Rye Fairway; 35 g/m²	-	-	-	-	50.81	100 m²	-
Golf courses; roughs							
"DLF Trifolium"; ref Promaster 25; 17 - 35 g/m²	-	-	-	-	13.86	100 m²	-
"Rigby Taylor"; ref Mascot R127 Golf Links & Rough; 35 g/m²	-	-	-	-	4.54	100 m²	-
Waste land; spoil heaps; quarries							
"British Seed Houses"; ref A15 Reclamation; 15 - 20 g/m²	-	-	-	-	8.94	100 m²	-
"DLF Trifolium"; ref Promaster 95; 12 - 35 g/m²	-	-	-	-	16.10	100 m²	-
"DLF Trifolium"; ref Promaster 105; 5 g/m²	-	-	-	-	4.41	100m²	-
Low maintenance; housing estates; amenity grassed areas							
"British Seed Houses"; ref A19 Housing Estate; 25 - 35 g/m²	-	-	-	-	16.08	100 m²	-
"British Seed Houses"; ref A22 Low Maintenance; 25 - 35 g/m²	-	-	-	-	18.85	100 m²	-
"DLF Trifolium"; ref Promaster 120; 25 - 35 g/m²	-	-	-	-	14.21	100 m²	-
Saline coastal; roadside areas							
"British Seed Houses"; ref A21 Coastal/ Saline Restoration; 15 - 20 g/m²	-	-	-	-	11.18	100 m²	-
"DLF Trifolium"; ref Promaster 90; 15 - 35 g/m²	-	-	-	-	14.70	100 m²	-
Turf production							
"British Seed Houses"; ref A25 Ley Mixture; 160 kg/ha	-	-	-	-	741.60	ha	-
"British Seed Houses"; ref A24 Wear & Tear; 185 kg/ha	-	-	-	-	861.05	ha	-
Market prices of wild flora seed mixtures							
Acid soils							
"British Seed Houses"; ref WF1 (Annual Flowering); 1.00 - 2.00 g/m²	-	-	-	-	21.38	100 m²	-
Neutral soils							
"British Seed Houses"; ref WF3 (Neutral Soils); 0.50 - 1.00 g/m²	-	-	-	-	10.69	100 m²	-
Market prices of wild flora and grass seed mixtures							
General purpose							
"DLF Trifolium"; ref Pro Flora 8 Old English Country Meadow Mix; 5 g/m²	-	-	-	-	12.50	100 m²	-
"DLF Trifolium"; ref Pro Flora 9 General purpose; 5 g/m²	-	-	-	-	10.00	100 m²	-

Q PAVING/PLANTING/FENCING/SITE FURNITURE

Item Excluding site overheads and profit	PC £	Labour hours	Labour £	Plant £	Material £	Unit	Total rate £
Q30 SEEDING/TURFING - cont'd							
Market prices of wild flora and grass seed mixtures - cont'd							
Acid soils							
"British Seed Houses"; ref WFG2 (Annual Meadow); 5 g/m^2	-	-	-	-	25.00	100 m^2	-
"DLF Trifolium"; ref Pro Flora 2 Acidic soils; 5 g/m^2	-	-	-	-	11.25	100 m^2	-
Neutral soils							
"British Seed Houses"; ref WFG4 (Neutral Meadow); 5 g/m^2	-	-	-	-	27.20	100 m^2	-
"British Seed Houses"; ref WFG13 (Scotland); 5 g/m^2	-	-	-	-	24.30	100 m^2	-
"DLF Trifolium"; ref Pro Flora 3 Damp loamy soils; 5 g/m^2	-	-	-	-	15.75	100 m^2	-
Calcareous soils							
"British Seed Houses"; ref WFG5 (Calcareous Soils); 5 g/m^2	-	-	-	-	25.60	100 m^2	-
"DLF Trifolium"; ref Pro Flora 4 Calcareous soils; 5 g/m^2	-	-	-	-	13.50	100 m^2	-
Heavy clay soils							
"British Seed Houses"; ref WFG6 (Clay Soils); 5 g/m^2	-	-	-	-	30.25	100 m^2	-
"British Seed Houses"; ref WFG12 (Ireland); 5 g/m^2	-	-	-	-	24.38	100 m^2	-
"DLF Trifolium"; ref Pro Flora 5 Wet loamy soils; 5 g/m^2	-	-	-	-	33.00	100 m^2	-
Sandy soils							
"British Seed Houses"; ref WFG7 (Free Draining Soils); 5 g/m^2	-	-	-	-	35.00	100 m^2	-
"British Seed Houses"; ref WFG11 (Ireland); 5 g/m^2	-	-	-	-	25.50	100 m^2	-
"British Seed Houses"; ref WFG14 (Scotland); 5 g/m^2	-	-	-	-	27.75	100 m^2	-
"DLF Trifolium"; ref Pro Flora 6 Dry free draining loamy soils; 5 g/m^2	-	-	-	-	29.25	100 m^2	-
Shaded areas							
"British Seed Houses"; ref WFG8 (Woodland and Hedgerow); 5 g/m^2	-	-	-	-	26.75	100 m^2	-
"DLF Trifolium"; ref Pro Flora 7 Hedgerow and light shade; 5 g/m^2	-	-	-	-	33.75	100 m^2	-
Educational							
"British Seed Houses"; ref WFG15 (Schools and Colleges); 5 g/m^2	-	-	-	-	36.50	100 m^2	-
Wetlands							
"British Seed Houses"; ref WFG9 (Wetlands and Ponds); 5 g/m^2	-	-	-	-	32.50	100 m^2	-
"DLF Trifolium"; ref Pro Flora 5 Wet loamy soils; 5 g/m^2	-	-	-	-	33.00	100 m^2	-
Scrub and moorland							
"British Seed Houses"; ref WFG10 (Cornfield Annuals); 5 g/m^2	-	-	-	-	35.55	100 m^2	-
Hedgerow							
"DLF Trifolium"; ref Pro Flora 7 Hedgerow and light shade; 5 g/m^2	-	-	-	-	33.75	100 m^2	-
Vacant sites							
"DLF Trifolium"; ref Pro Flora 1 Cornfield annuals; 5 g/m^2	-	-	-	-	30.00	100 m^2	-

Q PAVING/PLANTING/FENCING/SITE FURNITURE

Item Excluding site overheads and profit	PC £	Labour hours	Labour £	Plant £	Material £	Unit	Total rate £
Regional Environmental mixes							
"British Seed Houses"; ref RE1 (Traditional Hay); 5 g/m²	-	-	-	-	27.45	100 m²	-
"British Seed Houses"; ref RE2 (Lowland Meadow); 5 g/m²	-	-	-	-	35.25	100 m²	-
"British Seed Houses"; ref RE3 (Riverflood Plain/Water Meadow); 5 g/m²	-	-	-	-	34.75	100 m²	-
"British Seed Houses"; ref RE4 (Lowland Limestone); 5 g/m²	-	-	-	-	34.25	100 m²	-
"British Seed Houses"; ref RE5 (Calcareous Sub-mountain Restoration); 5 g/m²	-	-	-	-	38.60	100 m²	-
"British Seed Houses"; ref RE6 (Upland Limestone); 5 g/m²	-	-	-	-	48.95	100 m²	-
"British Seed Houses"; ref RE7 (Acid Sub-mountain Restoration); 5 g/m²	-	-	-	-	33.75	100 m²	-
"British Seed Houses"; ref RE8 (Coastal Reclamation); 5 g/m²	-	-	-	-	42.95	100 m²	-
"British Seed Houses"; ref RE9 (Farmland Mixture); 5 g/m²	-	-	-	-	30.35	100 m²	-
"British Seed Houses"; ref RE10 (Marginal Land); 5 g/m²	-	-	-	-	44.25	100 m²	-
"British Seed Houses"; ref RE11 (Heath Scrubland); 5 g/m²	-	-	-	-	15.30	100 m²	-
"British Seed Houses"; ref RE12 (Drought Land); 5 g/m²	-	-	-	-	42.00	100 m²	-
Cultivation							
Ripping up subsoil; using approved subsoiling machine; minimum depth 250 mm below topsoil; at 1.20 m centres; in							
gravel or sandy clay	-	-	-	2.40	-	100 m²	2.40
soil compacted by machines	-	-	-	2.80	-	100 m²	2.80
clay	-	-	-	3.00	-	100 m²	3.00
chalk or other soft rock	-	-	-	6.01	-	100 m²	6.01
Extra for subsoiling at 1 m centres	-	-	-	0.60	-	100 m²	0.60
Breaking up existing ground; using pedestrian operated tine cultivator or rotavator							
100 mm deep	-	0.22	4.01	2.12	-	100 m²	6.14
150 mm deep	-	0.28	5.02	2.65	-	100 m²	7.67
200 mm deep	-	0.37	6.69	3.54	-	100 m²	10.23
As above but in heavy clay or wet soils							
100 mm deep	-	0.44	8.03	4.25	-	100 m²	12.28
150 mm deep	-	0.66	12.04	6.37	-	100 m²	18.41
200 mm deep	-	0.82	15.06	7.96	-	100 m²	23.02
Breaking up existing ground; using tractor drawn tine cultivator or rotavator Single pass							
100 mm deep	-	-	-	0.28	-	100 m²	0.28
150 mm deep	-	-	-	0.36	-	100 m²	0.36
200 mm deep	-	-	-	0.47	-	100 m²	0.47
600 mm deep	-	-	-	1.42	-	100 m²	1.42
Cultivating ploughed ground; using disc, drag or chain harrow							
4 passes	-	-	-	1.71	-	100 m²	1.71
Rolling cultivated ground lightly; using self-propelled agricultural roller	-	0.06	1.01	0.60	-	100 m²	1.62

Q PAVING/PLANTING/FENCING/SITE FURNITURE

Item Excluding site overheads and profit	PC £	Labour hours	Labour £	Plant £	Material £	Unit	Total rate £
Q30 SEEDING/TURFING - cont'd							
Cultivation - cont'd							
Importing and storing selected and approved topsoil; from source not exceeding 13 km from site; inclusive of settlement							
small quantities (less than 15 m³)	40.00	-	-	-	48.00	m³	-
over 15 m³	26.00	-	-	-	31.20	m³	-
Spreading and lightly consolidating approved topsoil (imported or from spoil heaps); in layers not exceeding 150 mm; travel distance from spoil heaps not exceeding 100 m; by machine (imported topsoil not included)							
minimum depth 100 mm	-	1.55	28.29	36.62	-	100 m²	64.91
minimum depth 150 mm	-	2.33	42.58	55.11	-	100 m²	97.69
minimum depth 300 mm	-	4.67	85.16	110.21	-	100 m²	195.37
minimum depth 450 mm	-	6.99	127.57	165.14	-	100 m²	292.71
Spreading and lightly consolidating approved topsoil (imported or from spoil heaps); in layers not exceeding 150 mm; travel distance from spoil heaps not exceeding 100 m; by hand (imported topsoil not included)							
minimum depth 100 mm	-	20.00	365.07	-	-	100 m²	365.07
minimum depth 150 mm	-	30.01	547.61	-	-	100 m²	547.61
minimum depth 300 mm	-	60.01	1095.22	-	-	100 m²	1095.22
minimum depth 450 mm	-	90.02	1642.83	-	-	100 m²	1642.83
Extra over for spreading topsoil to slopes 15 - 30°; by machine or hand	-	-	-	-	-	10%	-
Extra over for spreading topsoil to slopes over 30°; by machine or hand	-	-	-	-	-	25%	-
Extra over for spreading topsoil from spoil heaps; travel exceeding 100 m; by machine							
100 - 150 m	-	0.01	0.22	0.06	-	m³	0.28
150 - 200 m	-	0.02	0.34	0.09	-	m³	0.43
200 - 300 m	-	0.03	0.51	0.14	-	m³	0.64
Extra over spreading topsoil for travel exceeding 100 m; by hand							
100 m	-	0.83	15.21	-	-	m³	15.21
200 m	-	1.67	30.42	-	-	m³	30.42
300 m	-	2.50	45.63	-	-	m³	45.63
Evenly grading; to general surfaces to bring to finished levels							
by machine (tractor mounted rotavator)	-	-	-	0.01	-	m²	0.01
by pedestrian operated rotavator	-	-	0.07	0.04	-	m²	0.12
by hand	-	0.01	0.18	-	-	m²	0.18
Extra over grading for slopes 15 - 30°; by machine or hand	-	-	-	-	-	10%	-
Extra over grading for slopes over 30°; by machine or hand	-	-	-	-	-	25%	-
Apply screened topdressing to grass surfaces; spread using Tru-Lute							
sand soil mixes 90/10 to 50/50	-	-	0.04	0.03	0.14	m²	0.20
Spread only existing cultivated soil to final levels using Tru-Lute							
cultivated soil	-	-	0.04	0.03	-	m²	0.07

Prices for Measured Works – Major Works

Q PAVING/PLANTING/FENCING/SITE FURNITURE

Item Excluding site overheads and profit	PC £	Labour hours	Labour £	Plant £	Material £	Unit	Total rate £
Clearing stones; disposing off site; to distance not exceeding 13 km							
by hand; stones not exceeding 50 mm in any direction; loading to skip 4.6 m³	-	0.01	0.18	0.04	-	m²	0.22
by mechanical stone rake; stones not exceeding 50 mm in any direction; loading to 15 m³ truck by mechanical loader	-	-	0.04	0.08	-	m²	0.11
Lightly cultivating; weeding; to fallow areas; disposing debris off site; to distance not exceeding 13 km							
by hand	-	0.01	0.26	-	0.10	m²	0.36
Surface applications; soil additives; pre-seeding; material delivered to a maximum of 25 m from area of application; applied; by machine							
Soil conditioners; to cultivated ground; ground limestone; PC £27.54/tonne; including turning in							
0.25 kg/m² = 2.50 tonnes/ha	0.69	-	-	2.92	0.69	100 m²	3.61
0.50 kg/m² = 5.00 tonnes/ha	1.38	-	-	2.92	1.38	100 m²	4.30
0.75 kg/m² = 7.50 tonnes/ha	2.07	-	-	2.92	2.07	100 m²	4.99
1.00 kg/m² = 10.00 tonnes/ha	2.75	-	-	2.92	2.75	100 m²	5.68
Soil conditioners; to cultivated ground; medium bark; based on deliveries of 25 m³ loads; PC £35.65/m³; including turning in							
1 m³ per 40 m² = 25 mm thick	0.89	-	-	0.09	0.89	m²	0.98
1 m³ per 20 m² = 50 mm thick	1.78	-	-	0.14	1.78	m²	1.92
1 m³ per 13.33 m² = 75 mm thick	2.67	-	-	0.19	2.67	m²	2.87
1 m³ per 10 m² = 100 mm thick	3.56	-	-	0.25	3.56	m²	3.82
Soil conditioners; to cultivated ground; mushroom compost; delivered in 25 m³ loads; PC £21.20/m³; including turning in							
1 m³ per 40 m² = 25 mm thick	0.53	0.02	0.30	0.09	0.53	m²	0.92
1 m³ per 20 m² = 50 mm thick	1.06	0.03	0.56	0.14	1.06	m²	1.76
1 m³ per 13.33 m² = 75 mm thick	1.59	0.04	0.73	0.19	1.59	m²	2.51
1 m³ per 10 m² = 100 mm thick	2.12	0.05	0.91	0.25	2.12	m²	3.28
Soil conditioners; to cultivated ground; mushroom compost; delivered in 35 m³ loads; PC £11.75/m³; including turning in							
1 m³ per 40 m² = 25 mm thick	0.29	0.02	0.30	0.09	0.29	m²	0.68
1 m³ per 20 m² = 50 mm thick	0.59	0.03	0.56	0.14	0.59	m²	1.29
1 m³ per 13.33 m² = 75 mm thick	0.88	0.04	0.73	0.19	0.88	m²	1.81
1 m³ per 10 m² = 100 mm thick	1.18	0.05	0.91	0.25	1.18	m²	2.34
Surface applications and soil additives; pre-seeding; material delivered to a maximum of 25 m from area of application; applied; by hand							
Soil conditioners; to cultivated ground; ground limestone; PC £27.54/tonne; including turning in							
0.25 kg/m² = 2.50 tonnes/ha	0.69	1.20	21.90	-	0.69	100 m²	22.59
0.50 kg/m² = 5.00 tonnes/ha	1.38	1.33	24.33	-	1.38	100 m²	25.71
0.75 kg/m² = 7.50 tonnes/ha	2.07	1.50	27.38	-	2.07	100 m²	29.44
1.00 kg/m² = 10.00 tonnes/ha	2.75	1.71	31.28	-	2.75	100 m²	34.04

Q PAVING/PLANTING/FENCING/SITE FURNITURE

Item Excluding site overheads and profit	PC £	Labour hours	Labour £	Plant £	Material £	Unit	Total rate £
Q30 SEEDING/TURFING - cont'd							
Surface applications and soil additives - cont'd							
Soil conditioners; to cultivated ground; medium bark; based on deliveries of 25 m³ loads; PC £35.65/m³; including turning in							
1 m³ per 40 m² = 25 mm thick	0.89	0.02	0.41	-	0.89	m²	1.30
1 m³ per 20 m² = 50 mm thick	1.78	0.04	0.81	-	1.78	m²	2.59
1 m³ per 13.33 m² = 75 mm thick	2.67	0.07	1.22	-	2.67	m²	3.89
1 m³ per 10 m² = 100 mm thick	3.56	0.08	1.46	-	3.56	m²	5.03
Soil conditioners; to cultivated ground; mushroom compost; delivered in 25 m³ loads; PC £21.20/m³; including turning in							
1 m³ per 40 m² = 25 mm thick	0.53	0.02	0.41	-	0.53	m²	0.94
1 m³ per 20 m² = 50 mm thick	1.06	0.04	0.81	-	1.06	m²	1.87
1 m³ per 13.33 m² = 75 mm thick	1.59	0.07	1.22	-	1.59	m²	2.81
1 m³ per 10 m² = 100 mm thick	2.12	0.08	1.46	-	2.12	m²	3.58
Soil conditioners; to cultivated ground; mushroom compost; delivered in 60 m³ loads; PC £9.50/m³; including turning in							
1 m³ per 40 m² = 25 mm thick	0.24	0.02	0.41	-	0.24	m²	0.64
1 m³ per 20 m²= 50 mm thick	0.42	0.04	0.81	-	0.42	m²	1.23
1 m³ per 13.33 m² = 75 mm thick	0.71	0.07	1.22	-	0.71	m²	1.93
1 m³ per 10 m² = 100 mm thick	0.95	0.08	1.46	-	0.95	m²	2.41
Preparation of seedbeds - General							
Preamble: For preliminary operations see "Cultivation" section.							
Preparation of seedbeds; soil preparation							
Lifting selected and approved topsoil from spoil heaps; passing through 6 mm screen; removing debris	-	0.08	1.52	4.88	0.01	m³	6.41
Topsoil; supply only; PC £26.00/m³; allowing for 20% settlement							
25 mm	-	-	-	-	0.78	m²	-
50 mm	-	-	-	-	1.56	m²	-
100 mm	-	-	-	-	3.12	m²	-
150 mm	-	-	-	-	4.68	m²	-
200 mm	-	-	-	-	6.24	m²	-
250 mm	-	-	-	-	7.80	m²	-
300 mm	-	-	-	-	9.36	m²	-
400 mm	-	-	-	-	12.48	m²	-
450 mm	-	-	-	-	14.04	m²	-
Spreading topsoil to form seedbeds (topsoil not included); by machine							
25 mm deep	-	-	0.05	0.10	-	m²	0.14
50 mm deep	-	-	0.06	0.12	-	m²	0.18
75 mm deep	-	-	0.07	0.14	-	m²	0.21
100 mm deep	-	0.01	0.09	0.19	-	m²	0.28
150 mm deep	-	0.01	0.14	0.29	-	m²	0.42
Spreading only topsoil to form seedbeds (topsoil not included); by hand							
25 mm deep	-	0.03	0.46	-	-	m²	0.46
50 mm deep	-	0.03	0.61	-	-	m²	0.61
75 mm deep	-	0.04	0.78	-	-	m²	0.78
100 mm deep	-	0.05	0.91	-	-	m²	0.91
150 mm deep	-	0.08	1.37	-	-	m²	1.37

Q PAVING/PLANTING/SITE FURNITURE

Item Excluding site overheads and profit	PC £	Labour hours	Labour £	Plant £	Material £	Unit	Total rate £
Bringing existing topsoil to a fine tilth for seeding; by raking or harrowing; stones not to exceed 6 mm; by machine	-	-	0.07	0.04	-	m²	0.12
Bringing existing topsoil to a fine tilth for seeding; by raking or harrowing; stones not to exceed 6 mm; by hand	-	0.01	0.16	-	-	m²	0.16
Preparation of seedbeds; soil treatments							
For the following operations add or subtract the following amounts for every £0.10 difference in the material cost price							
35 g/m²	-	-	-	-	0.35	100 m²	-
50 g/m²	-	-	-	-	0.50	100 m²	-
70 g/m²	-	-	-	-	0.70	100 m²	-
100 g/m²	-	-	-	-	1.00	100 m²	-
125 kg/ha	-	-	-	-	12.50	ha	-
150 kg/ha	-	-	-	-	15.00	ha	-
175 kg/ha	-	-	-	-	17.50	ha	-
200 kg/ha	-	-	-	-	20.00	ha	-
225 kg/ha	-	-	-	-	22.50	ha	-
250 kg/ha	-	-	-	-	25.00	ha	-
300 kg/ha	-	-	-	-	30.00	ha	-
350 kg/ha	-	-	-	-	35.00	ha	-
400 kg/ha	-	-	-	-	40.00	ha	-
500 kg/ha	-	-	-	-	50.00	ha	-
700 kg/ha	-	-	-	-	70.00	ha	-
1000 kg/ha	-	-	-	-	100.00	ha	-
1250 kg/ha	-	-	-	-	125.00	ha	-
Pre-seeding fertilizers (12+00+09); PC £0.85/kg; to seedbeds; by machine							
35 g/m²	2.38	-	-	0.19	2.38	100 m²	2.57
50 g/m²	3.40	-	-	0.19	3.40	100 m²	3.59
70 g/m²	4.76	-	-	0.19	4.76	100 m²	4.95
100 g/m²	6.80	-	-	0.19	6.80	100 m²	6.99
Pre-seeding fertilizers (12+00+09); PC £0.85/kg; to seedbeds; by hand							
35 g/m²	2.38	0.17	3.04	-	2.38	100 m²	5.42
50 g/m²	3.40	0.17	3.04	-	3.40	100 m²	6.44
70 g/m²	4.76	0.17	3.04	-	4.76	100 m²	7.80
100 g/m²	6.80	0.20	3.65	-	6.80	100 m²	10.45
Seeding							
Seeding labours only in two operations; by machine (for seed prices see above)							
35 g/m²	-	-	-	0.51	-	100 m²	0.51

Q PAVING/PLANTING/FENCING/SITE FURNITURE

Item Excluding site overheads and profit	PC £	Labour hours	Labour £	Plant £	Material £	Unit	Total rate £
Q30 SEEDING/TURFING - cont'd							
Seeding - cont'd							
Grass seed; spreading in two operations; PC £4.50/kg; (for changes in material prices please refer to table above); by machine							
35 g/m²	-	-	-	0.51	15.75	100 m²	16.26
50 g/m²	-	-	-	0.51	22.50	100 m²	23.01
70 g/m²	-	-	-	0.51	31.50	100 m²	32.01
100 g/m²	-	-	-	0.51	45.00	100 m²	45.51
125 kg/ha	-	-	-	50.95	562.50	ha	613.45
150 kg/ha	-	-	-	50.95	675.00	ha	725.95
200 kg/ha	-	-	-	50.95	900.00	ha	950.95
250 kg/ha	-	-	-	50.95	1125.00	ha	1175.95
300 kg/ha	-	-	-	50.95	1350.00	ha	1400.95
350 kg/ha	-	-	-	50.95	1575.00	ha	1625.95
400 kg/ha	-	-	-	50.95	1800.00	ha	1850.95
500 kg/ha	-	-	-	50.95	2250.00	ha	2300.95
700 kg/ha	-	-	-	50.95	3150.00	ha	3200.95
Extra over seeding by machine for slopes over 30° (allowing for the actual area but measured in plan)							
35 g/m²	-	-	-	0.08	2.36	100 m²	2.44
50 g/m²	-	-	-	0.08	3.38	100 m²	3.45
70 g/m²	-	-	-	0.08	4.72	100 m²	4.80
100 g/m²	-	-	-	0.08	6.75	100 m²	6.83
125 kg/ha	-	-	-	7.64	84.38	ha	92.02
150 kg/ha	-	-	-	7.64	101.25	ha	108.89
200 kg/ha	-	-	-	7.64	135.00	ha	142.64
250 kg/ha	-	-	-	7.64	168.75	ha	176.39
300 kg/ha	-	-	-	7.64	202.50	ha	210.14
350 kg/ha	-	-	-	7.64	236.25	ha	243.89
400 kg/ha	-	-	-	7.64	270.00	ha	277.64
500 kg/ha	-	-	-	7.64	337.50	ha	345.14
700 kg/ha	-	-	-	7.64	472.50	ha	480.14
Seeding labours only in two operations; by machine (for seed prices see above)							
35 g/m²	-	0.17	3.04	-	-	100 m²	3.04
Grass seed; spreading in two operations; PC £4.50/kg (for changes in material prices please refer to table above); by hand							
35 g/m²	-	0.17	3.04	-	15.75	100 m²	18.79
50 g/m²	-	0.17	3.04	-	22.50	100 m²	25.54
70 g/m²	-	0.17	3.04	-	31.50	100 m²	34.54
100 g/m²	-	0.20	3.65	-	45.00	100 m²	48.65
125 g/m²	-	0.20	3.65	-	56.25	100 m²	59.90
Extra over seeding by hand for slopes over 30° (allowing for the actual area but measured in plan)							
35 g/m²	2.34	-	0.07	-	2.34	100 m²	2.41
50 g/m²	3.38	-	0.07	-	3.38	100 m²	3.44
70 g/m²	4.72	-	0.07	-	4.72	100 m²	4.79
100 g/m²	6.75	-	0.08	-	6.75	100 m²	6.83
125 g/m²	8.41	-	0.08	-	8.41	100 m²	8.49
Harrowing seeded areas; light chain harrow	-	-	-	0.09	-	100 m²	0.09
Raking over seeded areas							
by mechanical stone rake	-	-	-	2.18	-	100 m²	2.18
by hand	-	0.80	14.60	-	-	100 m²	14.60

Q PAVING/PLANTING/FENCING/SITE FURNITURE

Item Excluding site overheads and profit	PC £	Labour hours	Labour £	Plant £	Material £	Unit	Total rate £
Rolling seeded areas; light roller							
by tractor drawn roller	-	-	-	0.54	-	100 m²	0.54
by pedestrian operated mechanical roller	-	0.08	1.52	0.54	-	100 m²	2.06
by hand drawn roller	-	0.17	3.04	-	-	100 m²	3.04
Extra over harrowing, raking or rolling seeded areas for slopes over 30°; by machine or hand	-	-	-	-	-	25%	-
Turf edging; to seeded areas; 300 mm wide	-	0.05	0.87	-	1.99	m²	2.86
Liquid sod; Turf Management Systems							
Spray on grass system of grass plantlets fertilizer, bio-degradable mulch carrier, root enhancer and water							
to prepared ground	-	-	-	-	-	m²	1.80
Preparation of turf beds							
Rolling turf to be lifted; lifting by hand or mechanical turf stripper; stacks to be not more than 1 m high							
cutting only preparing to lift; pedestrian turf cutter	-	0.75	13.69	9.96	-	100 m²	23.64
lifting and stacking; by hand	-	8.33	152.08	-	-	100 m²	152.08
Rolling up; moving to stacks							
distance not exceeding 100 m	-	2.50	45.63	-	-	100 m²	45.63
extra over rolling and moving turf to stacks to transport per additional 100 m	-	0.83	15.21	-	-	100 m²	15.21
Lifting selected and approved topsoil from spoil heaps							
passing through 6 mm screen; removing debris	-	0.17	3.04	9.77	-	m³	12.81
Extra over lifting topsoil and passing through screen for imported topsoil; plus 20% allowance for settlement	26.00	-	-	-	31.20	m³	-
Topsoil; PC £26.00/m³; plus 20% allowance for settlement							
25 mm deep	-	-	-	-	0.78	m²	-
50 mm deep	-	-	-	-	1.56	m²	-
100 mm deep	-	-	-	-	3.12	m²	-
150 mm deep	-	-	-	-	4.68	m²	-
200 mm deep	-	-	-	-	6.24	m²	-
250 mm deep	-	-	-	-	7.80	m²	-
300 mm deep	-	-	-	-	9.36	m²	-
400 mm deep	-	-	-	-	12.48	m²	-
450 mm deep	-	-	-	-	14.04	m²	-
Spreading topsoil to form turfbeds (topsoil not included); by machine							
25 mm deep	-	-	0.05	0.10	-	m²	0.14
50 mm deep	-	-	0.06	0.12	-	m²	0.18
75 mm deep	-	-	0.07	0.14	-	m²	0.21
100 mm deep	-	0.01	0.09	0.19	-	m²	0.28
150 mm deep	-	0.01	0.14	0.29	-	m²	0.42
Spreading topsoil to form turfbeds (topsoil not included); by hand							
25 mm deep	-	0.03	0.46	-	-	m²	0.46
50 mm deep	-	0.03	0.61	-	-	m²	0.61
75 mm deep	-	0.04	0.78	-	-	m²	0.78
100 mm deep	-	0.05	0.91	-	-	m²	0.91
150 mm deep	-	0.08	1.37	-	-	m²	1.37

Q PAVING/PLANTING/FENCING/SITE FURNITURE

Item Excluding site overheads and profit	PC £	Labour hours	Labour £	Plant £	Material £	Unit	Total rate £
Q30 SEEDING/TURFING - cont'd							
Preparation of turf beds - cont'd							
Bringing existing topsoil to a fine tilth for turfing by raking or harrowing; stones not to exceed 6 mm; by machine	-	-	0.07	0.04	-	m^2	0.12
Bringing existing topsoil to a fine tilth for turfing by raking or harrowing; stones not to exceed 6 mm; by hand	-	0.01	0.17	-	-	m^2	0.17
Turfing							
Turfing; laying only; to stretcher bond; butt joints; including providing and working from barrow plank runs where necessary to surfaces not exceeding 30° from horizontal							
specially selected lawn turves from previously lifted stockpile	-	0.08	1.37	-	-	m^2	1.37
cultivated lawn turves; to large open areas	-	0.06	1.06	-	-	m^2	1.06
cultivated lawn turves; to domestic or garden areas	-	0.08	1.42	-	-	m^2	1.42
road verge quality turf	-	0.04	0.73	-	-	m^2	0.73
Industrially grown turf; PC prices listed represent the general range of industrial turf prices for sportsfields and amenity purposes; prices will vary with quantity and site location							
"Rolawn"							
ref RB Medallion; sports fields, domestic lawns, general landscape; full loads 1720 m^2	1.99	0.07	1.28	-	1.99	m^2	3.27
ref RB Medallion; sports fields, domestic lawns, general landscape; part loads	1.99	0.07	1.28	-	1.99	m^2	3.27
"Tensar Ltd"							
"Tensar Turf Mat" reinforced turf for embankments; laid to embankments 3.3 m^2 turves	3.50	0.11	2.01	-	3.50	m^2	5.51
"Inturf"							
ref Inturf 1; fine lawns, golf greens, bowling greens	2.42	0.10	1.74	-	2.42	m^2	4.16
ref Inturf 2; football grounds, parks, hardwearing areas	1.67	0.05	0.96	-	1.67	m^2	2.62
ref Inturf 2 Bargold; fine turf; football grounds, parks, hardwearing areas	1.92	0.05	0.96	-	1.92	m^2	2.87
ref Inturf 3; hockey grounds, polo, medium wearing areas	1.92	0.05	0.99	-	1.92	m^2	2.91
Custom Grown Turf; "green" specific seed mixtures to suit soil or site conditions	6.42	0.08	1.46	-	6.42	m^2	7.88
"RTF" Rhizomatous Tall Fescue; deep rooted self repairing for difficult site or drought conditions or poor soils	2.42	0.08	1.46	-	2.42	m^2	3.88

Q PAVING/PLANTING/FENCING/SITE FURNITURE

Item Excluding site overheads and profit	PC £	Labour hours	Labour £	Plant £	Material £	Unit	Total rate £
Reinforced turf; Boddingtons Netlon Advanced Turf; Boddingtons Ltd; blended mesh elements incorporated into root zone; root zone spread and levelled over cultivated, prepared and reduced and levelled ground (not included); compacted with vibratory roller							
"Boddingtons ATS 400/B" with selected turf and fertilizer 100 mm thick							
100 - 500 m^2	23.00	0.03	0.61	0.72	23.00	m^2	24.33
over 500 m^2	21.50	0.03	0.61	0.72	21.50	m^2	22.83
"Boddingtons ATS 400/B" with selected turf and fertilizer 150 mm thick							
100 - 500 m^2	29.00	0.04	0.73	0.81	29.00	m^2	30.54
over 500 m^2	28.00	0.04	0.73	0.81	28.00	m^2	29.54
"Boddingtons ATS 400/B" with selected turf and fertilizer 200 mm thick							
100 - 500 m^2	32.00	0.05	0.91	0.93	32.00	m^2	33.84
over 500 m^2	30.00	0.05	0.91	0.93	30.00	m^2	31.84
Firming turves with wooden beater	-	0.01	0.18	-	-	m^2	0.18
Rolling turfed areas; light roller							
by tractor with turf tyres and roller	-	-	-	0.54	-	100 m^2	0.54
by pedestrian operated mechanical roller	-	0.08	1.52	0.54	-	100 m^2	2.06
by hand drawn roller	-	0.17	3.04	-	-	100 m^2	3.04
Dressing with finely sifted topsoil; brushing into joints	0.02	0.05	0.91	-	0.03	m^2	0.94
Turfing; laying only							
to slopes over 30°; to diagonal bond (measured as plan area - add 15% to these rates for the incline area of 30° slopes)	-	0.12	2.19	-	-	m^2	2.19
Extra over laying turfing for pegging down turves							
wooden or galvanized wire pegs; 200 mm long; 2 pegs per 0.50 m^2	1.44	0.01	0.24	-	1.44	m^2	1.68
Artificial grass; Artificial Lawn Company; laid to sharp sand bed priced separately 15 kg kiln sand brushed in per m^2							
"Leisure Lawn"; 24 mm thick artificial sports turf; sand filled	-	-	-	-	-	m^2	27.80
"Budget Grass"; for general use; budget surface; sand filled	-	-	-	-	-	m^2	21.95
"Multi Grass"; patios, conservatories and pool surrounds; sand filled	-	-	-	-	-	m^2	23.90
"Premier"; lawns and patios	-	-	-	-	-	m^2	30.00
"Play Lawn"; grass/sand and rubber filled	-	-	-	-	-	m^2	33.85
"Grassflex"; safety surfacing for play areas	-	-	-	-	-	m^2	49.50
Maintenance operations (Note: the following rates apply to aftercare maintenance executed as part of a landscaping contract only)							
Initial cutting; to turfed areas							
20 mm high; using pedestrian guided power driven cylinder mower; including boxing off cuttings (stone picking and rolling not included)	-	0.18	3.29	0.29	-	100 m^2	3.58

Q PAVING/PLANTING/FENCING/SITE FURNITURE

Item Excluding site overheads and profit	PC £	Labour hours	Labour £	Plant £	Material £	Unit	Total rate £
Q30 SEEDING/TURFING - cont'd							
Maintenance operations - cont'd							
Repairing damaged grass areas							
scraping out; removing slurry; from ruts and holes; average 100 mm deep	-	0.13	2.43	-	-	m²	2.43
100 mm topsoil	-	0.13	2.43	-	3.12	m²	5.55
Repairing damaged grass areas; sowing grass seed to match existing or as specified; to individually prepared worn patches							
35 g/m²	0.16	0.01	0.18	-	0.18	m²	0.36
50 g/m²	0.23	0.01	0.18	-	0.26	m²	0.44
Sweeping leaves; disposing off site; motorized vacuum sweeper or rotary brush sweeper							
areas of maximum 2500 m² with occasional large tree and established boundary planting; 4.6 m³ (1 skip of material to be removed)	-	0.40	7.30	3.29	-	100 m²	10.59
Leaf clearance; clearing grassed area of leaves and other extraneous debris Using equipment towed by tractor							
large grassed areas with perimeters of mature trees such as sports fields and amenity areas	-	0.01	0.23	0.04	-	100 m²	0.27
large grassed areas containing ornamental trees and shrub beds	-	0.03	0.46	0.06	-	100 m²	0.52
Using pedestrian operated mechanical equipment and blowers							
grassed areas with perimeters of mature trees such as sports fields and amenity areas	-	0.04	0.73	0.05	-	100 m²	0.78
grassed areas containing ornamental trees and shrub beds	-	0.10	1.82	0.13	-	100 m²	1.96
verges	-	0.07	1.22	0.09	-	100 m²	1.31
By hand							
grassed areas with perimeters of mature trees such as sports fields and amenity areas	-	0.05	0.91	0.10	-	100 m²	1.01
grassed areas containing ornamental trees and shrub beds	-	0.08	1.52	0.17	-	100 m²	1.69
verges	-	1.00	18.25	1.99	-	100 m²	20.24
Removal of arisings							
areas with perimeters of mature trees	-	0.01	0.10	0.09	1.20	100 m²	1.39
areas containing ornamental trees and shrub beds	-	0.02	0.30	0.34	3.00	100 m²	3.64
Cutting grass to specified height; per cut							
multi unit gang mower	-	0.59	10.73	18.30	-	ha	29.03
ride-on triple cylinder mower	-	0.01	0.26	0.14	-	100 m²	0.39
ride-on triple rotary mower	-	0.01	0.26	-	-	100 m²	0.26
pedestrian mower	-	0.18	3.29	0.69	-	100 m²	3.98
Cutting grass to banks; per cut							
side arm cutter bar mower	-	0.02	0.43	0.34	-	100 m²	0.76
Cutting rough grass; per cut							
power flail or scythe cutter	-	0.04	0.64	-	-	100 m²	0.64
Extra over cutting grass for slopes not exceeding 30°	-	-	-	-	-	10%	-
Extra over cutting grass for slopes exceeding 30°	-	-	-	-	-	40%	-

Q PAVING/PLANTING/FENCING/SITE FURNITURE

Item Excluding site overheads and profit	PC £	Labour hours	Labour £	Plant £	Material £	Unit	Total rate £
Cutting fine sward							
pedestrian operated seven-blade cylinder lawn mower	-	0.14	2.56	0.22	-	100 m²	**2.78**
Extra over cutting fine sward for boxing off cuttings							
pedestrian mower	-	0.03	0.51	0.04	-	100 m²	**0.56**
Cutting areas of rough grass							
scythe	-	1.00	18.25	-	-	100 m²	**18.25**
sickle	-	2.00	36.50	-	-	100 m²	**36.50**
petrol operated strimmer	-	0.30	5.48	0.42	-	100 m²	**5.90**
Cutting areas of rough grass which contain trees or whips							
petrol operated strimmer	-	0.40	7.30	0.56	-	100 m²	**7.86**
Extra over cutting rough grass for on site raking up and dumping	-	0.33	6.08	-	-	100 m²	**6.08**
Trimming edge of grass areas; edging tool							
with petrol powered strimmer	-	0.13	2.43	0.19	-	100 m	**2.62**
by hand	-	0.67	12.17	-	-	100 m	**12.17**
Marking out pitches using approved line marking compound; including initial setting out and marking							
discus, hammer, javelin or shot putt area	2.85	2.00	36.50	-	2.85	nr	**39.35**
cricket square	1.90	2.00	36.50	-	1.90	nr	**38.40**
cricket boundary	6.65	8.00	146.00	-	6.65	nr	**152.65**
grass tennis court	2.85	4.00	73.00	-	2.85	nr	**75.85**
hockey pitch	9.50	8.00	146.00	-	9.50	nr	**155.50**
football pitch	9.50	8.00	146.00	-	9.50	nr	**155.50**
rugby pitch	9.50	8.00	146.00	-	9.50	nr	**155.50**
eight lane running track; 400 m	19.00	16.00	292.00	-	19.00	nr	**311.00**
Re-marking out pitches using approved line marking compound							
discus, hammer, javelin or shot putt area	1.90	0.50	9.13	-	1.90	nr	**11.03**
cricket square	1.90	0.50	9.13	-	1.90	nr	**11.03**
grass tennis court	6.65	1.00	18.25	-	6.65	nr	**24.90**
hockey pitch	6.65	1.00	18.25	-	6.65	nr	**24.90**
football pitch	6.65	1.00	18.25	-	6.65	nr	**24.90**
rugby pitch	6.65	1.00	18.25	-	6.65	nr	**24.90**
eight lane running track; 400 m	19.00	2.50	45.63	-	19.00	nr	**64.63**
Rolling grass areas; light roller							
by tractor drawn roller	-	-	-	0.54	-	100 m²	**0.54**
by pedestrian operated mechanical roller	-	0.08	1.52	0.54	-	100 m²	**2.06**
by hand drawn roller	-	0.17	3.04	-	-	100 m²	**3.04**
Aerating grass areas; to a depth of 100 mm							
using tractor-drawn aerator	-	0.06	1.06	1.01	-	100 m²	**2.07**
using pedestrian-guided motor powered solid or slitting tine turf aerator	-	0.18	3.20	2.96	-	100 m²	**6.16**
using hollow tine aerator; including sweeping up and dumping corings	-	0.50	9.13	5.92	-	100 m²	**15.04**
using hand aerator or fork	-	1.67	30.42	-	-	100 m²	**30.42**
Extra over aerating grass areas for on site sweeping up and dumping corings	-	0.17	3.04	-	-	100 m²	**3.04**
Switching off dew; from fine turf areas	-	0.20	3.65	-	-	100 m²	**3.65**
Scarifying grass areas to break up thatch; removing dead grass							
using tractor-drawn scarifier	-	0.07	1.28	0.31	-	100 m²	**1.59**
using self-propelled scarifier; including removing and disposing of grass on site	-	0.33	6.08	0.15	-	100 m²	**6.23**

Q PAVING/PLANTING/FENCING/SITE FURNITURE

Item Excluding site overheads and profit	PC £	Labour hours	Labour £	Plant £	Material £	Unit	Total rate £
Q30 SEEDING/TURFING - cont'd							
Maintenance operations - cont'd							
Harrowing grass areas							
using drag mat	-	0.03	0.51	0.30	-	100 m²	0.81
using chain harrow	-	0.04	0.64	0.38	-	100 m²	1.02
using drag mat	-	2.80	51.11	30.25	-	ha	81.35
using chain harrow	-	3.50	63.88	37.81	-	ha	101.69
Extra for scarifying and harrowing grass areas for disposing excavated material off site; to tip not exceeding 13 km; loading by machine							
slightly contaminated	-	-	-	1.65	21.33	m³	22.98
rubbish	-	-	-	1.65	21.33	m³	22.98
inert material	-	-	-	1.10	7.00	m³	8.10
For the following topsoil improvement and seeding operations add or subtract the following amounts for every £0.10 difference in the material cost price							
35 g/m²	-	-	-	-	0.35	100 m²	-
50 g/m²	-	-	-	-	0.50	100 m²	-
70 g/m²	-	-	-	-	0.70	100 m²	-
100 g/m²	-	-	-	-	1.00	100 m²	-
125 kg/ha	-	-	-	-	12.50	ha	-
150 kg/ha	-	-	-	-	15.00	ha	-
175 kg/ha	-	-	-	-	17.50	ha	-
200 kg/ha	-	-	-	-	20.00	ha	-
225 kg/ha	-	-	-	-	22.50	ha	-
250 kg/ha	-	-	-	-	25.00	ha	-
300 kg/ha	-	-	-	-	30.00	ha	-
350 kg/ ha	-	-	-	-	35.00	ha	-
400 kg/ha	-	-	-	-	40.00	ha	-
500 kg/ha	-	-	-	-	50.00	ha	-
700 kg/ha	-	-	-	-	70.00	ha	-
1000 kg/ha	-	-	-	-	100.00	ha	-
1250 kg/ha	-	-	-	-	125.00	ha	-
Top dressing fertilizers (7+7+7); PC £1.13/ kg; to seedbeds; by machine							
35 g/m²	3.95	-	-	0.19	3.95	100 m²	4.14
50 g/m²	5.65	-	-	0.19	5.65	100 m²	5.84
300 kg/ha	338.85	-	-	18.76	338.85	ha	357.61
350 kg/ha	395.32	-	-	18.76	395.32	ha	414.08
400 kg/ha	451.80	-	-	18.76	451.80	ha	470.56
500 kg/ha	564.75	-	-	30.01	564.75	ha	594.76
Top dressing fertilizers (7+7+7); PC £1.13/ kg; to seedbeds; by hand							
35 g/m²	3.95	0.17	3.04	-	3.95	100 m²	7.00
50 g/m²	5.65	0.17	3.04	-	5.65	100 m²	8.69
70 g/m²	7.91	0.17	3.04	-	7.91	100 m²	10.95

Q PAVING/PLANTING/FENCING/SITE FURNITURE

Item Excluding site overheads and profit	PC £	Labour hours	Labour £	Plant £	Material £	Unit	Total rate £
Watering turf; evenly; at a rate of 5 l/m² using movable spray lines powering 3 nr sprinkler heads with a radius of 15 m and allowing for 60% overlap (irrigation machinery costs not included)	-	0.02	0.29	-	-	100 m²	**0.29**
using sprinkler equipment and with sufficient water pressure to run 1 nr 15 m radius sprinkler	-	0.02	0.36	-	-	100 m²	**0.36**
using hand-held watering equipment	-	0.25	4.56	-	-	100 m²	**4.56**
Q31 PLANTING							
Note: For market prices of landscape chemicals please see sections Q30 or Q35.							
Market prices of planting materials (Note: the rates shown generally reflect the manufacturer's recommended retail prices; trade and bulk discounts are often available on the prices shown)							
Topsoil							
general purpose screened	-	-	-	-	26.00	m³	-
50/50; as dug/10 mm screened	-	-	-	-	32.00	m³	-
Market prices of mulching materials							
Melcourt Industries Ltd; 25 m³ loads (items labelled FSC are Forest Stewardship Council certified, items marked "FT" are certified firetested)							
"Ornamental Bark Mulch" FT	-	-	-	-	52.90	m³	-
"Bark Nuggets"® FT	-	-	-	-	50.20	m³	-
"Graded Bark Flakes" FT	-	-	-	-	52.80	m³	-
"Amenity Bark Mulch"; FSC; FT	-	-	-	-	35.65	m³	-
"Contract Bark Mulch"; FSC	-	-	-	-	33.40	m³	-
"Spruce Ornamental"; FSC; FT	-	-	-	-	36.40	m³	-
"Decorative Biomulch"®	-	-	-	-	35.05	m³	-
"Rustic Biomulch"®	-	-	-	-	38.80	m³	-
"Mulch 2000"	-	-	-	-	27.70	m³	-
"Forest BioMulch"®	-	-	-	-	32.55	m³	-
Melcourt Industries Ltd; 50 m³ loads							
"Mulch 2000"	-	-	-	-	17.20	m³	-
Melcourt Industries Ltd; 70 m³ loads							
"Contract Bark Mulch"; FSC	-	-	-	-	21.00	m³	-
Melcourt Industries Ltd; 80 m³ loads							
"Ornamental Bark Mulch" FT	-	-	-	-	39.90	m³	-
"Bark Nuggets"® FT	-	-	-	-	37.20	m³	-
"Graded Bark Flakes" FT	-	-	-	-	37.80	m³	-
"Amenity Bark Mulch"; FSC	-	-	-	-	22.65	m³	-
"Spruce Ornamental"; FSC, FT	-	-	-	-	23.40	m³	-
"Decorative Biomulch"®	-	-	-	-	22.05	m³	-
"Rustic Biomulch"®	-	-	-	-	25.80	m³	-
"Forest BioMulch"®	-	-	-	-	19.55	m³	-
Mulch; Melcourt Industries Ltd; 25 m³ loads							
"Composted Fine Bark"; FSC	-	-	-	-	31.65	m³	-
"Humus 2000"	-	-	-	-	26.85	m³	-
"Spent Mushroom Compost"	-	-	-	-	21.20	m³	-
"Topgrow"	-	-	-	-	31.05	m³	-

Q PAVING/PLANTING/FENCING/SITE FURNITURE

Item Excluding site overheads and profit	PC £	Labour hours	Labour £	Plant £	Material £	Unit	Total rate £
Q31 PLANTING - cont'd							
Market prices of mulching materials - cont'd							
Mulch; Melcourt Industries Ltd; 50 m³ loads							
"Humus 2000"	-	-	-	-	16.25	m³	-
Mulch; Melcourt Industries Ltd; 60 m³ loads							
"Spent Mushroom Compost"	-	-	-	-	9.50	m³	-
Mulch; Melcourt Industries Ltd; 65 m³ loads							
"Composted Fine Bark"	-	-	-	-	18.65	m³	-
"Super Humus"	-	-	-	-	18.60	m³	-
"Topgrow"	-	-	-	-	20.55	m³	-
Fertilizers; British Seed Houses							
Floranid permanent NPK; 16%+7%+15%; slow release fertilizer	36.00	-	-	-	36.00	kg	-
Floranid Eagle NK-NPK; 20%+0%+18%; slow release fertilizer	43.25	-	-	-	43.25	kg	-
Floranid Master Extra NPK; 19%+5% +10%; slow release fertilizer	40.25	-	-	-	40.25	kg	-
Basotop Fair NPK; 235+6%+10%; slow release fertilizer	37.00	-	-	-	37.00	kg	-
Fertilizers; Scotts UK							
"Enmag"; controlled release fertilizer (8-9 months); 11+22+09; 70 g/m²	-	-	-	-	14.71	100 m²	-
Fertilizers; Scotts UK; granular "Osmocote Flora"; 15+09+11; controlled release fertilizer; costs for recommended application rates							
transplant	-	-	-	-	0.08	nr	-
whip	-	-	-	-	0.12	nr	-
feathered	-	-	-	-	0.15	nr	-
light standard	-	-	-	-	0.15	nr	-
standard	-	-	-	-	0.19	nr	-
selected standard	-	-	-	-	0.27	nr	-
heavy standard	-	-	-	-	0.31	nr	-
extra heavy standard	-	-	-	-	0.38	nr	-
16 - 18 cm girth	-	-	-	-	0.42	nr	-
18 - 20 cm girth	-	-	-	-	0.46	nr	-
20 - 22 cm girth	-	-	-	-	0.54	nr	-
22 - 24 cm girth	-	-	-	-	0.58	nr	-
24 - 26 cm girth	-	-	-	-	0.62	nr	-
Fertilizers; Farmura Environmental Ltd; Seanure Soilbuilder							
soil amelioration; 70 g/m²	-	-	-	-	7.75	100 m²	-
to plant pits; 300 x 300 x 300 mm	-	-	-	-	0.03	nr	-
to plant pits; 600 x 600 x 600 mm	-	-	-	-	0.36	nr	-
to tree pits; 1.00 x 1.00 x 1.00 m	-	-	-	-	1.66	nr	-
Fertilizers; Scotts UK							
"TPMC" tree planting and mulching compost	-	-	-	-	4.86	bag	-

Q PAVING/PLANTING/FENCING/SITE FURNITURE

Item Excluding site overheads and profit	PC £	Labour hours	Labour £	Plant £	Material £	Unit	Total rate £
Fertilizers; Rigby Taylor Ltd; fertilizer application rates 35 g/m² unless otherwise shown							
straight fertilizer; "Bone Meal"; at 70 g/m²	-	-	-	-	7.76	100 m²	-
straight fertilizer; "Sulphate of Ammonia"	-	-	-	-	2.77	100 m²	-
straight fertilizer; "Sulphate of Iron"	-	-	-	-	2.68	100 m²	-
straight fertilizer; "Sulphate of Potash"	-	-	-	-	6.20	100 m²	-
straight fertilizer; "Super Phosphate Powder"	-	-	-	-	1.22	100 m²	-
liquid fertilizer; "Mascot Microflow-CX"; 4 +2+18 + trace elements; 200-1400 ml /100 m²	-	-	-	-	6.03	100 m²	-
liquid fertilizer; "Mascot Microflow-CX"; 12 +0+8 + trace elements; 200-1400 ml /100 m²	-	-	-	-	5.47	100 m²	-
liquid fertilizer; "Mascot Microflow-C"; 25 +0+0 + trace elements; 200-1400 ml /100 m²	-	-	-	-	5.70	100 m²	-
Wetting agents; Rigby Taylor Ltd							
wetting agent; "Breaker Advance Liquid"; 10 l	-	-	-	-	1.55	100 m²	-
wetting agent; "Breaker Advance Granules"; 20 kg	-	-	-	-	12.13	100 m²	-
Planting - General							
Preamble: Prices for all planting work are deemed to include carrying out planting in accordance with all good horticultural practice.							
Site protection; temporary protective fencing							
Cleft chestnut rolled fencing; to 100 mm diameter chestnut posts; driving into firm ground at 3 m centres; pales at 50 mm centres							
900 mm high	2.55	0.11	1.95	-	3.15	m	**5.10**
1100 mm high	2.95	0.11	1.95	-	3.42	m	**5.36**
1500 mm high	4.91	0.11	1.95	-	5.39	m	**7.33**
Extra over temporary protective fencing for removing and making good (no allowance for re-use of material)	-	0.07	1.22	0.20	-	m	**1.42**
Cultivation							
Ripping up subsoil; using approved subsoiling machine; minimum depth 250 mm below topsoil; at 1.20 m centres; in							
gravel or sandy clay	-	-	-	2.40	-	100 m²	**2.40**
soil compacted by machines	-	-	-	2.80	-	100 m²	**2.80**
clay	-	-	-	3.00	-	100 m²	**3.00**
chalk or other soft rock	-	-	-	6.01	-	100 m²	**6.01**
Extra for subsoiling at 1 m centres	-	-	-	0.60	-	100 m²	**0.60**
Breaking up existing ground; using pedestrian operated tine cultivator or rotavator							
100 mm deep	-	0.22	4.01	2.12	-	100 m²	**6.14**
150 mm deep	-	0.28	5.02	2.65	-	100 m²	**7.67**
200 mm deep	-	0.37	6.69	3.54	-	100 m²	**10.23**

Q PAVING/PLANTING/FENCING/SITE FURNITURE

Item Excluding site overheads and profit	PC £	Labour hours	Labour £	Plant £	Material £	Unit	Total rate £
Q31 PLANTING - cont'd							
Cultivation - cont'd							
As above but in heavy clay or wet soils							
100 mm deep	-	0.44	8.03	4.25	-	100 m²	12.28
150 mm deep	-	0.66	12.04	6.37	-	100 m²	18.41
200 mm deep	-	0.82	15.06	7.96	-	100 m²	23.02
Breaking up existing ground; using tractor drawn tine cultivator or rotavator							
100 mm deep	-	-	-	0.28	-	100 m²	0.28
150 mm deep	-	-	-	0.36	-	100 m²	0.36
200 mm deep	-	-	-	0.47	-	100 m²	0.47
600 mm deep	-	-	-	1.42	-	100 m²	1.42
Cultivating ploughed ground; using disc, drag or chain harrow							
4 passes	-	-	-	1.71	-	100 m²	1.71
Rolling cultivated ground lightly; using self-propelled agricultural roller	-	0.06	1.01	0.60	-	100 m²	1.62
Importing only selected and approved topsoil; from source not exceeding 13 km from site							
1 - 14 m³	26.00	-	-	-	78.00	m³	-
over 15 m³	26.00	-	-	-	26.00	m³	-
Spreading and lightly consolidating approved topsoil (imported or from spoil heaps); in layers not exceeding 150 mm; travel distance from spoil heaps not exceeding 100 m; by machine (imported topsoil not included)							
minimum depth 100 mm	-	1.55	28.29	36.62	-	100 m²	64.91
minimum depth 150 mm	-	2.33	42.58	55.11	-	100 m²	97.69
minimum depth 300 mm	-	4.67	85.16	110.21	-	100 m²	195.37
minimum depth 450 mm	-	6.99	127.57	165.14	-	100 m²	292.71
Spreading and lightly consolidating approved topsoil (imported or from spoil heaps); in layers not exceeding 150 mm; travel distance from spoil heaps not exceeding 100 m; by hand (imported topsoil not included)							
minimum depth 100 mm	-	20.00	365.07	-	-	100 m²	365.07
minimum depth 150 mm	-	30.01	547.61	-	-	100 m²	547.61
minimum depth 300 mm	-	60.01	1095.22	-	-	100 m²	1095.22
minimum depth 450 mm	-	90.02	1642.83	-	-	100 m²	1642.83
Extra over for spreading topsoil to slopes 15 - 30°; by machine or hand	-	-	-	-	-	10%	-
Extra over for spreading topsoil to slopes over 30°; by machine or hand	-	-	-	-	-	25%	-
Extra over spreading topsoil for travel exceeding 100 m; by machine							
100 - 150 m	-	0.02	0.36	0.10	-	m³	0.46
150 - 200 m	-	0.03	0.49	0.13	-	m³	0.62
200 - 300 m	-	0.04	0.65	0.17	-	m³	0.82
Extra over spreading topsoil for travel exceeding 100 m; by hand							
100 m	-	2.50	45.63	-	-	m³	45.63
200 m	-	3.50	63.88	-	-	m³	63.88
300 m	-	4.50	82.13	-	-	m³	82.13

Q PAVING/PLANTING/FENCING/SITE FURNITURE

Item Excluding site overheads and profit	PC £	Labour hours	Labour £	Plant £	Material £	Unit	Total rate £
Evenly grading; to general surfaces to bring to finished levels							
by machine (tractor mounted rotavator)	-	-	-	0.01	-	m²	0.01
by pedestrian operated rotavator	-	-	0.07	0.04	-	m²	0.12
by hand	-	0.01	0.18	-	-	m²	0.18
Extra over grading for slopes 15 - 30°; by machine or hand	-	-	-	-	-	10%	-
Extra over grading for slopes over 30°; by machine or hand	-	-	-	-	-	25%	-
Clearing stones; disposing off site; to distance not exceeding 13 km							
by hand; stones not exceeding 50 mm in any direction; loading to skip 4.6 m³	-	0.01	0.18	0.04	-	m²	0.22
by mechanical stone rake; stones not exceeding 50 mm in any direction; loading to 15 m³ truck by mechanical loader	-	-	0.04	0.08	-	m²	0.11
Lightly cultivating; weeding; to fallow areas; disposing debris off site; to distance not exceeding 13 km							
by hand	-	0.01	0.26	-	0.10	m²	0.36
Preparation of planting operations							
For the following topsoil improvement and planting operations add or subtract the following amounts for every £0.10 difference in the material cost price							
35 g/m²	-	-	-	-	0.35	100 m²	-
50 g/m²	-	-	-	-	0.50	100 m²	-
70 g/m²	-	-	-	-	0.70	100 m²	-
100 g/m²	-	-	-	-	1.00	100 m²	-
150 kg/ha	-	-	-	-	15.00	ha	-
200 kg/ha	-	-	-	-	20.00	ha	-
250 kg/ha	-	-	-	-	25.00	ha	-
300 kg/ha	-	-	-	-	30.00	ha	-
400 kg/ha	-	-	-	-	40.00	ha	-
500 kg/ha	-	-	-	-	50.00	ha	-
700 kg/ha	-	-	-	-	70.00	ha	-
1000 kg/ha	-	-	-	-	100.00	ha	-
1250 kg/ha	-	-	-	-	125.00	ha	-
Fertilizers; in top 150 mm of topsoil; at 35 g/m²							
"Mascot Microfine"; controlled release turf fertilizer; 8+0+6 + 2% Mg + 4% Fe	6.25	0.12	2.24	-	6.25	100 m²	8.48
"Enmag"; controlled release fertilizer (8-9 months); 11+22+09	7.35	0.12	2.24	-	7.72	100 m²	9.96
"Mascot Sports Amenity"; turf fertilizer; 7+7+7	3.95	0.12	2.24	-	3.95	100 m²	6.19
"Super Phosphate Powder"	4.27	0.12	2.24	-	4.48	100 m²	6.72
"Mascot Sports Amenity"; turf fertilizer; 15+5+10	4.72	0.12	2.24	-	4.72	100 m²	6.96
"Bone meal"	3.88	0.12	2.24	-	4.07	100 m²	6.31

Q PAVING/PLANTING/FENCING/SITE FURNITURE

Item Excluding site overheads and profit	PC £	Labour hours	Labour £	Plant £	Material £	Unit	Total rate £
Q31 PLANTING - cont'd							
Preparation of planting operations - cont'd							
Fertilizers; in top 150 mm of topsoil at 70 g/m^2							
"Mascot Microfine"; controlled release turf fertilizer; 8+0+6 + 2% Mg + 4% Fe	12.50	0.12	2.24	-	12.50	100 m^2	14.73
"Enmag"; controlled release fertilizer (8-9 months); 11+22+09	14.71	0.12	2.24	-	15.44	100 m^2	17.68
"Mascot Sports Amenity"; turf fertilizer; 7+7+7	7.91	0.12	2.24	-	7.91	100 m^2	10.14
"Super Phosphate Powder"	8.54	0.12	2.24	-	8.97	100 m^2	11.20
"Mascot Sports Amenity"; turf fertilizer; 15+5+10	9.44	0.12	2.24	-	9.44	100 m^2	11.68
"Bone meal"	7.76	0.12	2.24	-	8.14	100 m^2	10.38
Spreading topsoil; from dump not exceeding 100 m distance; by machine (topsoil not included)							
at 1 m^3 per 13 m^2; 75 mm thick	-	0.56	10.14	27.46	-	100 m^2	37.60
at 1 m^3 per 10 m^2; 100 mm thick	-	0.89	16.22	28.05	-	100 m^2	44.28
at 1 m^3 per 6.50 m^2; 150 mm thick	-	1.11	20.26	49.37	-	100 m^2	69.62
at 1 m^3 per 5 m^2; 200 mm thick	-	1.48	27.01	65.63	-	100 m^2	92.64
Spreading topsoil; from dump not exceeding 100 m distance; by hand (topsoil not included)							
at 1 m^3 per 13 m^2; 75 mm thick	-	15.00	273.80	-	-	100 m^2	273.80
at 1 m^3 per 10 m^2; 100 mm thick	-	20.00	365.07	-	-	100 m^2	365.07
at 1 m^3 per 6.50 m^2; 150 mm thick	-	30.01	547.61	-	-	100 m^2	547.61
at 1 m^3 per 5 m^2; 200 mm thick	-	36.67	669.30	-	-	100 m^2	669.30
Imported topsoil; tipped 100 m from area of application; by machine							
at 1 m^3 per 13 m^2; 75 mm thick	195.00	0.56	10.14	27.46	234.00	100 m^2	271.60
at 1 m^3 per 10 m^2; 100 mm thick	260.00	0.89	16.22	28.05	312.00	100 m^2	356.28
at 1 m^3 per 6.50 m^2; 150 mm thick	390.00	1.11	20.26	49.37	468.00	100 m^2	537.62
at 1 m^3 per 5 m^2; 200 mm thick	520.00	1.48	27.01	65.63	624.00	100 m^2	716.63
Imported topsoil; tipped 100 m from area of application; by hand							
at 1 m^3 per 13 m^2; 75 mm thick	195.00	15.00	273.80	-	234.00	100 m^2	507.80
at 1 m^3 per 10 m^2; 100 mm thick	260.00	20.00	365.07	-	312.00	100 m^2	677.07
at 1 m^3 per 6.50 m^2; 150 mm thick	390.00	30.01	547.61	-	468.00	100 m^2	1015.61
at 1 m^3 per 5 m^2; 200 mm thick	520.00	36.67	669.30	-	624.00	100 m^2	1293.30
Composted bark and manure soil conditioner (20 m^3 loads); from not further than 25 m from location; cultivating into topsoil by machine							
50 mm thick	151.50	2.86	52.14	1.77	159.07	100 m^2	212.99
100 mm thick	303.00	6.05	110.39	1.77	318.15	100 m^2	430.31
150 mm thick	454.50	8.90	162.51	1.77	477.23	100 m^2	641.51
200 mm thick	606.00	12.90	235.51	1.77	636.30	100 m^2	873.58
Composted bark soil conditioner (20 m^3 loads); placing on beds by mechanical loader; spreading and rotavating into topsoil by machine							
50 mm thick	151.50	-	-	6.20	151.50	100 m^2	157.70
100 mm thick	303.00	-	-	10.45	318.15	100 m^2	328.60
150 mm thick	454.50	-	-	14.87	477.23	100 m^2	492.09
200 mm thick	606.00	-	-	19.20	636.30	100 m^2	655.50

Q PAVING/PLANTING/SITE FURNITURE

Item Excluding site overheads and profit	PC £	Labour hours	Labour £	Plant £	Material £	Unit	Total rate £
Mushroom compost (20 m³ loads); from not further than 25 m from location; cultivating into topsoil by machine							
50 mm thick	106.00	2.86	52.14	1.77	106.00	100 m²	159.91
100 mm thick	212.00	6.05	110.39	1.77	212.00	100 m²	324.16
150 mm thick	318.00	8.90	162.51	1.77	318.00	100 m²	482.28
200 mm thick	424.00	12.90	235.51	1.77	424.00	100 m²	661.28
Mushroom compost (20 m³ loads); placing on beds by mechanical loader; spreading and rotavating into topsoil by machine							
50 mm thick	106.00	-	-	6.20	106.00	100 m²	112.20
100 mm thick	212.00	-	-	10.45	212.00	100 m²	222.45
150 mm thick	318.00	-	-	14.87	318.00	100 m²	332.87
200 mm thick	424.00	-	-	19.20	424.00	100 m²	443.20
Manure (60 m³ loads); from not further than 25 m from location; cultivating into topsoil by machine							
50 mm thick	114.90	2.86	52.14	1.77	114.90	100 m²	168.81
100 mm thick	229.80	6.05	110.39	1.77	241.29	100 m²	353.45
150 mm thick	344.70	8.90	162.51	1.77	361.94	100 m²	526.22
200 mm thick	459.60	12.90	235.51	1.77	482.58	100 m²	719.86
Surface applications and soil additives; pre-planting; from not further than 25 m from location; by machine							
Medium bark soil conditioner; AHS Ltd; including turning in to cultivated ground; delivered in 15 m³ loads							
1 m³ per 40 m² = 25 mm thick	0.89	-	-	0.09	0.89	m²	0.98
1 m³ per 20 m² = 50 mm thick	1.78	-	-	0.14	1.78	m²	1.92
1 m³ per 13.33 m² = 75 mm thick	2.67	-	-	0.19	2.67	m²	2.87
1 m³ per 10 m² = 100 mm thick	3.56	-	-	0.25	3.56	m²	3.82
Mushroom compost soil conditioner; AHS Ltd; including turning in to cultivated ground; delivered in 20 m³ loads							
1 m³ per 40 m² = 25 mm thick	0.53	0.02	0.30	-	0.53	m²	0.83
1 m³ per 20 m² = 50 mm thick	1.06	0.03	0.56	-	1.06	m²	1.62
1 m³ per 13.33 m² = 75 mm thick	1.59	0.04	0.73	-	1.59	m²	2.32
1 m³ per 10 m² = 100 mm thick	2.12	0.05	0.91	-	2.12	m²	3.03
Mushroom compost soil conditioner; AHS Ltd; including turning in to cultivated ground; delivered in 35 m³ loads							
1 m³ per 40 m² = 25 mm thick	0.29	0.02	0.30	-	0.29	m²	0.60
1 m³ per 20 m² = 50 mm thick	0.59	0.03	0.56	-	0.59	m²	1.15
1 m³ per 13.33 m² = 75 mm thick	0.88	0.04	0.73	-	0.88	m²	1.61
1 m³ per 10 m² = 100 mm thick	1.18	0.05	0.91	-	1.18	m²	2.09
Surface applications and soil additives; pre-planting; from not further than 25 m from location; by hand							
Ground limestone soil conditioner; including turning in to cultivated ground							
0.25 kg/m² = 2.50 tonnes/ha	0.69	1.20	21.90	-	0.69	100 m²	22.59
0.50 kg/m² = 5.00 tonnes/ha	1.38	1.33	24.33	-	1.38	100 m²	25.71
0.75 kg/m² = 7.50 tonnes/ha	2.07	1.50	27.38	-	2.07	100 m²	29.44
1.00 kg/m² = 10.00 tonnes/ha	2.75	1.71	31.28	-	2.75	100 m²	34.04

Q PAVING/PLANTING/FENCING/SITE FURNITURE

Item Excluding site overheads and profit	PC £	Labour hours	Labour £	Plant £	Material £	Unit	Total rate £
Q31 PLANTING - cont'd							
Surface applications and soil additives - cont'd							
Medium bark soil conditioner; AHS Ltd; including turning in to cultivated ground; delivered in 15 m³ loads							
1 m³ per 40 m² = 25 mm thick	0.89	0.02	0.41	-	0.89	m²	1.30
1 m³ per 20 m² = 50 mm thick	1.78	0.04	0.81	-	1.78	m²	2.59
1 m³ per 13.33 m² = 75 mm thick	2.67	0.07	1.22	-	2.67	m²	3.89
1 m³ per 10 m² = 100 mm thick	3.56	0.08	1.46	-	3.56	m²	5.03
Mushroom compost soil conditioner; Melcourt Industries Ltd; including turning in to cultivated ground; delivered in 25 m³ loads							
1 m³ per 40 m² = 25 mm thick	0.53	0.02	0.41	-	0.53	m²	0.94
1 m³ per 20 m² = 50 mm thick	1.06	0.04	0.81	-	1.06	m²	1.87
1 m³ per 13.33 m² = 75 mm thick	1.59	0.07	1.22	-	1.59	m²	2.81
1 m³ per 10 m² = 100 mm thick	2.12	0.08	1.46	-	2.12	m²	3.58
Mushroom compost soil conditioner; AHS Ltd; including turning in to cultivated ground; delivered in 35 m³ loads							
1 m³ per 40 m² = 25 mm thick	0.29	0.02	0.41	-	0.29	m²	0.70
1 m³ per 20 m² = 50 mm thick	0.59	0.04	0.81	-	0.59	m²	1.40
1 m³ per 13.33 m² = 75 mm thick	0.88	0.07	1.22	-	0.88	m²	2.10
1 m³ per 10 m² = 100 mm thick	1.18	0.08	1.46	-	1.18	m²	2.63
"Super Humus" mixed bark and manure conditioner; Melcourt Industries Ltd; delivered in 25 m³ loads							
1 m³ per 40 m² = 25 mm thick	0.76	0.02	0.41	-	0.80	m²	1.20
1 m³ per 20 m² = 50 mm thick	1.51	0.04	0.81	-	1.59	m²	2.40
1 m³ per 13.33 m² = 75 mm thick	2.27	0.07	1.22	-	2.39	m²	3.60
1 m³ per 10 m² = 100 mm thick	3.03	0.08	1.46	-	3.18	m²	4.64
Tree planting; pre-planting operations							
Excavating tree pits; depositing soil alongside pits; by machine							
600 x 600 x 600 mm deep	-	0.15	2.69	0.63	-	nr	3.32
900 x 900 x 600 mm deep	-	0.33	6.02	1.42	-	nr	7.45
1.00 x 1.00 m x 600 mm deep	-	0.61	11.20	1.76	-	nr	12.96
1.25 x 1.25 m x 600 mm deep	-	0.96	17.53	2.76	-	nr	20.29
1.00 x 1.00 x 1.00 m deep	-	1.02	18.66	2.94	-	nr	21.60
1.50 x 1.50 x 750 mm deep	-	1.73	31.50	4.96	-	nr	36.46
1.50 x 1.50 x 1.00 m deep	-	2.30	41.89	6.60	-	nr	48.49
1.75 x 1.75 x 1.00 m deep	-	3.13	57.16	9.01	-	nr	66.17
2.00 x 2.00 x 1.00 m deep	-	4.09	74.66	11.76	-	nr	86.42
Excavating tree pits; depositing soil alongside pits; by hand							
600 x 600 x 600 mm deep	-	0.44	8.03	-	-	nr	8.03
900 x 900 x 600 mm deep	-	1.00	18.25	-	-	nr	18.25
1.00 x 1.00 m x 600 mm deep	-	1.13	20.53	-	-	nr	20.53
1.25 x 1.25 m x 600 mm deep	-	1.93	35.22	-	-	nr	35.22
1.00 x 1.00 x 1.00 m deep	-	2.06	37.59	-	-	nr	37.59
1.50 x 1.50 x 750 mm deep	-	3.47	63.33	-	-	nr	63.33
1.75 x 1.50 x 750 mm deep	-	4.05	73.91	-	-	nr	73.91
1.50 x 1.50 x 1.00 m deep	-	4.63	84.50	-	-	nr	84.50
2.00 x 2.00 m x 750 mm deep	-	6.17	112.60	-	-	nr	112.60
2.00 x 2.00 x 1.00 m deep	-	8.23	150.21	-	-	nr	150.21

Q PAVING/PLANTING/FENCING/SITE FURNITURE

Item Excluding site overheads and profit	PC £	Labour hours	Labour £	Plant £	Material £	Unit	Total rate £
Breaking up subsoil in tree pits; to a depth of 200 mm	-	0.03	0.61	-	-	m^2	0.61
Spreading and lightly consolidating approved topsoil (imported or from spoil heaps); in layers not exceeding 150 mm; distance from spoil heaps not exceeding 100 m (imported topsoil not included); by machine							
minimum depth 100 mm	-	1.55	28.29	36.62	-	100 m^2	64.91
minimum depth 150 mm	-	2.33	42.58	55.11	-	100 m^2	97.69
minimum depth 300 mm	-	4.67	85.16	110.21	-	100 m^2	195.37
minimum depth 450 mm	-	6.99	127.57	165.14	-	100 m^2	292.71
Spreading and lightly consolidating approved topsoil (imported or from spoil heaps); in layers not exceeding 150 mm; distance from spoil heaps not exceeding 100 m (imported topsoil not included); by hand							
minimum depth 100 mm	-	20.00	365.07	-	-	100 m^2	365.07
minimum depth 150 mm	-	30.01	547.61	-	-	100 m^2	547.61
minimum depth 300 mm	-	60.01	1095.22	-	-	100 m^2	1095.22
minimum depth 450 mm	-	90.02	1642.83	-	-	100 m^2	1642.83
Extra for filling tree pits with imported topsoil; PC £26.00/m^3; plus allowance for 20% settlement							
depth 100 mm	-	-	-	-	3.12	m^2	-
depth 150 mm	-	-	-	-	4.68	m^2	-
depth 200 mm	-	-	-	-	6.24	m^2	-
depth 300 mm	-	-	-	-	9.36	m^2	-
depth 400 mm	-	-	-	-	12.48	m^2	-
depth 450 mm	-	-	-	-	14.04	m^2	-
depth 500 mm	-	-	-	-	15.60	m^2	-
depth 600 mm	-	-	-	-	18.72	m^2	-
Add or deduct the following amounts for every £0.50 change in the material price of topsoil							
depth 100 mm	-	-	-	-	0.06	m^2	-
depth 150 mm	-	-	-	-	0.09	m^2	-
depth 200 mm	-	-	-	-	0.12	m^2	-
depth 300 mm	-	-	-	-	0.18	m^2	-
depth 400 mm	-	-	-	-	0.24	m^2	-
depth 450 mm	-	-	-	-	0.27	m^2	-
depth 500 mm	-	-	-	-	0.30	m^2	-
depth 600 mm	-	-	-	-	0.36	m^2	-
Structural soils (Amsterdam tree sand); Heicom; non compressive soil mixture for tree planting in areas to receive compressive surface treatments Backfilling and lightly compacting in layers; excavation, disposal, moving of material from delivery position and surface treatments not included; by machine							
individual treepits	61.05	0.50	9.13	2.16	61.05	m^3	72.33
in trenches	61.05	0.42	7.60	1.80	61.05	m^3	70.45

Q PAVING/PLANTING/FENCING/SITE FURNITURE

Item Excluding site overheads and profit	PC £	Labour hours	Labour £	Plant £	Material £	Unit	Total rate £
Q31 PLANTING - cont'd							
Structural soils (Amsterdam tree sand) - cont'd							
Backfilling and lightly compacting in layers; excavation, disposal, moving of material from delivery position and surface treatments not included; by hand							
individual treepits	61.05	1.60	29.20	-	61.05	m³	**90.25**
in trenches	61.05	1.33	24.27	-	61.05	m³	**85.32**
Tree staking							
J Toms Ltd; extra over trees for tree stake(s); driving 500 mm into firm ground; trimming to approved height; including two tree ties to approved pattern							
one stake; 1.52 m long x 32 x 32 mm	1.47	0.20	3.65	-	1.47	nr	**5.12**
two stakes; 1.21 m long x 25 x 25 mm	1.60	0.30	5.47	-	1.60	nr	**7.08**
two stakes; 1.52 m long x 32 x 32 mm	2.16	0.30	5.47	-	2.16	nr	**7.63**
three stakes; 1.52 m long x 32 x 32 mm	3.24	0.36	6.57	-	3.24	nr	**9.81**
Tree anchors							
Platipus Anchors Ltd; extra over trees for tree anchors							
ref RF1P rootball kit; for 75 to 220 mm girth, 2 to 4.5 m high; inclusive of "Plati-Mat" PM1	28.25	1.00	18.25	-	28.25	nr	**46.50**
ref RF2P; rootball kit; for 220 to 450 mm girth, 4.5 to 7.5 m high; inclusive of "Plati-Mat" PM2	48.09	1.33	24.27	-	48.09	nr	**72.36**
ref RF3P; rootball kit; for 450 to 750 mm girth, 7.5 to 12 m high; inclusive of "Plati-Mat" PM3	101.82	1.50	27.38	-	101.82	nr	**129.19**
ref CG1; guy fixing kit; 75 to 220 mm girth, 2 to 4.5 m high	18.17	1.67	30.42	-	18.17	nr	**48.59**
ref CG2; guy fixing kit; 220 to 450 mm girth, 4.5 to 7.5 m high	36.75	2.00	36.50	-	36.75	nr	**73.25**
installation tools; drive rod for RF1/CG1 kits	-	-	-	-	58.90	nr	**-**
installation tools; drive rod for RF2/CG2 kits	-	-	-	-	88.76	nr	**-**
Extra over trees for land drain to tree pits; 100 mm diameter perforated flexible agricultural drain; including excavating drain trench; laying pipe; backfilling	1.10	1.00	18.25	-	1.21	m	**19.46**
Tree planting; tree pit additives							
Melcourt Industries Ltd; "Topgrow"; incorporating into topsoil at 1 part "Topgrow" to 3 parts excavated topsoil; supplied in 75 l bags; pit size							
600 x 600 x 600 mm	1.44	0.02	0.36	-	1.44	nr	**1.80**
900 x 900 x 900 mm	4.86	0.06	1.09	-	4.86	nr	**5.96**
1.00 x 1.00 x 1.00 m	6.66	0.24	4.38	-	6.66	nr	**11.04**
1.25 x 1.25 x 1.25 m	13.02	0.40	7.30	-	13.02	nr	**20.32**
1.50 x 1.50 x 1.50 m	22.50	0.90	16.43	-	22.50	nr	**38.92**

Q PAVING/PLANTING/FENCING/SITE FURNITURE

Item Excluding site overheads and profit	PC £	Labour hours	Labour £	Plant £	Material £	Unit	Total rate £
Melcourt Industries Ltd; "Topgrow"; incorporating into topsoil at 1 part "Topgrow" to 3 parts excavated topsoil; supplied in 60 m³ loose loads; pit size							
600 x 600 x 600 mm	1.11	0.02	0.30	-	1.11	nr	1.41
900 x 900 x 900 mm	3.75	0.05	0.91	-	3.75	nr	4.66
1.00 x 1.00 x 1.00 m	5.14	0.20	3.65	-	5.14	nr	8.79
1.25 x 1.25 x 1.25 m	10.03	0.33	6.08	-	10.03	nr	16.12
1.50 x 1.50 x 1.50 m	17.34	0.75	13.69	-	17.34	nr	31.03
Tree planting; root barriers							
English Woodlands; "Root Director"; one-piece root control planters for installation at time of planting to divert root growth down away from pavements and out for anchorage; excavation measured separately							
RD 1050; 1050 x 1050 mm to 1300 x 1300 mm at base	89.53	0.25	4.56	-	89.53	nr	94.09
RD 640; 640 x 640 mm to 870 x 870 mm at base	58.15	0.25	4.56	-	58.15	nr	62.71
Greenleaf Horticulture; linear root deflection barriers; installed to trench measured separately							
Re-Root 2000; 1.5 mm thick	5.36	0.05	0.91	-	5.36	m	6.27
Re-Root 2000; 1.0 mm thick	4.53	0.05	0.91	-	4.53	m	5.44
Re-Root 600; 1.0 mm thick	6.97	0.05	0.91	-	6.97	m	7.88
Greenleaf Horticulture; irrigation systems; "Root Rain" tree pit irrigation systems							
"Metro"; small; 35 mm pipe diameter x 1.25 m long; for specimen shrubs and standard trees							
plastic	6.29	0.25	4.56	-	6.29	nr	10.85
plastic; with chain	8.24	0.25	4.56	-	8.24	nr	12.80
metal; with chain	10.21	0.25	4.56	-	10.21	nr	14.77
"Metro"; medium; 35 mm pipe diameter x 1.75 m long; for standard and selected standard trees							
plastic	6.53	0.29	5.21	-	6.53	nr	11.74
plastic; with chain	8.65	0.29	5.21	-	8.65	nr	13.86
metal; with chain	10.58	0.29	5.21	-	10.58	nr	15.79
"Metro"; large; 35 mm pipe diameter x 2.50 m long; for selected standards and extra heavy standards							
plastic	7.73	0.33	6.08	-	7.73	nr	13.81
plastic; with chain	9.68	0.33	6.08	-	9.68	nr	15.76
metal; with chain	12.94	0.33	6.08	-	12.94	nr	19.02

Q PAVING/PLANTING/FENCING/SITE FURNITURE

Item Excluding site overheads and profit	PC £	Labour hours	Labour £	Plant £	Material £	Unit	Total rate £
Q31 PLANTING - cont'd							
Greenleaf Horticulture - cont'd							
"Urban"; for large capacity general purpose irrigation to parkland and street verge planting							
RRUrb1; 3.0 m pipe	-	0.29	5.21	-	15.51	nr	**20.72**
RRUrb2; 5.0 m pipe	17.82	0.33	6.08	-	17.82	nr	**23.90**
RRUrb3; 8.0 m pipe	21.28	0.40	7.30	-	21.28	nr	**28.58**
"Civic"; heavy cast aluminium inlet; for heavily trafficked locations							
5.0 m pipe	30.63	0.33	6.08	-	30.63	nr	**36.71**
8.0 m pipe	34.03	0.40	7.30	-	34.03	nr	**41.33**
Mulching of tree pits; Melcourt Industries Ltd; FSC (Forest Stewardship Council) certified and fire tested							
Spreading mulch; to individual trees; maximum distance 25 m (mulch not included)							
50 mm thick	-	0.05	0.89	-	-	m²	**0.89**
75 mm thick	-	0.07	1.33	-	-	m²	**1.33**
100 mm thick	-	0.10	1.77	-	-	m²	**1.77**
Mulch; "Bark Nuggets®"; to individual trees; delivered in 80 m³ loads; maximum distance 25 m							
50 mm thick	1.86	0.05	0.89	-	1.95	m²	**2.84**
75 mm thick	2.79	0.07	1.33	-	2.93	m²	**4.26**
100 mm thick	3.72	0.10	1.77	-	3.91	m²	**5.68**
Mulch; "Bark Nuggets®"; to individual trees; delivered in 25 m³ loads; maximum distance 25 m							
50 mm thick	2.51	0.05	0.89	-	2.64	m²	**3.52**
75 mm thick	3.77	0.05	0.91	-	3.96	m²	**4.87**
100 mm thick	5.02	0.07	1.22	-	5.27	m²	**6.49**
Mulch; "Amenity Bark Mulch"; to individual trees; delivered in 80 m³ loads; maximum distance 25 m							
50 mm thick	1.13	0.05	0.89	-	1.19	m²	**2.08**
75 mm thick	1.70	0.07	1.33	-	1.78	m²	**3.12**
100 mm thick	2.27	0.10	1.77	-	2.38	m²	**4.15**
Mulch; "Amenity Bark Mulch"; to individual trees; delivered in 25 m³ loads; maximum distance 25 m							
50 mm thick	1.78	0.05	0.89	-	1.87	m²	**2.76**
75 mm thick	2.67	0.07	1.33	-	2.81	m²	**4.14**
100 mm thick	3.56	0.07	1.22	-	3.74	m²	**4.96**
Trees; planting labours only							
Bare root trees; including backfilling with previously excavated material (all other operations and materials not included)							
light standard; 6 - 8 cm girth	-	0.35	6.39	-	-	nr	**6.39**
standard; 8 - 10 cm girth	-	0.40	7.30	-	-	nr	**7.30**
selected standard; 10 - 12 cm girth	-	0.58	10.59	-	-	nr	**10.59**
heavy standard; 12 - 14 cm girth	-	0.83	15.21	-	-	nr	**15.21**
extra heavy standard; 14 - 16 cm girth	-	1.00	18.25	-	-	nr	**18.25**

Prices for Measured Works – Major Works

Q PAVING/PLANTING/FENCING/SITE FURNITURE

Item Excluding site overheads and profit	PC £	Labour hours	Labour £	Plant £	Material £	Unit	Total rate £
Root balled trees; including backfilling with previously excavated material (all other operations and materials not included)							
standard; 8 - 10 cm girth	-	0.50	9.13	-	-	nr	**9.13**
selected standard; 10 - 12 cm girth	-	0.60	10.95	-	-	nr	**10.95**
heavy standard; 12 - 14 cm girth	-	0.80	14.60	-	-	nr	**14.60**
extra heavy standard;14 - 16 cm girth	-	1.50	27.38	-	-	nr	**27.38**
16 - 18 cm girth	-	1.30	23.65	21.38	-	nr	**45.04**
18 - 20 cm girth	-	1.60	29.20	26.40	-	nr	**55.60**
20 - 25 cm girth	-	4.50	82.13	85.19	-	nr	**167.32**
25 - 30 cm girth	-	6.00	109.50	112.13	-	nr	**221.63**
30 - 35 cm girth	-	11.00	200.75	197.92	-	nr	**398.67**
Tree planting; containerized trees; nursery stock; James Coles & Sons (Nurseries) Ltd							
"Acer platanoides"; including backfillling with excavated material (other operations not included)							
standard; 8 - 10 cm girth	35.00	0.48	8.76	-	35.00	nr	**43.76**
selected standard; 10 - 12 cm girth	55.00	0.56	10.23	-	55.00	nr	**65.23**
heavy standard; 12 - 14 cm girth	65.00	0.76	13.95	-	65.00	nr	**78.95**
extra heavy standard; 14 - 16 cm girth	75.00	1.20	21.90	-	75.00	nr	**96.90**
"Carpinus betulus"; including backfillling with excavated material (other operations not included)							
standard; 8 - 10 cm girth	40.00	0.48	8.76	-	40.00	nr	**48.76**
selected standard; 10 - 12 cm girth	60.00	0.56	10.23	-	60.00	nr	**70.23**
heavy standard; 12 - 14 cm girth	75.00	0.76	13.95	-	75.00	nr	**88.95**
extra heavy standard; 14 - 16 cm girth	85.00	1.20	21.90	-	85.00	nr	**106.90**
"Fraxinus excelsior"; including backfillling with excavated material (other operations not included)							
standard; 8 - 10 cm girth	35.00	0.48	8.76	-	35.00	nr	**43.76**
selected standard; 10 - 12 cm girth	55.00	0.56	10.23	-	55.00	nr	**65.23**
heavy standard; 12 - 14 cm girth	65.00	0.76	13.95	-	65.00	nr	**78.95**
extra heavy standard; 14 - 16 cm girth	75.00	1.20	21.90	-	75.00	nr	**96.90**
"Prunus avium Plena"; including backfillling with excavated material (other operations not included)							
standard; 8 - 10 cm girth	35.00	0.40	7.30	-	35.00	nr	**42.30**
selected standard; 10 - 12 cm girth	60.00	0.56	10.23	-	60.00	nr	**70.23**
heavy standard; 12 - 14 cm girth	70.00	0.76	13.95	-	70.00	nr	**83.95**
extra heavy standard; 14 - 16 cm girth	80.00	1.20	21.90	-	80.00	nr	**101.90**
"Quercus robur"; including backfillling with excavated material (other operations not included)							
standard; 8 - 10 cm girth	40.00	0.48	8.76	-	40.00	nr	**48.76**
selected standard; 10 - 12 cm girth	60.00	0.56	10.23	-	60.00	nr	**70.23**
heavy standard; 12 - 14 cm girth	75.00	0.76	13.95	-	75.00	nr	**88.95**
extra heavy standard; 14 - 16 cm girth	85.00	1.20	21.90	-	85.00	nr	**106.90**
"Betula utilis jaquemontii"; multistemmed; including backfillling with excavated material (other operations not included)							
175/200 mm high; feathered	23.00	0.48	8.76	-	23.00	nr	**31.76**
200/250 mm high; feathered	30.00	0.56	10.23	-	30.00	nr	**40.23**
250/300 mm high	35.00	0.76	13.95	-	35.00	nr	**48.95**
300/350 mm high	55.00	1.20	21.90	-	55.00	nr	**76.90**

Q PAVING/PLANTING/FENCING/SITE FURNITURE

Item Excluding site overheads and profit	PC £	Labour hours	Labour £	Plant £	Material £	Unit	Total rate £
Q31 PLANTING - cont'd							
Tree planting; root balled trees; advanced nursery stock and semi-mature - General Preamble: The cost of planting semi-mature trees will depend on the size and species, and on the access to the site for tree handling machines. Prices should be obtained for individual trees and planting.							
Tree planting; bare root trees; nursery stock; James Coles & Sons (Nurseries) Ltd							
"Acer platanoides"; including backfillling with excavated material (other operations not included)							
light standard; 6 - 8 cm girth	9.00	0.35	6.39	-	9.00	nr	**15.39**
standard; 8 - 10 cm girth	11.25	0.40	7.30	-	11.25	nr	**18.55**
selected standard; 10 - 12 cm girth	18.25	0.58	10.59	-	18.25	nr	**28.84**
heavy standard; 12 - 14 cm girth	33.50	0.83	15.21	-	33.50	nr	**48.71**
extra heavy standard; 14 - 16 cm girth	42.00	1.00	18.25	-	42.00	nr	**60.25**
"Carpinus betulus"; including backfillling with excavated material (other operations not included)							
light standard; 6 - 8 cm girth	13.25	0.35	6.39	-	13.25	nr	**19.64**
standard; 8 - 10 cm girth	26.50	0.40	7.30	-	26.50	nr	**33.80**
selected standard; 10 - 12 cm girth	37.75	0.58	10.59	-	37.75	nr	**48.34**
heavy standard; 12 - 14 cm girth	39.25	0.83	15.21	-	39.25	nr	**54.47**
extra heavy standard; 14 - 16 cm girth	46.25	1.00	18.25	-	46.25	nr	**64.50**
"Fraxinus excelsior"; including backfillling with excavated material (other operations not included)							
light standard; 6 - 8 cm girth	9.75	0.35	6.39	-	9.75	nr	**16.14**
standard; 8 - 10 cm girth	16.00	0.40	7.30	-	16.00	nr	**23.30**
selected standard; 10 - 12 cm girth	22.50	0.58	10.59	-	22.50	nr	**33.09**
heavy standard; 12 - 14 cm girth	39.25	0.83	15.15	-	39.25	nr	**54.40**
extra heavy standard; 14 - 16 cm girth	47.50	1.00	18.25	-	47.50	nr	**65.75**
"Prunus avium Plena"; including backfillling with excavated material (other operations not included)							
light standard; 6 - 8 cm girth	9.75	0.36	6.64	-	9.75	nr	**16.39**
standard; 8 - 10 cm girth	16.00	0.40	7.30	-	16.00	nr	**23.30**
selected standard; 10 - 12 cm girth	29.50	0.58	10.59	-	29.50	nr	**40.09**
heavy standard; 12 - 14 cm girth	47.50	0.83	15.21	-	47.50	nr	**62.71**
extra heavy standard; 14 - 16 cm girth	63.00	1.00	18.25	-	63.00	nr	**81.25**
"Quercus robur"; including backfillling with excavated material (other operations not included)							
light standard; 6 - 8 cm girth	21.00	0.35	6.39	-	21.00	nr	**27.39**
standard; 8 - 10 cm girth	30.75	0.40	7.30	-	30.75	nr	**38.05**
selected standard; 10 - 12 cm girth	42.00	0.58	10.59	-	42.00	nr	**52.59**
heavy standard; 12 - 14 cm girth	58.75	0.83	15.21	-	58.75	nr	**73.96**
"Robinia pseudoacacia Frisia"; including backfillling with excavated material (other operations not included)							
light standard; 6 - 8 cm girth	21.00	0.35	6.39	-	21.00	nr	**27.39**
standard; 8 - 10 cm girth	28.00	0.40	7.30	-	28.00	nr	**35.30**
selected standard; 10 - 12 cm girth	44.75	0.58	10.59	-	44.75	nr	**55.34**

Q PAVING/PLANTING/FENCING/SITE FURNITURE

Item Excluding site overheads and profit	PC £	Labour hours	Labour £	Plant £	Material £	Unit	Total rate £
Tree planting; root balled trees; nursery stock; James Coles & Sons (Nurseries) Ltd							
"Acer platanoides"; including backfillling with excavated material (other operations not included)							
standard; 8 - 10 cm girth	18.75	0.48	8.76	-	18.75	nr	**27.51**
selected standard; 10 - 12 cm girth	28.25	0.56	10.23	-	28.25	nr	**38.48**
heavy standard; 12 - 14 cm girth	48.50	0.76	13.95	-	48.50	nr	**62.45**
extra heavy standard; 14 - 16 cm girth	57.00	1.20	21.90	-	57.00	nr	**78.90**
"Carpinus betulus"; including backfillling with excavated material (other operations not included)							
standard; 8 - 10 cm girth	34.00	0.48	8.76	-	34.00	nr	**42.76**
selected standard; 10 - 12 cm girth	47.75	0.56	10.23	-	47.75	nr	**57.98**
heavy standard; 12 - 14 cm girth	79.50	0.76	13.95	-	79.50	nr	**93.45**
extra heavy standard; 14 - 16 cm girth	106.00	1.20	21.90	-	106.00	nr	**127.90**
"Fraxinus excelsior"; including backfillling with excavated material (other operations not included)							
standard; 8 - 10 cm girth	23.50	0.48	8.76	-	23.50	nr	**32.26**
selected standard; 10 - 12 cm girth	32.50	0.56	10.23	-	32.50	nr	**42.73**
heavy standard; 12 - 14 cm girth	54.25	0.76	13.95	-	54.25	nr	**68.20**
extra heavy standard; 14 - 16 cm girth	62.50	1.20	21.90	-	62.50	nr	**84.40**
"Prunus avium Plena"; including backfillling with excavated material (other operations not included)							
standard; 8 - 10 cm girth	30.50	0.40	7.30	-	30.50	nr	**37.80**
selected standard; 10 - 12 cm girth	30.50	0.56	10.23	-	30.50	nr	**40.73**
heavy standard; 12 - 14 cm girth	62.50	0.76	13.95	-	62.50	nr	**76.45**
extra heavy standard; 14 - 16 cm girth	78.00	1.20	21.90	-	78.00	nr	**99.90**
"Quercus robur"; including backfillling with excavated material (other operations not included)							
standard; 8 - 10 cm girth	38.25	0.48	8.76	-	38.25	nr	**47.01**
selected standard; 10 - 12 cm girth	52.00	0.56	10.23	-	52.00	nr	**62.23**
heavy standard; 12 - 14 cm girth	73.75	0.76	13.95	-	73.75	nr	**87.70**
extra heavy standard; 14 - 16 cm girth	107.75	1.20	21.90	-	107.75	nr	**129.65**
"Robinia pseudoacacia Frisia"; including backfillling with excavated material (other operations not included)							
standard; 8 - 10 cm girth	35.50	0.48	8.76	-	35.50	nr	**44.26**
selected standard; 10 - 12 cm girth	54.75	0.56	10.23	-	54.75	nr	**64.98**
heavy standard; 12 - 14 cm girth	98.00	0.76	13.95	-	98.00	nr	**111.95**
Tree planting; "Airpot" container grown trees; advanced nursery stock and semi-mature; Deepdale Trees Ltd							
"Acer platanoides Emerald Queen"; including backfilling with excavated material (other operations not included)							
16 - 18 cm girth	95.00	1.98	36.13	39.14	95.00	nr	**170.27**
18 - 20 cm girth	130.00	2.18	39.75	41.96	130.00	nr	**211.70**
20 - 25 cm girth	190.00	2.38	43.36	46.96	190.00	nr	**280.33**
25 - 30 cm girth	250.00	2.97	54.20	71.31	250.00	nr	**375.51**
30 - 35 cm girth	450.00	3.96	72.27	78.28	450.00	nr	**600.55**

Q PAVING/PLANTING/FENCING/SITE FURNITURE

Item Excluding site overheads and profit	PC £	Labour hours	Labour £	Plant £	Material £	Unit	Total rate £
Q31 PLANTING - cont'd							
Tree planting - cont'd							
"Aesculus briotti"; including backfilling with excavated material (other operations not included)							
16 - 18 cm girth	90.00	1.98	36.13	39.14	90.00	nr	165.27
18 - 20 cm girth	130.00	1.60	29.20	41.96	130.00	nr	201.16
20 - 25 cm girth	180.00	2.38	43.36	46.96	180.00	nr	270.33
25 - 30 cm girth	240.00	2.97	54.20	71.31	240.00	nr	365.51
30 - 35 cm girth	330.00	3.96	72.27	78.28	330.00	nr	480.55
"Prunus avium Flora Plena"; including backfilling with excavated material (other operations not included)							
16 - 18 cm girth	95.00	1.98	36.13	39.14	95.00	nr	170.27
18 - 20 cm girth	130.00	1.60	29.20	41.96	130.00	nr	201.16
20 - 25 cm girth	190.00	2.38	43.36	46.96	190.00	nr	280.33
25 - 30 cm girth	250.00	2.97	54.20	71.31	250.00	nr	375.51
30 - 35 cm girth	350.00	3.96	72.27	78.28	350.00	nr	500.55
"Quercus palustris - Pin Oak"; including backfilling with excavated material (other operations not included)							
16 - 18 cm girth	99.75	1.98	36.13	39.14	99.75	nr	175.02
18 - 20 cm girth	136.50	1.60	29.20	41.96	136.50	nr	207.66
20 - 25 cm girth	199.50	2.38	43.36	46.96	199.50	nr	289.83
25 - 30 cm girth	262.50	2.97	54.20	71.31	262.50	nr	388.01
30 - 35 cm girth	393.75	3.96	72.27	78.28	393.75	nr	544.30
Tree planting; "Airpot" container grown trees; semi-mature and mature trees; Deepdale Trees Ltd; planting and back filling; planted by telehandler or by crane; delivery included; all other operations priced separately							
Semi mature trees; indicative prices							
40 - 45 cm girth	550.00	4.00	73.00	48.48	550.00	nr	671.48
45 - 50 cm girth	750.00	4.00	73.00	48.48	750.00	nr	871.48
55 - 60 cm girth	1350.00	6.00	109.50	48.48	1350.00	nr	1507.98
60 - 70 cm girth	2500.00	7.00	127.75	65.60	2500.00	nr	2693.35
70 - 80 cm girth	3500.00	7.50	136.88	82.72	3500.00	nr	3719.59
80 - 90 cm girth	4500.00	8.00	146.00	96.95	4500.00	nr	4742.95
Tree planting; rootballed trees; advanced nursery stock and semi-mature; Lorenz von Ehren							
"Acer platanoides Emerald Queen"; including backfilling with excavated material (other operations not included)							
16 - 18 cm girth	85.00	1.30	23.65	3.59	85.00	nr	112.25
18 - 20 cm girth	100.00	1.60	29.20	3.59	100.00	nr	132.79
20 - 25 cm girth	125.00	4.50	82.13	16.33	125.00	nr	223.46
25 - 30 cm girth	165.00	6.00	109.50	20.32	165.00	nr	294.82
30 - 35 cm girth	300.00	11.00	200.75	27.20	300.00	nr	527.95

Q PAVING/PLANTING/SITE FURNITURE

Item Excluding site overheads and profit	PC £	Labour hours	Labour £	Plant £	Material £	Unit	Total rate £
"Aesculus carnea Briotti"; including backfilling with excavated material (other operations not included)							
16 - 18 cm girth	130.00	1.30	23.65	3.59	130.00	nr	157.25
18 - 20 cm girth	160.00	1.60	29.20	3.59	160.00	nr	192.79
20 - 25 cm girth	200.00	4.50	82.13	16.33	200.00	nr	298.46
25 - 30 cm girth	220.00	6.00	109.50	20.32	220.00	nr	349.82
30 - 35 cm girth	290.00	11.00	200.75	27.20	290.00	nr	517.95
"Prunus avium Plena"; including backfilling with excavated material (other operations not included)							
16 - 18 cm girth	90.00	1.30	23.65	3.59	90.00	nr	117.25
18 - 20 cm girth	110.00	1.60	29.20	3.59	110.00	nr	142.79
20 - 25 cm girth	140.00	4.50	82.13	16.33	140.00	nr	238.46
25 - 30 cm girth	170.00	6.00	109.50	20.32	170.00	nr	299.82
30 - 35 cm girth	270.00	11.00	200.75	23.91	270.00	nr	494.66
"Quercus palustris - Pin Oak"; including backfilling with excavated material (other operations not included)							
16 - 18 cm girth	110.00	1.30	23.65	3.59	110.00	nr	137.25
18 - 20 cm girth	130.00	1.60	29.20	3.59	130.00	nr	162.79
20 - 25 cm girth	140.00	4.50	82.13	16.33	140.00	nr	238.46
25 - 30 cm girth	220.00	6.00	109.50	20.32	220.00	nr	349.82
30 - 35 cm girth	340.00	11.00	200.75	27.20	340.00	nr	567.95
"Tilia cordata Greenspire"; including backfilling with excavated material (other operations not included)							
16 - 18 cm girth	95.00	1.30	23.65	3.59	95.00	nr	122.25
18 - 20 cm girth	100.00	1.60	29.20	3.59	100.00	nr	132.79
20 - 25 cm girth	120.00	4.50	82.13	16.33	120.00	nr	218.46
25 - 30 cm girth; 5 x transplanted; 4.0 - 5.0 m tall	155.00	6.00	109.50	20.32	155.00	nr	284.82
30 - 35 cm girth; 5 x transplanted; 5.0 - 7.0 m tall	260.00	11.00	200.75	27.20	260.00	nr	487.95
"Betula pendula (3 stems)"; including backfilling with excavated material (other operations not included)							
3.0 - 3.5 m high	70.00	1.98	36.13	39.14	70.00	nr	145.27
3.5 - 4.0 m high	90.00	1.60	29.20	41.96	90.00	nr	161.16
4.0 - 4.5 m high	140.00	2.38	43.36	46.96	140.00	nr	230.33
4.5 - 5.0 m high	170.00	2.97	54.20	71.31	170.00	nr	295.51
5.0 - 6.0 m high	210.00	3.96	72.27	78.28	210.00	nr	360.55
6.0 - 7.0 m high	400.00	4.50	82.13	92.64	400.00	nr	574.76
"Pinus sylvestris"; including backfilling with excavated material (other operations not included)							
3.0 - 3.5 m high	400.00	1.98	36.13	39.14	400.00	nr	475.27
3.5 - 4.0 m high	500.00	1.60	29.20	41.96	500.00	nr	571.16
4.0 - 4.5 m high	800.00	2.38	43.36	46.96	800.00	nr	890.33
4.5 - 5.0 m high	1100.00	2.97	54.20	71.31	1100.00	nr	1225.51
5.0 - 6.0 m high	1500.00	3.96	72.27	78.28	1500.00	nr	1650.55
6.0 - 7.0 m high	2500.00	4.50	82.13	95.48	2500.00	nr	2677.61

Q PAVING/PLANTING/FENCING/SITE FURNITURE

Item Excluding site overheads and profit	PC £	Labour hours	Labour £	Plant £	Material £	Unit	Total rate £
Q31 PLANTING - cont'd							
Tree planting; containerized trees; Lorenz von Ehren; to the tree prices above add for Airpot containerization only							
Tree size							
20 - 25 cm	-	-	-	-	60.00	nr	-
25 - 30 cm	-	-	-	-	100.00	nr	-
30 - 35 cm	-	-	-	-	150.00	nr	-
35 - 40 cm	-	-	-	-	200.00	nr	-
40 - 50 cm	-	-	-	-	250.00	nr	-
50 - 70 cm	-	-	-	-	300.00	nr	-
70 - 80 cm	-	-	-	-	400.00	nr	-
80 - 90 cm	-	-	-	-	500.00	nr	-
Tree planting; rootballed trees; semi-mature and mature trees; Lorenz von Ehren; planting and back filling; planted by telehandler or by crane; delivery included; all other operations priced separately							
Semi-mature trees							
40 - 45 cm girth	500.00	8.00	146.00	62.17	500.00	nr	708.17
45 - 50 cm girth	700.00	8.00	146.00	83.35	700.00	nr	929.35
50 - 60 cm girth	1200.00	10.00	182.50	100.54	1200.00	nr	1483.04
60 - 70 cm girth	2200.00	15.00	273.75	194.00	2200.00	nr	2667.75
70 - 80 cm girth	3200.00	18.00	328.50	186.00	3200.00	nr	3714.50
80 - 90 cm girth	4500.00	18.00	328.50	186.00	4500.00	nr	5014.50
90 - 100 cm girth	5900.00	18.00	328.50	186.00	5900.00	nr	6414.50
Tree planting; tree protection - General Preamble: Care must be taken to ensure that tree grids and guards are removed when trees grow beyond the specified diameter of guard.							
Tree planting; tree protection Crowders Nurseries; "Crowders Tree Tube"; olive green							
1200 mm high x 80 x 80 mm stakes; 1500 mm high for "Crowders Tree Tube"; driving into ground	0.84	0.07	1.22	-	0.84	nr	2.06
	0.64	0.05	0.91	-	0.64	nr	1.55
Crowders Nurseries; expandable plastic tree guards; including 25 mm softwood stakes							
500 mm high	0.66	0.17	3.04	-	0.66	nr	3.70
1.00 m high	0.70	0.17	3.04	-	0.70	nr	3.74
English Woodlands; Weldmesh Tree Guards; nailing to tree stakes (tree stakes not included)							
1800 mm high x 200 mm diameter	17.50	0.33	6.08	-	17.50	nr	23.58
1800 mm high x 225 mm diameter	15.00	0.33	6.08	-	15.00	nr	21.08
1800 mm high x 250 mm diameter	15.50	0.25	4.56	-	15.50	nr	20.06
1800 mm high x 300 mm diameter	16.50	0.33	6.08	-	16.50	nr	22.58
J. Toms Ltd; spiral rabbit guards; clear or brown							
450 x 38 mm	0.19	0.03	0.61	-	0.19	nr	0.80
610 x 38 mm	0.21	0.03	0.61	-	0.21	nr	0.82

Q PAVING/PLANTING/FENCING/SITE FURNITURE

Item Excluding site overheads and profit	PC £	Labour hours	Labour £	Plant £	Material £	Unit	Total rate £
English Woodlands; Plastic Mesh Tree Guards; black; supplied in 50 m rolls							
13 x 13 mm small mesh; roll width 60 cm	0.97	0.04	0.65	-	1.45	nr	2.10
13 x 13 mm small mesh; roll width 120 cm	1.73	0.06	1.07	-	2.38	nr	3.46
English Woodlands; Plastic Mesh Tree Guard in pre-cut pieces; 600 x 150 mm; supplied flat packed	0.81	0.06	1.07	-	1.29	nr	2.36
Tree guards of 3 nr 2.40 m x 100 mm stakes; driving 600 mm into firm ground; bracing with timber braces at top and bottom; including 3 strands barbed wire	0.43	1.00	18.25	-	0.43	nr	18.68
English Woodlands; strimmer guard in heavy duty black plastic; 225 mm high	2.72	0.07	1.22	-	2.72	nr	3.94
Tubex Ltd; "Standard Treeshelter" inclusive of 25 mm stake; prices shown for quantities of 500 nr							
0.6 m high	0.56	0.05	0.91	-	0.76	each	1.67
0.75 m high	0.67	0.05	0.91	-	0.98	each	1.89
1.2 m high	2.24	0.05	0.91	-	2.57	each	3.48
1.5 m high	1.29	0.05	0.91	-	1.69	each	2.60
Tubex Ltd; "Shrubshelter" inclusive of 25 mm stake; prices shown for quantities of 500 nr							
"Ecostart" shelter for forestry transplants and seedlings	0.67	0.07	1.22	-	0.67	nr	1.89
0.6 m high	1.26	0.07	1.22	-	1.46	nr	2.68
0.75 m high	1.38	0.07	1.22	-	1.62	nr	2.84
Extra over trees for spraying with antidesiccant spray; "Wiltpruf"							
selected standards; standards; light standards	2.62	0.20	3.65	-	2.62	nr	6.27
standards; heavy standards	4.36	0.25	4.56	-	4.36	nr	8.93
Hedges							
Excavating trench for hedges; depositing soil alongside trench; by machine							
300 mm deep x 300 mm wide	-	0.03	0.55	0.26	-	m	0.81
300 mm deep x 450 mm wide	-	0.05	0.82	0.39	-	m	1.21
Excavating trench for hedges; depositing soil alongside trench; by hand							
300 mm deep x 300 mm wide	-	0.12	2.19	-	-	m	2.19
300 mm deep x 450 mm wide	-	0.23	4.11	-	-	m	4.11
Setting out; notching out; excavating trench; breaking up subsoil to minimum depth 300 mm							
minimum 400 mm deep	-	0.25	4.56	-	-	m	4.56
Hedge planting; including backfill with excavated topsoil; PC £0.33/nr							
single row; 200 mm centres	1.65	0.06	1.14	-	1.65	m	2.79
single row; 300 mm centres	1.10	0.06	1.01	-	1.10	m	2.11
single row; 400 mm centres	0.82	0.04	0.76	-	0.82	m	1.59
single row; 500 mm centres	0.66	0.03	0.61	-	0.66	m	1.27
double row; 200 mm centres	3.30	0.17	3.04	-	3.30	m	6.34
double row; 300 mm centres	2.20	0.13	2.43	-	2.20	m	4.63
double row; 400 mm centres	1.65	0.08	1.52	-	1.65	m	3.17
double row; 500 mm centres	1.32	0.07	1.22	-	1.32	m	2.54
Extra over hedges for incorporating manure; at 1 m^3 per 30 m	0.77	0.03	0.46	-	0.77	m	1.22

Q PAVING/PLANTING/FENCING/SITE FURNITURE

Item Excluding site overheads and profit	PC £	Labour hours	Labour £	Plant £	Material £	Unit	Total rate £
Q31 PLANTING - cont'd							
Topiary **Clipped topiary; Lorenz von Ehren; German field grown clipped and transplanted as detailed; planted to plantpit; including backfilling with excavated material and TPMC**							
Buxus sempervirens (box); balls							
300 mm diameter; 3 x transplanted; container grown or rootballed	13.00	0.25	4.56	-	22.22	nr	**26.78**
500 mm diameter; 4 x transplanted; wire rootballed	43.00	1.50	27.38	5.47	111.86	nr	**144.71**
900 mm diameter; 5 x transplanted; wire rootballed	195.00	2.75	50.19	5.47	309.02	nr	**364.67**
1300 mm diameter; 6 x transplanted; wire rootballed	650.00	3.10	56.58	6.56	761.58	nr	**824.72**
Buxus sempervirens (box); pyramids							
500 mm high; 3 x transplanted; container grown or rootballed	30.00	1.50	27.38	5.47	98.86	nr	**131.71**
900 mm high; 4 x transplanted; wire rootballed	125.00	2.75	50.19	5.47	239.02	nr	**294.67**
1300 mm high; 5 x transplanted; wire rootballed	390.00	3.10	56.58	6.56	501.58	nr	**564.72**
Buxus sempervirens (box); truncated pyramids							
500 mm high; 3 x transplanted; container grown or rootballed	78.00	1.50	27.38	5.47	146.86	nr	**179.71**
900 mm high; 4 x transplanted; wire rootballed	300.00	2.75	50.19	5.47	414.02	nr	**469.67**
Buxus sempervirens (box); truncated cone							
1300 mm high; 5 x transplanted; wire rootballed	950.00	3.10	56.58	6.56	1061.58	nr	**1124.72**
Buxus sempervirens (box); cubes							
500 mm square; 4 x transplanted; rootballed	80.00	1.50	27.38	5.47	148.86	nr	**181.71**
900 mm square; 5 x transplanted; wire rootballed	360.00	2.75	50.19	5.47	474.02	nr	**529.67**
Taxus baccata (yew); balls							
500 mm diameter; 4 x transplanted; rootballed	52.00	1.50	27.38	5.47	120.86	nr	**153.71**
900 mm diameter; 5 x transplanted; wire rootballed	170.00	2.75	50.19	5.47	284.02	nr	**339.67**
1300 mm diameter; 7 x transplanted; wire rootballed	560.00	3.10	56.58	6.56	671.58	nr	**734.72**
Taxus baccata (yew); cones							
800 mm high; 4 x transplanted; wire rootballed	55.00	1.50	27.38	5.47	123.86	nr	**156.71**
1500 mm high; 5 x transplanted; wire rootballed	140.00	2.00	36.50	5.47	224.86	nr	**266.83**
2500 mm high; 7 x transplanted; wire rootballed	470.00	3.10	56.58	6.56	581.58	nr	**644.72**
Taxus baccata (yew); cubes							
500 mm square; 4 x transplanted; rootballed	65.00	1.50	27.38	5.47	133.86	nr	**166.71**
900 mm square; 5 x transplanted; wire rootballed	310.00	2.75	50.19	5.47	424.02	nr	**479.67**

Q PAVING/PLANTING/FENCING/SITE FURNITURE

Item Excluding site overheads and profit	PC £	Labour hours	Labour £	Plant £	Material £	Unit	Total rate £
Taxus baccata (yew); pyramids							
900 mm high; 4 x transplanted; wire rootballed	120.00	1.50	27.38	5.47	188.86	nr	221.71
1500 mm high; 5 x transplanted; wire rootballed	300.00	3.10	56.58	6.56	411.58	nr	474.72
2500 mm high; 7 x transplanted; wire rootballed	870.00	4.00	73.00	10.94	1013.01	nr	1096.95
Carpinus betulus (common hornbeam); columns; round base							
800 mm wide x 2000 mm high; 4 x transplanted; wire rootballed	150.00	3.10	56.58	6.56	261.58	nr	324.72
800 mm wide x 2750 mm high; 4 x transplanted; wire rootballed	300.00	4.00	73.00	10.94	443.01	nr	526.95
Carpinus betulus (common hornbeam); cones							
3 m high; 5 x transplanted; wire rootballed	270.00	4.00	73.00	10.94	413.01	nr	496.95
4 m high; 6 x transplanted; wire rootballed	560.00	5.00	91.25	10.94	729.73	nr	831.92
Carpinus betulus 'Fastigiata'; pyramids							
4 m high; 6 x transplanted; wire rootballed	960.00	5.00	91.25	10.94	1129.73	nr	1231.92
7 m high; 7 x transplanted; wire rootballed	2600.00	7.50	136.88	16.41	2809.73	nr	2963.01
Shrub planting - General Preamble: For preparation of planting areas see "Cultivation" at the beginning of the section on planting.							
Shrub planting Setting out; selecting planting from holding area; loading to wheelbarrows; planting as plan or as directed; distance from holding area maximum 50 m; plants 2 - 3 litre containers							
single plants not grouped	-	0.04	0.73	-	-	nr	0.73
plants in groups of 3 - 5 nr	-	0.03	0.46	-	-	nr	0.46
plants in groups of 10 - 100 nr	-	0.02	0.30	-	-	nr	0.30
plants in groups of 100 nr minimum	-	0.01	0.21	-	-	nr	0.21
Forming planting holes; in cultivated ground (cultivating not included); by mechanical auger; trimming holes by hand; depositing excavated material alongside holes							
250 mm diameter	-	0.03	0.61	0.04	-	nr	0.65
250 x 250 mm	-	0.04	0.73	0.08	-	nr	0.81
300 x 300 mm	-	0.08	1.37	0.10	-	nr	1.46
Hand excavation; forming planting holes; in cultivated ground (cultivating not included); depositing excavated material alongside holes							
100 x 100 x 100 mm deep; with mattock or hoe	-	0.01	0.12	-	-	nr	0.12
250 x 250 x 300 mm deep	-	0.04	0.73	-	-	nr	0.73
300 x 300 x 300 mm deep	-	0.06	1.01	-	-	nr	1.01
400 x 400 x 400 mm deep	-	0.13	2.28	-	-	nr	2.28
500 x 500 x 500 mm deep	-	0.25	4.56	-	-	nr	4.56
600 x 600 x 600 mm deep	-	0.43	7.90	-	-	nr	7.90
900 x 900 x 600 mm deep	-	1.00	18.25	-	-	nr	18.25
1.00 x 1.00 m x 600 mm deep	-	1.23	22.45	-	-	nr	22.45
1.25 x 1.25 m x 600 mm deep	-	1.93	35.22	-	-	nr	35.22

Q PAVING/PLANTING/FENCING/SITE FURNITURE

Item Excluding site overheads and profit	PC £	Labour hours	Labour £	Plant £	Material £	Unit	Total rate £
Q31 PLANTING - cont'd							
Shrub planting - cont'd							
Hand excavation; forming planting holes; in uncultivated ground; depositing excavated material alongside holes							
100 x 100 x 100 mm deep; with mattock or hoe	-	0.03	0.46	-	-	nr	**0.46**
250 x 250 x 300 mm deep	-	0.06	1.01	-	-	nr	**1.01**
300 x 300 x 300 mm deep	-	0.06	1.14	-	-	nr	**1.14**
400 x 400 x 400 mm deep	-	0.25	4.56	-	-	nr	**4.56**
500 x 500 x 500 mm deep	-	0.33	5.94	-	-	nr	**5.94**
600 x 600 x 600 mm deep	-	0.55	10.04	-	-	nr	**10.04**
900 x 900 x 600 mm deep	-	1.25	22.81	-	-	nr	**22.81**
1.00 x 1.00 m x 600 mm deep	-	1.54	28.06	-	-	nr	**28.06**
1.25 x 1.25 m x 600 mm deep	-	2.41	44.03	-	-	nr	**44.03**
Bare root planting; to planting holes (forming holes not included); including backfilling with excavated material (bare root plants not included)							
bare root 1+1; 30 - 90 mm high	-	0.02	0.30	-	-	nr	**0.30**
bare root 1+2; 90 - 120 mm high	-	0.02	0.30	-	-	nr	**0.30**
Containerized planting; to planting holes (forming holes not included); including backfilling with excavated material (shrub or ground cover not included)							
9 cm pot	-	0.01	0.18	-	-	nr	**0.18**
2 litre container	-	0.02	0.36	-	-	nr	**0.36**
3 litre container	-	0.02	0.41	-	-	nr	**0.41**
5 litre container	-	0.03	0.61	-	-	nr	**0.61**
10 litre container	-	0.05	0.91	-	-	nr	**0.91**
15 litre container	-	0.07	1.22	-	-	nr	**1.22**
20 litre container	-	0.08	1.52	-	-	nr	**1.52**
Shrub planting; 2 litre containerized plants; in cultivated ground (cultivating not included); PC £2.65/nr							
average 2 plants per m^2	-	0.06	1.02	-	5.30	m^2	**6.32**
average 3 plants per m^2	-	0.08	1.53	-	7.95	m^2	**9.48**
average 4 plants per m^2	-	0.11	2.05	-	10.60	m^2	**12.65**
average 6 plants per m^2	-	0.17	3.07	-	15.90	m^2	**18.97**
Extra over shrubs for stakes	0.64	0.02	0.30	-	0.64	nr	**0.94**
Composted bark soil conditioners; 20 m^3 loads; on beds by mechanical loader; spreading and rotavating into topsoil; by machine							
50 mm thick	151.50	-	-	6.20	151.50	100 m^2	**157.70**
100 mm thick	303.00	-	-	10.45	318.15	100 m^2	**328.60**
150 mm thick	454.50	-	-	14.87	477.23	100 m^2	**492.09**
200 mm thick	606.00	-	-	19.20	636.30	100 m^2	**655.50**
Mushroom compost; 25 m^3 loads; delivered not further than 25 m from location; cultivating into topsoil by pedestrian operated machine							
50 mm thick	106.00	2.86	52.14	1.77	106.00	100 m^2	**159.91**
100 mm thick	212.00	6.05	110.39	1.77	212.00	100 m^2	**324.16**
150 mm thick	318.00	8.90	162.51	1.77	318.00	100 m^2	**482.28**
200 mm thick	424.00	12.90	235.51	1.77	424.00	100 m^2	**661.28**

Q PAVING/PLANTING/FENCING/SITE FURNITURE

Item Excluding site overheads and profit	PC £	Labour hours	Labour £	Plant £	Material £	Unit	Total rate £
Mushroom compost; 25 m³ loads; on beds by mechanical loader; spreading and rotavating into topsoil by tractor drawn rotavator							
50 mm thick	106.00	-	-	6.20	106.00	100 m²	112.20
100 mm thick	212.00	-	-	10.45	212.00	100 m²	222.45
150 mm thick	318.00	-	-	14.87	318.00	100 m²	332.87
200 mm thick	424.00	-	-	19.20	424.00	100 m²	443.20
Manure; 20 m³ loads; delivered not further than 25 m from location; cultivating into topsoil by pedestrian operated machine							
50 mm thick	225.00	2.86	52.14	1.77	225.00	100 m²	278.91
100 mm thick	229.80	6.05	110.39	1.77	241.29	100 m²	353.45
150 mm thick	344.70	8.90	162.51	1.77	361.94	100 m²	526.22
200 mm thick	459.60	12.90	235.51	1.77	482.58	100 m²	719.86
Fertilizer (7+7+7); PC £1.13/kg; to beds; by hand							
35 g/m²	3.95	0.17	3.04	-	3.95	100 m²	7.00
50 g/m²	5.65	0.17	3.04	-	5.65	100 m²	8.69
70 g/m²	7.91	0.17	3.04	-	7.91	100 m²	10.95
Fertilizers; "Enmag"; PC £2.63/kg; controlled release fertilizer; to beds; by hand							
35 g/m²	7.35	0.17	3.04	-	7.35	100 m²	10.40
50 g/m²	10.51	0.17	3.04	-	10.51	100 m²	13.55
70 g/m²	15.76	0.17	3.04	-	15.76	100 m²	18.80
Note: For machine incorporation of fertilizers and soil conditioners see "Cultivation".							
Herbaceous and groundcover planting							
Herbaceous plants; PC £1.20/nr; including forming planting holes in cultivated ground (cultivating not included); backfilling with excavated material; 1 litre containers							
average 4 plants per m² - 500 mm centres	-	0.09	1.70	-	4.80	m²	6.50
average 6 plants per m² - 408 mm centres	-	0.14	2.56	-	7.20	m²	9.76
average 8 plants per m² - 354 mm centres	-	0.19	3.41	-	9.60	m²	13.01
Note: For machine incorporation of fertilizers and soil conditioners see "Cultivation". Plant support netting; Bridport Gundry; on 50 mm diameter stakes; 750 mm long; driving into ground at 1.50 m centres							
green extruded plastic mesh; 125 mm square	0.49	0.04	0.73	-	0.49	m²	1.22
Bulb planting							
Bulbs; including forming planting holes in cultivated area (cultivating not included); backfilling with excavated material							
small	13.00	0.83	15.21	-	13.00	100 nr	28.21
medium	22.00	0.83	15.21	-	22.00	100 nr	37.21
large	25.00	0.91	16.59	-	25.00	100 nr	41.59
Bulbs; in grassed area; using bulb planter; including backfilling with screened topsoil or peat and cut turf plug							
small	13.00	1.67	30.42	-	13.00	100 nr	43.42
medium	22.00	1.67	30.42	-	22.00	100 nr	52.42
large	25.00	2.00	36.50	-	25.00	100 nr	61.50

Q PAVING/PLANTING/FENCING/SITE FURNITURE

Item Excluding site overheads and profit	PC £	Labour hours	Labour £	Plant £	Material £	Unit	Total rate £
Q31 PLANTING - cont'd							
Aquatic planting							
Aquatic plants; in prepared growing medium in pool; plant size 2 - 3 litre containerized (plants not included)	-	0.04	0.73	-	-	nr	0.73
MARKET PRICES OF NATIVE SPECIES							
Oakover Nurseries Ltd; the following prices are typically some of the more popular species used in native plantations							
1+1; 1 year seedling transplanted and grown for a year							
30–40 cm	-	-	-	-	0.45	nr	-
40–60 cm	-	-	-	-	0.55	nr	-
1+2; 1 year seedling transplanted and grown for 2 years							
60–80 cm	-	-	-	-	0.65	nr	-
Whips							
80–100 cm	-	-	-	-	0.90	nr	-
100–125 cm	-	-	-	-	1.30	nr	-
Feathered trees							
125–150 cm	-	-	-	-	2.10	nr	-
150–180 cm	-	-	-	-	2.50	nr	-
180–200 cm	-	-	-	-	4.00	nr	-
200–250 cm	-	-	-	-	6.50	nr	-
250–300 cm	-	-	-	-	8.00	nr	-
Native species planting							
Forming planting holes; in cultivated ground (cultivating not included); by mechanical auger; trimming holes by hand; depositing excavated material alongside holes							
250 mm diameter	-	0.03	0.61	0.04	-	nr	0.65
250 x 250 mm	-	0.04	0.73	0.08	-	nr	0.81
300 x 300 mm	-	0.08	1.37	0.10	-	nr	1.46
Hand excavation; forming planting holes; in cultivated ground (cultivating not included); depositing excavated material alongside holes							
100 x 100 x 100 mm deep; with mattock or hoe	-	0.01	0.12	-	-	nr	0.12
250 x 250 x 300 mm deep	-	0.04	0.73	-	-	nr	0.73
300 x 300 x 300 mm deep	-	0.06	1.01	-	-	nr	1.01
Operations after planting							
Initial cutting back to shrubs and hedge plants; including disposal of all cuttings	-	1.00	18.25	-	-	100 m²	18.25
Mulch; Melcourt Industries Ltd; "Bark Nuggets®"; to plant beds; delivered in 80 m³ loads; maximum distance 25 m							
50 mm thick	1.86	0.04	0.81	-	1.95	m²	2.76
75 mm thick	2.79	0.07	1.22	-	2.93	m²	4.15
100 mm thick	3.72	0.09	1.62	-	3.91	m²	5.53

Q PAVING/PLANTING/FENCING/SITE FURNITURE

Item Excluding site overheads and profit	PC £	Labour hours	Labour £	Plant £	Material £	Unit	Total rate £
Mulch; Melcourt Industries Ltd; "Bark Nuggets®"; to plant beds; delivered in 25 m^3 loads; maximum distance 25 m							
50 mm thick	2.23	0.03	0.53	-	2.34	m^2	2.87
75 mm thick	3.77	0.07	1.22	-	3.96	m^2	5.17
100 mm thick	5.02	0.09	1.62	-	5.27	m^2	6.89
Mulch; Melcourt Industries Ltd; "Amenity Bark Mulch"; FSC; to plant beds; delivered in 80 m^3 loads; maximum distance 25 m							
50 mm thick	1.13	0.04	0.81	-	1.19	m^2	2.00
75 mm thick	1.70	0.07	1.22	-	1.78	m^2	3.00
100 mm thick	2.27	0.09	1.62	-	2.38	m^2	4.00
Mulch; Melcourt Industries Ltd; "Amenity Bark Mulch"; FSC; to plant beds; delivered in 25 m^3 loads; maximum distance 25 m							
50 mm thick	1.78	0.04	0.81	-	1.87	m^2	2.68
75 mm thick	2.67	0.07	1.22	-	2.81	m^2	4.03
100 mm thick	3.56	0.09	1.62	-	3.74	m^2	5.37
Mulch mats; English Woodlands; lay mulch mat to planted area or plant station; mat secured with metal J pins 240 mm x 3 mm							
Mats to individual plants and plant stations							
Bonflora Biodegradable; 45 cm diameter; circular mat	1.60	0.05	0.91	-	1.98	ea	2.89
Bonflora Biodegradable; 60 cm diameter; circular mat	2.10	0.05	0.91	-	2.48	ea	3.39
Hemcore Biodegradable; 50 x 50 cm; square mat	0.96	0.05	0.91	-	1.34	ea	2.25
Woven Polypropylene; 50 x 50 cm; square mat	0.29	0.05	0.91	-	0.67	ea	1.58
Woven Polypropylene; 1 x 1 m; square mat	0.78	0.07	1.22	-	1.16	ea	2.38
Hedging mats							
Woven Polypropylene; 1 x 100 m roll; hedge planting	0.47	0.01	0.23	-	0.66	m^2	0.89
General planting areas							
Plantex membrane; 1 x 14 m roll; weed control	1.14	0.01	0.18	-	1.33	m^2	1.52
Fertilizers; in top 150 mm of topsoil at 35 g/m^2							
"Mascot Microfine"; turf fertilizer; 8+0+6 + 2% Mg + 4% Fe	6.25	0.12	2.24	-	6.25	100 m^2	8.48
"Enmag"; controlled release fertilizer (8-9 months); 11+22+09	7.35	0.12	2.24	-	7.72	100 m^2	9.96
"Mascot Sports Amenity"; turf fertilizer; 7 +7+7	3.95	0.12	2.24	-	3.95	100 m^2	6.19
"Mascot Sports Amenity"; turf fertilizer; 15 +5+10	4.72	0.12	2.24	-	4.72	100 m^2	6.96
"Super Phosphate Powder"	4.27	0.12	2.24	-	4.48	100 m^2	6.72
"Bone Meal"	3.88	0.12	2.24	-	4.07	100 m^2	6.31

Q PAVING/PLANTING/FENCING/SITE FURNITURE

Item Excluding site overheads and profit	PC £	Labour hours	Labour £	Plant £	Material £	Unit	Total rate £
Q31 PLANTING - cont'd							
Mulch mats - cont'd							
Fertilizers; in top 150 mm of topsoil at 70 g/m^2							
"Mascot Microfine"; turf fertilizer; 8+0+6 + 2% Mg + 4% Fe	12.50	0.12	2.24	-	12.50	100 m^2	14.73
"Enmag"; controlled release fertilizer (8-9 months); 11+22+09	14.71	0.12	2.24	-	15.44	100 m^2	17.68
"Mascot Sports Amenity"; turf fertilizer; 7+7+7	7.91	0.12	2.24	-	7.91	100 m^2	10.14
"Mascot Sports Amenity"; turf fertilizer; 15+5+10	9.44	0.12	2.24	-	9.44	100 m^2	11.68
"Super Phosphate Powder"	8.54	0.12	2.24	-	8.97	100 m^2	11.20
"Bone meal"	7.76	0.12	2.24	-	8.14	100 m^2	10.38
Maintenance operations (Note: the following rates apply to aftercare maintenance executed as part of a landscaping contract only - Please see section Q35 of this publication for further maintenance rates)							
Weeding and hand forking planted areas; including disposing weeds and debris on site; areas maintained weekly	-	-	0.07	-	-	m^2	0.07
Weeding and hand forking planted areas; including disposing weeds and debris on site; areas maintained monthly	-	0.01	0.18	-	-	m^2	0.18
Extra over weeding and hand forking planted areas for disposing excavated material off site; to tip not exceeding 13 km; mechanically loaded							
slightly contaminated	-	-	-	1.65	21.33	m^3	22.98
rubbish	-	-	-	1.65	21.33	m^3	22.98
inert material	-	-	-	1.10	7.00	m^3	8.10
Mulch; Melcourt Industries Ltd; "Bark Nuggets®"; to plant beds; delivered in 80 m^3 loads; maximum distance 25 m							
50 mm thick	1.86	0.04	0.81	-	1.95	m^2	2.76
75 mm thick	2.79	0.07	1.22	-	2.93	m^2	4.15
100 mm thick	3.72	0.09	1.62	-	3.91	m^2	5.53
Mulch; Melcourt Industries Ltd; "Bark Nuggets®"; to plant beds; delivered in 25 m^3 loads; maximum distance 25 m							
50 mm thick	2.51	0.04	0.81	-	2.64	m^2	3.45
75 mm thick	3.77	0.07	1.22	-	3.96	m^2	5.17
100 mm thick	5.02	0.09	1.62	-	5.27	m^2	6.89
Mulch; Melcourt Industries Ltd; "Amenity Bark Mulch"; FSC; to plant beds; delivered in 80 m^3 loads; maximum distance 25 m							
50 mm thick	1.13	0.04	0.81	-	1.19	m^2	2.00
75 mm thick	1.70	0.07	1.22	-	1.78	m^2	3.00
100 mm thick	2.27	0.09	1.62	-	2.38	m^2	4.00

Q PAVING/PLANTING/FENCING/SITE FURNITURE

Item Excluding site overheads and profit	PC £	Labour hours	Labour £	Plant £	Material £	Unit	Total rate £
Mulch; Melcourt Industries Ltd; "Amenity Bark Mulch"; FSC; to plant beds; delivered in 25 m³ loads; maximum distance 25 m							
50 mm thick	1.78	0.04	0.81	-	1.87	m²	2.68
75 mm thick	2.67	0.07	1.22	-	2.81	m²	4.03
100 mm thick	3.56	0.09	1.62	-	3.74	m²	5.37
Fertilizers; at 35 g/m²							
"Mascot Microfine"; turf fertilizer; 8+0+6 + 2% Mg + 4% Fe	6.25	0.12	2.24	-	6.25	100 m²	8.48
"Enmag"; controlled release fertilizer (8-9 months); 11+22+09	7.35	0.12	2.24	-	7.72	100 m²	9.96
"Mascot Sports Amenity"; turf fertilizer; 7+7+7	3.95	0.12	2.24	-	3.95	100 m²	6.19
"Mascot Sports Amenity"; turf fertilizer; 15+5+10	4.72	0.12	2.24	-	4.72	100 m²	6.96
"Super Phosphate Powder"	4.27	0.12	2.24	-	4.48	100 m²	6.72
"Bone meal"	3.88	0.12	2.24	-	4.07	100 m²	6.31
Fertilizers; at 70 g/m²							
"Mascot Microfine"; turf fertilizer; 8+0+6 + 2% Mg + 4% Fe	12.50	0.12	2.24	-	12.50	100 m²	14.73
"Enmag"; controlled release fertilizer (8-9 months); 11+22+09	14.71	0.12	2.24	-	15.44	100 m²	17.68
"Mascot Sports Amenity"; turf fertilizer; 7+7+7	7.91	0.12	2.24	-	7.91	100 m²	10.14
"Mascot Sports Amenity"; turf fertilizer; 15+5+10	9.44	0.12	2.24	-	9.44	100 m²	11.68
"Super Phosphate Powder"	8.54	0.12	2.24	-	8.97	100 m²	11.20
"Bone meal"	7.76	0.12	2.24	-	8.14	100 m²	10.38
Watering planting; evenly; at a rate of 5 l/m²							
using hand-held watering equipment	-	0.25	4.56	-	-	100 m²	4.56
using sprinkler equipment and with sufficient water pressure to run one 15 m radius sprinkler	-	0.14	2.54	-	-	100 m²	2.54
using movable spray lines powering three sprinkler heads with a radius of 15 m and allowing for 60% overlap (irrigation machinery costs not included)	-	0.02	0.29	-	-	100 m²	0.29
Forestry planting							
Deep ploughing rough ground to form planting ridges at							
2.00 m centres	-	0.63	11.41	7.51	-	100 m²	18.92
3.00 m centres	-	0.59	10.73	7.07	-	100 m²	17.80
4.00 m centres	-	0.40	7.30	4.81	-	100 m²	12.11
Notching plant forestry seedlings; "T" or "L" notch	33.00	0.75	13.69	-	33.00	100 nr	46.69
Turf planting forestry seedlings	33.00	2.00	36.50	-	33.00	100 nr	69.50
Tree tubes; to young trees	148.00	0.30	5.47	-	148.00	100 nr	153.47
Cleaning and weeding around seedlings; once	-	0.50	9.13	-	-	100 nr	9.13
Treading in and firming ground around seedlings planted; at 2500 per ha after frost or other ground disturbance; once	-	0.33	6.08	-	-	100 nr	6.08
Beating up initial planting; once (including supply of replacement seedlings at 10% of original planting)	3.30	0.25	4.56	-	3.30	100 nr	7.86

Q PAVING/PLANTING/FENCING/SITE FURNITURE

Item Excluding site overheads and profit	PC £	Labour hours	Labour £	Plant £	Material £	Unit	Total rate £
Q35 LANDSCAPE MAINTENANCE							
Preamble: Long-term landscape maintenance							
Maintenance on long-term contracts differs in cost from that of maintenance as part of a landscape contract. In this section the contract period is generally 3 - 5 years. Staff are generally allocated to a single project only and therefore productivity is higher whilst overhead costs are lower. Labour costs in this section are lower than the costs used in other parts of the book. Machinery is assumed to be leased over a five year period and written off over the same period. The costs of maintenance and consumables for the various machinery types have been included in the information that follows. Finance costs for the machinery have not been allowed for. The rates shown below are for machines working in unconfined contiguous areas. Users should adjust the times and rates if working in smaller spaces or spaces with obstructions.							
LANDSCAPE CHEMICALS							
Chemical applications labours only							
Controlled droplet application (CDA)							
plants at 1.50 m centres; 4444 nr/ha	-	18.52	337.99	-	-	ha	337.99
plants at 1.50 m centres; 44.44 nr/100 m^2	-	1.85	33.80	-	-	100 m^2	33.80
plants at 1.75 m centres; 3265 nr/ha	-	13.60	248.20	-	-	ha	248.20
plants at 1.75 m centres; 3265 nr/ha	-	1.36	24.82	-	-	100 m^2	24.82
plants at 2.00 m centres; 2500 nr/ha	-	10.42	190.16	-	-	ha	190.16
plants at 2.00 m centres; 2500 nr/ha	-	1.04	19.02	-	-	100 m^2	19.02
spot spraying to shrub beds	-	0.40	7.30	-	-	100 m^2	7.30
spot spraying to hard surfaces	-	0.25	4.56	-	-	100 m^2	4.56
mass spraying	-	5.00	91.25	-	-	ha	91.25
mass spraying	-	0.50	9.13	-	-	100 m^2	9.13
Power spraying							
mass spraying to areas containing vegetation	-	4.00	73.00	60.02	-	ha	133.02
mass spraying to hard surfaces	-	0.01	0.15	0.12	-	100 m^2	0.28
spraying to kerbs and channels along roadways 300 mm wide	-	2.50	45.63	37.51	-	1 km	83.14
General herbicides; in accordance with manufacturer's instructions; "Knapsack" spray application; see "Market rates of chemicals" for costs of specific applications							
knapsack sprayer; selective spraying around bases of plants	-	0.35	6.40	-	-	100 m^2	6.40
knapsack sprayer; spot spraying	-	0.20	3.65	-	-	100 m^2	3.65
knapsack sprayer; mass spraying	-	0.20	3.65	-	-	100 m^2	3.65
granular distribution by hand or hand applicator	-	0.40	7.30	-	-	100 m^2	7.30

Q PAVING/PLANTING/FENCING/SITE FURNITURE

Item Excluding site overheads and profit	PC £	Labour hours	Labour £	Plant £	Material £	Unit	Total rate £
Backpack spraying; keeping weed free at bases of plants							
plants at 1.50 m centres; 4444 nr/ha	-	27.78	506.99	-	-	ha	**506.99**
plants at 1.75 m centres; 3265 nr/ha	-	20.40	372.30	-	-	ha	**372.30**
plants at 2.00 m centres; 2500 nr/ha	-	15.63	285.25	-	-	ha	**285.25**
mass spraying	-	15.00	273.75	-	-	ha	**273.75**
The following table provides the areas of spraying in each planting scenario for plants which require 1.00 m diameter weed free circles							
Area of weed free circles required in each planting density							
plants at 500 mm centres; 400 nr/100 m²	-	-	-	-	400.00	m²	-
plants at 600 m centres; 278 nr/100 m²	-	-	-	-	278.00	m²	-
plants at 750 mm centres; 178 nr/100 m²	-	-	-	-	178.00	m²	-
plants at 1.00 m centres; 100 nr/100 m²	-	-	-	-	100.00	m²	-
plants at 1.50 m centres; 4444 nr/ha	-	-	-	-	3490.00	m²	-
plants at 1.75 m centres; 3265 nr/ha	-	-	-	-	2564.00	m²	-
plants at 2.00 m centres; 2500 nr/ha	-	-	-	-	1963.00	m²	-
MARKET PRICES OF LANDSCAPE CHEMICALS AT SUGGESTED APPLICATION RATES							
Note: All chemicals are standard knapsack or backpack applied unless specifically stated as CDA or TDA.							
Total herbicides							
Herbicide applications; CDA; chemical application via low pressure specialized wands to landscape planting; application to maintain 1.00 m diameter clear circles (0.79 m²) around new planting							
Vanquish Biactive; ALS Ltd; Bayer; enhanced movement glyphosate; application rate 15 litre/ha							
plants at 1.50 m centres; 4444 nr/ha	-	-	-	-	46.02	ha	-
plants at 1.75 m centres; 3265 nr/ha	-	-	-	-	33.97	ha	-
plants at 2.00 m centres; 2500 nr/ha	-	-	-	-	26.05	ha	-
mass spraying	-	-	-	-	132.00	ha	-
spot spraying	-	-	-	-	1.32	ha	-

Q PAVING/PLANTING/FENCING/SITE FURNITURE

Item Excluding site overheads and profit	PC £	Labour hours	Labour £	Plant £	Material £	Unit	Total rate £
Q35 LANDSCAPE MAINTENANCE - cont'd							
Total herbicides; CDA; low pressure sustainable applications							
"Hilite"; Nomix Enviro Ltd; glyphosate oil-emulsion; Total Droplet Control (TDC formerly CDA); low volume sustainable chemical applications; formulation controls annual and perennial grasses and non-woody broad-leaved weeds in amenity and industrial areas, parks, highways, paved areas and factory sites; also effective for clearing vegetation prior to planting of ornamental species, around the bases of trees, shrubs and roses, as well as in forestry areas							
10 l/ha	-	-	-	-	1.58	100 m²	-
7.5 l/ha	-	-	-	-	1.18	100 m²	-
"Tribute Plus"; Nomix Enviro Ltd ; TDA; MCPA/dicamba/mecaprop-p; TDC; for selective control of broad-leaved weeds in turf							
6 l/ha; white clover, plantains, creeping buttercup, thistle	-	-	-	-	0.26	100 m²	-
8 l/ha; yarrow, ragwort, parsley, piert, cinquefoil, mouse eared chickweed, selfheal, pearlwort, sorrel, black meddick, dandelion and related weeds, bulbous buttercup	-	-	-	-	0.35	100 m²	-
"Stirrup"; Rigby Taylor Ltd; glyphosate; CDA; 192 g							
10 l/ha	-	-	-	-	1.58	100 m²	-
7.5 l/ha	-	-	-	-	1.58	100 m²	-
"Gallup Biograde Amenity"; Rigby Taylor Ltd; glyphosate 360 g/l formulation							
general use; woody weeds; ash, beech, bracken, bramble; 3 l/ha	-	-	-	-	0.38	100 m²	-
annual and perennial grasses; heather (peat soils); 4 l/ha	-	-	-	-	0.51	100 m²	-
pre-planting; general clearance; 5 l/ha	-	-	-	-	0.63	100 m²	-
heather; mineral soils; 6 l/ha	-	-	-	-	0.76	100 m²	-
rhododendron; 10 l/ha	-	-	-	-	1.26	100 m²	-
"Gallup Hi-aktiv Amenity"; Rigby Taylor Ltd; 490 g/l formulation							
general use; woody weeds; ash, beech, bracken, bramble; 2.2 l/ha	-	-	-	-	0.34	100 m²	-
annual and perennial grasses; heather (peat soils); 2.9 l/ha	-	-	-	-	0.45	100 m²	-
pre-planting; general clearance; 3.7 l/ha	-	-	-	-	0.57	100 m²	-
heather; mineral soils; 4.4 l/ha	-	-	-	-	0.68	100 m²	-
rhododendron; 7.3 l/ha	-	-	-	-	1.12	100 m²	-

Q PAVING/PLANTING/FENCING/SITE FURNITURE

Item Excluding site overheads and profit	PC £	Labour hours	Labour £	Plant £	Material £	Unit	**Total rate £**
"Discman" CDA Biograde; Rigby Taylor Ltd							
application rate; 7 l/ha; light vegetation and annual weeds	-	-	-	-	1.01	100 m²	-
application rate; 8.5 l/ha; forestry/ woodlands; before planting, to control broad-leaved and grass weeds	-	-	-	-	1.23	100 m²	-
application rate; 10 l/ha; established deep-rooted annuals, perennial grasses and broad-leaved weeds	-	-	-	-	1.45	100 m²	-
Total herbicides; standard backpack or power sprayer applications							
"Roundup Pro Biactive 360"; Scotts UK							
application rate; 5 l/ha; land not intended for cropping	-	-	-	-	0.61	100 m²	-
application rate; 10 l/ha; forestry; weed control	-	-	-	-	1.21	100 m²	-
"Roundup Pro Biactive 450"; Scotts UK; 450 g/litre glyphosate							
application rate 4 l/ha; natural surfaces not intended to bear vegetation; amenity vegetation control; permeable surfaces overlaying soil; hard surfaces	-	-	-	-	0.58	100 m²	-
application rate 8 l/ha; forest; forest nursery weed control	-	-	-	-	1.15	100 m²	-
application rate 16 mm/l; stump application	-	-	-	-	0.23	litre	-
"Roundup Pro Biactive 450" Scotts UK; enhanced movement glyphosate; application rate 4 litre/ha							
plants at 1.50 m centres; 4444 nr/ha	75.74	-	-	-	83.32	ha	-
plants at 1.75 m centres; 3265 nr/ha	55.71	-	-	-	61.28	ha	-
plants at 2.00 m centres; 2500 nr/ha	42.62	-	-	-	46.89	ha	-
mass spraying	57.60	-	-	-	63.36	ha	-
spot spraying	2.16	-	-	-	2.38	100 m²	-
Kerb							
"Kerb Flowable"; propyzamide	-	-	-	-	2.16	100 m²	-
"Kerb Granules"; 2 x 2000 tree pack	-	-	-	-	0.06	tree	-
"Proliance Quattro"; Scotts UK; 360 g/l glyphosate; advanced glyphosate formulation which incorporates a fully optimized surfactant system and water conditioner; to hazard classification; offers a high standard of operator and environmental safety							
application rate 6 l/ha; perennial broad-leaved weeds	-	-	-	-	1.05	100 m²	-
application rate 4 l/ha; perennial grasses	-	-	-	-	0.70	100 m²	-
application rate 1.5 l/ha; annual weeds	-	-	-	-	0.26	100 m²	-
"Dual"; Nomix Enviro Ltd; glyphosate/ sulfosulfuron; TDC; low volume sustainable chemical applications; residual effect lasting up to 6 months; shrub beds, gravelled areas, tree bases, parks, industrial areas and around obstacles							
application rate 9 l/ha	-	-	-	-	1.98	100 m²	-

Q PAVING/PLANTING/FENCING/SITE FURNITURE

Item Excluding site overheads and profit	PC £	Labour hours	Labour £	Plant £	Material £	Unit	Total rate £
Q35 LANDSCAPE MAINTENANCE - cont'd							
Total herbicides - cont'd							
"Glyfos Dakar Pro"; Headland Amenity Ltd; 68% glyphosate as water dispersible granules; broad-leaved weeds and grasses							
application rate 2.5 kg/ha	-	-	-	-	0.40	100 m^2	-
Total herbicides with residual pre-emergent action							
"Pistol"; ALS; Bayer; barrier prevention for emergence control and glyphosate for control of emerged weeds at a single application per season; application rate 4.5 litre/ha; 40 g/l diflufenican and 250 g/l glyphosate							
plants at 1.50 m centres; 4444 nr/ha	-	-	-	-	32.81	ha	-
plants at 1.75 m centres; 3265 nr/ha	-	-	-	-	24.04	ha	-
plants at 2.00 m centres; 2500 nr/ha	-	-	-	-	23.51	ha	-
Residual herbicides							
"Chikara"; Rigby Taylor Ltd; "Belchim Crop Protection"; flazasulfuron; systemic pre-emergent and early post-emergent herbicide for the control of annual and perennial weeds on natural surfaces not intended to bear vegetation and permeable surfaces over-lying soil							
150 g/ha	-	-	-	-	2.49	100 m^2	-
Contact herbicides							
"Jewel"; Scotts UK; carfentrazone and mecoprop-p; selective weed and mosskiller							
application rate; 1.5 kg/ha	-	-	-	-	1.25	100 m^2	-
"Natural Weed and Moss spray"; Headland Amenity Ltd; acetic acid							
application rate; 100 l/ha	-	-	-	-	4.50	100 m^2	-
Selective herbicides CDA							
"Hilite"; Nomix Enviro; glyphosate oil-emulsion; TD ; low volume sustainable chemical applications; formulation controls annual and perennial grasses and non-woody broad-leaved weeds in amenity and industrial areas, parks, highways, paved areas and factory sites; also effective for clearing vegetation prior to planting of ornamental species, around the bases of trees, shrubs and roses, as well as in forestry areas							
10 l/ha	-	-	-	-	1.58	100 m^2	-
7.5 l/ha	-	-	-	-	1.18	100 m^2	-

Q PAVING/PLANTING/FENCING/SITE FURNITURE

Item Excluding site overheads and profit	PC £	Labour hours	Labour £	Plant £	Material £	Unit	Total rate £
"Tribute Plus"; Nomix Enviro; MCPA/ dicamba/mecaprop-p; TDC; for selective control of broad-leaved weeds in turf							
6 l/ha; white clover, plantains, creeping buttercup, thistle	-	-	-	-	0.26	100 m²	-
8 l/ha; yarrow, ragwort, parsley, piert, cinquefoil, mouse eared chickweed, selfheal, pearlwort, sorrel, black meddick, dandelion and related weeds, bulbous buttercup	-	-	-	-	0.35	100 m²	-
Contact and systemic combined herbicides							
"Jewel"; Scotts UK; carfentrazone and mecoprop-p; selective weed and mosskiller							
application rate; 1.5 kg/ha	-	-	-	-	1.25	100 m²	-
"Finale"; ALS; Bayer; glufosinate-ammonium; controls grasses and broad-leaved weeds; it can also be used for weed control around non-edible ornamental crops including trees and shrubs and in tree nurseries; may be used to "burn off" weeds in shrub beds prior to treatment with a residual herbicide; is also ideal for preparing sports turf for linemarking							
application rate; 5 l/ha; seedlings of all species; established annual weeds and grasses as specified	-	-	-	-	0.79	100 m²	-
application rate; 8 l/ha; established annual and perennial weeds and grasses as specified in the product guide	-	-	-	-	1.26	100 m²	-
Selective herbicides; power or knapsack sprayer							
"Junction"; Rigby Taylor Ltd; selective post-emergence turf herbicide containing 6.25 g/litre florasulam plus 452 g/litre 2,4-D; specially designed for use by contractors for maintenance treatments of a wide range of broad-leaved weeds whilst giving excellent safety to turfgrass							
application rate; 1.2 l/ha	-	-	-	-	0.50	100 m²	-
"Greenor"; Rigby Taylor Ltd; fluroxypyr, clopyralid and MCPA; systemic selective treatment of weeds in turfgrass							
application rate; 4 l/ha	-	-	-	-	1.00	100 m²	-
"Super Selective Plus"; Rigby Taylor Ltd; combination of three active ingredients (mecoprop-p, MCPA and dicamba) enables this formulation to give good control of most commonly occurring broad-leaved weeds found in sports and amenity turf							
application rate; 3.5 l/ha	-	-	-	-	0.21	100 m²	-
"Crossbar"; Rigby Taylor Ltd; fluroxypyr, 2, 4-D and dicamba for control of weeds in amenity turf including fine turf							
application rate; 2 l/ha	-	-	-	-	0.76	100 m²	-

Q PAVING/PLANTING/FENCING/SITE FURNITURE

Item Excluding site overheads and profit	PC £	Labour hours	Labour £	Plant £	Material £	Unit	Total rate £
Q35 LANDSCAPE MAINTENANCE - cont'd							
Selective herbicides - cont'd "Barclay Greengard"; Rigby Taylor Ltd; 2, 4-D, triclopyr, and dicamba; selective herbicide for the control of certain broad-leaved weeds in amenity grassland and scrub control in forestry and uncropped areas including Japanese Knotweed and weeds listed under the Weeds Act of 1959							
application rate; 5 l/ha	-	-	-	-	2.48	100 m^2	-
application rate; 2 l/ha	-	-	-	-	0.99	100 m^2	-
"Intrepid 2"; Scotts UK; for broad-leaved weed control in grass							
application rate; 7700 m^2/5 l	-	-	-	-	0.51	100 m^2	-
Re-Act; Scotts UK; MCPA, mecoprop-p and dicamba; selective herbicide; established managed amenity turf and newly seeded grass; controls many annual and perennial weeds; will not vaporise in hot conditions							
application rate; 9090 - 14,285 m^2/5 l	-	-	-	-	0.46	100 m^2	-
"Praxys"; Scotts UK; fluroxypyr-meptyl, clopyralid and florasulam; selective, systemic post-emergence herbicide for managed amenity turf, lawns and amenity grassland with high selectivity to established and young turf							
application rate; 9090 - 14,285 m^2/5 l	-	-	-	-	1.39	100 m^2	-
"Asulox"; ALS; Bayer; post-emergence translocated herbicide for the control of bracken and docks							
application rate; docks; 17857 m^2/5 l	-	-	-	-	0.34	100 m^2	-
application rate; bracken; 4545 m^2/5 l	-	-	-	-	1.32	100 m^2	-
"Spearhead"; ALS; Bayer; selective herbicide for the control of broad-leaved weeds in established turf							
application rate; 4.5 l/ha	-	-	-	-	0.85	100 m^2	-
"Cabadex"; Headland Amenity Ltd; fluroxypyr and florasulam; newly sown and established turf (particularly speedwell, yarrow and yellow suckling clover)							
application rate 2 l/ha	-	-	-	-	0.65	100 m^2	-
"Relay Turf"; Headland Amenity Ltd; mecoprop-p, MCPA and dicamba; many broad-leaved weeds in managed amenity turf							
application rate; 5 l/ha	-	-	-	-	0.45	100 m^2	-
"Blaster"; Headland Amenity Ltd; triclopyr clopyralid; nettles, docks, thistles, brambles, woody weeds in amenity grassland							
application rate; 4 l/ha	-	-	-	-	1.40	100 m^2	-

Q PAVING/PLANTING/FENCING/SITE FURNITURE

Item Excluding site overheads and profit	PC £	Labour hours	Labour £	Plant £	Material £	Unit	Total rate £
Woody weed herbicides; backpack or power sprayers							
"Timbrel"; Rigby Taylor Ltd; Bayer; summer applied; water based; selective scrub and brushwood herbicide							
2.0 l/ha; bramble, briar, broom, gorse, nettle	-	-	-	-	1.24	100 m²	-
4.0 l/ha; alder, birch, blackthorn, dogwood, elder, poplar, rosebay willowherb, sycamore	-	-	-	-	2.49	100 m²	-
6.0 l/ha; beech, box, buckthorn, elm, hazel, hornbeam, horse chestnut, lime, maple, privet, rowan, Spanish chestnut, willow, wild pear	-	-	-	-	3.73	100 m²	-
8.0 l/ha; ash, oak, rhododendron	-	-	-	-	4.98	100 m²	-
"Timbrel"; Rigby Taylor Ltd; Bayer; winter applied; paraffin or diesel based; selective scrub and brushwood herbicide							
3.0 l/ha; bramble, briar, broom, gorse, nettle	-	-	-	-	1.87	100 m²	-
6.0 l/ha; alder, ash, beech, birch, blackthorn, box, buckthorn, dogwood, elder, elm, hazel, hornbeam, horse chestnut, lime, maple, oak, poplar, privet, rowan, Spanish chestnut, sycamore, willow, wild pear	-	-	-	-	3.73	100 m²	-
10.0 l/ha; hawthorn, laurel, rhododendron	-	-	-	-	6.22	100 m²	-
"Garlon 4"; Nomix Enviro Ltd; 667 g/litre triclopyr; control of brushwood, scrub, woody and perennial herbaceous weeds in amenity, forestry, rail and industrial areas; also suitable for forestry production							
2 l/ha; bramble, briar, broom, gorse, nettle	-	-	-	-	0.98	100 m²	-
4 l/ha; alder, birch, blackthorn, dogwood, elder, poplar, rosebay, willow-herb, sycamore	-	-	-	-	1.96	100 m²	-
6 l/ha; beech, box, elm, buckthorn, hazel, hornbeam, horse chestnut, lime, maple, privet, rowan, Spanish chestnut, willow, wild pear	-	-	-	-	2.94	100 m²	-
8 l/ha; rhododendron, ash, oak	-	-	-	-	3.92	100 m²	-
"Tordon 22K"; Nomix Enviro; picloram; selective action; control of tough woody weeds, invasive species, scrub and deep-rooted perennials; residual action grass cover is maintained							
2.8 l/ha; white clover, creeping thistle, hogweed, plantains	-	-	-	-	1.19	100 m²	-
4.3 - 5.6 l/ha; common bird s-foot trefoil, field bindweed, creeping buttercup, burdock, cat s ear, common chickweed, colt s foot, creeping cinquefoil, dandelion, docks, hawkweed, Japanese knotweed, mugwort, common nettle, silverweed, tansy, cotton thistle, rosebay willowherb, yarrow	-	-	-	-	2.12	100 m²	-
8.4 l/ha; wild mignonettte, common ragwort	-	-	-	-	3.56	100 m²	-
11.2 l/ha; bracken, brambles	-	-	-	-	7.72	100 m²	-

Q PAVING/PLANTING/FENCING/SITE FURNITURE

Item Excluding site overheads and profit	PC £	Labour hours	Labour £	Plant £	Material £	Unit	Total rate £
Q35 LANDSCAPE MAINTENANCE - cont'd							
Japanese knotweed and ragwort control "Loram 24"; ALS Ltd; Bayer; picloram							
4.2 l/ha; Japanese Knotweed and other weeds as listed in the suppliers literature	-	-	-	-	1.60	100 m^2	-
5.6 l/ha; common ragwort, wild mignonette	-	-	-	-	2.13	100 m^2	-
Aquatic herbicides; backpack or power sprayers "Roundup Pro Biactive 360"; Scotts UK							
application rate; 6 l/ha; aquatic weed control	-	-	-	-	0.73	100 m^2	-
"Roundup Pro Biactive 450"; Scotts UK; 450 g/litre glyphosate							
application rate 4.8 l/ha; enclosed waters, open waters, land immediately adjacent to aquatic areas	-	-	-	-	0.69	100 m^2	-
Aquatic herbicides; TDC "Conqueror"; Nomix Enviro; low volume sustainable chemical applications glyphosate with aquatic approval							
10 l/ha; established annuals, deep-rooted perennials and broad-leaved weeds	-	-	-	-	0.16	100 m^2	-
7.5 l/ha; light vegetation and small annual weeds aquatic situations	-	-	-	-	1.18	100 m^2	-
12.5 l/ha; aquatic situations emergent weeds e.g. reed, grasses and watercress	-	-	-	-	0.20	100 m^2	-
15 l/ha; aquatic situations floating weeds such as water lilies	-	-	-	-	0.24	100 m^2	-
Fungicides "Masalon"; Rigby Taylor Ltd; systemic							
application rate; 8 l/ha	-	-	-	-	7.54	100 m^2	-
"Defender"; Rigby Taylor Ltd; broad spectrum fungicide displaying protective and curative activity against fusarium patch and red thread							
application rate; 700 g/ha	-	-	-	-	6.23	100 m^2	-
"Fusion"; Rigby Taylor Ltd; contact and systemic fungicide with curative and preventative activity against major turf diseases such as fusarium, red thread, dollar spot, anthracnose and rust							
application rate; 700 g/ha	-	-	-	-	6.09	100 m^2	-
"Rayzor"; Rigby Taylor Ltd; broad spectrum contact fungicide with both protectant and eradicant activity for the control of fusarium patch and red thread in managed amenity turf							
application rate; 5 l/2000 m^2	-	-	-	-	8.69	100 m^2	-
"Mascot Contact"; multi-site protectant fungicide for the control of a range of diseases on turf							
application rate; 5 l/1667 m^2	-	-	-	-	6.59	100 m^2	-
"Daconil Weatherstik"; Scotts UK; turf							
application rate; 333 m^2/l	-	-	-	-	5.07	100 m^2	-

Q PAVING/PLANTING/FENCING/SITE FURNITURE

Item Excluding site overheads and profit	PC £	Labour hours	Labour £	Plant £	Material £	Unit	Total rate £
Moss control "Jewel"; Scotts UK; carfentrazone and mecoprop-p; selective weed and moss killer							
application rate; 1.5 kg/ha	-	-	-	-	1.25	100 m²	-
Insect control "Crossfire 480"; Rigby Taylor Ltd; control of leatherjackets and fruit fly in sports and amenity turf							
application rate; 1.5 l/ha	-	-	-	-	0.56	100 m²	-
"Merit Turf"; Bayer; insecticide for the control of chafer grubs and leatherjackets							
application rate; 10 kg/3333 m²	-	-	-	-	5.95	100 m²	-
Grass cutting; ride-on or tractor drawn equipment; Norris & Gardiner Ltd; works carried out on one site only such as large fields or amenity areas where labour and machinery is present for a full day Using multiple-gang mower with cylindrical cutters; contiguous areas such as playing fields and the like larger than 3000 m²							
3 gang; 2.13 m cutting width	-	0.01	0.16	0.11	-	100 m²	**0.27**
5 gang; 3.40 m cutting width	-	0.01	0.14	0.15	-	100 m²	**0.29**
7 gang; 4.65 m cutting width	-	0.01	0.09	0.10	-	100 m²	**0.19**
Using multiple-gang mower with cylindrical cutters; non-contiguous areas such as verges and general turf areas							
3 gang	-	0.02	0.32	0.11	-	100 m²	**0.43**
5 gang	-	0.01	0.20	0.15	-	100 m²	**0.35**
Using multiple-rotary mower with vertical drive shaft and horizontally rotating bar or disc cutters; contiguous areas larger than 3000 m²							
cutting grass, overgrowth or the like using flail mower or reaper	-	0.02	0.30	0.25	-	100 m²	**0.55**
Grass cutting; ride-on or tractor drawn equipment; Norris & Gardiner Ltd; works carried out on multiple sites; machinery moved by trailer between sites Using multiple-gang mower with cylindrical cutters; contiguous areas such as playing fields and the like larger than 3000 m²							
3 gang; 2.13 m cutting width	-	0.02	0.26	0.11	-	100 m²	**0.37**
5 gang; 3.40 m cutting width	-	0.02	0.23	0.15	-	100 m²	**0.38**
7 gang; 4.65 m cutting width	-	0.01	0.14	0.10	-	100 m²	**0.24**
Using multiple-gang mower with cylindrical cutters; non-contiguous areas such as verges and general turf areas							
3 gang	-	0.03	0.51	0.11	-	100 m²	**0.62**
5 gang	-	0.02	0.33	0.15	-	100 m²	**0.48**
Using multiple-rotary mower with vertical drive shaft and horizontally rotating bar or disc cutters; contiguous areas larger than 3000 m²							
cutting grass, overgrowth or the like using flail mower or reaper	-	0.03	0.50	0.25	-	100 m²	**0.75**

Q PAVING/PLANTING/FENCING/SITE FURNITURE

Item Excluding site overheads and profit	PC £	Labour hours	Labour £	Plant £	Material £	Unit	Total rate £
Q35 LANDSCAPE MAINTENANCE - cont'd							
Grass cutting - cont'd							
Cutting grass, overgrowth or the like; using tractor-mounted side-arm flail mower; in areas inaccessible to alternative machine; on surface							
not exceeding 30° from horizontal	-	0.02	0.36	0.32	-	100 m²	**0.68**
30° to 50° from horizontal	-	0.05	0.71	0.32	-	100 m²	**1.03**
Grass cutting; ride-on self-propelled equipment; Norris & Gardiner Ltd; works carried out on one site only such as large fields or amenity areas where labour and machinery is present for a full day							
Using ride-on multiple-cylinder mower							
3 gang; 2.13 m cutting width	-	0.01	0.16	0.11	-	100 m²	**0.27**
5 gang; 3.40 m cutting width	-	0.01	0.10	0.15	-	100 m²	**0.25**
Using ride-on multiple-rotary mower with horizontally rotating bar, disc or chain cutters							
cutting width 1.52 m	-	0.01	0.19	0.17	-	100 m²	**0.36**
cutting width 1.82 m	-	0.01	0.16	0.12	-	100 m²	**0.28**
cutting width 2.97 m	-	0.01	0.10	0.14	-	100 m²	**0.24**
Grass cutting; ride-on self-propelled equipment; Norris & Gardiner Ltd; works carried out on multiple sites; machinery moved by trailer between sites							
Using ride-on multiple-cylinder mower							
3 gang; 2.13 m cutting width	-	0.02	0.26	0.11	-	100 m²	**0.37**
5 gang; 3.40 m cutting width	-	0.01	0.16	0.15	-	100 m²	**0.31**
Using ride-on multiple-rotary mower with horizontally rotating bar, disc or chain cutters							
cutting width 1.52 m	-	0.02	0.32	0.17	-	100 m²	**0.49**
cutting width 1.82 m	-	0.02	0.27	0.12	-	100 m²	**0.39**
cutting width 2.97 m	-	0.01	0.17	0.14	-	100 m²	**0.31**
Add for using grass box/collector for removal and depositing of arisings	-	0.04	0.58	0.06	-	100 m²	**0.64**
Grass cutting; pedestrian operated equipment; Norris & Gardiner Ltd							
Using cylinder lawn mower fitted with not less than five cutting blades, front and rear rollers; on surface not exceeding 30° from horizontal; arisings let fly; width of cut							
51 cm	-	0.06	0.93	0.19	-	100 m	**1.12**
61 cm	-	0.05	0.78	0.17	-	100 m	**0.95**
71 cm	-	0.04	0.67	0.17	-	100 m	**0.84**
91 cm	-	0.03	0.52	0.20	-	100 m	**0.72**
Using rotary self-propelled mower; width of cut							
45 cm	-	0.04	0.60	0.08	-	100 m²	**0.68**
81 cm	-	0.02	0.33	0.08	-	100 m²	**0.41**
91 cm	-	0.02	0.30	0.11	-	100 m²	**0.41**
120 cm	-	0.01	0.23	0.36	-	100 m²	**0.59**
Add for using grass box for collecting and depositing arisings							
removing and depositing arisings	-	0.05	0.76	-	-	100 m²	**0.76**
Add for 30° to 50° from horizontal	-	-	-	-	-	33%	**-**

Q PAVING/PLANTING/FENCING/SITE FURNITURE

Item Excluding site overheads and profit	PC £	Labour hours	Labour £	Plant £	Material £	Unit	Total rate £
Add for slopes exceeding 50°	-	-	-	-	-	100%	-
Cutting grass or light woody undergrowth; using trimmer with nylon cord or metal disc cutter; on surface							
not exceeding 30° from horizontal	-	0.20	3.05	0.28	-	100 m²	3.33
30 - 50° from horizontal	-	0.40	6.10	0.56	-	100 m²	6.66
exceeding 50° from horizontal	-	0.50	7.63	0.70	-	100 m²	8.33
Grass cutting; collecting arisings; Norris & Gardener Ltd							
Extra over for tractor drawn and self-propelled machinery using attached grass boxes; depositing arisings							
22 cuts per year	-	0.05	0.76	-	-	100 m²	0.76
18 cuts per year	-	0.08	1.14	-	-	100 m²	1.14
12 cuts per year	-	0.10	1.52	-	-	100 m²	1.52
4 cuts per year	-	0.25	3.81	-	-	100 m²	3.81
bedding plant	-	-	-	-	0.29	each	-
Arisings collected by trailed sweepers							
22 cuts per year	-	0.02	0.25	0.24	-	100 m²	0.49
18 cuts per year	-	0.02	0.34	0.24	-	100 m²	0.58
12 cuts per year	-	0.03	0.51	0.24	-	100 m²	0.75
4 cuts per year	-	0.07	1.02	0.24	-	100 m²	1.26
Disposing arisings on site; 100 m distance maximum							
22 cuts per year	-	0.01	0.12	0.02	-	100 m²	0.14
18 cuts per year	-	0.01	0.15	0.03	-	100 m²	0.18
12 cuts per year	-	0.01	0.22	0.04	-	100 m²	0.26
4 cuts per year	-	0.04	0.66	0.13	-	100 m²	0.79
Disposal of arisings off site							
22 cuts per year	-	0.01	0.13	0.04	0.75	100 m²	0.92
18 cuts per year	-	0.01	0.15	0.05	1.00	100 m²	1.20
12 cuts per year	-	0.01	0.10	0.14	1.80	100 m²	2.03
4 cuts per year	-	0.02	0.30	0.82	3.00	100 m²	4.12
Harrowing							
Harrowing grassed area with							
drag harrow	-	0.01	0.19	0.14	-	100 m²	0.33
chain or light flexible spiked harrow	-	0.02	0.25	0.14	-	100 m²	0.39
Scarifying							
Mechanical							
Sisis ARP4; including grass collection box; towed by tractor; area scarified annually	-	0.02	0.36	0.10	-	100 m²	0.46
Sisis ARP4; including grass collection box; towed by tractor; area scarified two years previously	-	0.03	0.43	0.12	-	100 m²	0.55
pedestrian operated self-powered equipment	-	0.07	1.07	0.22	-	100 m²	1.29
add for disposal of arisings	-	0.03	0.38	1.02	14.98	100 m²	16.38
By hand							
hand implement	-	0.50	7.63	-	-	100 m²	7.63
add for disposal of arisings	-	0.03	0.38	1.02	14.98	100 m²	16.38

Q PAVING/PLANTING/FENCING/SITE FURNITURE

Item Excluding site overheads and profit	PC £	Labour hours	Labour £	Plant £	Material £	Unit	Total rate £
Q35 LANDSCAPE MAINTENANCE - cont'd							
Rolling							
Rolling grassed area; equipment towed by tractor; once over; using							
smooth roller	-	0.01	0.22	0.15	-	100 m²	**0.37**
Turf aeration							
By machine							
Vertidrain turf aeration equipment towed by tractor to effect a minimum penetration of 100 to 250 mm at 100 mm centres	-	0.04	0.59	1.29	-	100 m²	**1.89**
Ryan GA 30; self-propelled turf aerating equipment; to effect a minimum penetration of 100 mm at varying centres	-	0.04	0.59	1.51	-	100 m²	**2.10**
Groundsman; pedestrian operated; self-powered solid or slitting tine turf aerating equipment to effect a minimum penetration of 100 mm	-	0.14	2.14	1.43	-	100 m²	**3.56**
Cushman core harvester; self-propelled; for collection of arisings	-	0.02	0.30	0.66	-	100 m²	**0.96**
By hand							
hand fork; to effect a minimum penetration of 100 mm and spaced 150 mm apart	-	1.33	20.33	-	-	100 m²	**20.33**
hollow tine hand implement; to effect a minimum penetration of 100 mm and spaced 150 mm apart	-	2.00	30.50	-	-	100 m²	**30.50**
collection of arisings by hand	-	3.00	45.75	-	-	100 m²	**45.75**
Turf areas; surface treatments and top dressing; Boughton Loam Ltd							
Apply screened topdressing to grass surfaces; spread using Tru-Lute							
Sand soil mixes 90/10 to 50/50	0.14	-	0.03	0.03	0.14	m²	**0.20**
Apply screened soil 3 mm, Kettering loam to goal mouths and worn areas							
20 mm thick	1.10	0.01	0.15	-	1.32	m²	**1.47**
10 mm thick	0.55	0.01	0.15	-	0.66	m²	**0.81**
Leaf clearance; clearing grassed area of leaves and other extraneous debris							
Using equipment towed by tractor							
large grassed areas with perimeters of mature trees such as sports fields and amenity areas	-	0.01	0.19	0.07	-	100 m²	**0.26**
large grassed areas containing ornamental trees and shrub beds	-	-	0.06	0.98	-	100 m²	**1.04**
Using pedestrian operated mechanical equipment and blowers							
grassed areas with perimeters of mature trees such as sports fields and amenity areas	-	0.02	0.25	1.57	-	100 m²	**1.82**
grassed areas containing ornamental trees and shrub beds	-	0.10	1.52	0.67	-	100 m²	**2.19**
verges	-	0.07	1.02	0.09	-	100 m²	**1.11**
By hand							
grassed areas with perimeters of mature trees such as sports fields and amenity areas	-	0.10	1.52	0.20	-	100 m²	**1.72**

Q PAVING/PLANTING/FENCING/SITE FURNITURE

Item Excluding site overheads and profit	PC £	Labour hours	Labour £	Plant £	Material £	Unit	Total rate £
grassed areas containing ornamental trees and shrub beds	-	0.17	2.54	0.33	-	100 m^2	2.87
verges	-	0.33	5.08	0.66	-	100 m^2	5.75
Removal of arisings							
areas with perimeters of mature trees	-	0.01	0.10	0.09	1.20	100 m^2	1.39
areas containing ornamental trees and shrub beds	-	0.02	0.30	0.34	3.00	100 m^2	3.64
Litter clearance							
Collection and disposal of litter from grassed area	-	0.01	0.15	-	0.06	100 m^2	0.22
Collection and disposal of litter from isolated grassed area not exceeding 1000 m^2	-	0.04	0.61	-	0.06	100 m^2	0.67
Edge maintenance							
Maintain edges where lawn abuts pathway or hard surface using							
strimmer	-	0.01	0.08	0.01	-	m	0.08
shears	-	0.02	0.25	-	-	m	0.25
Maintain edges where lawn abuts plant bed using							
mechanical edging tool	-	0.01	0.10	0.03	-	m	0.13
shears	-	0.01	0.17	-	-	m	0.17
half moon edging tool	-	0.02	0.30	-	-	m	0.30
Tree guards, stakes and ties etc.							
Adjusting existing tree tie	-	0.03	0.51	-	-	nr	0.51
Taking up single or double tree stake and ties; removing and disposing	-	0.05	0.91	-	-	nr	0.91
Pruning shrubs							
Trimming ground cover planting							
soft groundcover; vinca ivy and the like	-	1.00	15.25	-	-	100 m^2	15.25
woody groundcover; cotoneaster and the like	-	1.50	22.88	-	-	100 m^2	22.88
Pruning massed shrub border (measure ground area)							
shrub beds pruned annually	-	0.01	0.15	-	-	m^2	0.15
shrub beds pruned hard every 3 years	-	0.03	0.42	-	-	m^2	0.42
Cutting off dead heads							
bush or standard rose	-	0.05	0.76	-	-	nr	0.76
climbing rose	-	0.08	1.27	-	-	nr	1.27
Pruning roses							
bush or standard rose	-	0.05	0.76	-	-	nr	0.76
climbing rose or rambling rose; tying in as required	-	0.07	1.02	-	-	nr	1.02
Pruning ornamental shrub; height before pruning (increase these rates by 50% if pruning work has not been executed during the previous two years)							
not exceeding 1m	-	0.04	0.61	-	-	nr	0.61
1 to 2 m	-	0.06	0.85	-	-	nr	0.85
exceeding 2 m	-	0.13	1.91	-	-	nr	1.91

Q PAVING/PLANTING/FENCING/SITE FURNITURE

Item Excluding site overheads and profit	PC £	Labour hours	Labour £	Plant £	Material £	Unit	Total rate £
Q35 LANDSCAPE MAINTENANCE - cont'd							
Pruning shrubs - cont'd							
Removing excess growth etc from face of building etc.; height before pruning							
not exceeding 2 m	-	0.03	0.44	-	-	nr	**0.44**
2 to 4 m	-	0.05	0.76	-	-	nr	**0.76**
4 to 6 m	-	0.08	1.27	-	-	nr	**1.27**
6 to 8 m	-	0.13	1.91	-	-	nr	**1.91**
8 to 10 m	-	0.14	2.18	-	-	nr	**2.18**
Removing epicormic growth from base of shrub or trunk and base of tree: any height; any diameter; number of growths							
not exceeding 10	-	0.05	0.76	-	-	nr	**0.76**
10 to 20	-	0.07	1.02	-	-	nr	**1.02**
Beds, borders and planters							
Lifting							
bulbs	-	0.50	7.63	-	-	100 nr	**7.63**
tubers or corms	-	0.40	6.10	-	-	100 nr	**6.10**
established herbaceous plants; hoeing and depositing for replanting	-	2.00	30.50	-	-	100 nr	**30.50**
Temporary staking and tying in herbaceous plant	-	0.03	0.51	-	0.11	nr	**0.62**
Cutting down spent growth of herbaceous plant; clearing arisings							
unstaked	-	0.02	0.30	-	-	nr	**0.30**
staked; not exceeding 4 stakes per plant; removing stakes and putting into store	-	0.03	0.38	-	-	nr	**0.38**
Hand weeding							
newly planted areas	-	2.00	30.50	-	-	100 m^2	**30.50**
established areas	-	0.50	7.63	-	-	100 m^2	**7.63**
Removing grasses from groundcover areas	-	3.00	45.80	-	-	100 m^2	**45.80**
Hand digging with fork; not exceeding 150 mm deep; breaking down lumps; leaving surface with a medium tilth	-	1.33	20.33	-	-	100 m^2	**20.33**
Hand digging with fork or spade to an average depth of 230 mm; breaking down lumps; leaving surface with a medium tilth	-	2.00	30.50	-	-	100 m^2	**30.50**
Hand hoeing; not exceeding 50 mm deep; leaving surface with a medium tilth	-	0.40	6.10	-	-	100 m^2	**6.10**
Hand raking to remove stones etc.; breaking down lumps: leaving surface with a fine tilth prior to planting	-	0.67	10.17	-	-	100 m^2	**10.17**
Hand weeding; planter, window box; not exceeding 1.00 m^2							
ground level box	-	0.05	0.76	-	-	nr	**0.76**
box accessed by stepladder	-	0.08	1.27	-	-	nr	**1.27**
Spreading only compost, mulch or processed bark to a depth of 75 mm							
on shrub bed with existing mature planting	-	0.09	1.39	-	-	m^2	**1.39**
recently planted areas	-	0.07	1.02	-	-	m^2	**1.02**
groundcover and herbaceous areas	-	0.08	1.14	-	-	m^2	**1.14**

Q PAVING/PLANTING/FENCING/SITE FURNITURE

Item Excluding site overheads and profit	PC £	Labour hours	Labour £	Plant £	Material £	Unit	Total rate £
Clearing cultivated area of leaves, litter and other extraneous debris; using hand implement							
weekly maintenance	-	0.13	1.91	-	-	100 m²	1.91
daily maintenance	-	0.02	0.25	-	-	100 m²	0.25
Bedding							
Lifting							
bedding plants; hoeing and depositing for disposal	-	3.00	45.75	-	-	100 m²	45.75
Hand digging with fork; not exceeding 150 mm deep: breaking down lumps; leaving surface with a medium tilth	-	0.75	11.44	-	-	100 m²	11.44
Hand weeding							
newly planted areas	-	2.00	30.50	-	-	100 m²	30.50
established areas	-	0.50	7.63	-	-	100 m²	7.63
Hand digging with fork or spade to an average depth of 230 mm; breaking down lumps; leaving surface with a medium tilth	-	0.50	7.63	-	-	100 m²	7.63
Hand hoeing: not exceeding 50 mm deep; leaving surface with a medium tilth	-	0.40	6.10	-	-	100 m²	6.10
Hand raking to remove stones etc.; breaking down lumps; leaving surface with a fine tilth prior to planting	-	0.67	10.17	-	-	100 m²	10.17
Hand weeding; planter, window box; not exceeding 1.00 m²							
ground level box	-	0.05	0.76	-	-	nr	0.76
box accessed by stepladder	-	0.08	1.27	-	-	nr	1.27
Spreading only; compost, mulch or processed bark to a depth of 75 mm							
on shrub bed with existing mature planting	-	0.09	1.39	-	-	m²	1.39
recently planted areas	-	0.07	1.02	-	-	m²	1.02
groundcover and herbaceous areas	-	0.08	1.14	-	-	m²	1.14
Collecting bedding from nursery	-	3.00	45.75	13.64	-	100 m²	59.39
Setting out							
mass planting single variety	-	0.13	1.91	-	-	m²	1.91
pattern	-	0.33	5.08	-	-	m²	5.08
Planting only							
massed bedding plants	-	0.20	3.05	-	-	m²	3.05
Clearing cultivated area of leaves, litter and other extraneous debris; using hand implement							
weekly maintenance	-	0.13	1.91	-	-	100 m²	1.91
daily maintenance	-	0.02	0.25	-	-	100 m²	0.25
Irrigation and watering							
Hand held hosepipe; flow rate 25 litres per minute; irrigation requirement							
10 litres/m²	-	0.74	11.24	-	-	100 m²	11.24
15 litres/m²	-	1.10	16.77	-	-	100 m²	16.77
20 litres/m²	-	1.46	22.31	-	-	100 m²	22.31
25 litres/m²	-	1.84	28.01	-	-	100 m²	28.01
Hand held hosepipe; flow rate 40 litres per minute; irrigation requirement							
10 litres/m²	-	0.46	7.05	-	-	100 m²	7.05
15 litres/m²	-	0.69	10.57	-	-	100 m²	10.57
20 litres/m²	-	0.91	13.92	-	-	100 m²	13.92
25 litres/m²	-	1.15	17.48	-	-	100 m²	17.48

Q PAVING/PLANTING/FENCING/SITE FURNITURE

Item Excluding site overheads and profit	PC £	Labour hours	Labour £	Plant £	Material £	Unit	Total rate £
Q35 LANDSCAPE MAINTENANCE - cont'd							
Hedge cutting; field hedges cut once or twice annually							
Trimming sides and top using hand tool or hand held mechanical tools							
not exceeding 2 m high	-	0.10	1.52	0.14	-	10 m²	**1.67**
2 to 4 m high	-	0.33	5.08	0.47	-	10 m²	**5.55**
Hedge cutting; ornamental							
Trimming sides and top using hand tool or hand held mechanical tools							
not exceeding 2 m high	-	0.13	1.91	0.18	-	10 m²	**2.08**
2 to 4 m high	-	0.50	7.63	0.70	-	10 m²	**8.33**
Hedge cutting; reducing width; hand tool or hand held mechanical tools							
Not exceeding 2 m high							
average depth of cut not exceeding 300 mm	-	0.03	0.38	0.04	-	m²	**0.42**
average depth of cut 300 to 600 mm	-	0.06	0.85	0.08	-	m²	**0.93**
average depth of cut 600 to 900 mm	-	0.06	0.95	0.09	-	m²	**1.04**
2 to 4 m high							
average depth of cut not exceeding 300 mm	-	0.04	0.61	0.06	-	m²	**0.67**
average depth of cut 300 to 600 mm	-	0.06	0.95	0.09	-	m²	**1.04**
average depth of cut 600 to 900 mm	-	0.10	1.52	0.14	-	m²	**1.67**
4 to 6 m high							
average depth of cut not exceeding 300 mm	-	0.10	1.52	0.14	-	m²	**1.67**
average depth of cut 300 to 600 mm	-	0.17	2.54	0.23	-	m²	**2.78**
average depth of cut 600 to 900 mm	-	0.50	7.63	0.70	-	m²	**8.33**
Hedge cutting; reducing width; tractor mounted hedge cutting equipment							
Not exceeding 2 m high							
average depth of cut not exceeding 300 mm	-	0.04	0.61	0.67	-	10 m²	**1.28**
average depth of cut 300 to 600 mm	-	0.05	0.76	0.83	-	10 m²	**1.59**
average depth of cut 600 to 900 mm	-	0.20	3.05	3.33	-	10 m²	**6.38**
2 to 4 m high							
average depth of cut not exceeding 300 mm	-	0.01	0.19	0.21	-	10 m²	**0.40**
average depth of cut 300 to 600 mm	-	0.03	0.38	0.42	-	10 m²	**0.80**
average depth of cut 600 to 900 mm	-	0.02	0.30	0.33	-	10 m²	**0.64**
Hedge cutting; reducing height; hand tool or hand held mechanical tools							
Not exceeding 2 m high							
average depth of cut not exceeding 300 mm	-	0.07	1.02	0.09	-	10 m²	**1.11**
average depth of cut 300 to 600 mm	-	0.13	2.03	0.19	-	10 m²	**2.22**
average depth of cut 600 to 900 mm	-	0.40	6.10	0.56	-	10 m²	**6.66**
2 to 4 m high							
average depth of cut not exceeding 300 mm	-	0.03	0.51	0.05	-	m²	**0.55**
average depth of cut 300 to 600 mm	-	0.07	1.02	0.09	-	m²	**1.11**
average depth of cut 600 to 900 mm	-	0.20	3.05	0.28	-	m²	**3.33**

Q PAVING/PLANTING/FENCING/SITE FURNITURE

Item Excluding site overheads and profit	PC £	Labour hours	Labour £	Plant £	Material £	Unit	Total rate £
4 to 6 m high							
average depth of cut not exceeding							
300 mm	-	0.07	1.02	0.09	-	m²	1.11
average depth of cut 300 to 600 mm	-	0.13	1.91	0.18	-	m²	2.08
average depth of cut 600 to 900 mm	-	0.25	3.81	0.35	-	m²	4.16
Hedge cutting; removal and disposal of arisings							
Sweeping up and depositing arisings							
300 mm cut	-	0.05	0.76	-	-	10 m²	0.76
600 mm cut	-	0.20	3.05	-	-	10 m²	3.05
900 mm cut	-	0.40	6.10	-	-	10 m²	6.10
Chipping arisings							
300 mm cut	-	0.02	0.30	0.11	-	10 m²	0.42
600 mm cut	-	0.08	1.27	0.48	-	10 m²	1.75
900 mm cut	-	0.20	3.05	1.14	-	10 m²	4.19
Disposal of unchipped arisings							
300 mm cut	-	0.02	0.25	0.51	2.00	10 m²	2.76
600 mm cut	-	0.03	0.51	1.02	3.00	10 m²	4.53
900 mm cut	-	0.08	1.27	2.56	7.49	10 m²	11.32
Disposal of chipped arisings							
300 mm cut	-	-	0.05	0.17	3.00	10 m²	3.22
600 mm cut	-	0.02	0.25	0.17	5.99	10 m²	6.42
900 mm cut	-	0.03	0.51	0.34	14.98	10 m²	15.83
Herbicide applications; standard backpack spray applicators; application to maintain 1.00 m diameter clear circles (0.79m²) around new planting							
Scotts UK; Roundup Pro Biactive 360; glyphosate; enhanced movement glyphosate; application rate 5 litre/ha							
plants at 1.50 m centres; 4444 nr/ha	21.21	24.69	450.59	-	23.33	ha	473.92
plants at 1.75 m centres; 3265 nr/ha	15.51	18.14	331.06	-	17.07	ha	348.12
plants at 2.00 m centres; 2500 nr/ha	11.97	13.89	253.49	-	13.17	ha	266.66
mass spraying	60.60	13.00	237.25	-	66.66	ha	303.91
Japanese Knotweed Management Plan; PBA Solutions Ltd							
Formulate Japanese Knotweed Management Plan (KMP) as recommended by the Environment Agency Code of Practice 2006 (EA CoP)	-	-	-	-	-	nr	1200.00

Q PAVING/PLANTING/FENCING/SITE FURNITURE

Item Excluding site overheads and profit	PC £	Labour hours	Labour £	Plant £	Material £	Unit	Total rate £
Q35 LANDSCAPE MAINTENANCE - cont'd							
Treatment of Japanese Knotweed; PBA Solutions Ltd; removal and/or treatment in accordance with 'The Knotweed Code of Practice' published by the Environment Agency 2006 (EACoP); Japanese knotweed must be treated by qualified practitioners							
Commercial chemical treatments based on areas less than 50 m² knotweed area; allows for the provision of contractual arrangements including method statements and spraying records; also allows for the provision of risk, environment and COSHH assessments							
rate per visit	-	-	-	-	-	nr	225.00
rate per year	-	-	-	-	-	nr	900.00
rate for complete treatment programme lasting 3 years	-	-	-	-	-	nr	2700.00
Commercial chemical treatments based on areas between 50-300 m² knotweed area; allows for the provision of contractual arrangements including method statements and spraying recordsl also allows for the provision of risk, environment and COSHH assessments							
rate per visit	-	-	-	-	-	nr	300.00
rate per year	-	-	-	-	-	nr	1200.00
rate for complete treatment programme lasting 3 years	-	-	-	-	-	nr	3600.00
Extra over minimum rate above; for areas 300 - 900 m²							
rate per visit	-	-	-	-	-	m²	0.48
rate per year	-	-	-	-	-	m²	1.92
rate for complete treatment programme lasting 3 years	-	-	-	-	-	m²	5.76
Extra over minimum rate above; for areas over 900 m²							
rate per visit	-	-	-	-	-	m²	0.40
rate per year	-	-	-	-	-	m²	1.60
rate for complete treatment programme lasting 3 years	-	-	-	-	-	m²	4.80

Q PAVING/PLANTING/FENCING/SITE FURNITURE

Item Excluding site overheads and profit	PC £	Labour hours	Labour £	Plant £	Material £	Unit	Total rate £
Note: Taking waste off site will require pre-treatment. Landfill operators are required by the Environment Agency to obtain evidence of pre-treatment to show that the waste has been treated to either reduce its volume or hazardousness nature, or facilitate its handling, or enhance its recovery by the waste produce (site of origin) prior being taken to landfill. Generally pre-treatment of Japanese Knotweed can include chemical treatment, cutting or burning. The cost to implement this operation would be similar to item 4/5 with a minimum charge of £300.00. To enable waste to go to landfill soil testing would need to be completed and consultation with a landfill operator or invasive weed specialist is recommended in the first instance.							
Japanese knotweed contaminated waste removal; PBA Solutions Ltd (see note above)							
Removal offsite to a licensed landfill site using 20 tonne gross vehicle weight lorries; price is for haulage and tipping only; the mileage stated below is the distance from point of collection to licensed landfill site; price excludes landfill tax and loading of material into lorries; rates will fluctuate between landfill operators and the figures given are for initial guidance only							
less than 20 miles	-	-	-	-	-	load	575.00
20 - 40 miles	-	-	-	-	-	load	637.00
40 - 60 miles	-	-	-	-	-	load	835.00
Rootbarriers; PBA Solutions Ltd; used in conjunctiont with on site control methods such as bund treatment, cell burial, vertical boundary protection and capping. The following barriers have been reliably used in connection with Japanese knotweed control and as such are effective when correctly installed as part of a control programme which would usually include for chemical control							
Linear root deflection barriers; installed to trench measured separately							
"Flexiroot" non permeable; reinforced LDPE coated polypropylene geomembrane	5.36	0.05	0.91	-	5.89	m	6.80
"bioroot-barrier x" permeable root barrier; mechanically bonded geocomposite consisting of a cooper-foil bonded between 2 layers of geotextile	10.76	0.05	0.91	-	10.76	m	11.67

Q PAVING/PLANTING/FENCING/SITE FURNITURE

Item Excluding site overheads and profit	PC £	Labour hours	Labour £	Plant £	Material £	Unit	Total rate £
Q35 LANDSCAPE MAINTENANCE - cont'd							
Landfill tax rates; based on waste being within landfill operator thresholds for controlled waste and with Japanese knotweed being the only form of contamination; the rates below are based on visual inspections of the load at the gate of the tip							
Rates to April 2010							
higher rate (where knotweed is visible at more than 5% on inspection at landfill)	-	-	-	-	40.00	tonne	-
lower rate (where knotweed is visible at less than 5% on inspection at landfill)	-	-	-	-	2.50	tonne	-
Rates from April 2010 - April 2011							
higher rate (where knotweed is visible at more than 5% on inspection at landfill)	-	-	-	-	48.00	tonne	-
lower rate (where knotweed is visible at less than 5% on inspection at landfill)	-	-	-	-	2.50	tonne	-
Q40 FENCING AND GATES							
Temorary security fence; HSS Hire; mesh framed unclimbable fencing; including precast concrete supports and couplings							
Weekly hire; 2.85 x 2.00 m high							
weekly hire rate	-	-	-	1.28	-	m	1.28
erection of fencing; labour only	-	0.10	1.82	-	-	m	1.82
removal of fencing loading to collection vehicle	-	0.07	1.22	-	-	m	1.22
delivery charge	-	-	-	0.80	-	m	0.80
return haulage charge	-	-	-	0.60	-	m	0.60
Protective fencing; AVS Fencing Supplies Ltd							
Cleft chestnut rolled fencing; fixing to 100 mm diameter chestnut posts; driving into firm ground at 3 m centres							
900 mm high	2.55	0.11	1.95	-	3.15	m	5.10
1200 mm high	2.95	0.16	2.92	-	3.57	m	6.49
1500 mm high; 3 strand	4.91	0.21	3.90	-	5.39	m	9.28
Enclosures; Earth Anchors							
"Rootfast" anchored galvanized steel enclosures post ref ADP 20-1000; 1000 mm high x 20 mm diameter with ref AA25-750 socket and padlocking ring	35.00	0.10	1.83	-	37.19	nr	39.01
steel cable, orange plastic coated	1.25	-	0.04	-	1.25	m	1.29
Timber fencing; AVS Fencing Supplies Ltd; tanalized softwood fencing; all timber posts are "kiln dried redwood" with 15 year guarantee							
Timber lap panels; pressure treated; fixed to timber posts 75 x 75 mm in 1:3:6 concrete; at 1.90 m centres							
900 mm high	7.49	0.67	12.17	-	14.21	m	26.38
1200 mm high	7.63	0.75	13.69	-	13.91	m	27.60
1500 mm high	7.63	0.80	14.60	-	14.20	m	28.80
1800 mm high	7.65	0.90	16.43	-	15.06	m	31.49

Q PAVING/PLANTING/FENCING/SITE FURNITURE

Item Excluding site overheads and profit	PC £	Labour hours	Labour £	Plant £	Material £	Unit	Total rate £
Timber lap panels; fixed to slotted concrete posts 100 x 100 mm 1:3:6 concrete at 1.88 m centres							
900 mm high	7.49	0.75	13.69	-	17.34	m	**31.03**
1200 mm high	7.62	0.80	14.60	-	17.47	m	**32.07**
1500 mm high	10.28	0.85	15.51	-	21.75	m	**37.26**
1800 mm high	10.30	1.00	18.25	-	21.95	m	**40.20**
extra for corner posts	19.55	0.90	16.43	-	26.03	nr	**42.46**
extra over panel fencing for 300 mm high trellis tops; slats at 100 mm centres; including additional length of posts	4.25	1.00	18.25	-	4.25	m	**22.50**
Close boarded fencing; to concrete posts 100 x 100 mm; 2 nr softwood arris rails; 100 x 22 mm softwood pales lapped 13 mm; including excavating and backfilling into firm ground at 3.00 m centres; setting in concrete 1:3:6; concrete gravel board 150 x 150 mm							
900 mm high	12.68	1.00	18.25	-	14.98	m	**33.23**
1050 mm high	12.92	1.10	20.07	-	15.22	m	**35.29**
1500 mm high	14.72	1.15	20.99	-	17.15	m	**38.14**
Close boarded fencing; to concrete posts 100 x 100 mm; 3 nr softwood arris rails; 100 x 22 mm softwood pales lapped 13 mm; including excavating and backfilling into firm ground at 3.00 m centres; setting in concrete 1:3:6							
1650 mm high	17.65	1.25	22.81	-	19.94	m	**42.75**
1800 mm high	17.77	1.30	23.73	-	20.20	m	**43.92**
Close boarded fencing; to timber redwood posts 100 x 100 mm; 2 nr softwood arris rails; 100 x 22 mm softwood pales lapped 13 mm; including excavating and backfilling into firm ground at 3.00 m centres; setting in concrete 1:3:6							
1350 mm high	14.13	1.00	18.25	-	16.56	m	**34.81**
1650 mm high	14.97	1.25	22.81	-	17.40	m	**40.21**
1800 mm high	15.43	1.30	23.73	-	17.86	m	**41.58**
Close boarded fencing; to timber posts 100 x 100 mm; 3 nr softwood arris rails; 100 x 22 mm softwood pales lapped 13 mm; including excavating and backfilling into firm ground at 3.00 m centres; setting in concrete 1:3:6							
1350 mm high	16.70	0.95	17.38	-	18.99	m	**36.37**
1650 mm high	17.54	1.30	23.73	-	19.97	m	**43.69**
1800 mm high	18.00	1.35	24.64	-	20.43	m	**45.06**
extra over for post 125 x 100 mm; for 1800 mm high fencing	7.67	-	-	-	7.67	m	-
extra over to the above for counter rail	0.59	-	-	-	0.59	m	-
extra over to the above for capping rail	0.59	0.10	1.82	-	0.59	m	**2.42**
extra over to the above for concrete gravel board 150 x 50 mm	4.52	-	-	-	4.52	m	-

Q PAVING/PLANTING/FENCING/SITE FURNITURE

Item Excluding site overheads and profit	PC £	Labour hours	Labour £	Plant £	Material £	Unit	Total rate £
Q40 FENCING AND GATES - cont'd							
Timber fencing - cont'd							
Feather edge board fencing; to timber posts 100 x 100 mm; 2 nr softwood cant rails; 125 x 22 mm feather edge boards lapped 13 mm; including excavating and backfilling into firm ground at 3.00 m centres; setting in concrete 1:3:6							
1350 mm high	13.39	0.56	10.22	-	14.48	m	24.70
1650 mm high	14.63	0.56	10.22	-	15.72	m	25.94
1800 mm high	14.99	0.56	10.22	-	16.08	m	26.30
Feather edge board fencing; to timber posts 100 x 100 mm; 3 nr softwood cant rails; 125 x 22 mm feather edge boards lapped 13 mm; including excavating and backfilling into firm ground at 3.00 m centres; setting in concrete 1:3:6							
1350 mm high	16.18	0.57	10.43	-	17.27	m	27.70
1650 mm high	17.42	0.57	10.43	-	18.51	m	28.94
1800 mm high	16.81	0.57	10.43	-	18.87	m	29.30
extra over for post 125 x 100 mm; for 1800 mm high fencing	7.67	-	-	-	7.67	nr	-
extra over to the above for counter rail and capping	1.70	0.10	1.82	-	1.70	m	3.53
Palisade fencing; 22 x 75 mm softwood vertical palings with pointed tops; nailing to two 50 x 100 mm horizontal softwood arris rails; morticed to 100 x 100 mm softwood posts with weathered tops at 3.00 m centres; setting in concrete							
900 mm high	11.56	0.90	16.43	-	13.99	m	30.41
1050 mm high	12.29	0.90	16.43	-	14.72	m	31.14
1200 mm high	12.62	0.95	17.34	-	15.05	m	32.39
extra over for rounded tops	0.66	-	-	-	0.66	m	-
Post-and-rail fencing; 90 x 38 mm softwood horizontal rails; fixing with galvanized nails to 150 x 75 mm softwood posts; including excavating and backfilling into firm ground at 1.80 m centres; all treated timber							
1200 mm high; 3 horizontal rails	7.49	0.35	6.39	-	7.57	m	13.97
1200 mm high; 4 horizontal rails	8.67	0.35	6.39	-	8.67	m	15.07
Cleft rail fencing; oak or chestnut adze tapered rails 2.80 m long; morticed into joints; to 125 x 100 mm softwood posts 1.95 m long; including excavating and backfilling into firm ground at 2.80 m centres							
two rails	9.04	0.25	4.56	-	9.04	m	13.60
three rails	11.39	0.28	5.11	-	11.39	m	16.51
four rails	13.74	0.35	6.39	-	13.74	m	20.13
"Hit and miss" horizontal rail fencing; 87 x 38 mm top and bottom rails; 100 x 22 mm vertical boards arranged alternately on opposite side of rails; to 100 x 100 mm posts; including excavating and backfilling into firm ground; setting in concrete at 1.8 m centres							
treated softwood; 1600 mm high	25.18	1.20	21.90	-	27.37	m	49.27
treated softwood; 1800 mm high	27.70	1.33	24.33	-	30.11	m	54.44
treated softwood; 2000 mm high	30.23	1.40	25.55	-	32.64	m	58.19

Q PAVING/PLANTING/FENCING/SITE FURNITURE

Item Excluding site overheads and profit	PC £	Labour hours	Labour £	Plant £	Material £	Unit	Total rate £
Trellis tops to fencing (see more detailed items in the Minor Works section of this book)							
Extra over screen fencing for 300 mm high trellis tops; slats at 100 mm centres; including additional length of posts	4.25	0.10	1.82	-	4.25	m	6.07
Boundary fencing; strained wire and wire mesh; AVS Fencing Supplies Ltd							
Strained wire fencing; concrete inter posts only at 2750 mm centres; 610 mm below ground; excavating holes; filling with concrete; replacing topsoil; disposing surplus soil off site							
900 mm high	2.89	0.56	10.14	0.59	4.14	m	14.88
1200 mm high	3.62	0.56	10.14	0.59	4.87	m	15.61
1800 mm high	4.95	0.78	14.19	1.48	6.20	m	21.88
Extra over strained wire fencing for concrete straining posts with one strut; posts and struts 610 mm below ground; struts, cleats, stretchers, winders, bolts and eye bolts; excavating holes; filling to within 150 mm of ground level with concrete (1:12) - 40 mm aggregate; replacing topsoil; disposing surplus soil off site							
900 mm high	23.35	0.67	12.23	1.33	49.69	nr	63.24
1200 mm high	25.97	0.67	12.17	1.33	53.25	nr	66.75
1800 mm high	36.96	0.67	12.17	1.33	66.58	nr	80.08
Extra over strained wire fencing for concrete straining posts with two struts; posts and struts 610 mm below ground; excavating holes; filling to within 150 mm of ground level with concrete (1:12) - 40 mm aggregate; replacing topsoil; disposing surplus soil off site							
900 mm high	33.68	0.91	16.61	1.71	73.02	nr	91.33
1200 mm high	37.54	0.85	15.51	1.33	76.87	nr	93.71
1800 mm high	53.66	0.85	15.51	1.33	102.38	nr	119.22
Strained wire fencing; galvanized steel angle posts only at 2750 mm centres; 610 mm below ground; driving in							
900 mm high; 40 x 40 x 5 mm	2.65	0.03	0.55	-	2.65	m	3.20
1200 mm high; 40 x 40 x 5 mm	3.01	0.03	0.61	-	3.01	m	3.62
1500 mm high; 40 x 40 x 5 mm	4.01	0.06	1.11	-	4.01	m	5.12
1800 mm high; 40 x 40 x 5 mm	4.32	0.07	1.33	-	4.32	m	5.65
1800 mm high; 45 x 45 x 5 mm with extention for three rows barbed wire	5.54	0.12	2.21	-	5.54	m	7.75
Galvanized steel straining posts with two struts for strained wire fencing; setting in concrete							
900 mm high; 50 x 50 x 6 mm	47.20	1.00	18.25	-	49.51	nr	67.76
1200 mm high; 50 x 50 x 6 mm	57.01	1.00	18.25	-	59.32	nr	77.57
1500 mm high; 50 x 50 x 6 mm	71.23	1.00	18.25	-	73.54	nr	91.79
1800 mm high; 50 x 50 x 6 mm	73.83	1.00	18.25	-	76.14	nr	94.39
2400 mm high; 60 x 60 x 6 mm	83.59	1.00	18.25	-	85.90	nr	104.15

Q PAVING/PLANTING/FENCING/SITE FURNITURE

Item Excluding site overheads and profit	PC £	Labour hours	Labour £	Plant £	Material £	Unit	Total rate £
Q40 FENCING AND GATES - cont'd							
Boundary fencing - cont'd							
Strained wire; to posts (posts not included); 3 mm galvanized wire; fixing with galvanized stirrups							
900 mm high; 2 wire	0.16	0.03	0.61	-	0.31	m	0.92
1200 mm high; 3 wire	0.24	0.05	0.85	-	0.39	m	1.24
1400 mm high; 3 wire	0.24	0.05	0.85	-	0.39	m	1.24
1800 mm high; 3 wire	0.24	0.05	0.85	-	0.39	m	1.24
Barbed wire; to posts (posts not included); 3 mm galvanized wire; fixing with galvanized stirrups							
900 mm high; 2 wire	0.25	0.07	1.22	-	0.40	m	1.61
1200 mm high; 3 wire	0.37	0.09	1.70	-	0.52	m	2.22
1400 mm high; 3 wire	0.37	0.09	1.70	-	0.52	m	2.22
1800 mm high; 3 wire	0.37	0.09	1.70	-	0.52	m	2.22
Chain link fencing; AVS Fencing Supplies Ltd; to strained wire and posts priced separately; 3 mm galvanized wire; 51 mm mesh; galvanized steel components; fixing to line wires threaded through posts and strained with eye-bolts; posts (not included)							
900 mm high	3.89	0.07	1.22	-	4.04	m	5.25
1200 mm high	5.44	0.07	1.22	-	5.59	m	6.81
1800 mm high	7.54	0.10	1.82	-	7.69	m	9.51
Chain link fencing; to strained wire and posts priced separately; 3.15 mm plastic coated galvanized wire (wire only 2.50 mm); 51 mm mesh; galvanized steel components; fencing with line wires threaded through posts and strained with eye-bolts; posts (not included) (Note: plastic coated fencing can be cheaper than galvanized finish as wire of a smaller cross-sectional area can be used)							
900 mm high	2.75	0.07	1.22	-	3.05	m	4.27
1200 mm high	3.79	0.07	1.22	-	4.09	m	5.31
1800 mm high	5.40	0.13	2.28	-	6.60	m	8.88
Extra over strained wire fencing for cranked arms and galvanized barbed wire							
1 row	3.47	0.02	0.30	-	3.47	m	3.78
2 row	3.60	0.05	0.91	-	3.60	m	4.51
3 row	3.72	0.05	0.91	-	3.72	m	4.64

Q PAVING/PLANTING/FENCING/SITE FURNITURE

Item Excluding site overheads and profit	PC £	Labour hours	Labour £	Plant £	Material £	Unit	Total rate £
Field fencing; AVS Fencing Supplies Ltd; welded wire mesh; fixed to posts and straining wires measured separately							
Cattle fence; 1200 m high; 114 x 300 mm at bottom to 230 x 300 mm at top	0.76	0.10	1.82	-	0.96	m	2.78
Sheep fence; 800 mm high; 140 x 300 mm at bottom to 230 x 300 mm at top	0.77	0.10	1.82	-	0.97	m	2.80
Deer fence; 2900 mm high; 89 x 150 mm at bottom to 267 x 300 mm at top	1.53	0.13	2.28	-	1.73	m	4.01
Extra for concreting in posts	-	0.50	9.13	-	3.89	nr	13.01
Extra for straining post	8.25	0.75	13.69	-	8.25	nr	21.94
Rabbit netting; AVS Fencing Supplies Ltd Timber stakes; peeled kiln dried pressure treated; pointed; 1.8 m posts driven 900 mm into ground at 3 m centres (line wires and netting priced separately)							
75 - 100 mm stakes	1.92	0.25	4.56	-	1.92	m	6.48
Corner posts or straining posts 150 mm diameter x 2.3 m high set in concrete; centres to suit local conditions or changes of direction							
1 strut	10.19	1.00	18.25	3.80	16.14	each	38.19
2 strut	12.51	1.00	18.25	3.80	18.46	each	40.51
Strained wire; to posts (posts not included); 3 mm galvanized wire; fixing with galvanized stirrups							
900 mm high; 2 wire	0.16	0.03	0.61	-	0.31	m	0.92
1200 mm high; 3 wire	0.24	0.05	0.85	-	0.39	m	1.24
Rabbit netting; 31 mm 19 gauge 1050 mm high netting fixed to posts line wires and straining posts or corner posts all priced separately							
900 mm high; turned in	0.84	0.04	0.73	-	0.86	m	1.59
900 mm high; buried 150 mm in trench	0.84	0.08	1.52	-	0.86	m	2.38
Boundary fencing; strained wire and wire mesh; Jacksons Fencing Tubular chain link fencing; galvanized; plastic coated; 60.3 mm diameter posts at 3.0 m centres; setting 700 mm into ground; choice of 10 mesh colours; including excavating holes, backfilling and removing surplus soil; with top rail only							
900 mm high	-	-	-	-	-	m	36.91
1200 mm high	-	-	-	-	-	m	38.61
1800 mm high	-	-	-	-	-	m	41.92
2000 mm high	-	-	-	-	-	m	42.81
Tubular chain link fencing; galvanized; plastic coated; 60.3 mm diameter posts at 3.0 m centres; cranked arms and three lines barbed wire; setting 700 mm into ground; including excavating holes; backfilling and removing surplus soil; with top rail only							
2000 mm high	-	-	-	-	-	m	44.12
1800 mm high	-	-	-	-	-	m	43.23

Q PAVING/PLANTING/FENCING/SITE FURNITURE

Item Excluding site overheads and profit	PC £	Labour hours	Labour £	Plant £	Material £	Unit	Total rate £
Q40 FENCING AND GATES - cont'd							
Boundary fencing; Steelway-Fensecure Ltd							
"Classic" two rail tubular fencing; top and bottom with stretcher bars and straining wires in between; comprising 60.3 mm tubular posts at 3.00 m centres; setting in concrete; 35 mm top rail tied with aluminium and steel fittings; 50 x 50 x 355/2.5 mm PVC coated chain link; all components galvanized and coated in green nylon							
964 mm high	-	-	-	-	-	m	36.90
1269 mm high	-	-	-	-	-	m	40.63
1574 mm high	-	-	-	-	-	m	46.19
1878 mm high	-	-	-	-	-	m	48.51
2188 mm high	-	-	-	-	-	m	51.71
2458 mm high	-	-	-	-	-	m	57.82
2948 mm high	-	-	-	-	-	m	73.28
3562 mm high	-	-	-	-	-	m	73.09
End posts; Classic range; 60.3 mm diameter; setting in concrete							
964 mm high	-	-	-	-	-	nr	51.19
1269 mm high	-	-	-	-	-	nr	55.19
1574 mm high	-	-	-	-	-	nr	57.83
1878 mm high	-	-	-	-	-	nr	67.19
2188 mm high	-	-	-	-	-	nr	73.08
2458 mm high	-	-	-	-	-	nr	83.92
2948 mm high	-	-	-	-	-	nr	91.24
3562 mm high	-	-	-	-	-	nr	109.80
Corner posts; 60.3 mm diameter; setting in concrete							
964 mm high	-	-	-	-	-	nr	41.42
1269 mm high	-	-	-	-	-	nr	24.83
1574 mm high	-	-	-	-	-	nr	48.45
1878 mm high	-	-	-	-	-	nr	50.77
2188 mm high	-	-	-	-	-	nr	59.05
2458 mm high	-	-	-	-	-	nr	62.90
2948 mm high	-	-	-	-	-	nr	75.54
3562 mm high	-	-	-	-	-	nr	80.43
Boundary fencing; McArthur Group							
"Paladin" welded mesh colour coated green fencing; fixing to metal posts at 2.975 m centres with manufacturer's fixings; setting 600 mm deep in firm ground; including excavating holes; backfilling and removing surplus excavated material; includes 10 year product guarantee							
1800 mm high	53.29	0.66	12.04	-	53.29	m	65.34
2000 mm high	54.40	0.66	12.04	-	62.45	m	74.50
2400 mm high	62.69	0.66	12.04	-	72.06	m	84.10
Extra over welded galvanized plastic coated mesh fencing for concreting in posts	7.78	0.11	2.03	-	7.78	m	9.81

Q PAVING/PLANTING/FENCING/SITE FURNITURE

Item Excluding site overheads and profit	PC £	Labour hours	Labour £	Plant £	Material £	Unit	Total rate £
Boundary fencing							
"Orsogril" rectangular steel bar mesh fence panels; pleione pattern; bolting to 60 x 8 mm uprights at 2 m centres; mesh 62 x 66 mm; setting in concrete							
930 mm high panels	-	-	-	-	-	m	118.00
1326 mm high panels	-	-	-	-	-	m	151.00
1722 mm high panels	-	-	-	-	-	m	198.00
"Orsogril" rectangular steel bar mesh fence panels; pleione pattern; bolting to 8 x 80 mm uprights at 2 m centres; mesh 62 x 66 mm; setting in concrete							
930 mm high panel	-	-	-	-	-	m	122.00
1326 mm high panel	-	-	-	-	-	m	157.00
1722 mm high panel	-	-	-	-	-	m	202.00
"Orsogril" rectangular steel bar mesh fence panels; sterope pattern; bolting to 60 x 80 mm uprights at 2.00 m centres; mesh 62 x 132 mm; setting in concrete paving (paving not included)							
930 mm high panels	-	-	-	-	-	m	149.03
1326 mm high panels	-	-	-	-	-	m	195.34
1722 mm high panels	-	-	-	-	-	m	245.68
Security fencing; Jacksons Fencing							
"Barbican" galvanized steel paling fencing; on 60 x 60 mm posts at 3 m centres; setting in concrete							
1250 mm high	-	-	-	-	-	m	81.31
1500 mm high	-	-	-	-	-	m	87.55
2000 mm high	-	-	-	-	-	m	95.04
2500 mm high	-	-	-	-	-	m	121.79
Gates; to match "Barbican" galvanized steel paling fencing							
width 1m	-	-	-	-	-	nr	994.00
width 2 m	-	-	-	-	-	nr	1005.48
width 3 m	-	-	-	-	-	nr	1033.00
width 4 m	-	-	-	-	-	nr	1067.75
width 8 m	-	-	-	-	-	pair	2220.00
width 9m	-	-	-	-	-	pair	2785.00
width 10 m	-	-	-	-	-	pair	3016.00
Security fencing; Jacksons Fencing; intruder guards							
"Viper Spike Intruder Guards"; to existing structures; including fixing bolts							
ref Viper 1; 40 x 5 mm x 1.1 m long; with base plate	31.05	1.00	18.25	-	31.05	nr	49.30
ref Viper 3; 160 x 190 mm wide; U shape; to prevent intruders climbing pipes	35.05	0.50	9.13	-	35.05	nr	44.18

Q PAVING/PLANTING/FENCING/SITE FURNITURE

Item Excluding site overheads and profit	PC £	Labour hours	Labour £	Plant £	Material £	Unit	Total rate £
Q40 FENCING AND GATES - cont'd							
Security fencing; Jacksons Fencing "Razor Barb Concertina"; spiral wire security barriers; fixing to 600 mm steel ground stakes							
ref 3275; 450 mm diameter roll; medium barb; galvanized	2.14	0.02	0.36	-	2.29	m	2.65
ref 3276; 730 mm diameter roll; medium barb; galvanized	2.91	0.02	0.36	-	3.06	m	3.43
ref 3277; 950 mm diameter roll; medium barb; galvanized	2.90	0.02	0.36	-	3.05	m	3.42
Three lines "Barbed Tape"; medium barb; on 50 x 50 mm mild steel angle posts; setting in concrete							
ref 3283; barbed tape; medium barb; galvanized	0.85	0.21	3.84	-	10.51	m	14.36
Five lines "Barbed Tape"; medium barb; on 50 x 50 mm mild steel angle posts; setting in concrete							
ref 3283; barbed tape; medium barb; galvanized	1.42	0.22	4.06	-	11.08	m	15.14
Trip rails; metal; Broxap Ltd Double row 48.3 mm internal diameter mild steel tubular rails with sleeved joints; to cast iron posts 525 mm above ground; setting in concrete at 1.20 m centres; all standard painted							
BX 201; cast iron with single rail; 42 mm diameter rail	65.66	1.00	18.25	-	70.93	m	89.18
Heavy duty; Anti Ram 540 mm high with 100 x 100 mm rail; Birdsmouth style	46.06	1.00	18.25	-	46.31	m	64.56
Trip rails; Birdsmouth; Jacksons Fencing Diamond rail fencing (Birdsmouth); posts 100 x 100 mm softwood planed at 1.35 m centres set in 1:3:6 concrete; rail 75 x 75 mm nominal secured with galvanized straps nailed to posts							
posts 900 mm (600 mm above ground); at 1.35 m centres	8.30	0.50	9.13	0.25	9.92	m	19.30
posts 1.20 m (900 mm above ground); at 1.35 m centres	9.67	0.50	9.13	0.25	11.29	m	20.67
posts 900 mm (600 mm above ground); at 1.80 m centres	6.90	0.38	6.86	0.19	8.13	m	15.18
posts 1.20 m (900 mm above ground); at 1.80 m centres	7.93	0.38	6.86	0.19	9.14	m	16.19
Trip rails; Townscape Products Ltd Hollow steel section knee rails; galvanized; 500 mm high; setting in concrete							
1000 mm bays	87.16	2.00	36.50	-	93.49	m	130.00
1200 mm bays	18.77	2.00	36.50	-	81.23	m	117.73

Q PAVING/PLANTING/FENCING/SITE FURNITURE

Item Excluding site overheads and profit	PC £	Labour hours	Labour £	Plant £	Material £	Unit	Total rate £
Concrete fencing							
Panel fencing; to precast concrete posts; in 2 m bays; setting posts 600 mm into ground; sandfaced finish							
900 mm high	11.74	0.25	4.56	-	13.27	m	17.84
1200 mm high	15.68	0.25	4.56	-	17.21	m	21.77
Panel fencing; to precast concrete posts; in 2 m bays; setting posts 750 mm into ground; sandfaced finish							
1500 mm high	20.46	0.33	6.08	-	21.99	m	28.07
1800 mm high	24.52	0.36	6.64	-	26.05	m	32.68
2100 mm high	28.39	0.40	7.30	-	29.92	m	37.22
2400 mm high	32.63	0.40	7.30	-	34.16	m	41.46
extra over capping panel to all of the above	6.72	-	-	-	6.72	m	-
Windbreak fencing							
Fencing; English Woodlands; "Shade and Shelter Netting" windbreak fencing; green; to 100 mm diameter treated softwood posts; setting 450 mm into ground; fixing with 50 x 25 mm treated softwood battens nailed to posts; including excavating and backfilling into firm ground; setting in concrete at 3 m centres							
1200 mm high	1.24	0.16	2.84	-	4.84	m	7.68
1800 mm high	1.76	0.16	2.84	-	5.36	m	8.20
Ball stop fencing; Steelway-Fensecure Ltd							
Ball stop net; 30 x 30 mm netting fixed to 60.3 mm diameter 12 mm solid bar lattice; galvanized; dual posts; top, middle and bottom rails							
4.5 m high	-	-	-	-	-	m	102.56
5.0 m high	-	-	-	-	-	m	129.49
6.0 m high	-	-	-	-	-	m	144.27
7.0 m high	-	-	-	-	-	m	159.07
8.0 m high	-	-	-	-	-	m	173.83
9.0 m high	-	-	-	-	-	m	188.63
10.0 m high	-	-	-	-	-	m	172.36
Ball stop net; corner posts							
4.5 m high	-	-	-	-	-	m	70.37
5.0 m high	-	-	-	-	-	m	77.02
6.0 m high	-	-	-	-	-	m	119.65
7.0 m high	-	-	-	-	-	m	106.59
8.0 m high	-	-	-	-	-	m	121.39
9.0 m high	-	-	-	-	-	m	126.17
10 m high	-	-	-	-	-	m	150.95
Ball stop net; end posts							
4.5 m high	-	-	-	-	-	m	58.54
5.0 m high	-	-	-	-	-	m	65.20
6.0 m high	-	-	-	-	-	m	79.98
7.0 m high	-	-	-	-	-	m	94.78
8.0 m high	-	-	-	-	-	m	109.54
9.0 m high	-	-	-	-	-	m	124.35
10 m high	-	-	-	-	-	m	126.17

Q PAVING/PLANTING/FENCING/SITE FURNITURE

Item Excluding site overheads and profit	PC £	Labour hours	Labour £	Plant £	Material £	Unit	Total rate £
Q40 FENCING AND GATES - cont'd							
Railings; Steelway-Fensecure Ltd							
Mild steel bar railings of balusters at 115 mm centres welded to flat rail top and bottom; bays 2.00 m long; bolting to 51 x 51 mm hollow square section posts; setting in concrete							
galvanized; 900 mm high	-	-	-	-	-	m	74.80
galvanized; 1200 mm high	-	-	-	-	-	m	90.60
galvanized; 1500 mm high	-	-	-	-	-	m	100.69
galvanized; 1800 mm high	-	-	-	-	-	m	120.79
primed; 900 mm high	-	-	-	-	-	m	70.50
primed; 1200 mm high	-	-	-	-	-	m	85.60
primed; 1500 mm high	-	-	-	-	-	m	94.63
primed; 1800 mm high	-	-	-	-	-	m	112.78
Mild steel blunt top railings of 19 mm balusters at 130 mm centres welded to bottom rail; passing through holes in top rail and welded; top and bottom rails 40 x 10 mm; bolting to 51 x 51 mm hollow square section posts; setting in concrete							
galvanized; 800 mm high	-	-	-	-	-	m	80.54
galvanized; 1000 mm high	-	-	-	-	-	m	84.58
galvanized; 1300 mm high	-	-	-	-	-	m	100.69
galvanized; 1500 mm high	-	-	-	-	-	m	108.74
primed; 800 mm high	-	-	-	-	-	m	77.53
primed; 1000 mm high	-	-	-	-	-	m	80.54
primed; 1300 mm high	-	-	-	-	-	m	95.65
primed; 1500 mm high	-	-	-	-	-	m	102.72
Railings; traditional pattern; 16 mm diameter verticals at 127 mm intervals with horizontal bars near top and bottom; balusters with spiked tops; 51 x 20 mm standards; including setting 520 mm into concrete at 2.75 m centres							
primed; 1200 mm high	-	-	-	-	-	m	90.60
primed; 1500 mm high	-	-	-	-	-	m	102.72
primed; 1800 mm high	-	-	-	-	-	m	110.75
Interlaced bow-top mild steel railings; traditional park type; 16 mm diameter verticals at 80 mm intervals; welded at bottom to 50 x 10 mm flat and slotted through 38 x 8 mm top rail to form hooped top profile; 50 x 10 mm standards; setting 560 mm into concrete at 2.75 m centres							
galvanized; 900 mm high	-	-	-	-	-	m	136.94
galvanized; 1200 mm high	-	-	-	-	-	m	167.15
galvanized; 1500 mm high	-	-	-	-	-	m	199.36
galvanized; 1800 mm high	-	-	-	-	-	m	231.58
primed; 900 mm high	-	-	-	-	-	m	132.90
primed; 1200 mm high	-	-	-	-	-	m	161.10
primed; 1500 mm high	-	-	-	-	-	m	191.32
primed; 1800 mm high	-	-	-	-	-	m	221.50

Q PAVING/PLANTING/FENCING/SITE FURNITURE

Item Excluding site overheads and profit	PC £	Labour hours	Labour £	Plant £	Material £	Unit	Total rate £
Metal estate fencing; Jacksons Fencing; mild steel flat bar angular fencing; galvanized; main posts at 5.00 m centres; fixed with 1:3:6 concrete 430 mm deep; intermediate posts fixed with fixing claw at 1.00 m centres							
Five plain, flat or round bars							
1200 high; under 500 m	60.00	0.28	5.07	-	62.19	m	67.26
1200 high; over 500 m	55.00	0.28	5.07	-	57.19	m	62.26
Five bar horizontal steel fencing; 1.20 m high; 38 x 10 mm joiner standards at 4.50 m centres; 38 x 8 mm intermediate standards at 900 mm centres; including excavating and backfilling into firm ground; setting in concrete							
galvanized	-	-	-	-	-	m	106.94
primed	-	-	-	-	-	m	87.91
Extra over five bar horizontal steel fencing for 76 mm diameter end and corner posts							
galvanized	-	-	-	-	-	each	116.92
primed	-	-	-	-	-	each	68.41
Five bar horizontal steel fencing with mild steel copings; 38 mm square hollow section mild steel standards at 1.80 m centres; including excavating and backfilling into firm ground; setting in concrete							
galvanized; 900 mm high	-	-	-	-	-	m	139.96
galvanized; 1200 mm high	-	-	-	-	-	m	147.23
galvanized; 1500 mm high	-	-	-	-	-	m	157.62
galvanized; 1800 mm high	-	-	-	-	-	m	168.01
primed; 900 mm high	-	-	-	-	-	m	124.19
primed; 1200 mm high	-	-	-	-	-	m	127.48
primed; 1500 mm high	-	-	-	-	-	m	134.40
primed; 1800 mm high	-	-	-	-	-	m	144.64
Traditional trellis panels; The Garden Trellis Company; bespoke trellis panels for decorative, screening or security applications; timber planed all round; height of trellis 1800 mm							
Freestanding panels; posts 70 x 70 mm set in concrete; timber frames mitred and grooved 45 x 34 mm; heights and widths to suit; slats 32 x 10 mm joinery quality tanalized timber at 100 mm ccs; elements fixed by galvanized staples; capping rail 70 x 34 mm							
HV68 horizontal and vertical slats; softwood	72.00	1.00	18.25	-	77.49	m	95.74
D68 diagonal slats; softwood	86.40	1.00	18.25	-	91.89	m	110.14
HV68 horizontal and vertical slats; hardwood iroko	176.40	1.00	18.25	-	181.89	m	200.14
D68 diagonal slats; hardwood iroko or Western Red Cedar	201.60	1.00	18.25	-	207.09	m	225.34

Q PAVING/PLANTING/FENCING/SITE FURNITURE

Item Excluding site overheads and profit	PC £	Labour hours	Labour £	Plant £	Material £	Unit	Total rate £
Q40 FENCING AND GATES - cont'd							
Traditional trellis panels - cont'd							
Trellis panels fixed to face of existing wall or railings; timber frames mitred and grooved 45 x 34 mm; heights and widths to suit; slats 32 x 10 mm joinery quality tanalized timber at 100 mm ccs; elements fixed by galvanized staples; capping rail 70 x 34 mm							
HV68 horizontal and vertical slats; softwood	68.40	0.50	9.13	-	70.17	m	**79.30**
D68 diagonal slats; softwood	81.00	0.50	9.13	-	82.77	m	**91.89**
HV68 horizontal and vertical slats; iroko or Western Red Cedar	117.00	0.50	9.13	-	118.77	m	**127.89**
D68 diagonal slats; iroko or Western Red Cedar	189.00	0.50	9.13	-	190.77	m	**199.90**
Contemporary style trellis panels; The Garden Trellis Company; bespoke trellis panels for decorative, screening or security applications; timber planed all round							
Freestanding panels 30/15; posts 70 x 70 mm set in concrete with 90 x 30 mm top capping; slats 30 x 14 mm with 15 mm gaps; vertical support at 450 mm ccs							
joinery treated softwood	117.00	1.00	18.25	-	122.49	m	**140.74**
hardwood iroko or Western Red Cedar	198.00	1.00	18.25	-	203.49	m	**221.74**
Panels 30/15 face fixed to existing wall or fence; posts 70 x 70 mm set in concrete with 90 x 30 mm top capping; slats 30 x 14 mm with 15 mm gaps; vertical support at 450 mm ccs							
joinery treated softwood	108.00	0.50	9.13	-	109.77	m	**118.89**
hardwood iroko or Western Red Cedar	180.00	0.50	9.13	-	181.77	m	**190.90**
Integral arches to ornamental trellis panels; The Garden Trellis Company							
Arches to trellis panels in 45 x 34 mm grooved timbers to match framing; fixed to posts							
R450 1/4 circle; joinery treated softwood	35.00	1.50	27.38	-	40.49	nr	**67.86**
R450 1/4 circle; hardwood iroko or Western Red Cedar	45.00	1.50	27.38	-	50.49	nr	**77.86**
Arches to span 1800 mm wide							
joinery treated softwood	45.00	1.50	27.38	-	50.49	nr	**77.86**
hardwood iroko or Western Red Cedar	60.00	1.50	27.38	-	65.49	nr	**92.86**
Painting or staining of trellis panels; high quality coatings							
microporous opaque paint or spirit based stain	-	-	-	-	24.00	m²	-

Q PAVING/PLANTING/FENCING/SITE FURNITURE

Item Excluding site overheads and profit	PC £	Labour hours	Labour £	Plant £	Material £	Unit	Total rate £
Gates - General Preamble: Gates in fences; see specification for fencing as gates in traditional or proprietary fencing systems are usually constructed of the same materials and finished as the fencing itself.							
Gates; hardwood; AVS Fencing Supplies Ltd Hardwood entrance gate; five bar diamond braced; curved hanging stile; planed iroko; fixed to 150 x 150 mm softwood posts; inclusive of hinges and furniture							
ref 1100 040; 0.9 m wide	189.97	5.00	91.25	-	240.06	nr	**331.31**
ref 1100 041; 1.2 m wide	202.55	5.00	91.25	-	252.64	nr	**343.89**
ref 1100 042; 1.5 m wide	264.02	5.00	91.25	-	314.11	nr	**405.36**
ref 1100 043; 1.8 m wide	279.57	5.00	91.25	-	329.66	nr	**420.91**
ref 1100 044; 2.1 m wide	306.11	5.00	91.25	-	356.20	nr	**447.45**
ref 1100 045; 2.4 m wide	288.09	5.00	91.25	-	338.18	nr	**429.43**
ref 1100 047; 3.0 m wide	351.86	5.00	91.25	-	401.95	nr	**493.20**
ref 1100 048; 3.3 m wide	367.51	5.00	91.25	-	417.60	nr	**508.85**
ref 1100 049; 3.6 m wide	380.05	5.00	91.25	-	430.14	nr	**521.39**
Hardwood field gate; five bar diamond braced; planed iroko; fixed to 150 x 150 mm softwood posts; inclusive of hinges and furniture							
ref 1100 100; 0.9 m wide	129.10	4.00	73.00	-	179.19	nr	**252.19**
ref 1100 101; 1.2 m wide	140.26	4.00	73.00	-	190.35	nr	**263.35**
ref 1100 102; 1.5 m wide	167.22	4.00	73.00	-	217.31	nr	**290.31**
ref 1100 103; 1.8 m wide	179.74	4.00	73.00	-	229.83	nr	**302.83**
ref 1100 104; 2.1 m wide	221.32	4.00	73.00	-	271.41	nr	**344.41**
ref 1100 105; 2.4 m wide	234.49	4.00	73.00	-	284.58	nr	**357.58**
ref 1100 106; 2.7 m wide	247.69	4.00	73.00	-	297.78	nr	**370.78**
ref 1100 108; 3.3 m wide	274.06	4.00	73.00	-	324.15	nr	**397.15**
ref 1100 109; 3.6 m wide	287.92	4.00	73.00	-	338.01	nr	**411.01**
Gates; softwood; Jacksons Fencing Timber field gates; including wrought iron ironmongery; five bar type; diamond braced; 1.80 m high; to 200 x 200 mm posts; setting 750 mm into firm ground							
treated softwood; width 2400 mm	81.63	10.00	182.50	-	200.91	nr	**383.41**
treated softwood; width 2700 mm	90.63	10.00	182.50	-	209.91	nr	**392.41**
treated softwood; width 3000 mm	97.56	10.00	182.50	-	216.84	nr	**399.34**
treated softwood; width 3300 mm	103.32	10.00	182.50	-	222.60	nr	**405.10**
Featherboard garden gates; including ironmongery; to 100 x 120 mm posts; one diagonal brace							
treated softwood; 1.0 x 1.2 m high	25.11	3.00	54.75	-	105.99	nr	**160.74**
treated softwood; 1.0 x 1.5 m high	32.94	3.00	54.75	-	116.02	nr	**170.77**
treated softwood; 1.0 x 1.8 m high	79.92	3.00	54.75	-	124.35	nr	**179.10**
Picket garden gates; including ironmongery; to match picket fence; width 1000 mm; to 100 x 120 mm posts; one diagonal brace							
treated softwood; 950 mm high	68.22	3.00	54.75	-	100.86	nr	**155.61**
treated softwood; 1200 mm high	71.91	3.00	54.75	-	104.55	nr	**159.30**
treated softwood; 1800 mm high	84.38	3.00	54.75	-	128.80	nr	**183.55**

Q PAVING/PLANTING/FENCING/SITE FURNITURE

Item Excluding site overheads and profit	PC £	Labour hours	Labour £	Plant £	Material £	Unit	Total rate £
Q40 FENCING AND GATES - cont'd							
Gates; tubular steel; Jacksons Fencing							
Tubular mild steel field gates; including ironmongery; diamond braced; 1.80 m high; to tubular steel posts; setting in concrete							
galvanized; width 3000 mm	125.37	5.00	91.25	-	227.37	nr	**318.62**
galvanized; width 3300 mm	132.57	5.00	91.25	-	234.57	nr	**325.82**
galvanized; width 3600 mm	139.91	5.00	91.25	-	241.90	nr	**333.15**
galvanized; width 4200 mm	154.57	5.00	91.25	-	256.57	nr	**347.82**
Gates; sliding; Jacksons Fencing							
"Sliding Gate"; including all galvanized rails and vertical rail infill panels; special guide and shutting frame posts (Note: foundations installed by suppliers)							
access width 4.00 m; 1.5 m high gates	-	-	-	-	-	nr	**4401.28**
access width 4.00 m; 2.0 m high gates	-	-	-	-	-	nr	**4521.28**
access width 4.00 m; 2.5 m high gates	-	-	-	-	-	nr	**4641.28**
access width 6.00 m; 1.5 m high gates	-	-	-	-	-	nr	**5174.28**
access width 6.00 m; 2.0 m high gates	-	-	-	-	-	nr	**5329.28**
access width 6.00 m; 2.5 m high gates	-	-	-	-	-	nr	**5485.28**
access width 8.00 m; 1.5 m high gates	-	-	-	-	-	nr	**5895.28**
access width 8.00 m; 2.0 m high gates	-	-	-	-	-	nr	**6088.28**
access width 8.00 m; 2.5 m high gates	-	-	-	-	-	nr	**6280.28**
access width 10.00 m; 2.0 m high gates	-	-	-	-	-	nr	**7564.12**
access width 10.00 m; 2.5 m high gates	-	-	-	-	-	nr	**7670.18**
Kissing gates							
Kissing gates; Jacksons Fencing; in galvanized metal bar; fixing to fencing posts (posts not included); 1.65 x 1.30 x 1.00 m high	221.40	5.00	91.25	-	221.40	nr	**312.65**
Stiles							
Stiles; two posts; setting into firm ground; three rails; two treads	72.00	3.00	54.75	-	82.37	nr	**137.12**
Pedestrian guard rails and barriers							
Mild steel pedestrian guard rails; Broxap Street Furniture; 1.00 m high with 150 mm toe space; to posts at 2.00 m centres; galvanized finish							
vertical "in-line" bar infill panel	44.00	2.00	36.50	-	49.95	m	**86.45**
vertical staggered bar infill with 230 mm visibility gap at top	46.00	2.00	36.50	-	51.95	m	**88.45**

Prices for Measured Works – Major Works

Q PAVING/PLANTING/FENCING/SITE FURNITURE

Item Excluding site overheads and profit	PC £	Labour hours	Labour £	Plant £	Material £	Unit	Total rate £
Q50 SITE/STREET FURNITURE/ EQUIPMENT							
Note: The costs of delivery from the supplier are generally not included in these items. Readers should check with the supplier to verify the delivery costs.							
Standing stones; CED Ltd; erect standing stones; vertical height above ground; in concrete base; including excavation setting in concrete to 1/3 depth and crane offload into position							
Purple schist							
1.00 m high	85.87	1.50	27.38	14.24	100.40	nr	**142.01**
1.25 m high	108.77	1.50	27.38	14.24	117.77	nr	**159.38**
1.50 m high	174.02	2.00	36.50	17.09	185.55	nr	**239.13**
2.00 m high	286.23	3.00	54.75	28.48	319.53	nr	**402.75**
2.50 m high	435.06	1.50	27.38	56.95	499.61	nr	**583.94**
Barriers - General							
Preamble: The provision of car and lorry control barriers may form part of the landscape contract. Barriers range from simple manual counterweighted poles to fully automated remote-control security gates, and the exact degree of control required must be specified. Complex barriers may need special maintenance and repair.							
Barriers; Autopa Ltd							
Manually operated pole barriers; counterbalance; to tubular steel supports; bolting to concrete foundation (foundation not included); aluminium boom; various finishes							
clear opening up to 3.00 m	817.00	6.00	109.50	-	837.08	nr	**946.58**
clear opening up to 4.00 m	879.00	6.00	109.50	-	899.08	nr	**1008.58**
clear opening 5.00 m	941.00	6.00	109.50	-	961.08	nr	**1070.58**
clear opening 6.00 m	1003.00	6.00	109.50	-	1023.08	nr	**1132.58**
clear opening 7.00 m	1063.00	6.00	109.50	-	1083.08	nr	**1192.58**
catch pole; arm rest for all manual barriers	116.00	-	-	-	118.51	nr	**-**
Electrically operated pole barriers; enclosed fan-cooled motor; double worm reduction gear; overload clutch coupling with remote controls; aluminium boom; various finishes (exclusive of electrical connections by electrician)							
Autopa AU 3.0; 3 m boom	2365.00	6.00	109.50	-	2385.08	nr	**2494.58**
Autopa AU 4.5; 4.5 m boom with catchpost	2465.00	6.00	109.50	-	2485.08	nr	**2594.58**
Autopa AU 6.0; 6.0 m boom with catchpost	2565.00	6.00	109.50	-	2585.08	nr	**2694.58**

Q PAVING/PLANTING/FENCING/SITE FURNITURE

Item Excluding site overheads and profit	PC £	Labour hours	Labour £	Plant £	Material £	Unit	Total rate £
Q50 SITE/STREET FURNITURE/ EQUIPMENT - cont'd							
Vehicle crash barriers - General Preamble: See Department of Environment Technical Memorandum BE5.							
Vehicle crash barriers Steel corrugated beams; untensioned; Broxap Street Furniture; effective length 3.20 m x 310 mm deep x 85 mm corrugations							
steel posts; Z-section; roadside posts	75.94	0.33	6.08	0.67	83.62	m	**90.38**
steel posts; Z-section; off highway	68.44	0.33	6.08	0.67	76.12	m	**82.88**
steel posts RSJ 760 high; for anchor fixing	88.44	0.33	6.08	-	102.50	m	**108.58**
steel posts RSJ 560 high; for anchor fixing to car parks	85.31	0.33	6.08	-	99.38	m	**105.46**
extra over for curved rail 6.00 m radius	67.19	-	-	-	67.19	m	**-**
Bases for street furniture Excavating; filling with concrete 1:3:6; bases for street furniture							
300 x 450 x 500 mm deep	-	0.75	13.69	-	6.04	nr	**19.72**
300 x 600 x 500 mm deep	-	1.11	20.26	-	8.05	nr	**28.31**
300 x 900 x 500 mm deep	-	1.67	30.48	-	12.07	nr	**42.55**
1750 x 900 x 300 mm deep	-	2.33	42.52	-	42.25	nr	**84.77**
2000 x 900 x 300 mm deep	-	2.67	48.73	-	48.29	nr	**97.01**
2400 x 900 x 300 mm deep	-	3.20	58.40	-	57.94	nr	**116.34**
2400 x 1000 x 300 mm deep	-	3.56	64.97	-	64.38	nr	**129.35**
Precast concrete flags; to concrete bases (not included); bedding and jointing in cement:mortar (1:4)							
450 x 600 x 50 mm	10.44	1.17	21.29	-	15.15	m²	**36.44**
Precast concrete paving blocks; to concrete bases (not included); bedding in sharp sand; butt joints							
200 x 100 x 65 mm	7.63	0.50	9.13	-	9.47	m²	**18.59**
200 x 100 x 80 mm	8.49	0.50	9.13	-	10.20	m²	**19.33**
Engineering paving bricks; to concrete bases (not included); bedding and jointing in sulphate-resisting cement:lime:sand mortar (1:1:6)							
over 300 mm wide	9.66	0.56	10.25	-	19.87	m²	**30.13**
Edge restraints to pavings; haunching in concrete (1:3:6)							
200 x 300 mm	4.86	0.10	1.82	-	4.86	m	**6.69**

Q PAVING/PLANTING/FENCING/SITE FURNITURE

Item Excluding site overheads and profit	PC £	Labour hours	Labour £	Plant £	Material £	Unit	Total rate £
Bases; Earth Anchors Ltd							
"Rootfast" ancillary anchors; ref A1; 500 mm long x 25 mm diameter and top strap ref F2; including bolting to site furniture (site furniture not included)	9.50	0.50	9.13	-	9.50	set	**18.63**
installation tool for above	23.50	-	-	-	23.50	nr	**-**
"Rootfast" ancillary anchors; ref A4; heavy duty 40 mm square fixed head anchors; including bolting to site furniture (site furniture not included)	29.00	0.33	6.08	-	29.00	set	**35.08**
"Rootfast" ancillary anchors; F4; vertical socket; including bolting to site furniture (site furniture not included)	12.75	0.05	0.91	-	12.75	set	**13.66**
"Rootfast" ancillary anchors; ref F3; horizontal socket; including bolting to site furniture (site furniture not included)	14.00	0.05	0.91	-	14.00	set	**14.91**
installation tools for the above	65.00	-	-	-	65.00	nr	**-**
"Rootfast" ancillary anchors; ref A3; anchored bases; including bolting to site furniture (site furniture not included)	45.00	0.33	6.08	-	45.00	set	**51.08**
installation tools for the above	65.00	-	-	-	65.00	nr	**-**
Furniture/equipment - General Preamble: The following items include fixing to manufacturer's instructions; holding down bolts or other fittings and making good (excavating, backfilling and tarmac, concrete or paving bases not included).							
Dog waste bins; all-steel Earth Anchors Ltd							
HG45A; 45 l; earth anchored; post mounted	173.00	0.33	6.08	-	173.00	nr	**179.08**
HG45A; 45 l; as above with pedal operation	203.00	0.33	6.08	-	203.00	nr	**209.08**
Dog waste bins; cast iron Bins; Furnitubes International Ltd							
ref PED 701; "Pedigree"; post mounted cast iron dog waste bins; 1250 mm total height above ground; 400 mm square bin	397.00	0.75	13.69	-	402.95	nr	**416.64**
Litter bins; precast concrete in textured white or exposed aggregate finish; with wire baskets and drainage holes Bins; Marshalls Plc							
"Boulevard 700"; concrete circular litter bin	390.00	0.50	9.13	-	390.00	nr	**399.13**
Bins; Neptune Outdoor Furniture Ltd							
ref SF 16; 42 l	211.00	0.50	9.13	-	211.00	nr	**220.13**
ref SF 14; 100 l	295.00	0.50	9.13	-	295.00	nr	**304.13**
Bins; Townscape Products Ltd							
"Sutton"; 750 mm high x 500 mm diameter; 70 l capacity; including GRP canopy	166.56	0.33	6.08	-	285.58	nr	**291.66**
"Braunton"; 750 mm high x 500 mm diameter; 70 l capacity; including GRP canopy	176.57	1.50	27.38	-	295.59	nr	**322.96**

Q PAVING/PLANTING/FENCING/SITE FURNITURE

Item Excluding site overheads and profit	PC £	Labour hours	Labour £	Plant £	Material £	Unit	Total rate £
Q50 SITE/STREET FURNITURE/ EQUIPMENT - cont'd							
Litter bins; metal; stove-enamelled perforated metal for holder and container							
Bins; Townscape Products Ltd							
"Metro"; 440 x 420 x 800 mm high; 62 l capacity	573.00	0.33	6.08	-	573.00	nr	579.08
"Voltan"; large round; 460 mm diameter x 780 mm high; 56 l capacity	493.08	0.33	6.08	-	493.08	nr	499.16
"Voltan"; small round with pedestal; 410 mm diameter x 760 mm high; 31 l capacity	477.42	0.33	6.08	-	477.42	nr	483.50
Litter bins; all-steel							
Bins; Marshall Plc							
"MSF Central"; steel litter bin	250.00	1.00	18.25	-	250.00	nr	268.25
"MSF Central"; stainless steel litter bin	500.00	0.67	12.17	-	500.00	nr	512.17
Bins; Earth Anchors Ltd							
"Ranger"; 100 l; pedestal mounted	427.00	0.50	9.13	-	427.00	nr	436.13
"Big Ben"; 82 l; steel frame and liner; colour coated finish; earth anchored	276.00	1.00	18.25	-	276.00	nr	294.25
"Beau"; 42 l; steel frame and liner; colour coated finish; earth anchored	224.00	1.00	18.25	-	224.00	nr	242.25
Bins; Townscape Products Ltd							
"Baltimore Major" with GRP canopy; 560 mm diameter x 960 mm high; 140 l capacity	827.95	1.00	18.25	-	827.95	nr	846.20
Litter bins; cast iron							
Bins; Marshalls Plc							
"MSF5501" Heritage; cast iron litter bin	465.00	1.00	18.25	3.65	465.00	nr	486.90
Bins; Furnitubes International Ltd							
"Wave Bin"; ref WVB 440; free standing 55 l liners; 440 mm diameter x 850 mm high	336.00	0.50	9.13	-	339.98	nr	349.11
"Wave Bin" ref WVB 520; free standing 85 l liners; 520 mm diameter x 850 mm high; cast iron plinth	388.00	0.50	9.13	-	391.98	nr	401.11
"Covent Garden"; ref COV 702; side opening, 500 mm square x 1050 mm high; 105 l capacity	270.00	0.50	9.13	-	276.49	nr	285.61
"Covent Garden"; ref COV 803; side opening, 500 mm diameter x 1025 mm high; 85 l capacity	248.00	0.50	9.13	-	254.49	nr	263.61
"Covent Garden"; ref COV 912; open top; 500 mm A/F octagonal x 820 mm high; 85 l capacity	360.00	0.50	9.13	-	366.49	nr	375.61
"Albert"; ref ALB 800; open top; 400 mm diameter x 845 mm high; 55 l capacity	510.00	0.50	9.13	-	516.49	nr	525.61
Bins; Broxap Street Furniture							
"Derby Hercules"; post mounted; steel; 60 l	149.00	1.00	18.25	-	155.49	nr	173.74
Bins; Townscape Products Ltd							
"York Major"; 650 mm diameter x 1060 mm high; 140 l capacity	991.29	1.00	18.25	-	993.15	nr	1011.40

Q PAVING/PLANTING/FENCING/SITE FURNITURE

Item Excluding site overheads and profit	PC £	Labour hours	Labour £	Plant £	Material £	Unit	Total rate £
Litter bins; timber faced; hardwood slatted casings with removable metal litter containers; ground or wall fixing							
Bins; Lister Lutyens Co Ltd							
"Monmouth"; 675 mm high x 450 mm wide; freestanding	123.00	0.33	6.08	-	123.00	nr	129.08
"Monmouth"; 675 mm high x 450 mm wide; bolting to ground (without legs)	114.00	1.00	18.25	-	122.48	nr	140.73
Bins; Woodscape Ltd							
square; 580 x 580 x 950 mm high; with lockable lid	656.25	0.50	9.13	-	656.25	nr	665.38
round; 580 mm diameter x 950 mm high; with lockable lid	656.25	0.50	9.13	-	656.25	nr	665.38
Plastic litter and grit bins; glassfibre reinforced polyester grit bins; yellow body; hinged lids							
Bins; Wybone Ltd; Victoriana glass fibre; cast iron effect litter bins; including lockable liner							
ref LBV/2; 521 x 521 x 673 mm high; open top; square shape; 0.078 m³ capacity	199.52	1.00	18.25	-	206.01	nr	224.26
ref LVC/3; 457 mm diameter x 648 mm high; open top; drum shape; 0.084 m³ capacity; with lockable liner	217.65	1.00	18.25	-	224.14	nr	242.39
Grit bins; Furnitubes International Ltd							
Grit and salt bin; yellow glass fibre; hinged lid; 170 l	222.00	-	-	-	222.00	ea	-
Lifebuoy stations							
Lifebuoy stations; Earth Anchors Ltd "Rootfast" lifebuoy station complete with post ref AP44; SOLAS approved lifebouys 590 mm and lifeline	165.00	2.00	36.50	-	174.30	nr	210.80
installation tool for above	-	-	-	-	81.00	nr	-
Outdoor seats							
CED Ltd; stone bench; "Sinuous bench"; stone type bench to organic "S" pattern; laid to concrete base; 500 mm high x 500 mm wide (not included)							
2.00 m long	1177.00	4.00	73.00	-	1177.00	nr	1250.00
5.00 m long	2889.00	8.00	146.00	-	2889.00	nr	3035.00
10.00 m long	5778.00	11.00	200.75	-	5778.00	nr	5978.75

Q PAVING/PLANTING/FENCING/SITE FURNITURE

Item Excluding site overheads and profit	PC £	Labour hours	Labour £	Plant £	Material £	Unit	Total rate £
Q50 SITE/STREET FURNITURE/ EQUIPMENT - cont'd							
Outdoor seats; concrete framed - General Preamble: Prices for the following concrete framed seats and benches with hardwood slats include for fixing (where necessary) by bolting into existing paving or concrete bases (bases not included) or building into walls or concrete foundations (walls and foundations not included). Delivery generally not included.							
Outdoor seats; concrete framed							
Outdoor seats; Townscape Products Ltd							
"Oxford" benches; 1800 x 430 x 440 mm	242.95	2.00	36.50	-	254.85	nr	291.35
"Maidstone" seats; 1800 x 610 x 785 mm	361.23	2.00	36.50	-	373.13	nr	409.63
Outdoor seats; concrete							
Outdoor seats; Marshalls Ltd							
"Boulevard 2000" concrete seat	725.00	2.00	36.50	-	744.86	nr	781.36
Outdoor seats; metal framed - General Preamble: Metal framed seats with hardwood backs to various designs can be bolted to ground anchors.							
Outdoor seats; metal framed							
Outdoor seats; Furnitubes International Ltd							
ref NS 6; "Newstead"; steel standards with iroko slats; 1.80 m long	215.00	2.00	36.50	-	234.86	nr	271.36
ref NEB 6; "New Forest Single Bench"; cast iron standards with iroko slats; 1.83 m long	199.00	2.00	36.50	-	218.86	nr	255.36
ref EA 6; "Eastgate"; cast iron standards with iroko slats; 1.86 m long	260.00	2.00	36.50	-	279.86	nr	316.36
ref NE 6; "New Forest Seat"; cast iron standards with iroko slats; 1.83 m long	294.00	2.00	36.50	-	313.86	nr	350.36
Outdoor seats; Orchard Street Furniture Ltd; "Bramley"; broad iroko slats to steel frame							
1.20 m long	186.46	2.00	36.50	-	199.44	nr	235.94
1.80 m long	229.76	2.00	36.50	-	242.74	nr	279.24
2.40 m long	265.39	2.00	36.50	-	278.37	nr	314.87
Outdoor seats; Orchard Street Furniture Ltd; "Laxton"; narrow iroko slats to steel frame							
1.20 m long	210.72	2.00	36.50	-	223.70	nr	260.20
1.80 m long	258.95	2.00	36.50	-	271.93	nr	308.43
2.40 m long	294.27	2.00	36.50	-	307.25	nr	343.75
Outdoor seats; Orchard Street Furniture Ltd; "Lambourne"; iroko slats to cast iron frame							
1.80 m long	547.39	2.00	36.50	-	560.37	nr	596.87
2.40 m long	644.14	2.00	36.50	-	657.12	nr	693.62
Outdoor seats; Broxap Street Furniture; metal and timber							
"Eastgate"	384.00	2.00	36.50	-	396.98	nr	433.48
"Rotherham"	639.00	2.00	36.50	-	651.98	nr	688.48

Q PAVING/PLANTING/FENCING/SITE FURNITURE

Item Excluding site overheads and profit	PC £	Labour hours	Labour £	Plant £	Material £	Unit	Total rate £
Outdoor seats; Earth Anchors Ltd; "Forest-Saver"; steel frame and recycled slats							
bench; 1.8 m	199.00	1.00	18.25	-	199.00	nr	217.25
seat; 1.8 m	312.00	1.00	18.25	-	312.00	nr	330.25
Outdoor seats; Earth Anchors Ltd; "Evergreen"; cast iron frame and recycled slats							
bench; 1.8 m	426.00	1.00	18.25	-	426.00	nr	444.25
seat; 1.8 m	522.00	1.00	18.25	-	522.00	nr	540.25
Outdoor seats; all steel							
Outdoor seats; Marshalls Plc							
"MSF Central"; stainless steel seat	1000.00	0.50	9.13	-	1000.00	nr	1009.13
"MSFCentral"; steel seat	500.00	2.00	36.50	-	500.00	nr	536.50
"MSF 502"; cast iron Heritage seat	467.00	2.00	36.50	-	467.00	nr	503.50
Outdoor seats; Earth Anchors Ltd; "Ranger"							
bench; 1.8 m	275.00	1.00	18.25	-	275.00	nr	293.25
seat; 1.8 m	431.00	1.00	18.25	-	431.00	nr	449.25
Outdoor seats; all timber							
Outdoor seats; Broxap Street Furniture; teak seats with back and armrests							
"Cambridge" seat; three seater; 1.8m	390.00	2.00	36.50	-	409.86	nr	446.36
"Milano" seat; three seater; 1.8 m	420.00	2.00	36.50	-	439.86	nr	476.36
Outdoor seats; Lister Lutyens Co Ltd; "Mendip"; teak							
1.524 m long	465.00	1.00	18.25	-	488.58	nr	506.83
1.829 m long	561.00	1.00	18.25	-	584.58	nr	602.83
2.438 m long (including centre leg)	703.00	1.00	18.25	-	726.58	nr	744.83
Outdoor seats; Lister Lutyens Co Ltd; "Sussex"; hardwood							
1.5 m	185.00	2.00	36.50	-	208.58	nr	245.08
Outdoor seats; Woodscape Ltd; solid hardwood							
seat type "3" with back; 2.00 m long; freestanding	745.50	2.00	36.50	-	769.08	nr	805.58
seat type "3" with back; 2.00 m long; building in	787.50	4.00	73.00	-	811.08	nr	884.08
seat type "4" with back; 2.00 m long; fixing to wall	462.00	3.00	54.75	-	473.94	nr	528.69
seat type "5" with back; 2.50 m long; freestanding	861.00	2.00	36.50	-	884.58	nr	921.08
seat type "4"; 2.00 m long; fixing to wall	346.50	2.00	36.50	-	370.08	nr	406.58
seat type "5" with back; 2.00 m long; freestanding	761.25	2.00	36.50	-	784.83	nr	821.33
seat type "4" with back; 2.50 m long; fixing to wall	540.75	3.00	54.75	-	552.69	nr	607.44
seat type "5" with back; building in	803.25	2.00	36.50	-	826.83	nr	863.33
seat type "5" with back; 2.50 m long; building in	903.00	3.00	54.75	-	926.58	nr	981.33
bench type "1"; 2.00 m long; freestanding	593.25	2.00	36.50	-	616.83	nr	653.33
bench type "1"; 2.00 m long; building in	635.25	4.00	73.00	-	658.83	nr	731.83
bench type "2"; 2.00 m long; freestanding	567.00	2.00	36.50	-	590.58	nr	627.08
bench type "2"; 2.00 m long; building in	609.00	4.00	73.00	-	632.58	nr	705.58
bench type "2"; 2.50 m long; freestanding	645.75	2.00	36.50	-	669.33	nr	705.83
bench type "2"; 2.50 m long; building in	687.75	4.00	73.00	-	711.33	nr	784.33
bench type "2"; 2.00 m long overall; curved to 5 m radius; building in	777.00	4.00	73.00	-	800.58	nr	873.58

Q PAVING/PLANTING/FENCING/SITE FURNITURE

Item Excluding site overheads and profit	PC £	Labour hours	Labour £	Plant £	Material £	Unit	Total rate £
Q50 SITE/STREET FURNITURE/ EQUIPMENT - cont'd							
Outdoor seats - cont'd							
Outdoor seats; Orchard Street Furniture Ltd; "Allington"; all iroko							
1.20 m long	293.66	2.00	36.50	-	317.24	nr	**353.74**
1.80 m long	337.59	2.00	36.50	-	361.17	nr	**397.67**
2.40 m long	410.69	2.00	36.50	-	434.27	nr	**470.77**
Outdoor seats; tree benches/seats							
Tree bench; Neptune Outdoor Furniture Ltd; "Beaufort"; hexagonal; timber							
SF34-15A; 1500 mm diameter	1020.00	0.50	9.13	-	1020.00	nr	**1029.13**
Tree seat; Neptune Outdoor Furniture Ltd; "Beaufort"; hexagonal; timber; with back							
SF32-10A; 720 mm diameter	1224.00	0.50	9.13	-	1224.00	nr	**1233.13**
SF32-20A; 1720 mm diameter	1432.00	0.50	9.13	-	1432.00	nr	**1441.13**
Street furniture ranges							
Townscape Products Ltd; "Belgrave"; natural grey concrete							
bollards; 250 mm diameter x 500 mm high	76.90	1.00	18.25	-	80.02	nr	**98.27**
seats; 1800 x 600 x 736 mm high	282.96	2.00	36.50	-	282.96	nr	**319.46**
Picnic benches - General							
Preamble: The following items include for fixing to ground to manufacturer's instructions or concreting in.							
Picnic tables and benches							
Picnic tables and benches; Broxap Street Furniture							
"Eastgate" picnic unit	629.00	2.00	36.50	-	641.98	nr	**678.48**
Picnic tables and benches; Woodscape Ltd							
table and benches built in; 2 m long	1905.75	2.00	36.50	-	1918.73	nr	**1955.23**
Market prices of containers							
Plant containers; terracotta							
Capital Garden Products Ltd							
Large Pot LP63; weathered terracotta; 1170 x 1600 mm diameter	-	-	-	-	793.00	nr	-
Large Pot LP38; weathered terracotta; 610 x 970 mm diameter	-	-	-	-	342.75	nr	-
Large Pot LP23; weathered terracotta; 480 x 580 mm diameter	-	-	-	-	184.75	nr	-
Manhole Cover Planter 3022; 760 x 560 x 210 mm high	-	-	-	-	122.40	nr	-
Apple Basket 2215; 380 x 560 mm diameter	-	-	-	-	126.00	nr	-
Indian style Shimmer Pot 2322; 585 x 560 mm diameter	-	-	-	-	167.75	nr	-
Indian style Shimmer Pot 1717; 430 x 430 mm diameter	-	-	-	-	107.75	nr	-
Indian style Shimmer Pot 1314; 330 x 355 mm diameter	-	-	-	-	97.75	nr	-

Q PAVING/PLANTING/FENCING/SITE FURNITURE

Item Excluding site overheads and profit	PC £	Labour hours	Labour £	Plant £	Material £	Unit	Total rate £
Plant containers; faux lead							
Capital Garden Products Ltd							
Trough 2508 "Tudor Rose"; 620 x 220 x 230 mm high	-	-	-	-	61.75	nr	-
Tub 2004 "Elizabethan"; 510 mm square	-	-	-	-	97.75	nr	-
Tub 1513 "Elizabethan"; 380 mm square	-	-	-	-	65.75	nr	-
Tub 1601 "Tudor Rose"; 420 x 400 mm diameter	-	-	-	-	68.75	nr	-
Plant containers; window boxes							
Capital Garden Products Ltd							
Adam 5401; faux lead; 1370 x 270 x 210 mm high	-	-	-	-	107.75	nr	-
Wheatsheaf WH54; faux lead; 1370 x 270 x 210 mm high	-	-	-	-	115.20	nr	-
Oakleaf OAK24; terracotta; 610 x 230 x 240 mm high	-	-	-	-	86.75	nr	-
Swag 2402; faux lead; 610 x 200 x 210 mm high	-	-	-	-	53.75	nr	-
Plant containers; timber							
Plant containers; hardwood; Neptune Outdoor Furniture Ltd							
"Beaufort" T38-4D; 1500 x 1500 x 900 mm high	-	-	-	-	1166.00	nr	-
"Beaufort" T38-3C; 1000 x 1500 x 700 mm high	-	-	-	-	866.00	nr	-
"Beaufort" T38-2A; 1000 x 500 x 500 mm high	-	-	-	-	452.00	nr	-
"Kara" T42-4D; 1500 x 1500 x 900 mm high	-	-	-	-	1257.00	nr	-
"Kara" T42-3C; 1000 x 1500 x 700 mm high	-	-	-	-	938.00	nr	-
"Kara" T42-2A; 1000 x 500 x 500 mm high	-	-	-	-	489.00	nr	-
Plant containers; hardwood; Woodscape Ltd square; 900 x 900 x 420 mm high	-	-	-	-	297.36	nr	-
Measured works							
Plant containers; precast concrete							
Plant containers; Marshalls Plc							
"Boulevard 700"; circular base and ring	530.00	2.00	36.50	16.50	530.00	nr	**583.00**
"Boulevard 1200"; circular base and ring	665.00	1.00	18.25	16.50	665.00	nr	**699.75**
Cycle holders							
Cycle stands; Marshalls Plc							
"Sheffield"; steel cycle stand; RCS1	44.00	0.50	9.13	-	44.00	nr	**53.13**
"Sheffield"; stainless steel cycle stand; RSCS1	121.00	0.50	9.13	-	121.00	nr	**130.13**
Cycle holders; Autopa; "VELOPA"; galvanized steel							
ref R; fixing to wall or post; making good	34.00	1.00	18.25	-	46.98	nr	**65.23**
ref SR(V); fixing in ground; making good	43.00	1.00	18.25	-	55.98	nr	**74.23**
"Sheffield" cycle stands; ragged steel	54.00	1.00	18.25	-	66.98	nr	**85.23**

Q PAVING/PLANTING/FENCING/SITE FURNITURE

Item Excluding site overheads and profit	PC £	Labour hours	Labour £	Plant £	Material £	Unit	Total rate £
Q50 SITE/STREET FURNITURE/ EQUIPMENT - cont'd							
Cycle holders - cont'd							
Cycle holders; Townscape Products Ltd							
"Guardian" cycle holders; tubular steel frame; setting in concrete; 1250 x 550 x 775 mm high; making good	271.84	1.00	18.25	-	284.82	nr	**303.07**
"Penny" cycle stands; 600 mm diameter; exposed aggregate bollards with 8 nr cycle holders; in galvanized steel; setting in concrete; making good	541.93	1.00	18.25	-	554.91	nr	**573.16**
Cycle holders; Broxap Street Furniture							
"Neath" cycle rack; ref BX/MW/AG; for 6 nr cycles; semi-vertical; galvanized and polyester powder coated; 1.320 m wide x 2.542 m long x 1.80 m high	435.00	10.00	182.50	-	460.96	nr	**643.46**
"Premier Senior" combined shelter and rack; ref BX/MW/AW; for 10 nr cycles; horizontal; galvanized only; 2.13 m wide x 3.05 m long x 2.15 m high	1600.00	10.00	182.50	-	1625.96	nr	**1808.46**
"Toast Rack" double sided free standing cycle rack; ref BX/MW/GH; for 10 nr cycles; galvanized and polyester coated; 3.250 m long	380.00	4.00	73.00	-	390.04	nr	**463.04**
Directional signage; cast aluminium							
Signage; Furnitubes International Ltd							
ref FFL 1; "Lancer"; cast aluminium finials	60.90	0.07	1.22	-	60.90	nr	**62.12**
ref FAA IS; arrow end type cast aluminium directional arms; single line; 90 mm wide	129.15	2.00	36.50	-	129.15	nr	**165.65**
ref FAA ID; arrow end type cast aluminium directional arms; double line; 145 mm wide	156.45	0.13	2.43	-	156.45	nr	**158.88**
ref FAA IS; arrow end type cast aluminium directional arms; treble line; 200 mm wide	180.60	0.20	3.65	-	180.60	nr	**184.25**
ref FCK1 211 G; "Kingston"; composite standard root columns	341.25	2.00	36.50	-	357.02	nr	**393.52**
Park signage; Browse Bion Architectural Signs							
Entrance signs and map boards; vitreous enamel							
entrance map board; 1250 x 1000 mm high with two support posts; all associated works to post bases	-	-	-	-	-	nr	**2159.00**
information board with two locking cabinets; 1250 x 1000 mm high with two support posts; all associated works to post bases	-	-	-	-	-	nr	**2061.20**

Q PAVING/PLANTING/FENCING/SITE FURNITURE

Item Excluding site overheads and profit	PC £	Labour hours	Labour £	Plant £	Material £	Unit	Total rate £
Miscellaneous park signage; vitreous enamel; Browse Bion Architectural Signs							
"No dogs" sign; 200 x 150 mm; fixing to fencing or gates	-	-	-	-	-	nr	319.58
"Dog exercise area" sign; 300 x 400 mm; fixing to fencing or gates	-	-	-	-	-	nr	397.92
"Nature conservation area" sign; 900 x 400 mm high with two support posts; all associated works to post bases	-	-	-	-	-	nr	1140.00
Excavating; for bollards and barriers; by hand							
Holes for bollards							
400 x 400 x 400 mm; disposing off site	-	0.42	7.60	-	0.54	nr	8.14
600 x 600 x 600 mm; disposing off site	-	0.97	17.70	-	1.51	nr	19.21
Concrete bollards - General							
Preamble: Precast concrete bollards are available in a very wide range of shapes and sizes. The bollards listed here are the most commonly used sizes and shapes; manufacturer's catalogues should be consulted for the full range. Most manufacturers produce bollards to match their suites of street furniture, which may include planters, benches, litter bins and cycle stands. Most parallel sided bollards can be supplied in removable form, with a reduced shank, precast concrete socket and lifting hole to permit removal with a bar.							
Concrete bollards							
Marshalls Plc; cylinder; straight or tapered; 200 - 400 mm diameter; plain grey concrete; setting into firm ground (excavating and backfilling not included)							
"Bridgford"; 915 mm high above ground	80.00	2.00	36.50	-	90.13	nr	126.63
Marshalls Plc; cylinder; straight or tapered; 200 - 400 mm diameter; "Beadalite" reflective finish; setting into firm ground (excavating and backfilling not included)							
"Wexham" concrete bollard; exposed silver grey	145.00	2.00	36.50	-	150.44	nr	186.94
"Wexham Major" concrete bollard; exposed silver grey	235.00	2.00	36.50	-	245.13	nr	281.63
Precast concrete verge markers; various shapes; 450 mm high							
plain grey concrete	34.00	1.00	18.25	-	37.89	nr	56.14
white concrete	35.65	1.00	18.25	-	39.54	nr	57.79
exposed aggregate	37.75	1.00	18.25	-	41.64	nr	59.89

Q PAVING/PLANTING/FENCING/SITE FURNITURE

Item Excluding site overheads and profit	PC £	Labour hours	Labour £	Plant £	Material £	Unit	Total rate £
Q50 SITE/STREET FURNITURE/ EQUIPMENT - cont'd							
Other bollards							
Removable parking posts; Dee-Organ Ltd							

"Spacekeeper"; ref 3014203; folding
  plastic coated galvanized steel parking
  posts with key; reflective bands; including
  300 x 300 x 300 mm concrete foundations
  and fixing bolts; 850 mm high
	120.59	2.00	36.50	-	132.54	nr	**169.04**

"Spacesaver" ref 1506101; hardwood
  removable bollards including key and
  base-plate; 150 x 150 mm; 800 mm high
	140.75	2.00	36.50	-	144.74	nr	**181.24**

Removable parking posts; Marshalls Plc							
"RT\RD4"; domestic telescopic bollard	160.00	2.00	36.50	-	165.44	nr	**201.94**
"RT\R8"; heavy duty telescopic bollard	214.00	2.00	36.50	-	219.44	nr	**255.94**
Plastic bollards; Marshalls Plc							
"Lismore"; three ring recycled plastic bollard	75.00	2.00	36.50	-	80.44	nr	**116.94**
Cast iron bollards - General							
Preamble: The following bollards are particularly suitable for conservation areas. Logos for civic crests can be incorporated to order.							
Cast iron bollards							
Bollards; Furnitubes International Ltd (excavating and backfilling not included)							
"Doric Round"; 920 mm high x 170 mm diameter	80.00	2.00	36.50	-	85.19	nr	**121.69**
"Gunner round"; 750 mm high x 165 mm diameter	71.00	2.00	36.50	-	76.19	nr	**112.69**
"Manchester Round"; 975 mm high; 225 mm square base	75.00	2.00	36.50	-	80.19	nr	**116.69**
"Cannon"; 1140 mm x 210 mm diameter	134.00	2.00	36.50	-	139.19	nr	**175.69**
"Kenton"; heavy duty galvanized steel; 900 mm high; 350 mm diameter	160.00	2.00	36.50	-	165.19	nr	**201.69**
Bollards; Marshalls Plc (excavating and backfilling not included)							
"MSF103"; cast iron bollard; "Small Manchester"	110.00	2.00	36.50	-	115.44	nr	**151.94**
"MSF102"; cast iron bollard; "Manchester"	115.00	2.00	36.50	-	120.44	nr	**156.94**
Cast iron bollards with rails - General							
Preamble: The following cast iron bollards are suitable for conservation areas.							

Q PAVING/PLANTING/FENCING/SITE FURNITURE

Item Excluding site overheads and profit	PC £	Labour hours	Labour £	Plant £	Material £	Unit	Total rate £
Cast iron bollards with rails Cast iron posts with steel tubular rails; Broxap Street Furniture; setting into firm ground (excavating not included)							
"Sheffield Short"; 420 mm high; one rail, type A	70.00	1.50	27.38	-	72.51	nr	**99.88**
"Mersey"; 1085 mm high above ground; two rails, type D	99.00	1.50	27.38	-	101.51	nr	**128.88**
"Promenade"; 1150 mm high above ground; square; three rails, type C	124.00	1.50	27.38	-	126.51	nr	**153.88**
"Type A" mild steel tubular rail; including connector	6.05	0.03	0.61	-	7.44	m	**8.05**
"Type C" mild steel tubular rail; including connector	8.50	0.03	0.61	-	10.05	m	**10.65**
Steel bollards Steel bollards; Marshalls Plc (excavating and backfilling not included)							
"SSB01"; stainless steel bollard; 101 x 1250 mm	105.00	2.00	36.50	-	110.44	nr	**146.94**
"RS001"; stainless steel bollard; 114 x 1500 mm	120.00	2.00	36.50	-	125.44	nr	**161.94**
"RB119 Brunel"; steel bollard; 168 x 1500 mm	123.00	2.00	36.50	-	133.13	nr	**169.63**
Timber bollards Woodscape Ltd; durable hardwood							
RP 250/1500; 250 mm diameter x 1500 mm long	220.60	1.00	18.25	-	220.83	nr	**239.08**
SP 250/1500; 250 mm square x 1500 mm long	220.60	1.00	18.25	-	220.83	nr	**239.08**
SP 150/1200; 150 mm square x 1200 mm long	55.34	1.00	18.25	-	55.57	nr	**73.82**
SP 125/750; 125 mm square x 750 mm long	43.89	1.00	18.25	-	44.12	nr	**62.37**
RP 125/750; 125 mm diameter x 750 mm long	43.89	1.00	18.25	-	44.12	nr	**62.37**
Deterrent bollards Semi-mountable vehicle deterrent and kerb protection bollards; Furnitubes International Ltd (excavating and backfilling not included)							
"Bell decorative"	381.00	2.00	36.50	-	388.38	nr	**424.88**
"Half bell"	357.00	2.00	36.50	-	364.38	nr	**400.88**
"Full bell"	502.00	2.00	36.50	-	509.38	nr	**545.88**
"Three quarter bell"	387.00	2.00	36.50	-	394.38	nr	**430.88**
Security bollards Security bollards; Furnitubes International Ltd (excavating and backfilling not included)							
"Gunner"; reinforced with steel insert and tie bars; 750 mm high above ground; 600 mm below ground	124.00	2.00	36.50	-	131.78	nr	**168.28**
"Burr Bloc Type 6"; removable steel security bollard; 750 mm high above ground; 410 x 285 mm	408.00	2.00	36.50	-	425.50	nr	**462.00**

Q PAVING/PLANTING/FENCING/SITE FURNITURE

Item Excluding site overheads and profit	PC £	Labour hours	Labour £	Plant £	Material £	Unit	Total rate £
Q50 SITE/STREET FURNITURE/ EQUIPMENT - cont'd							
Tree grilles; cast iron							
Cast iron tree grilles; Furnitubes International Ltd							
ref GS 1070 Greenwich; two part; 1000 mm square x 700 mm diameter tree hole	71.00	2.00	36.50	-	71.00	nr	**107.50**
ref GC 1270 Greenwich; two part, 1200 mm diameter x 700 mm diameter tree hole	101.00	2.00	36.50	-	101.00	nr	**137.50**
Cast iron tree grilles; Marshalls Plc							
"Heritage"; cast iron grille plus frame; 1 x 1 m	370.00	3.00	54.75	-	380.05	nr	**434.80**
Cast iron tree grilles; Townscape Products Ltd							
"Baltimore"; 1200 mm square x 460 mm diameter tree hole	391.46	2.00	36.50	-	391.46	nr	**427.96**
"Baltimore"; hexagonal; maximum width 1440 mm nominal x 600 mm diameter tree hole	629.87	2.00	36.50	-	629.87	nr	**666.37**
Tree grilles; steel							
Steel tree grilles; Furnitubes International Ltd							
ref GSF 102G Greenwich; steel tree grille frame for GS 1045; one part	123.00	2.00	36.50	-	123.00	nr	**159.50**
ref GSF 122G Greenwich; steel tree grille frame for GS 1270 and GC 1245; one part	128.00	2.00	36.50	-	128.00	nr	**164.50**
Note: Care must be taken to ensure that tree grids and guards are removed when trees grow beyond the specified diameter of guard.							

Q PAVING/PLANTING/FENCING/SITE FURNITURE

Item Excluding site overheads and profit	PC £	Labour hours	Labour £	Plant £	Material £	Unit	Total rate £
Playground equipment - General Preamble: The range of equipment manufactured or available in the UK is so great that comprehensive coverage would be impossible, especially as designs, specifications and prices change fairly frequently. The following information should be sufficient to give guidance to anyone designing or equipping a playground. In comparing prices note that only outline specification details are given here and that other refinements which are not mentioned may be the reason for some difference in price between two apparently identical elements. The fact that a particular manufacturer does not appear under one item heading does not necessarily imply that they do not make it. Landscape designers are advised to check that equipment complies with ever more stringent safety standards before specifying.							
Playground equipment - Installation The rates below include for installation of the specified equipment by the manufacturers. Most manufacturers will offer an option to install the equipment they have supplied.							
Play systems; Kompan Ltd; "Galaxy"; multiple play activity systems for non-prescribed play; for children 6-14 years; galvanized steel and high density polyethylene							
Adara GXY 906; 14 different play activities	-	-	-	-	-	nr	15726.00
Sirius GXY 8011; 10 different play activities	-	-	-	-	-	nr	9717.50
Sports and social areas; multi-use games areas (MUGA); Kompan Ltd; enclosed sports areas; complete with surfacing boundary and goals and targets; galvanized steel framework with high density polyethylene panels, galvanized steel goals and equipment; surfacing priced separately							
FRE 1211; end sports wall; multigoal	-	-	-	-	-	nr	5534.00
Pitch; complete; suitable for multiple ball sports; suitable for use with natural artificial or hard landscape surfaces; fully enclosed including two end sports walls							
Cosmos FRE 2110; 12.0 x 20.0 m	-	-	-	-	-	nr	30469.00
Cosmos FRE 2116; 19.0 x 36.0 m	-	-	-	-	-	nr	44732.00
FRE 3000; meeting point shelter or social area	-	-	-	-	-	nr	4598.00

Q PAVING/PLANTING/FENCING/SITE FURNITURE

Item Excluding site overheads and profit	PC £	Labour hours	Labour £	Plant £	Material £	Unit	Total rate £
Q50 SITE/STREET FURNITURE/ EQUIPMENT - cont'd							
Swings - General							
Preamble: Prices for the following vary considerably. Those given represent the middle of the range and include multiple swings with tubular steel frames and timber or tyre seats; ground fixing and priming only.							
Swings							
Swings; Wicksteed Ltd							
traditional swings; 1850 mm high; 1 bay; 2 seat	-	-	-	-	-	nr	2227.00
traditional swings; 1850 mm high; 2 bay; 4 seat	-	-	-	-	-	nr	3578.00
traditional swings; 2450 mm high; 1 bay; 2 seat	-	-	-	-	-	nr	2151.00
traditional swings; 2450 mm high; 2 bay; 4 seat	-	-	-	-	-	nr	3396.00
traditional swings; 3050 mm high; 1 bay; 2 seat	-	-	-	-	-	nr	2284.00
traditional swings; 3050 mm high; 2 bay; 4 seat	-	-	-	-	-	nr	3524.00
double arch swing; cradle safety seats; 1850 mm high	-	-	-	-	-	nr	2259.00
twin double arch swing; cradle safety seat; 1850 mm high	-	-	-	-	-	nr	3783.00
single arch swing; flat rubber safety seat; 2450 mm high	-	-	-	-	-	nr	1739.00
double arch swing; flat rubber safety seats; 2450 mm high	-	-	-	-	-	nr	2133.00
Swings; Lappset UK Ltd							
ref 020414M; swing frame with two flat seats	-	-	-	-	-	nr	1460.00
Swings; Kompan Ltd							
ref M947; double swings	-	-	-	-	-	nr	2174.00
ref M961P; double swings	-	-	-	-	-	nr	2000.00
ref M951; "Sunflower" swings	-	-	-	-	-	nr	1590.00
Slides							
Slides; Wicksteed Ltd							
"Pedestal" slides; 3.40 m	-	-	-	-	-	nr	4022.00
"Pedestal" slides; 4.40 m	-	-	-	-	-	nr	3089.00
"Pedestal" slides; 5.80 m	-	-	-	-	-	nr	4680.00
"Embankment" slides; 3.40 m	-	-	-	-	-	nr	2367.00
"Embankment" slides; 4.40 m	-	-	-	-	-	nr	3077.00
"Embankment" slides; 5.80 m	-	-	-	-	-	nr	3916.00
"Embankment" slides; 7.30 m	-	-	-	-	-	nr	4976.00
"Embankment" slides; 9.10 m	-	-	-	-	-	nr	6233.00
"Embankment" slides; 11.00 m	-	-	-	-	-	nr	7454.00
Slides; Lappset UK Ltd							
ref 142015M; slide	-	-	-	-	-	nr	3540.00
ref 141115M; "Jumbo" slide	-	-	-	-	-	nr	5305.00
Slides; Kompan Ltd							
ref M351; slide	-	-	-	-	-	nr	2770.00
ref M326; Aladdin's cave slide	-	-	-	-	-	nr	2792.00

Q PAVING/PLANTING/FENCING/SITE FURNITURE

Item Excluding site overheads and profit	PC £	Labour hours	Labour £	Plant £	Material £	Unit	Total rate £
Moving equipment - General Preamble: The following standard items of playground equipment vary considerably in quality and price; the following prices are middle of the range.							
Moving equipment Roundabouts; Wicksteed Ltd							
"Turnstile"	-	-	-	-	-	nr	861.00
"Speedway" (without restrictor)	-	-	-	-	-	nr	3393.00
"Spiro Whirl" (without restrictor)	-	-	-	-	-	nr	3865.00
Roundabouts; Kompan Ltd Supernova GXY 916; multifunctional spinning and balancing disc; capacity approximately 15 children	-	-	-	-	-	nr	4180.00
Seesaws Seesaws; Lappset UK Ltd							
ref 010300; seesaws	-	-	-	-	-	nr	925.00
ref 010237; seesaws	-	-	-	-	-	nr	2345.00
Seesaws; Wicksteed Ltd							
"Seesaw"; non-bump	-	-	-	-	-	nr	2506.00
"Jolly Gerald"; non-bump	-	-	-	-	-	nr	2702.00
"Rocking Rockette"; with motion restrictor	-	-	-	-	-	nr	3625.00
"Rocking Horse"; with motion restrictor	-	-	-	-	-	nr	4135.00
Play sculptures - General Preamble: Many variants on the shapes of playground equipment are available, simulating spacecraft, trains, cars, houses etc., and these designs are frequently changed. The basic principles remain constant but manufacturer's catalogues should be checked for the latest styles.							
Climbing equipment and play structures - General Preamble: Climbing equipment generally consists of individually designed modules. Play structures generally consist of interlinked and modular pieces of equipment and sculptures. These may consist of climbing, play and skill based modules, nets and various other activities. Both are set into either safety surfacing or defined sand pit areas. The equipment below outlines a range from various manufacturers. Individual catalogues should be consulted in each instance. Safety areas should be allowed round all equipment.							

Q PAVING/PLANTING/FENCING/SITE FURNITURE

Item Excluding site overheads and profit	PC £	Labour hours	Labour £	Plant £	Material £	Unit	Total rate £
Q50 SITE/STREET FURNITURE/ EQUIPMENT - cont'd							
Climbing equipment and play structures							
Climbing equipment and play structures; Kompan Ltd							
ref M623; "Fairy Castle"	-	-	-	-	-	nr	4902.00
Climbing equipment; Wicksteed Ltd							
Funrun Fitness Trail "Under Starter's Orders"; set of 12 units	-	-	-	-	-	nr	15884.00
Climbing equipment; Lappset UK Ltd							
ref 138401M; "Storks Nest"	-	-	-	-	-	nr	4200.00
ref 122457M; "Playhouse"	-	-	-	-	-	nr	3620.00
ref 120100M; "Activity Tower"	-	-	-	-	-	nr	13320.00
ref 120124M; "Tower and Climbing Frame"	-	-	-	-	-	nr	12960.00
SMP playgrounds							
Action Pack - Cape Horn; 8 module	-	-	-	-	-	nr	13385.12
Spring equipment							
Spring based equipment for 1 - 8 year olds; Kompan Ltd							
ref M101; "Crazy Hen"	-	-	-	-	-	nr	714.00
ref M128; "Crazy Daisy"	-	-	-	-	-	nr	978.00
ref M141; "Crazy Seesaw"	-	-	-	-	-	nr	2096.00
ref M155; "Quartet Seesaw"	-	-	-	-	-	nr	1332.00
Spring based equipment for under 12's; Lappset UK Ltd							
ref 010501; "Horse Springer"	-	-	-	-	-	nr	1070.00
Sand pits							
Market prices							
play pit sand; Boughton Loam Ltd	-	-	-	-	54.00	m^3	-
Kompan Ltd							
"Basic 550"; 5760 x 4540 x 310 mm deep	-	-	-	-	-	nr	1335.00
Flagpoles							
Flagpoles; Harrison Flagpoles; in glass fibre; smooth white finish; terylene halyards with mounting accessories; setting in concrete; to manufacturer's recommendations (excavating not included)							
6.00 m high; plain	158.60	3.00	54.75	-	178.68	nr	233.43
6.00 m high; hinged baseplate; external halyard	177.00	4.00	73.00	-	197.08	nr	270.08
6.00 m high; hinged baseplate; internal halyard	255.00	4.00	73.00	-	275.08	nr	348.08
10.00 m high; plain	314.21	5.00	91.25	-	334.29	nr	425.54
10.00 m high; hinged baseplate; external halyard	322.00	5.00	91.25	-	342.08	nr	433.33
10.00 m high; hinged baseplate; internal halyard	416.00	5.00	91.25	-	436.08	nr	527.33
15.00 m high; plain	719.20	6.00	109.50	-	739.28	nr	848.78
15.00 m high; hinged baseplate; external halyard	769.07	6.00	109.50	-	789.15	nr	898.65
15.00 m high; hinged baseplate; internal halyard	967.58	6.00	109.50	-	987.66	nr	1097.16

Q PAVING/PLANTING/FENCING/SITE FURNITURE

Item Excluding site overheads and profit	PC £	Labour hours	Labour £	Plant £	Material £	Unit	Total rate £
Flagpoles; Harrison Flagpoles; in glass fibre; smooth white finish; terylene halyards for wall mounting							
3.0 m pole	187.00	2.00	36.50	-	187.00	nr	223.50
Flagpoles; Harrison Flagpoles; parallel galvanized steel banner flagpole; guide price							
6 m high	997.50	4.00	73.00	-	1017.58	nr	1090.58
Sports equipment; Edwards Sports Products Ltd							
Tennis courts; steel posts and sockets for hardcourt							
round	263.71	1.00	18.25	-	263.71	set	281.96
square	251.70	1.00	18.25	-	251.70	set	269.95
Tennis nets; not including posts or fixings							
ref 5001; "Championship"	133.41	3.00	54.75	-	140.10	set	194.85
ref 5021; "Matchplay"	95.61	3.00	54.75	-	102.30	set	157.05
ref 5075; "Club"	80.97	3.00	54.75	-	87.66	set	142.41
Football goals; senior; socketed; including backirons and 2.5 mm nets							
ref 2711/2197/20261; aluminium	1116.79	4.00	73.00	-	1123.48	set	1196.48
ref 2760/2190/20261; steel	746.48	0.14	2.63	-	932.40	set	935.03
Freestanding football goals; senior; including nets							
ref 2719/2000A; aluminium	1760.72	4.00	73.00	-	1767.41	set	1840.41
ref 2890/20261; steel	1002.34	4.00	73.00	-	1009.03	set	1082.03
Freestanding mini-soccer goals; including nets							
ref 2717/CN; aluminium	527.94	4.00	73.00	-	534.63	set	607.63
ref 2961/CN; steel	498.32	4.00	73.00	-	505.01	set	578.01
Rugby goal posts; including sockets							
ref 2800; 7.32 m high	719.20	6.00	109.50	-	737.05	set	846.55
ref 2802; 10.70 m high	993.23	6.00	109.50	-	1011.08	set	1120.58
Hockey goals; steel; including backboards and nets							
ref 2906/2310; freestanding	1269.28	1.00	18.25	-	1269.28	set	1287.53
ref 2851/2853/2841/2310; socketed	966.46	1.00	18.25	-	966.46	set	984.71
Cricket cages; steel; including netting							
ref 5454; freestanding	919.27	12.00	219.00	-	919.27	set	1138.27
ref 5454/W; wheelaway	1183.09	12.00	219.00	-	1183.09	set	1402.09

Q PAVING/PLANTING/FENCING/SITE FURNITURE

Item Excluding site overheads and profit	PC £	Labour hours	Labour £	Plant £	Material £	Unit	Total rate £
Q50 SITE/STREET FURNITURE/ EQUIPMENT - cont'd							
Pergolas; AVS Fencing Supplies Ltd; construct timber pergola; posts 150 x 150 mm x 2.40 m finished height in 600 mm deep minimum concrete pits; beams of 200 x 50 mm x 2.00 m wide; fixed to posts with dowels; rafters 200 x 38 mm notched to beams; inclusive of all mechanical excavation and disposal off site							
Pergola 2.00 m wide in green oak; posts at 2.00 m centres; dowel fixed							
rafters at 600 mm centres	-	8.40	172.20	3.59	240.04	m	**415.83**
rafters at 400 mm centres	-	10.80	221.40	3.59	290.47	m	**515.46**
Pergola 3.00 m wide in green oak; posts at 1.50 m centres							
rafters at 600 mm centres	-	7.20	155.70	4.31	318.79	m	**478.80**
rafters at 400 mm centres	-	8.00	173.00	4.31	394.29	m	**571.60**
Pergola 2.00 wide in prepared softwood; beams fixed with bolts; posts at 2.00 m centres							
rafters at 600 mm centres	-	5.00	108.13	3.59	141.38	m	**253.10**
rafters at 400 mm centres	-	6.00	129.75	3.59	162.84	m	**296.18**
Pergola 3.00 m wide in prepared softwood; posts at 1.50 m centres							
rafters at 600 mm centres	-	7.20	155.70	4.31	140.92	m	**300.94**
rafters at 400 mm centres	-	8.00	173.00	4.31	161.35	m	**338.67**

R DISPOSAL SYSTEMS

Item Excluding site overheads and profit	PC £	Labour hours	Labour £	Plant £	Material £	Unit	Total rate £
R12 DRAINAGE BELOW GROUND							
Silt pits and inspection chambers							
Excavating pits; starting from ground level; by machine							
maximum depth not exceeding 1.00 m	-	0.50	9.13	16.50	-	m^3	25.63
maximum depth not exceeding 2.00 m	-	0.50	9.13	24.75	-	m^3	33.88
maximum depth not exceeding 4.00 m	-	0.50	9.13	49.50	-	m^3	58.63
Disposal of excavated material; depositing on site in permanent spoil heaps; average 50 m	-	0.04	0.76	1.56	-	m^3	2.32
Filling to excavations; obtained from on site spoil heaps; average thickness not exceeding 0.25 m	-	0.13	2.43	4.40	-	m^3	6.83
Surface treatments; compacting; bottoms of excavations	-	0.05	0.91	-	-	m^2	0.91
Earthwork support; distance between opposing faces not exceeding 2.00 m							
maximum depth not exceeding 1.00 m	-	0.20	3.65	-	4.40	m^3	8.05
maximum depth not exceeding 2.00 m	-	0.30	5.47	-	4.40	m^3	9.87
maximum depth not exceeding 4.00 m	-	0.67	12.17	-	1.73	m^3	13.90
Silt pits and inspection chambers; in situ concrete							
Beds; plain in situ concrete; 11.50 N/mm^2 - 40 mm aggregate							
thickness not exceeding 150 mm	-	1.00	18.25	-	92.96	m^3	111.21
thickness 150 - 450 mm	-	0.67	12.17	-	92.96	m^3	105.13
Benchings in bottoms; plain in situ concrete; 25.50 N/mm^2 - 20 mm aggregate							
thickness 150 - 450 mm	-	2.00	36.50	-	92.96	m^3	129.46
Isolated cover slabs; reinforced In situ concrete; 21.00 N/mm^2 - 20 mm aggregate							
thickness not exceeding 150 mm	92.96	4.00	73.00	-	92.96	m^3	165.96
Fabric reinforcement; A193 (3.02 kg/m^2) in cover slabs	1.40	0.06	1.14	-	1.54	m^2	2.68
Formwork to reinforced in situ concrete; isolated cover slabs							
soffits; horizontal	-	3.28	59.86	-	4.58	m^2	64.44
height not exceeding 250 mm	-	0.97	17.70	-	1.98	m	19.68
Silt pits and inspection chambers; precast concrete units							
Precast concrete inspection chamber units; FP McCann Ltd; bedding, jointing and pointing in cement mortar (1:3); 600 x 450 mm internally							
600 mm deep	50.56	6.00	109.50	-	55.35	nr	164.85
900 mm deep	63.20	7.00	127.75	-	68.30	nr	196.05
Drainage chambers; FP McCann Ltd; 1200 x 750 mm reducing to 600 x 600 mm; no base unit; depth of invert							
1050 mm deep	383.00	9.00	164.25	-	398.44	nr	562.70
1650 mm deep	567.00	11.00	200.75	-	590.57	nr	791.32
2250 mm deep	751.00	12.50	228.13	-	781.51	nr	1009.64

R DISPOSAL SYSTEMS

Item Excluding site overheads and profit	PC £	Labour hours	Labour £	Plant £	Material £	Unit	Total rate £
R12 DRAINAGE BELOW GROUND - cont'd							
Silt pits and inspection chambers - cont'd							
Cover slabs for chambers or shaft sections; FP McCann Ltd; heavy duty							
900 mm diameter internally	62.00	0.67	12.15	-	62.00	nr	**74.15**
1050 mm diameter internally	68.00	2.00	36.50	11.00	68.00	nr	**115.50**
1200 mm diameter internally	82.00	1.00	18.25	11.00	82.00	nr	**111.25**
1500 mm diameter internally	139.04	1.00	18.25	11.00	139.04	nr	**168.29**
1800 mm diameter internally	208.47	2.00	36.50	24.75	208.47	nr	**269.72**
Brickwork							
Walls to manholes; bricks; PC £300.00/1000; in cement mortar (1:3)							
one brick thick	36.00	3.00	54.75	-	44.84	m²	**99.59**
one and a half brick thick	54.00	4.00	73.00	-	67.27	m²	**140.27**
two brick thick projection of footing or the like	72.00	4.80	87.60	-	89.69	m²	**177.29**
Walls to manholes; engineering bricks; PC £239.00/1000; in cement mortar (1:3)							
one brick thick	28.68	3.00	54.75	-	37.16	m²	**91.91**
one and a half brick thick	43.02	4.00	73.00	-	55.74	m²	**128.74**
two brick thick projection of footing or the like	57.36	4.80	87.60	-	74.32	m²	**161.92**
Extra over common or engineering bricks in any mortar for fair face; flush pointing as work proceeds; English bond walls or the like	-	0.13	2.43	-	-	m²	**2.43**
In situ finishings; cement:sand mortar (1:3); steel trowelled; 13 mm one coat work to manhole walls; to brickwork or blockwork base; over 300 mm wide	-	0.80	14.60	-	2.35	m²	**16.95**
Building into brickwork; ends of pipes; making good facings or renderings							
small	-	0.20	3.65	-	-	nr	**3.65**
large	-	0.30	5.47	-	-	nr	**5.47**
extra large	-	0.40	7.30	-	-	nr	**7.30**
extra large; including forming ring arch cover	-	0.50	9.13	-	-	nr	**9.13**
Inspection chambers; Hepworth Plc							
Polypropylene; up to 600 mm deep							
300 mm diameter x 960 mm deep; heavy duty round covers and frames; double seal recessed with 6 nr 110 mm outlets/inlets	228.14	3.00	54.75	-	230.57	nr	**285.32**
Polypropylene; up to 1200 mm deep							
475 mm diameter x 1030 mm deep; heavy duty round covers and frames; double seal recessed with 5 nr 110 mm outlets/inlets	92.81	4.00	73.00	-	295.32	nr	**368.32**
Step irons; Ashworth Ltd. drainage systems; malleable cast iron; galvanized; building into joints							
General purpose pattern; for one brick walls	4.14	0.17	3.10	-	4.14	nr	**7.24**

R DISPOSAL SYSTEMS

Item Excluding site overheads and profit	PC £	Labour hours	Labour £	Plant £	Material £	Unit	Total rate £
Best quality vitrified clay half section channels; Hepworth Plc; bedding and jointing in cement:mortar (1:2)							
Channels; straight							
100 mm	6.10	0.80	14.60	-	8.66	m	**23.26**
150 mm	10.15	1.00	18.25	-	12.72	m	**30.97**
225 mm	22.81	1.35	24.64	-	25.37	m	**50.01**
300 mm	46.82	1.80	32.85	-	49.38	m	**82.23**
Bends; 15, 30, 45 or 90°							
100 mm bends	5.49	0.75	13.69	-	6.77	nr	**20.46**
150 mm bends	9.49	0.90	16.43	-	11.42	nr	**27.84**
225 mm bends	36.81	1.20	21.90	-	39.37	nr	**61.27**
300 mm bends	75.05	1.10	20.07	-	78.26	nr	**98.34**
Best quality vitrified clay three quarter section channels; Hepworth Plc; bedding and jointing in cement:mortar (1:2)							
Branch bends; 115, 140 or 165°; left or right hand							
100 mm	5.49	0.75	13.69	-	6.77	nr	**20.46**
150 mm	9.49	0.90	16.43	-	11.42	nr	**27.84**
Intercepting traps							
Vitrified clay; inspection arms; brass stoppers; iron levers; chains and staples; galvanized; staples cut and pinned to brickwork; cement:mortar (1:2) joints to vitrified clay pipes and channels; bedding and surrounding in concrete; 11.50 N/mm² - 40 mm aggregate; cutting and fitting brickwork; making good facings							
100 mm inlet; 100 mm outlet	79.85	3.00	54.75	-	95.87	nr	**150.62**
150 mm inlet; 150 mm outlet	115.13	2.00	36.50	-	132.95	nr	**169.45**
Excavating trenches; using 3 tonne tracked excavator; to receive pipes; grading bottoms; earthwork support; filling with excavated material to within 150 mm of finished surfaces and compacting; completing fill with topsoil; disposal of surplus soil							
Services not exceeding 200 mm nominal size							
average depth of run not exceeding 0.50 m	1.17	0.12	2.19	0.99	1.40	m	**4.59**
average depth of run not exceeding 0.75 m	1.17	0.16	2.97	1.36	1.40	m	**5.74**
average depth of run not exceeding 1.00 m	1.17	0.28	5.17	2.37	1.40	m	**8.94**
average depth of run not exceeding 1.25 m	1.17	0.38	7.00	3.20	1.17	m	**11.36**

R DISPOSAL SYSTEMS

Item Excluding site overheads and profit	PC £	Labour hours	Labour £	Plant £	Material £	Unit	Total rate £
R12 DRAINAGE BELOW GROUND - cont'd							
Granular beds to trenches; lay granular material to trenches excavated separately; to receive pipes (not included)							
300 mm wide x 100 mm thick							
reject sand	-	0.05	0.91	0.22	1.58	m	**2.71**
reject gravel	-	0.05	0.91	0.22	3.76	m	**4.89**
shingle 40 mm aggregate	-	0.05	0.91	0.22	3.81	m	**4.94**
sharp sand	2.87	0.05	0.91	0.22	3.16	m	**4.29**
300 mm wide x 150 mm thick							
reject sand	-	0.08	1.37	0.32	2.38	m	**4.07**
reject gravel	-	0.08	1.37	0.32	5.64	m	**7.33**
shingle 40 mm aggregate	-	0.08	1.37	0.32	5.71	m	**7.40**
sharp sand	4.31	0.08	1.37	0.32	4.74	m	**6.43**
Excavating trenches; using 3 tonne tracked excavator; to receive pipes; grading bottoms; earthwork support; filling with imported granular material type 2 and compacting; disposal of surplus soil							
Services not exceeding 200 mm nominal size							
average depth of run not exceeding 0.50 m	1.44	0.09	1.58	0.70	2.70	m	**4.99**
average depth of run not exceeding 0.75 m	2.16	0.11	1.98	0.88	4.05	m	**6.91**
average depth of run not exceeding 1.00 m	2.88	0.14	2.55	1.14	5.40	m	**9.09**
average depth of run not exceeding 1.25 m	3.60	0.23	4.17	1.91	6.75	m	**12.83**
Excavating trenches; using 3 tonne tracked excavator; to receive pipes; grading bottoms; earthwork support; filling with concrete, ready mixed ST2; disposal of surplus soil							
Services not exceeding 200 mm nominal size							
average depth of run not exceeding 0.50 m	12.62	0.11	1.95	0.34	14.20	m	**16.49**
average depth of run not exceeding 0.75 m	18.93	0.13	2.37	0.43	21.30	m	**24.10**
average depth of run not exceeding 1.00 m	25.25	0.17	3.04	0.57	28.40	m	**32.01**
average depth of run not exceeding 1.25 m	31.56	0.23	4.11	0.86	35.50	m	**40.47**
Earthwork support; providing support to opposing faces of excavation; moving along as work proceeds							
Maximum depth not exceeding 2.00 m							
distance between opposing faces not exceeding 2.00 m	-	0.80	14.60	17.70	-	m	**32.30**

R DISPOSAL SYSTEMS

Item Excluding site overheads and profit	PC £	Labour hours	Labour £	Plant £	Material £	Unit	Total rate £
Clay pipes and fittings; Hepworth Plc;							
Supersleeve							
100 mm clay pipes; polypropylene slip							
coupling; in trenches (trenches not included)							
laid straight	2.61	0.25	4.56	-	4.54	m	9.10
short runs under 3.00 m	2.61	0.31	5.70	-	4.54	m	10.24
Extra over 100 mm clay pipes for							
bends; 15 - 90°; single socket	8.70	0.25	4.56	-	8.70	nr	13.26
junction; 45 or 90°; double socket	18.34	0.25	4.56	-	18.34	nr	22.90
slip couplings polypropylene	3.08	0.08	1.52	-	3.08	nr	4.60
gully with "P" trap; 100 mm; 154 x 154 mm							
plastic grating	37.24	1.00	18.25	-	51.04	nr	69.29
150 mm vitrified clay pipes; polypropylene							
slip coupling; in trenches (trenches not							
included)							
laid straight	10.43	0.30	5.47	-	10.43	m	15.90
short runs under 3.00 m	8.03	0.33	6.08	-	8.03	m	14.11
Extra over 150 mm vitrified clay pipes for							
bends; 15 - 90°	11.62	0.28	5.11	-	17.21	nr	22.32
junction; 45 or 90°; 100 x 150 mm	15.54	0.40	7.30	-	26.73	nr	34.03
junction; 45 or 90°; 150 x 150 mm	17.06	0.40	7.30	-	28.26	nr	35.56
slip couplings polypropylene	5.60	0.05	0.91	-	5.60	nr	6.51
taper pipe; 100 - 150 mm	17.47	0.50	9.13	-	17.47	nr	26.60
taper pipe; 150 - 225 mm	44.88	0.50	9.13	-	44.88	nr	54.00
socket adaptor; connection to traditional							
pipes and fittings	11.78	0.33	6.02	-	17.38	nr	23.40
Accessories in clay							
150 mm access pipe	43.82	-	-	-	55.01	nr	-
150 mm rodding eye	41.91	0.50	9.13	-	45.96	nr	55.09
gully with "P" traps; 150 mm; 154 x							
154 mm plastic grating	56.91	1.00	18.25	-	64.13	nr	82.38
PVC-u pipes and fittings; Wavin Plastics							
Ltd; OsmaDrain							
110 mm PVC-u pipes; in trenches (trenches							
not included)							
laid straight	7.34	0.08	1.46	-	7.34	m	8.80
short runs under 3.00 m	7.34	0.12	2.19	-	8.25	m	10.44
Extra over 110 mm PVC-u pipes for							
bends; short radius	14.31	0.25	4.56	-	14.31	nr	18.87
bends; long radius	26.82	0.25	4.56	-	26.82	nr	31.38
junctions; equal; double socket	17.07	0.25	4.56	-	17.07	nr	21.63
slip couplings	8.25	0.25	4.56	-	8.25	nr	12.81
adaptors to clay	16.13	0.50	9.13	-	16.13	nr	25.25
160 mm PVC-u pipes; in trenches (trenches							
not included)							
laid straight	17.24	0.08	1.46	-	17.24	m	18.70
short runs under 3.00 m	30.64	0.12	2.19	-	32.75	m	34.94

R DISPOSAL SYSTEMS

Item Excluding site overheads and profit	PC £	Labour hours	Labour £	Plant £	Material £	Unit	Total rate £
R12 DRAINAGE BELOW GROUND - cont'd							
PVC-u pipes and fittings - cont'd							
Extra over 160 mm PVC-u pipes for							
socket bend; double; 90 or 45°	54.13	0.20	3.65	-	54.13	nr	**57.78**
socket bend; double; 15 or 30°	50.60	0.20	3.65	-	50.60	nr	**54.25**
socket bend; single; 87.5 or 45°	30.85	0.20	3.65	-	30.85	nr	**34.50**
socket bend; single; 15 or 30°	27.35	0.20	3.65	-	27.35	nr	**31.00**
bends; short radius	40.20	0.25	4.56	-	54.51	nr	**59.07**
bends; long radius	104.84	0.25	4.56	-	104.84	nr	**109.40**
junctions; single	99.09	0.33	6.08	-	99.09	nr	**105.17**
pipe coupler	20.10	0.05	0.91	-	20.10	nr	**21.01**
slip couplings PVC-u	10.02	0.05	0.91	-	10.02	nr	**10.93**
adaptors to clay	43.82	0.50	9.13	-	43.82	nr	**52.95**
level invert reducer	16.50	0.50	9.13	-	16.50	nr	**25.63**
spiggot	36.60	0.20	3.65	-	36.60	nr	**40.25**
Accessories in PVC-u							
110 mm screwed access cover	15.11	-	-	-	15.11	nr	**-**
110 mm rodding eye	29.25	0.50	9.13	-	33.30	nr	**42.43**
gully with "P" traps; 110 mm; 154 x 154 mm grating	58.58	1.00	18.25	-	60.20	nr	**78.45**
Kerbs; to gullies; in one course Class B engineering bricks; to four sides; rendering in cement:mortar (1:3); dished to gully gratings	1.91	1.00	18.25	-	2.71	nr	**20.96**
Gullies; concrete; FP McCann							
Concrete road gullies; trapped with rodding eye and stoppers; 450 mm diameter x 1.07 m deep	46.00	6.00	109.50	-	68.40	nr	**177.90**
Gullies; vitrified clay; Hepworth Plc; bedding in concrete; 11.50 N/mm² - 40 mm aggregate							
Vitrified clay yard gullies (mud); trapped; domestic duty (up to 1 tonne)							
RGP5; 100 mm outlet; 100 mm diameter; 225 mm internal width; 585 mm internal depth	96.92	3.50	63.88	-	97.48	nr	**161.35**
RGP7; 150 mm outlet; 100 mm diameter; 225 mm internal width; 585 mm internal depth	96.92	3.50	63.88	-	97.48	nr	**161.35**
Vitrified clay yard gullies (mud); trapped; medium duty (up to 5 tonnes)							
100 mm outlet; 100 mm diameter; 225 mm internal width; 585 mm internal depth	137.01	3.50	63.88	-	137.57	nr	**201.45**
150 mm outlet; 100 mm diameter; 225 mm internal width; 585 mm internal depth	150.12	3.50	63.88	-	150.69	nr	**214.56**
Combined filter and silt bucket for yard gullies							
225 mm wide	35.04	-	-	-	35.04	nr	**-**

R DISPOSAL SYSTEMS

Item Excluding site overheads and profit	PC £	Labour hours	Labour £	Plant £	Material £	Unit	Total rate £
Vitrified clay road gullies; trapped with rodding eye							
100 mm outlet; 300 mm internal diameter; 600 mm internal depth	93.22	3.50	63.88	-	93.78	nr	**157.66**
150 mm outlet; 300 mm internal diameter; 600 mm internal depth	95.46	3.50	63.88	-	96.02	nr	**159.90**
150 mm outlet; 400 mm internal diameter; 750 mm internal depth	110.70	3.50	63.88	-	111.27	nr	**175.14**
150 mm outlet; 450 mm internal diameter; 900 mm internal depth	149.79	3.50	63.88	-	150.35	nr	**214.22**
Hinged gratings and frames for gullies; alloy							
193 mm for 150 mm diameter gully	-	-	-	-	23.22	nr	-
120 x 120 mm	-	-	-	-	8.25	nr	-
150 x 150 mm	-	-	-	-	14.94	nr	-
230 x 230 mm	-	-	-	-	27.31	nr	-
316 x 316 mm	-	-	-	-	72.42	nr	-
Hinged gratings and frames for gullies; cast iron							
265 mm for 225 mm diameter gully	-	-	-	-	46.40	nr	-
150 x 150 mm	-	-	-	-	14.94	nr	-
230 x 230 mm	-	-	-	-	27.31	nr	-
316 x 316 mm	-	-	-	-	72.42	nr	-
Universal gully trap PVC-u; Wavin Plastics Ltd; OsmaDrain system; bedding in concrete; 11.50 N/mm² - 40 mm aggregate							
Universal gully fitting; comprising gully trap only							
110 mm outlet; 110 mm diameter; 205 mm internal depth	12.14	3.50	63.88	-	12.70	nr	**76.58**
Vertical inlet hopper c\w plastic grate							
272 x 183 mm	16.31	0.25	4.56	-	16.31	nr	**20.87**
Sealed access hopper							
110 x 110 mm	38.35	0.25	4.56	-	38.35	nr	**42.91**
Universal gully PVC-u; Wavin Plastics Ltd; OsmaDrain system; accessories to universal gully trap							
Hoppers; backfilling with clean granular material; tamping; surrounding in lean mix concrete							
plain hopper; with 110 mm spigot; 150 mm long	13.33	0.40	7.30	-	13.67	nr	**20.97**
vertical inlet hopper; with 110 mm spigot; 150 mm long	16.31	0.40	7.30	-	16.31	nr	**23.61**
sealed access hopper; with 110 mm spigot; 150 mm long	38.35	0.40	7.30	-	38.35	nr	**45.65**
plain hopper; solvent weld to trap	9.10	0.40	7.30	-	9.10	nr	**16.40**
vertical inlet hopper; solvent wed to trap	15.64	0.40	7.30	-	15.64	nr	**22.94**
sealed access cover; PVC-u	19.53	0.10	1.82	-	19.53	nr	**21.36**

R DISPOSAL SYSTEMS

Item Excluding site overheads and profit	PC £	Labour hours	Labour £	Plant £	Material £	Unit	Total rate £
R12 DRAINAGE BELOW GROUND - cont'd							
Gullies PVC-u; Wavin Plastics Ltd; OsmaDrain system; bedding in concrete; 11.50 N/mm^2 - 40 mm aggregate							
Bottle gully; providing access to the drainage system for cleaning							
bottle gully; 228 x 228 x 317 mm deep	32.38	0.50	9.13	-	32.87	nr	**41.99**
sealed access cover; PVC-u 217 x 217 mm	25.17	0.10	1.82	-	25.17	nr	**27.00**
grating; ductile iron; 215 x 215 mm	19.40	0.10	1.82	-	19.40	nr	**21.22**
bottle gully riser; 325 mm	4.17	0.50	9.13	-	5.22	nr	**14.35**
Yard gully; trapped; 300 mm diameter x 600 mm deep; including catchment bucket and ductile iron cover and frame; medium duty loading							
305 mm diameter x 600 mm deep	196.34	2.50	45.63	-	199.37	nr	**244.99**
Kerbs to gullies							
One course Class B engineering bricks to four sides; rendering in cement:mortar (1:3); dished to gully gratings							
150 x 150 mm	0.96	0.33	6.08	-	1.49	nr	**7.57**
Linear drainage							
Marshalls Plc "Mini Beany" combined kerb and channel drainage system; to trenches (not included)							
Precast concrete drainage channel base; 185 - 385 mm deep; bedding, jointing and pointing in cement mortar (1:3); on 150 mm deep concrete (ready mixed) foundation; including haunching with in situ concrete; 11.50 N/mm^2 - 40 mm aggregate one side; channels 250 mm wide x 1.00 m long							
straight; 1.00 m long	20.75	1.00	18.25	-	24.28	m	**42.53**
straight; 500 mm long	22.17	1.05	19.16	-	25.70	m	**44.86**
radial; 30 - 10 m or 9 - 6 m internal or external	20.48	1.33	24.33	-	24.01	m	**48.34**
angles 45 or 90°	61.11	1.00	18.25	-	64.64	nr	**82.89**
Mini Beany Top Block; perforated kerb unit to drainage channel above; natural grey							
straight	11.00	0.33	6.08	-	12.22	m	**18.30**
radial; 30 - 10 m or 9 - 6 m internal or external	14.65	0.50	9.13	-	15.87	m	**24.99**
angles 45 or 90°	34.12	0.50	9.13	-	35.34	nr	**44.46**
Mini Beany; outfalls; two section concrete trapped outfall with Mini Beany cast iron access cover and frame; to concrete foundation							
high capacity outfalls; silt box 150/225 mm outlet; two section trapped outfall silt box and cast iron access cover	41.40	1.00	18.25	-	256.06	nr	**274.31**
inline side or end outlet; outfall 150 mm; 2 section concrete trapped outfall; cast iron Mini Beany access cover and frame	221.64	1.00	18.25	-	222.23	nr	**240.48**

R DISPOSAL SYSTEMS

Item Excluding site overheads and profit	PC £	Labour hours	Labour £	Plant £	Material £	Unit	Total rate £
Ancillaries to Mini Beany							
end cap	12.41	0.25	4.56	-	12.41	nr	**16.97**
end cap outlets	32.25	0.25	4.56	-	32.25	nr	**36.81**
Precast concrete channels; Charcon Hard Landscaping; on 150 mm deep concrete foundations; including haunching with in situ concrete; 21.00 N/mm^2 - 20 mm aggregate; both sides							
"Charcon Safeticurb"; slotted safety channels; for pedestrians and light vehicles							
ref DBJ; 305 x 305 mm	69.58	0.67	12.17	-	85.38	m	**97.55**
ref DBA; 250 x 250 mm	35.82	0.67	12.17	-	51.62	m	**63.79**
"Charcon Safeticurb"; slotted safety channels; for heavy vehicles							
ref DBM; 248 x 248 mm	63.78	0.80	14.60	-	79.58	m	**94.18**
ref Clearway; 324 x 257 mm	73.15	0.80	14.60	-	88.95	m	**103.55**
"Charcon Safeticurb"; inspection units; ductile iron lids; including jointing to drainage channels							
248 x 248 x 914 mm	86.79	1.50	27.38	-	87.72	nr	**115.09**
Silt box tops; concrete frame; cast iron grid lids; type 1; set over gully							
457 x 610 mm	389.03	2.00	36.50	-	389.96	nr	**426.46**
Manhole covers; type K; cast iron; providing inspection to blocks and back gullies	494.27	2.00	36.50	-	496.37	nr	**532.87**
Linear drainage							
Loadings for slot drains A15 - 1.5 tonne - pedestrian B125 - 12.5 tonne - domestic use C250 - 25 tonne - car parks, supermarkets, industrial units D400 - 40 tonne - highways E600 - 60 tonne - forklifts							
Slot drains; ACO Technologies; laid to concrete bed C25 on compacted granular base on 200 mm deep concrete bed; haunched with 200 mm concrete surround; all in 750 x 430 mm wide trench with compacted 200 mm granular base surround (excavation and sub-base not included)							
ACO MultiDrain M100PPD; recycled polypropylene drainage channel; range of gratings to complement installations which require discreet slot drainage							
142 mm wide x 150 mm deep	43.43	1.20	21.90	-	53.54	m	**75.44**
Accessories for M100PPD							
connectors; vertical outlet 110 mm	9.98	-	-	-	9.98	nr	**-**
connectors; vertical outlet 160 mm	9.98	-	-	-	9.98	nr	**-**
sump units; 110 mm	78.94	1.50	27.38	-	97.53	nr	**124.91**
universal gully and bucket; 440 x 440 x 1315 mm deep	379.35	3.00	54.75	-	414.21	nr	**468.96**

R DISPOSAL SYSTEMS

Item Excluding site overheads and profit	PC £	Labour hours	Labour £	Plant £	Material £	Unit	Total rate £
R12 DRAINAGE BELOW GROUND - cont'd							
Channel drains; ACO Technologies; ACO MultiDrain MD polymer concrete channel drainage system; traditional channel and grate drainage solution							
ACO MultiDrain MD Brickslot; offset galvanized slot drain grating for M100PPD; load class C250							
brickslot; galvanized steel; 1.00 mm	62.80	-	-	-	62.80	m	-
ACO MultiDrain MD Brickslot; offset slot drain grating; load class C250 - 400							
brickslot; galvanized steel; 1000 mm	106.23	-	-	-	106.23	m	-
brickslot; stainless steel; 1000 mm	125.95	-	-	-	251.90	m	-
Accessories for MultiDrain MD							
sump unit; complete with sediment bucket and access unit	78.94	1.00	18.25	-	180.39	nr	**198.64**
end cap; closing piece	8.50	-	-	-	8.50	nr	-
end cap; inlet/outlet	17.20	-	-	-	17.20	nr	-
ACO Drainlock Gratings for M100PPD and M100D system							
A15 loading							
slotted galvanized steel	12.75	0.15	2.74	-	12.75	m	**15.49**
perforated galvanized steel	23.40	-	-	-	23.40	m	-
C250 loading							
Heelguard composite black; 500 mm long with security locking	31.80	-	-	-	31.80	m	-
Intercept; ductile iron; 500 mm long	70.06	-	-	-	70.06	m	-
slotted galvanized steel; 1.00 mm long	33.00	-	-	-	33.00	m	-
perforated galvanized steel; 1.00 mm long	48.50	-	-	-	48.50	m	-
mesh galvanized steel; 1.00 mm long	32.95	-	-	-	32.95	m	-
Wade Ltd; stainless steel channel drains; specialized applications; bespoke manufacture							
Drain in stainless steel; c/w tie in lugs and inbuilt falls to 100 mm spigot outlet; stainless steel gratings							
"NE" channel; ref 12430; secured gratings	380.00	1.20	21.90	-	390.11	m	**412.01**
Fin drains; Cooper Clarke Civils and Lintels; Geofin Shallow Drain; drainage of sports fields and grassed landscaped areas							
Geofin Shallow Drain; to trenches; excavation by trenching machine; backfilled with single size aggregate 20 mm and covered with sharp sand rootzone 200 mm thick							
Geofin; 25 mm thick x 150 mm deep	4.42	0.05	0.91	0.71	7.66	m	**9.28**
Geofin; 25 mm thick x 450 mm deep	3.27	0.07	1.22	0.95	10.04	m	**12.20**
Geofin; 25 mm thick x 900 mm deep	7.18	0.10	1.82	0.95	20.53	m	**23.30**

R DISPOSAL SYSTEMS

Item Excluding site overheads and profit	PC £	Labour hours	Labour £	Plant £	Material £	Unit	Total rate £
Fin drains; Cooper Clarke Civils and Lintels; Geofin Geocomposite Finn Drain; laid to slabs							
Geofin fin drain laid horizontally to slab or blinded ground; covered with 20 mm shingle 200 mm thick							
Geofin; 25 mm thick x 900 mm wide; laid flat to falls	7.97	0.02	0.36	0.14	15.23	m²	15.74
Maxit Ltd; Leca (light expanded clay aggregate); drainage aggregate to roofdecks and planters							
Placed mechanically to planters; average 100 mm thick; by mechanical plant tipped into planters							
aggregate size 10 - 20 mm; delivered in 30 m³ loads	42.45	0.20	3.65	5.20	42.45	m³	51.30
aggregate size 10 - 20 mm; delivered in 70 m³ loads	37.19	0.20	3.65	5.20	37.19	m³	46.04
Placed by light aggregate blower (maximum 40 m)							
aggregate size 10 - 20 mm; delivered in 30 m³ loads	42.45	0.14	2.61	-	43.45	m³	46.06
aggregate size 10 - 20 mm; delivered in 70 m³ loads	37.19	0.14	2.61	-	38.19	m³	40.80
By hand							
aggregate size 10 - 20 mm; delivered in 30 m³ loads	42.45	1.33	24.33	-	42.45	m³	66.78
aggregate size 10 - 20 mm; delivered in 70 m³ loads	37.19	1.33	24.33	-	37.19	m³	61.52
Drainage boards laid to insulated slabs on roof decks; boards laid below growing medium and granulated drainage layer and geofabric (all not included) to collect and channel water to drainage outlets (not included)							
Alumasc Floradrain; polyethylene irrigation/ drainage layer; inclusive of geofabric laid over the surface of the drainage board							
Floradrain FD40; 0.96 x 2.08 m panels	10.66	0.05	0.91	-	11.68	m²	12.59
Floradrain FD60; 1.00 x 2.00 m panel	18.96	0.07	1.22	-	19.98	m²	21.20

R DISPOSAL SYSTEMS

Item Excluding site overheads and profit	PC £	Labour hours	Labour £	Plant £	Material £	Unit	Total rate £
R12 DRAINAGE BELOW GROUND - cont'd							
Access covers and frames FACTA (Fabricated Access Cover Trade Association) class: A - 0.5 tonne maximum slow moving wheel load AA - 1.5 tonne maximum slow moving wheel load AAA - 2.5 tonne maximum slow moving wheel load B - 5 tonne maximum slow moving wheel load C - 6.5 tonne maximum slow moving wheel load D - 11 tonne maximum slow moving wheel load							
Access covers and frames; solid top; galvanized; Steelway Brickhouse; "Bristeel"; bedding frame in cement mortar (1:3); cover in grease and sand; clear opening sizes; base size shown in brackets (50 mm depth)							
FACTA AA; single seal							
450 x 450 mm (520 x 520 mm)	69.71	1.50	27.38	-	78.15	nr	**105.53**
600 x 450 mm (670 x 520 mm)	78.38	1.80	32.85	-	87.76	nr	**120.61**
600 x 600 mm (670 x 670 mm)	84.48	2.00	36.50	-	95.73	nr	**132.23**
FACTA AA; double seal							
450 x 450 mm (560 x 560 mm)	119.99	1.50	27.38	-	128.43	nr	**155.80**
600 x 450 mm (710 x 560 mm)	132.96	1.80	32.85	-	142.33	nr	**175.18**
600 x 600 mm (710 x 710 mm)	146.60	2.00	36.50	-	157.85	nr	**194.35**
FACTA AAA; single seal							
450 x 450 mm (520 x 520 mm)	99.97	1.50	27.38	-	108.41	nr	**135.78**
600 x 450 mm (670 x 520 mm)	108.68	1.80	32.85	-	118.06	nr	**150.91**
600 x 600 mm (670 x 670 mm)	116.45	2.00	36.50	-	127.70	nr	**164.20**
FACTA AAA; double seal							
450 mm x 450 mm (560 x 560)	147.17	1.50	27.38	-	155.61	nr	**182.98**
600 mm x 450 mm (710 x 560)	161.82	1.80	32.85	-	171.19	nr	**204.04**
600 mm x 600 mm (710 x 710)	181.60	2.00	36.50	-	192.85	nr	**229.35**
FACTA B; single seal							
450 x 450 mm (520 x 520 mm)	87.30	1.50	27.38	-	95.74	nr	**123.11**
600 x 450 mm (670 x 520 mm)	105.24	1.80	32.85	-	114.61	nr	**147.46**
600 x 600 mm (670 x 670 mm)	123.95	2.00	36.50	-	135.20	nr	**171.70**
FACTA B; double seal							
450 x 450 mm (560 x 560 mm)	139.46	1.50	27.38	-	147.89	nr	**175.27**
600 x 450 mm (710 x 560 mm)	156.54	1.80	32.85	-	165.91	nr	**198.76**
600 x 600 mm (710 x 710 mm)	187.24	2.00	36.50	-	198.49	nr	**234.99**
Access covers and frames; recessed; galvanized; Steelway Brickhouse; "Bripave"; bedding frame in cement mortar (1:3); cover in grease and sand; clear opening sizes; base size shown in brackets (for block depths 50 mm, 65 mm, 80 mm or 100 mm)							
FACTA AA							
450 x 450 mm (562 x 562 mm)	188.99	1.50	27.38	-	197.43	nr	**224.80**
600 x 450 mm (712 x 562 mm)	211.69	1.80	32.85	-	221.06	nr	**253.91**
600 x 600 mm (712 x 712 mm)	221.09	2.00	36.50	-	231.88	nr	**268.38**
750 x 600 mm (862 x 712 mm)	251.95	2.40	43.80	-	264.13	nr	**307.93**

R DISPOSAL SYSTEMS

Item Excluding site overheads and profit	PC £	Labour hours	Labour £	Plant £	Material £	Unit	Total rate £
FACTA B							
450 x 450 mm (562 x 562 mm)	189.40	1.50	27.38	-	197.83	nr	225.21
600 x 450 mm (712 x 562 mm)	227.54	1.80	32.85	-	236.91	nr	269.76
600 x 600 mm (712 x 712 mm)	231.83	2.00	36.50	-	243.08	nr	279.58
750 x 600 mm (862 x 712 mm)	280.17	2.40	43.80	-	292.36	nr	336.16
FACTA D							
450 x 450 mm (562 x 562 mm)	217.75	1.50	27.38	-	226.18	nr	253.56
600 x 450 mm (712 x 562 mm)	234.17	1.80	32.85	-	243.55	nr	276.40
600 x 600 mm (712 x 712 mm)	243.44	2.00	36.50	-	254.69	nr	291.19
750 x 600 mm (862 x 712 mm)	306.10	2.40	43.80	-	318.29	nr	362.09
Access covers and frames; Jones of Oswestry; bedding frame in cement mortar (1:3); cover in grease and sand; clear opening sizes							
Access covers and frames; "Suprabloc"; to paved areas; filling with blocks cut and fitted to match surrounding paving							
pedestrian weight; 300 x 300 mm	121.47	2.50	45.63	-	127.11	nr	172.73
pedestrian weight; 450 x 450 mm	207.27	2.50	45.63	-	207.27	nr	252.90
pedestrian weight; 450 x 600 mm	221.01	2.50	45.63	-	226.88	nr	272.50
light vehicular weight; 300 x 300 mm	155.63	2.40	43.80	-	161.27	nr	205.07
light vehicular weight; 450 x 450 mm	119.98	3.00	54.75	-	126.44	nr	181.19
light vehicular weight; 450 x 600 mm	233.57	4.00	73.00	-	239.21	nr	312.21
heavy vehicular weight; 300 x 300 mm	155.97	3.00	54.75	-	161.84	nr	216.59
heavy vehicular weight; 450 x 600 mm	271.60	3.00	54.75	-	278.06	nr	332.81
heavy vehicular weight; 600 x 600 mm	290.00	4.00	73.00	-	299.39	nr	372.39
Extra over manhole frames and covers for							
filling recessed manhole covers with brick paviors; PC £305.00/1000	12.20	1.00	18.25	-	13.95	m²	32.20
filling recessed manhole covers with vehicular paving blocks; PC £7.63/m²	7.63	0.75	13.69	-	8.15	m²	21.84
filling recessed manhole covers with concrete paving flags; PC £12.54/m²	12.54	0.35	6.39	-	12.86	m²	19.25
R13 LAND DRAINAGE							
Market prices of backfilling materials							
Sand	14.40	-	-	-	17.28	m³	-
Gravel rejects	35.82	-	-	-	39.40	m³	-
Topsoil; allowing for 20% settlement	26.00	-	-	-	31.20	m³	-
Soakaways							
Excavating mechanical; to reduce levels							
maximum depth not exceeding 1.00 m; JCB sitemaster	-	0.05	0.91	1.65	-	m³	2.56
maximum depth not exceeding 1.00 m; 360 tracked excavator	-	0.04	0.73	1.65	-	m³	2.38
maximum depth not exceeding 2.00 m; 360 tracked excavator	-	0.06	1.09	2.48	-	m³	3.57
Disposal							
Excavated material; off site; to tip; mechanically loaded (JCB)							
inert	-	0.04	0.76	1.25	13.86	m³	15.87

R DISPOSAL SYSTEMS

Item Excluding site overheads and profit	PC £	Labour hours	Labour £	Plant £	Material £	Unit	Total rate £
R13 LAND DRAINAGE - cont'd							
In situ concrete ring beam foundations to base of soakaway; 300 mm wide x 250 mm deep; poured on or against earth or unblinded hardcore							
Internal diameters of rings							
900 mm	17.18	4.00	73.00	-	18.90	nr	91.90
1200 mm	22.90	4.50	82.13	-	25.20	nr	107.32
1500 mm	28.63	5.00	91.25	-	31.49	nr	122.74
2400 mm	45.82	5.50	100.38	-	50.39	nr	150.77
Concrete soakaway rings; Milton Pipes Ltd; perforations and step irons to concrete rings at manufacturers recommended centres; placing of concrete ring soakaways to in situ concrete ring beams (1:3:6) (not included); filling and surrounding base with gravel 225 mm deep (not included)							
Ring diameter 900 mm							
1.00 m deep; volume 636 litres	60.00	1.25	22.81	18.05	156.77	nr	197.63
1.50 m deep; volume 954 litres	90.00	1.65	30.11	47.64	232.27	nr	310.03
2.00 m deep; volume 1272 litres	120.00	1.65	30.11	47.64	307.77	nr	385.53
Ring diameter 1200 mm							
1.00 m deep; volume 1131 litres	75.00	3.00	54.75	43.31	187.25	nr	285.31
1.50 m deep; volume 1696 litres	112.50	4.50	82.13	64.97	275.75	nr	422.84
2.00 m deep; volume 2261 litres	150.00	4.50	82.13	64.97	364.25	nr	511.34
2.50 m deep; volume 2827 litres	187.50	6.00	109.50	86.63	414.25	nr	610.37
Ring diameter 1500 mm							
1.00 m deep; volume 1767 litres	125.00	4.50	82.13	43.31	265.02	nr	390.46
1.50 m deep; volume 2651 litres	187.50	6.00	109.50	60.64	389.52	nr	559.66
2.00 m deep; volume 3534 litres	250.00	6.00	109.50	60.64	514.02	nr	684.16
2.50 m deep; volume 4418 litres	187.50	7.50	136.88	108.28	513.52	nr	758.68
Ring diameter 2400 mm							
1.00 m deep; volume 4524 litres	465.00	6.00	109.50	43.31	654.52	nr	807.34
1.50 m deep; volume 6786 litres	697.50	7.50	136.88	60.64	971.02	nr	1168.54
2.00 m deep; volume 9048 litres	930.00	7.50	136.88	60.64	1293.02	nr	1490.54
2.50 m deep; volume 11310 litres	1162.50	9.00	164.25	108.28	1609.52	nr	1882.06
Extra over for							
250 mm depth chamber ring	-	-	-	-	-	100%	-
500 mm depth chamber ring	-	-	-	-	-	50%	-
Cover slabs to soakaways							
Heavy duty precast concrete							
900 mm diameter	95.00	1.00	18.25	21.66	95.00	nr	134.91
1200 mm diameter	115.00	1.00	18.25	21.66	115.00	nr	154.91
1500 mm diameter	175.00	1.00	18.25	21.66	175.00	nr	214.91
2400 mm diameter	670.00	1.00	18.25	21.66	670.00	nr	709.91
Step irons to concrete chamber rings	25.00	-	-	-	25.00	m	-

R DISPOSAL SYSTEMS

Item Excluding site overheads and profit	PC £	Labour hours	Labour £	Plant £	Material £	Unit	Total rate £
Extra over soakaways for filter wrapping with a proprietary filter membrane							
900 mm diameter x 1.00 m deep	1.19	1.00	18.25	-	1.49	nr	**19.74**
900 mm diameter x 2.00 m deep	2.39	1.50	27.38	-	2.99	nr	**30.36**
1050 mm diameter x 1.00 m deep	1.39	1.50	27.38	-	1.74	nr	**29.12**
1050 mm diameter x 2.00 m deep	2.79	2.00	36.50	-	3.48	nr	**39.98**
1200 mm diameter x 1.00 m deep	1.59	2.00	36.50	-	1.99	nr	**38.49**
1200 mm diameter x 2.00 m deep	3.18	2.50	45.63	-	3.98	nr	**49.60**
1500 mm diameter x 1.00 m deep	1.99	2.50	45.63	-	2.49	nr	**48.11**
1500 mm diameter x 2.00 m deep	3.97	2.50	45.63	-	4.96	nr	**50.59**
1800 mm diameter x 1.00 m deep	2.39	3.00	54.75	-	2.98	nr	**57.73**
1800 mm diameter x 2.00 m deep	4.77	3.25	59.31	-	5.96	nr	**65.28**
Gravel surrounding to concrete ring soakaway							
40 mm aggregate backfilled to vertical face of soakaway wrapped with geofabric (not included)							
250 mm thick	36.27	0.20	3.65	8.66	36.27	m³	**48.58**
Backfilling to face of soakaway; carefully compacting as work proceeds							
Arising from the excavations							
average thickness exceeding 0.25 m; depositing in layers 150 mm maximum thickness	-	0.03	0.61	4.33	-	m³	**4.94**
"Aquacell" soakaway; Wavin Plastics Ltd; preformed polypropylene soakaway infiltration crate units; to trenches; surrounded by geotextile and 40 mm aggregate laid 100 mm thick in trenches (excavation, disposal and backfilling not included)							
1.00 m x 500 x 400 mm; internal volume 190 litres							
4 crates; 2.00 x 1.00 m x 400 mm; 760 litres	173.64	1.60	29.20	13.86	222.95	nr	**266.01**
8 crates; 2.00 x 1.00 m x 800 mm; 1520 litres	347.28	3.20	58.40	17.82	408.25	nr	**484.47**
12 crates; 6.00 m x 500 x 800 mm; 2280 litres	520.92	4.80	87.60	36.96	638.25	nr	**762.81**
16 crates; 4.00 x 1.00 m x 800 mm; 3040 litres	694.56	6.40	116.80	33.00	800.23	nr	**950.03**
20 crates; 5.00 x 1.00 m x 800 mm; 3800 litres	868.20	8.00	146.00	32.67	981.39	nr	**1160.06**
30 crates; 15.00 x 1.00 m x 400 mm; 5700 litres	1302.30	12.00	219.00	89.10	1581.67	nr	**1889.77**
60 crates; 15.00 x 1.00 m x 800 mm; 11400 litres	2604.60	20.00	365.00	116.49	2981.61	nr	**3463.10**
Geofabric surround to Aquacell units; Terram Ltd							
"Terram" synthetic fibre filter fabric; to face of concrete rings (not included); anchoring whilst backfilling (not included)							
"Terram 1000"; 0.70 mm thick; mean water flow 50 l/m²/s	0.42	0.05	0.91	-	0.51	m²	**1.42**

R DISPOSAL SYSTEMS

Item Excluding site overheads and profit	PC £	Labour hours	Labour £	Plant £	Material £	Unit	Total rate £
R13 LAND DRAINAGE - cont'd							
Sand slitting; Agripower Ltd Drainage slits; at 1.00 m centres; using spinning disc trenching machine; backfilling to 100 mm of surface with pea gravel; finished surface with medim grade sand 100; arisings to be loaded, hauled and tipped onsite; minimum pitch size 6000 m^2							
250 mm depth	-	-	-	-	-	m	**1.80**
300 mm depth	-	-	-	-	-	m	**1.97**
400 mm depth	-	-	-	-	-	m	**3.01**
Sand grooving or banding; Agripower Ltd Drainage bands at 260 mm m centres; using Blec Sandmaster; to existing grass at 260 mm centres; minimum pitch size 4000 m^2							
sand banding or sand grooving	-	-	-	-	-	m^2	**1.17**
Land drainage; calculation table For calculation of drainage per hectare the following table can be used; rates show the lengths of drains per unit and not the value Lateral drains							
10.00 m centres	-	-	-	-	-	m/ha	**1000.00**
15.00 m centres	-	-	-	-	-	m/ha	**650.00**
25.00 m centres	-	-	-	-	-	m/ha	**400.00**
30.00 m centres	-	-	-	-	-	m/ha	**330.00**
Main drains							
1 nr (at 100 m centres)	-	-	-	-	-	m/ha	**100.00**
2 nr (at 50 m centres)	-	-	-	-	-	m/ha	**200.00**
3 nr (at 33.333 m centres)	-	-	-	-	-	m/ha	**300.00**
4 nr (at 25 m centres)	-	-	-	-	-	m/ha	**400.00**
Land drainage; excavating Removing 150 mm depth of topsoil; 300 mm wide; depositing beside trench; by machine	-	1.50	27.38	12.94	-	100 m	**40.31**
Removing 150 mm depth of topsoil; 300 mm wide; depositing beside trench; by hand	-	8.00	146.00	-	-	100 m	**146.00**
Disposing on site; to spoil heaps; by machine							
not exceeding 100 m distance	-	0.07	1.20	2.27	-	m^3	**3.48**
average 100 - 150 m distance	-	0.08	1.45	2.73	-	m^3	**4.17**
average 150 - 200 m distance	-	0.09	1.69	3.19	-	m^3	**4.88**
Removing excavated material from site to tip not exceeding 13 km; mechanically loaded excavated material and clean hardcore rubble	-	-	-	1.10	7.00	m^3	**8.10**

R DISPOSAL SYSTEMS

Item Excluding site overheads and profit	PC £	Labour hours	Labour £	Plant £	Material £	Unit	Total rate £
Land drainage; Agripower Ltd; excavating trenches (with minimum project size of 1000 m) by trenching machine; spreading arisings on site; laying perforated pipe; as shown below; backfilling with shingle to 350 mm from surface							
Width 150 mm; 80 mm perforated pipe							
depth 750 mm	-	-	-	-	-	m	**5.50**
Width 175 mm; 100 mm perforated pipe							
depth 800 mm	-	-	-	-	-	m	**6.65**
Width 250 mm; 150 mm perforated pipe							
depth 900 mm	-	-	-	-	-	m	**9.70**
Sports or amenity drainage; Agripower Ltd Ltd; excavating trenches (with minimum project size of 1000 m) by trenching machine; arisings to spoil heap on site maximum 100 m; backfill with shingle to within 125 mm from surface and with rootzone to ground level							
Width 145 mm; 80 mm perforated pipe							
depth 550 mm	-	-	-	-	-	m	**7.94**
Width 145 mm; 100 mm perforated pipe							
depth 650 mm	-	-	-	-	-	m	**9.12**
Width 145 mm; 150 mm perforated pipe							
depth 700 mm	-	-	-	-	-	100 m	**11.75**
Land drainage; excavating for drains; by backacter excavator JCB C3X Sitemaster; including disposing spoil to spoil heaps not exceeding 100 m; boning to levels by laser							
Width 150 - 225 mm							
depth 450 mm	-	9.67	176.42	252.50	-	100 m	**428.92**
depth 600 mm	-	13.00	237.25	378.75	-	100 m	**616.00**
depth 700 mm	-	15.50	282.88	473.44	-	100 m	**756.31**
depth 900 mm	-	23.00	419.75	757.50	-	100 m	**1177.25**
Width 300 mm							
depth 450 mm	-	9.67	176.42	252.50	-	100 m	**428.92**
depth 700 mm	-	14.00	255.50	416.63	-	100 m	**672.13**
depth 900 mm	-	23.00	419.75	757.50	-	100 m	**1177.25**
depth 1000 mm	-	25.00	456.25	833.25	-	100 m	**1289.50**
depth 1200 mm	-	25.00	456.25	946.88	-	100 m	**1403.13**
depth 1500 mm	-	31.00	565.75	1060.50	-	100 m	**1626.25**
Land drainage; excavating for drains; by 7 tonne tracked excavator; including disposing spoil to spoil heaps not exceeding 100 m							
Width 225 mm							
depth 450 mm	-	7.00	127.75	71.75	-	100 m	**199.50**
depth 600 mm	-	7.17	130.79	74.74	-	100 m	**205.53**
depth 700 mm	-	7.35	134.10	77.99	-	100 m	**212.09**
depth 900 mm	-	7.76	141.65	85.42	-	100 m	**227.07**
depth 1000 mm	-	8.00	146.00	89.69	-	100 m	**235.69**
depth 1200 mm	-	8.26	150.80	94.41	-	100 m	**245.21**

R DISPOSAL SYSTEMS

Item Excluding site overheads and profit	PC £	Labour hours	Labour £	Plant £	Material £	Unit	Total rate £
R13 LAND DRAINAGE - cont'd							
Land drainage - cont'd							
Width 300 mm							
depth 450 mm	-	7.35	134.10	77.99	-	100 m	**212.09**
depth 600 mm	-	7.76	141.65	85.42	-	100 m	**227.07**
depth 700 mm	-	8.26	150.80	94.41	-	100 m	**245.21**
depth 900 mm	-	9.25	168.81	112.11	-	100 m	**280.92**
depth 1000 mm	-	10.14	185.11	128.13	-	100 m	**313.23**
depth 1200 mm	-	11.00	200.75	143.50	-	100 m	**344.25**
Width 600 mm							
depth 450 mm	-	13.00	237.25	179.38	-	100 m	**416.63**
depth 600 mm	-	14.11	257.53	199.31	-	100 m	**456.83**
depth 700 mm	-	16.33	298.08	239.17	-	100 m	**537.25**
depth 900 mm	-	17.29	315.46	256.25	-	100 m	**571.71**
depth 1000 mm	-	18.38	335.52	275.96	-	100 m	**611.48**
depth 1200 mm	-	21.18	386.57	326.14	-	100 m	**712.70**
Land drainage; excavating for drains; by hand, including disposing spoil to spoil heaps not exceeding 100 m							
Width 150 mm							
depth 450 mm	-	22.03	402.05	-	-	100 m	**402.05**
depth 600 mm	-	29.38	536.18	-	-	100 m	**536.18**
depth 700 mm	-	34.27	625.43	-	-	100 m	**625.43**
depth 900 mm	-	44.06	804.10	-	-	100 m	**804.10**
Width 225 mm							
depth 450 mm	-	33.05	603.16	-	-	100 m	**603.16**
depth 600 mm	-	44.06	804.10	-	-	100 m	**804.10**
depth 700 mm	-	51.41	938.23	-	-	100 m	**938.23**
depth 900 mm	-	66.10	1206.33	-	-	100 m	**1206.33**
depth 1000 mm	-	73.44	1340.28	-	-	100 m	**1340.28**
Width 300 mm							
depth 450 mm	-	44.06	804.10	-	-	100 m	**804.10**
depth 600 mm	-	58.75	1072.19	-	-	100 m	**1072.19**
depth 700 mm	-	68.54	1250.86	-	-	100 m	**1250.86**
depth 900 mm	-	88.13	1608.37	-	-	100 m	**1608.37**
depth 1000 mm	-	97.92	1787.04	-	-	100 m	**1787.04**
Width 375 mm							
depth 450 mm	-	55.08	1005.21	-	-	100 m	**1005.21**
depth 600 mm	-	73.44	1340.28	-	-	100 m	**1340.28**
depth 700 mm	-	85.68	1563.66	-	-	100 m	**1563.66**
depth 900 mm	-	110.16	2010.42	-	-	100 m	**2010.42**
depth 1000 mm	-	122.40	2233.80	-	-	100 m	**2233.80**
Width 450 mm							
depth 450 mm	-	66.10	1206.33	-	-	100 m	**1206.33**
depth 600 mm	-	88.13	1608.37	-	-	100 m	**1608.37**
depth 700 mm	-	102.82	1876.46	-	-	100 m	**1876.46**
depth 900 mm	-	132.19	2412.47	-	-	100 m	**2412.47**
depth 1000 mm	-	146.88	2680.56	-	-	100 m	**2680.56**
Width 600 mm							
depth 450 mm	-	88.13	1608.37	-	-	100 m	**1608.37**
depth 600 mm	-	117.50	2144.38	-	-	100 m	**2144.38**
depth 700 mm	-	137.09	2501.89	-	-	100 m	**2501.89**
depth 900 mm	-	176.26	3216.74	-	-	100 m	**3216.74**
depth 1000 mm	-	195.84	3574.08	-	-	100 m	**3574.08**

R DISPOSAL SYSTEMS

Item Excluding site overheads and profit	PC £	Labour hours	Labour £	Plant £	Material £	Unit	Total rate £
Width 900 mm							
depth 450 mm	-	132.19	2412.47	-	-	100 m	2412.47
depth 600 mm	-	176.26	3216.74	-	-	100 m	3216.74
depth 700 mm	-	205.63	3752.75	-	-	100 m	3752.75
depth 900 mm	-	264.38	4824.94	-	-	100 m	4824.94
depth 1000 mm	-	293.76	5361.12	-	-	100 m	5361.12
Earthwork support; moving along as work proceeds							
Maximum depth not exceeding 2.00 m distance between opposing faces not exceeding 2.00 m	-	0.80	14.60	17.70	-	m	32.30
Land drainage; pipe laying							
Hepworth; agricultural clay drain pipes; 300 mm length; butt joints; in straight runs							
75 mm diameter	415.58	8.00	146.00	-	425.97	100 m	571.97
100 mm diameter	713.95	9.00	164.25	-	731.80	100 m	896.05
150 mm diameter	1459.87	10.00	182.50	-	1496.37	100 m	1678.87
Extra over clay drain pipes for filter-wrapping pipes with "Terram" or similar filter fabric							
"Terram 700"	0.23	0.04	0.73	-	0.23	m²	0.95
"Terram 1000"	0.21	0.04	0.73	-	0.21	m²	0.94
Junctions between drains in clay pipes							
75 x 75 mm	14.18	0.25	4.56	-	14.50	nr	19.06
100 x 100 mm	18.83	0.25	4.56	-	19.34	nr	23.91
150 x 150 mm	23.18	0.25	4.56	-	23.95	nr	28.52
Wavin Plastics Ltd; flexible plastic perforated pipes in trenches (not included); to a minimum depth of 450 mm (couplings not included)							
"OsmaDrain"; flexible plastic perforated pipes in trenches (not included); to a minimum depth of 450 mm (couplings not included)							
80 mm diameter; available in 100 m coil	68.00	2.00	36.50	-	69.70	100 m	106.20
100 mm diameter; available in 100 m coil	110.00	2.00	36.50	-	112.75	100 m	149.25
160 mm diameter; available in 35 m coil	266.50	2.00	36.50	-	273.16	100 m	309.66
"WavinCoil"; plastic pipe junctions							
80 x 80 mm	3.07	0.05	0.91	-	3.07	nr	3.98
100 x 100 mm	3.43	0.05	0.91	-	3.43	nr	4.34
100 x 60 mm	3.38	0.05	0.91	-	3.38	nr	4.29
100 x 80 mm	3.26	0.05	0.91	-	3.26	nr	4.17
160 x 160 mm	8.16	0.05	0.91	-	8.16	nr	9.07
"WavinCoil"; couplings for flexible pipes							
80 mm diameter	1.08	0.03	0.61	-	1.08	nr	1.69
100 mm diameter	1.19	0.03	0.61	-	1.19	nr	1.80
160 mm diameter	1.61	0.03	0.61	-	1.61	nr	2.22

R DISPOSAL SYSTEMS

Item Excluding site overheads and profit	PC £	Labour hours	Labour £	Plant £	Material £	Unit	Total rate £
R13 LAND DRAINAGE - cont'd							
Land drainage; backfilling trench after laying pipes with gravel rejects or similar; blind filling with ash or sand; topping with 150 mm topsoil from dumps not exceeding 100 m; by machine							
Width 150 mm							
depth 450 mm	-	3.30	60.23	53.29	219.16	100 m	**332.67**
depth 600 mm	-	4.30	78.48	69.79	302.35	100 m	**450.61**
depth 750 mm	-	4.96	90.52	80.68	353.48	100 m	**524.68**
depth 900 mm	-	6.30	114.98	102.79	460.94	100 m	**678.71**
Width 225 mm							
depth 450 mm	-	4.95	90.34	79.93	332.88	100 m	**503.15**
depth 600 mm	-	6.45	117.71	104.68	453.59	100 m	**675.98**
depth 750 mm	-	7.95	145.09	129.43	574.66	100 m	**849.18**
depth 900 mm	-	9.45	172.46	154.18	695.38	100 m	**1022.02**
Width 375 mm							
depth 450 mm	-	8.25	150.56	133.22	554.63	100 m	**838.41**
depth 600 mm	-	10.75	196.19	174.47	755.94	100 m	**1126.59**
depth 750 mm	-	13.25	241.81	215.72	957.60	100 m	**1415.13**
depth 900 mm	-	15.75	287.44	256.97	1158.91	100 m	**1703.32**
Land drainage; backfilling trench after laying pipes with gravel rejects or similar, blind filling with ash or sand; topping with 150 mm topsoil from dumps not exceeding 100 m; by hand							
Width 150 mm							
depth 450 mm	-	18.63	340.00	-	207.46	100 m	**547.46**
depth 600 mm	-	24.84	453.33	-	302.35	100 m	**755.68**
depth 750 mm	-	31.05	566.66	-	294.98	100 m	**861.65**
depth 900 mm	-	37.26	680.00	-	460.94	100 m	**1140.94**
Width 225 mm							
depth 450 mm	-	27.94	509.90	-	332.88	100 m	**842.78**
depth 600 mm	-	37.26	680.00	-	453.59	100 m	**1133.59**
depth 750 mm	-	46.57	849.90	-	574.66	100 m	**1424.57**
depth 900 mm	-	55.89	1019.99	-	695.38	100 m	**1715.37**
Width 375 mm							
depth 450 mm	-	46.57	849.90	-	554.63	100 m	**1404.53**
depth 600 mm	-	61.10	1115.08	-	755.94	100 m	**1871.01**
depth 750 mm	-	77.63	1416.75	-	957.60	100 m	**2374.35**
depth 900 mm	-	93.15	1699.99	-	1158.91	100 m	**2858.90**
Catchwater or french drains; laying pipes to drain excavated separately; 100 mm diameter non-coilable perforated plastic pipes; including straight jointing; pipes laid with perforations uppermost; lining trench; wrapping pipes with filter fabric; backfilling with shingle							
Width 300 mm							
depth 450 mm	1249.41	9.10	166.07	30.00	1737.92	100 m	**1933.99**
depth 600 mm	1253.13	9.84	179.58	46.86	1905.40	100 m	**2131.84**
depth 750 mm	1257.01	10.60	193.45	59.40	2073.09	100 m	**2325.94**
depth 900 mm	1260.81	11.34	206.96	71.61	2240.68	100 m	**2519.24**
depth 1000 mm	1263.35	11.84	216.08	79.86	2352.40	100 m	**2648.34**
depth 1200 mm	1268.41	12.84	234.33	96.36	2575.84	100 m	**2906.53**

R DISPOSAL SYSTEMS

Item Excluding site overheads and profit	PC £	Labour hours	Labour £	Plant £	Material £	Unit	Total rate £
Width 450 mm							
depth 450 mm	1255.11	10.44	190.53	86.46	1989.29	100 m	**2266.28**
depth 600 mm	1260.81	11.34	206.96	71.61	2240.68	100 m	**2519.24**
depth 750 mm	1266.51	12.46	227.40	90.09	2492.05	100 m	**2809.53**
depth 900 mm	1272.21	13.60	248.20	108.90	2743.43	100 m	**3100.53**
depth 1000 mm	1276.01	14.34	261.70	121.11	2911.01	100 m	**3293.83**
depth 1200 mm	1283.61	15.84	289.08	145.86	3246.18	100 m	**3681.12**
depth 1500 mm	1295.01	18.10	330.32	183.15	3748.93	100 m	**4262.41**
depth 2000 mm	1314.01	21.84	398.58	244.86	4586.86	100 m	**5230.30**
Width 600 mm							
depth 450 mm	1260.81	11.24	205.13	71.61	2240.68	100 m	**2517.41**
depth 600 mm	1268.41	12.84	234.33	96.36	2575.84	100 m	**2906.53**
depth 750 mm	1276.01	14.34	261.70	121.11	2911.01	100 m	**3293.83**
depth 900 mm	1283.61	15.84	289.08	145.86	3246.18	100 m	**3681.12**
depth 1000 mm	1290.85	16.84	307.33	162.36	3460.91	100 m	**3930.60**
depth 1200 mm	1298.81	18.84	343.83	195.36	3916.52	100 m	**4455.71**
depth 1500 mm	1314.01	21.84	398.58	514.86	4586.86	100 m	**5500.30**
depth 2000 mm	1339.35	19.84	362.08	442.86	5704.09	100 m	**6509.03**
Width 900 mm							
depth 450 mm	1272.21	13.60	248.20	108.90	2743.43	100 m	**3100.53**
depth 600 mm	1283.61	7.92	144.54	145.86	3246.18	100 m	**3536.58**
depth 750 mm	1295.01	9.05	165.16	183.15	3748.93	100 m	**4097.25**
depth 900 mm	1306.41	10.17	185.60	220.11	4251.69	100 m	**4657.40**
depth 1000 mm	1314.01	10.92	199.29	244.86	4586.86	100 m	**5031.01**
depth 1200 mm	1329.21	12.42	226.66	294.36	5257.20	100 m	**5778.22**
depth 1500 mm	1352.01	14.67	267.73	368.61	6262.71	100 m	**6899.04**
depth 2000 mm	1390.01	18.92	345.29	582.36	7938.55	100 m	**8866.20**
depth 2500 mm	1428.01	22.17	404.60	706.11	9614.40	100 m	**10725.11**
depth 3000 mm	1466.01	25.92	473.04	739.86	11290.25	100 m	**12503.15**
Catchwater or french drains; laying pipes to drains excavated separately; 160 mm diameter non-coilable perforated plastic pipes; including straight jointing; pipes laid with perforations uppermost; lining trench; wrapping pipes with filter fabric; backfilling with shingle							
Width 600 mm							
depth 450 mm	2293.55	11.20	204.40	69.30	3271.07	100 m	**3544.77**
depth 600 mm	2301.15	12.70	231.78	94.05	3606.24	100 m	**3932.07**
depth 750 mm	2308.75	14.20	259.15	118.80	3941.41	100 m	**4319.36**
depth 900 mm	2316.35	15.70	286.52	143.55	4276.58	100 m	**4706.65**
depth 1000 mm	2321.42	16.70	304.77	160.05	4500.03	100 m	**4964.85**
depth 1200 mm	2331.55	18.70	341.27	193.15	4946.92	100 m	**5481.34**
depth 1500 mm	2346.75	21.70	396.02	242.55	5617.26	100 m	**6255.83**
depth 2000 mm	2372.09	26.70	487.27	325.05	6734.49	100 m	**7546.81**
depth 2500 mm	2397.42	31.70	578.52	407.55	7851.72	100 m	**8837.80**
depth 3000 mm	2422.75	36.70	669.77	490.05	8968.95	100 m	**10128.78**
Width 900 mm							
depth 450 mm	1272.21	13.60	248.20	108.90	2743.43	100 m	**3100.53**
depth 600 mm	1283.61	15.84	289.08	145.86	3246.18	100 m	**3681.12**
depth 750 mm	1295.01	18.10	330.32	183.15	3748.93	100 m	**4262.41**
depth 900 mm	1306.41	20.34	371.20	220.11	4251.69	100 m	**4843.00**
depth 1000 mm	1314.01	21.84	398.58	244.86	4586.86	100 m	**5230.30**
depth 1200 mm	1329.21	24.84	453.33	294.36	5257.20	100 m	**6004.89**
depth 1500 mm	1352.01	29.34	535.46	368.61	6262.71	100 m	**7166.77**
depth 2000 mm	1390.01	36.84	672.33	492.36	7938.55	100 m	**9103.24**
depth 2500 mm	1428.01	44.34	809.21	616.11	9614.40	100 m	**11039.72**
depth 3000 mm	1466.01	51.84	946.08	739.86	11290.25	100 m	**12976.19**

R DISPOSAL SYSTEMS

Item Excluding site overheads and profit	PC £	Labour hours	Labour £	Plant £	Material £	Unit	Total rate £
R13 LAND DRAINAGE - cont'd							
Catchwater or french drains; Exxon Chemical Geopolymers Ltd; "Filtram" filter drain; in trenches (trenches not included); comprising filter fabric, liquid conducting core and 110 mm uPVC slitpipes; all in accordance with manufacturer's instructions; backfilling with shingle							
Width 600 mm							
depth 1000 mm	1288.68	16.84	307.33	162.36	3469.63	100 m	3939.32
depth 1200 mm	1298.81	18.84	343.83	195.36	3916.52	100 m	4455.71
depth 1500 mm	1314.01	21.84	398.58	514.86	4586.86	100 m	5500.30
depth 2000 mm	1339.35	19.84	362.08	442.86	5704.09	100 m	6509.03
Width 900 mm							
depth 450 mm	1272.21	13.60	248.20	108.90	2743.43	100 m	3100.53
depth 600 mm	1283.61	15.84	289.08	145.86	3246.18	100 m	3681.12
depth 750 mm	1295.01	18.10	330.32	183.15	3748.93	100 m	4262.41
depth 900 mm	1306.41	20.34	371.20	220.11	4251.69	100 m	4843.00
depth 1000 mm	1314.01	21.84	398.58	244.86	4586.86	100 m	5230.30
depth 1200 mm	1329.21	24.84	453.33	294.36	5257.20	100 m	6004.89
depth 1500 mm	1352.01	29.34	535.46	368.61	6262.71	100 m	7166.77
depth 2000 mm	1390.01	37.84	690.58	582.36	7938.55	100 m	9211.49
depth 2500 mm	1428.01	44.34	809.21	706.11	9614.40	100 m	11129.72
depth 3000 mm	1466.01	51.84	946.08	829.86	11290.25	100 m	13066.19
Ditching; clear silt and bottom ditch not exceeding 1.50 m deep; strim back vegetation; disposing to spoil heaps; by machine							
Up to 1.50 m wide at top	-	6.00	109.50	25.88	-	100 m	135.38
1.50 - 2.50 m wide at top	-	8.00	146.00	34.50	-	100 m	180.50
2.50 - 4.00 m wide at top	-	9.00	164.25	38.81	-	100 m	203.06
Ditching; clear only vegetation from ditch not exceeding 1.50 m deep; disposing to spoil heaps; by strimmer							
Up to 1.50 m wide at top	-	5.00	91.25	7.01	-	100 m	98.26
1.50 - 2.50 m wide at top	-	6.00	109.50	12.61	-	100 m	122.11
2.50 - 4.00 m wide at top	-	7.00	127.75	19.62	-	100 m	147.37
Ditching; clear silt from ditch not exceeding 1.50 m deep; trimming back vegetation; disposing to spoil heaps; by hand							
Up to 1.50 m wide at top	-	15.00	273.75	7.01	-	100 m	280.76
1.50 - 2.50 m wide at top	-	27.00	492.75	12.61	-	100 m	505.36
2.50 - 4.00 m wide at top	-	42.00	766.50	19.62	-	100 m	786.12
Ditching; excavating and forming ditch and bank to given profile (normally 45°); in loam or sandy loam; by machine							
Width 300 mm							
depth 600 mm	-	3.70	67.53	31.91	-	100 m	99.44
depth 900 mm	-	5.20	94.90	44.85	-	100 m	139.75
depth 1200 mm	-	7.20	131.40	62.10	-	100 m	193.50
depth 1500 mm	-	9.40	171.55	70.94	-	100 m	242.49

R DISPOSAL SYSTEMS

Item Excluding site overheads and profit	PC £	Labour hours	Labour £	Plant £	Material £	Unit	Total rate £
Width 600 mm							
depth 600 mm	-	-	-	80.99	-	100 m	80.99
depth 900 mm	-	-	-	120.17	-	100 m	120.17
depth 1200 mm	-	-	-	163.28	-	100 m	163.28
depth 1500 mm	-	-	-	235.13	-	100 m	235.13
Width 900 mm							
depth 600 mm	-	-	-	315.00	-	100 m	315.00
depth 900 mm	-	-	-	555.00	-	100 m	555.00
depth 1200 mm	-	-	-	745.80	-	100 m	745.80
depth 1500 mm	-	-	-	924.00	-	100 m	924.00
Width 1200 mm							
depth 600 mm	-	-	-	495.00	-	100 m	495.00
depth 900 mm	-	-	-	735.00	-	100 m	735.00
depth 1200 mm	-	-	-	973.50	-	100 m	973.50
depth 1500 mm	-	-	-	1247.40	-	100 m	1247.40
Width 1500 mm							
depth 600 mm	-	-	-	610.50	-	100 m	610.50
depth 900 mm	-	-	-	930.60	-	100 m	930.60
depth 1200 mm	-	-	-	1237.50	-	100 m	1237.50
depth 1500 mm	-	-	-	1551.00	-	100 m	1551.00
Extra for ditching in clay	-	-	-	-	-	20%	-
Ditching; excavating and forming ditch and bank to given profile (normal 45°); in loam or sandy loam; by hand							
Width 300 mm							
depth 600 mm	-	36.00	657.00	-	-	100 m	657.00
depth 900 mm	-	42.00	766.50	-	-	100 m	766.50
depth 1200 mm	-	56.00	1022.00	-	-	100 m	1022.00
depth 1500 mm	-	70.00	1277.50	-	-	100 m	1277.50
Width 600 mm							
depth 600 mm	-	56.00	1022.00	-	-	100 m	1022.00
depth 900 mm	-	84.00	1533.00	-	-	100 m	1533.00
depth 1200 mm	-	112.00	2044.00	-	-	100 m	2044.00
depth 1500 mm	-	140.00	2555.00	-	-	100 m	2555.00
Width 900 mm							
depth 600 mm	-	84.00	1533.00	-	-	100 m	1533.00
depth 900 mm	-	126.00	2299.50	-	-	100 m	2299.50
depth 1200 mm	-	168.00	3066.00	-	-	100 m	3066.00
depth 1500 mm	-	210.00	3832.50	-	-	100 m	3832.50
Width 1200 mm							
depth 600 mm	-	112.00	2044.00	-	-	100 m	2044.00
depth 900 mm	-	168.00	3066.00	-	-	100 m	3066.00
depth 1200 mm	-	224.00	4088.00	-	-	100 m	4088.00
depth 1500 mm	-	280.00	5110.00	-	-	100 m	5110.00
Extra for ditching in clay	-	-	-	-	-	50%	-
Extra for ditching in compacted soil	-	-	-	-	-	90%	-
Piped ditching Jointed concrete pipes; FP McCann Ltd; including bedding, haunching and topping with 150 mm concrete; 11.50 N/mm^2 - 40 mm aggregate; to existing ditch							
300 mm diameter	16.13	0.67	12.17	6.60	33.10	m	51.87
450 mm diameter	23.97	0.67	12.17	6.60	49.43	m	68.20
600 mm diameter	38.66	0.67	12.17	6.60	74.81	m	93.58
900 mm diameter	102.05	1.00	18.25	9.90	125.78	m	153.93

R DISPOSAL SYSTEMS

Item Excluding site overheads and profit	PC £	Labour hours	Labour £	Plant £	Material £	Unit	Total rate £
R13 LAND DRAINAGE - cont'd							
Piped ditching - cont'd							
Jointed concrete pipes; FP McCann Ltd; including bedding, haunching and topping with 150 mm concrete; 11.50 N/mm^2 - 40 mm aggregate; to existing ditch							
1200 mm diameter	175.39	1.00	18.25	9.90	206.44	m	**234.59**
extra over jointed concrete pipes for bends to 45°	161.30	0.67	12.17	8.25	170.77	nr	**191.18**
extra over jointed concrete pipes for single junctions 300 mm diameter	110.22	0.67	12.17	8.25	119.69	nr	**140.10**
extra over jointed concrete pipes for single junctions 450 mm diameter	163.39	0.67	12.17	7.50	172.86	nr	**192.52**
extra over jointed concrete pipes for single junctions 600 mm diameter	264.66	0.67	12.17	7.50	274.13	nr	**293.79**
extra over jointed concrete pipes for single junctions 900 mm diameter	447.98	0.67	12.17	7.50	457.45	nr	**477.11**
extra over jointed concrete pipes for single junctions 1200 mm diameter	684.30	0.67	12.17	7.50	693.77	nr	**713.43**
Concrete road gullies; trapped; cement: mortar (1:2) joints to concrete pipes; bedding and surrounding in concrete; 11.50 N/mm^2 - 40 mm aggregate; 450 mm diameter x 1.07 m deep; rodding eye; stoppers	46.00	6.00	109.50	-	67.16	nr	**176.66**
Preamble: Flat rate hourly rainfall = 50 mm/hr and assumes 100 impermeability of the run-off area. A storage capacity of the soakaway should be 1/3 of the hourly rainfall. Formulae for calculating soakaway depths are provided in the publications mentioned below and in the Tables and Memoranda section of this publication. The design of soakaways is dependent on, amongst other factors, soil conditions, permeability, groundwater level and runoff. The definitive documents for design of soakaways are CIRIA 156 and BRE Digest 365 dated September 1991. The suppliers of the systems below will assist through their technical divisions. Excavation earthwork support of pits and disposal not included.							
Outfalls							
Reinforced concrete outfalls to water course; flank walls; for 150 mm drain outlets; overall dimensions							
900 x 1050 x 900 mm high	-	-	-	-	-	m	**487.60**
GRC outfall headwalls	-	-	-	-	-	nr	**105.00**
JKH Unit; standard small	-	-	-	-	-	nr	**85.00**

S PIPED SUPPLY SYSTEMS

Item Excluding site overheads and profit	PC £	Labour hours	Labour £	Plant £	Material £	Unit	Total rate £
S10 COLD WATER							
Borehole drilling; Agripower Ltd							
Drilling operations to average 100 m depth; lining with 150 mm diameter sieve							
drilling to 100 m	-	-	-	-	-	nr	12000.00
rate per m over 100 m	-	-	-	-	-	m	120.00
pump, rising main, cable and kiosk	-	-	-	-	-	nr	5000.00
Blue MDPE polythene pipes; type 50; for cold water services; with compression fittings; bedding on 100 mm DOT type 1 granular fill material							
Pipes							
20 mm diameter	0.32	0.08	1.46	-	3.49	m	4.95
25 mm diameter	0.39	0.08	1.46	-	3.56	m	5.02
32 mm diameter	0.95	0.08	1.46	-	4.14	m	5.60
50 mm diameter	1.65	0.10	1.82	-	4.86	m	6.68
60 mm diameter	3.57	0.10	1.82	-	6.82	m	8.65
Hose union bib taps; including fixing to wall; making good surfaces							
15 mm	11.42	0.75	13.69	-	13.28	nr	26.97
22 mm	16.12	0.75	13.69	-	18.91	nr	32.60
Stopcocks; including fixing to wall; making good surfaces							
15 mm	6.78	0.75	13.69	-	8.64	nr	22.33
22 mm	10.56	0.75	13.69	-	12.42	nr	26.11
Standpipes; to existing 25 mm water mains							
1.00 m high	-	-	-	-	-	nr	225.00
Hose junction bib taps; to standpipes							
19 mm	-	-	-	-	-	nr	80.00
S14 IRRIGATION							
KAR UK; water recycling							
Fully automatic self cleaning water recycling system for use with domestic grey water; including subsurface tank, self flushing filter pump and control system; excludes plumbing of the drainage system from the house into the tank	-	-	-	-	-	nr	2987.00
KAR UK; main or ring main supply							
Excavate and lay mains supply pipe to supply irrigated area							
PE 80 20 mm	-	-	-	-	-	m	4.87
PE 80 25 mm	-	-	-	-	-	m	5.06
Pro-Flow PEHD 63 mm PE	-	-	-	-	-	m	8.68
Pro-Flow PEHD 50 mm PE	-	-	-	-	-	m	7.09
Pro-Flow PEHD 40 mm PE	-	-	-	-	-	m	6.11
Pro-Flow PEHD 32 mm PE	-	-	-	-	-	m	5.49

S PIPED SUPPLY SYSTEMS

Item Excluding site overheads and profit	PC £	Labour hours	Labour £	Plant £	Material £	Unit	Total rate £
S14 IRRIGATION - cont'd							
KAR UK; head control to sprinkler stations							
Valves installed at supply point on irrigation supply manifold; includes for pressure control and filtration							
2 valve unit	-	-	-	-	-	nr	132.03
4 valve unit	-	-	-	-	-	nr	206.11
6 valve unit	-	-	-	-	-	nr	291.49
12 valve unit	-	-	-	-	-	nr	558.92
Solenoid valve; 25 mm with chamber; extra over for each active station	-	-	-	-	-	nr	106.16
Cable to solenoid valves (alternative to valves on manifold as above)							
4 core	-	-	-	-	-	m	1.57
6 core	-	-	-	-	-	m	7.63
12 core	-	-	-	-	-	m	3.19
KAR UK; supply irrigation infrastructure for irrigation system							
Header tank and submersible pump and pressure stat							
1000 litre (500 gallon 25 mm/10 days to 1500 m²) 2 m³/hr pump	-	-	-	-	-	nr	489.67
4540 litre (1000 gallon 25 mm/10 days to 3500 m²)	-	-	-	-	-	nr	991.89
9080 litre (2000 gallon 25 mm/10 days to 7000 m²)	-	-	-	-	-	nr	1642.27
Electric multistation controllers 240 V							
WeatherMatic Smartline; 4 station; standard	-	-	-	-	-	nr	149.21
WeatherMatic Smartline; 6 station; standard	-	-	-	-	-	nr	167.62
WeatherMatic Smartline; 12 station; standard	-	-	-	-	-	nr	284.18
WeatherMatic Smartline; 12 station; radio controlled	-	-	-	-	-	nr	583.80
WeatherMatic Smartline; 12 station; with moisture sensor	-	-	-	-	-	nr	981.43
WeatherMatic Smartline; 12 station; radio controlled; with moisture sensor	-	-	-	-	-	nr	1178.61
KAR UK; station consisting of multiple sprinklers; inclusive of all trenching, wiring and connections to ring main or main supply							
Sprayheads; "WeatherMatic LX"; including nozzle; 21 m² coverage							
100 mm (4")	-	-	-	-	-	nr	15.52
150 mm (6")	-	-	-	-	-	nr	21.92
300 mm (12")	-	-	-	-	-	nr	25.28
Matched precipitation sprinklers; "WeatherMatic LX"							
MP 1000; 16 m² coverage	-	-	-	-	-	nr	35.57
MP 2000; 30 m² coverage	-	-	-	-	-	nr	35.57
MP 3000; 81 m² coverage	-	-	-	-	-	nr	35.57

S PIPED SUPPLY SYSTEMS

Item Excluding site overheads and profit	PC £	Labour hours	Labour £	Plant £	Material £	Unit	Total rate £
Gear drive sprinklers; "WeatherMatic"; placed to provide head to head (100%) overlap; average							
T3; 121 m² coverage	-	-	-	-	-	nr	46.72
CT70; 255 m² coverage	-	-	-	-	-	nr	82.36
CT70SS; 361 m² coverage	-	-	-	-	-	nr	91.46
KAR UK; driplines; inclusive of connections and draindown systems							
"Metzerplas TechLand" 30 cm drip emitter							
fixed on soil surface	-	-	-	-	-	m	1.07
subsurface	-	-	-	-	-	m	1.57
"Metzerplas TechLand" 50 cm drip emitter							
fixed on soil surface	-	-	-	-	-	m	0.90
subsurface	-	-	-	-	-	m	1.41
Commissioning and testing of irrigation system							
per station	-	-	-	-	-	nr	39.56
Annual maintenance costs of irrigation system							
Call out charge per visit	-	-	-	-	-	nr	329.59
Extra over per station	-	-	-	-	-	nr	13.19
KAR UK; "KAR Moisture Control" moisture sensing control system							
Large garden comprising 7000 m² and 24 stations with four moisture sensors							
turf only	-	-	-	-	-	nr	13333.33
turf/shrub beds; 70/30	-	-	-	-	-	nr	7777.78
Medium garden comprising 3500 m² and 12 stations with two moisture sensors							
turf only	-	-	-	-	-	nr	8888.89
turf/shrub beds; 70/30	-	-	-	-	-	nr	11555.56
Smaller gardens garden comprising 1000 m² and 6 stations with one moisture sensor							
turf only	-	-	-	-	-	nr	5444.44
turf/shrub beds; 70/30	-	-	-	-	-	nr	6000.00
turf/shrub beds; 50/50	-	-	-	-	-	nr	6444.44
Leaky Pipe Systems Ltd; "Leaky Pipe"; moisture leaking pipe irrigation system							
Main supply pipe inclusive of machine excavation; exclusive of connectors							
20 mm LDPE polytubing	1.09	0.05	0.91	0.61	1.12	m	2.64
16 mm LDPE polytubing	0.73	0.05	0.91	0.61	0.75	m	2.27
Water filters and cartridges							
No 10; 20 mm	-	-	-	-	54.47	nr	-
Big Blue and RR30 cartridge; 25 mm	-	-	-	-	139.36	nr	-
Water filters and pressure regulator sets; complete assemblies							
No 10; flow rate 3.1 - 82 litres per minute	-	-	-	-	92.95	nr	-

S PIPED SUPPLY SYSTEMS

Item Excluding site overheads and profit	PC £	Labour hours	Labour £	Plant £	Material £	Unit	Total rate £
S14 IRRIGATION - cont'd							
Leaky Pipe Systems Ltd - cont'd							
Leaky pipe hose; placed 150 mm subsurface for turf irrigation Distance between laterals 350 mm; excavation and backfilling priced separately							
LP12L low leak	4.73	0.04	0.73	-	4.73	m^2	**5.46**
LP12H high leak	4.10	0.04	0.73	-	4.10	m^2	**4.83**
LP12UH ultra high leak	5.02	0.04	0.73	-	5.02	m^2	**5.75**
Leaky pipe hose; laid to surface for landscape irrigation; distance between laterals 600 mm							
LP12L low leak	2.76	0.03	0.46	-	2.82	m^2	**3.28**
LP12H high leak	2.39	0.03	0.46	-	2.45	m^2	**2.91**
LP12UH ultra high leak	2.92	0.03	0.46	-	2.99	m^2	**3.45**
Leaky pipe hose; laid to surface for landscape irrigation; distance between laterals 900 mm							
LP12L low leak	1.84	0.02	0.30	-	1.89	m^2	**2.19**
LP12H high leak	1.60	0.02	0.30	-	1.64	m^2	**1.94**
LP12UH ultra high leak	1.95	0.02	0.30	-	2.00	m^2	**2.31**
Leaky pipe hose; laid to surface for tree irrigation laid around circumference of tree pit							
LP12L low leak	2.54	0.13	2.28	-	2.79	nr	**5.08**
LP12H high leak	2.20	0.13	2.28	-	2.42	nr	**4.70**
LP12UH ultra high leak	2.69	0.13	2.28	-	2.96	nr	**5.24**
Accessories							
automatic multi-station controller stations; inclusive of connections	375.70	2.00	36.50	-	375.70	nr	**412.20**
solenoid valves; inclusive of wiring and connections to a multi-station controller; nominal distance from controller 25 m	62.40	0.50	9.13	-	517.40	nr	**526.52**
Rainwater harvesting; Britannia Rainwater Recycling Systems; tanks for collection of water for use in landscape irrigation; inclusive of filtration of first stage particle and leaf matter; installed on surface; inclusive of base, gullies for water catchment pipework from catchment area and submersible pumps; excavated material to stockpile							
Tanks installed below ground; gullies for rainwater catchment and pipework not included							
23000 litre	6200.00	12.00	219.00	238.76	6657.27	nr	**7115.03**
13000 litre	5340.00	10.00	182.50	204.96	5590.00	nr	**5977.46**
6500 litre	3780.00	8.00	146.00	102.48	4237.27	nr	**4485.75**

S PIPED SUPPLY SYSTEMS

S15 FOUNTAINS/WATER FEATURES

Lakes and ponds - General
Preamble: The pressure of water against a retaining wall or dam is considerable, and where water retaining structures form part of the design of water features, the landscape architect is advised to consult a civil engineer. Artificially contained areas of water in raised reservoirs over 25,000 m^3 have to be registered with the local authority and their dams will have to be covered by a civil engineer's certificate of safety.

Typical linings - General
Preamble: In addition to the traditional methods of forming the linings of lakes and ponds in puddled clay or concrete, there are a number of lining materials available. They are mainly used for reservoirs but can also help to form comparatively economic water features especially in soil which is not naturally water retentive. Information on the construction of traditional clay puddle ponds can be obtained from the British Trust for Conservation Volunteers, 36 St. Mary's Street, Wallingford, Oxfordshire, OX10 0EU. Tel: (01491) 39766. The cost of puddled clay ponds depends on the availability of suitable clay, the type of hand or machine labour that can be used and the use to which the pond is to be put.

Item Excluding site overheads and profit	PC £	Labour hours	Labour £	Plant £	Material £	Unit	Total rate £
Lake liners; Fairwater Water Garden Design Consultants Ltd; to evenly graded surface of excavations (excavating not included); all stones over 75 mm; removing debris; including all welding and jointing of liner sheets							
Geotextile underlay; inclusive of spot welding to prevent dragging							
to water features	-	-	-	-	-	m^2	2.16
to lakes or large features	-	-	-	-	-	1000m^2	2157.80
Butyl rubber liners; "Varnamo" inclusive of site vulcanizing laid to geotextile above							
0.75 mm thick	-	-	-	-	-	m^2	7.57
0.75 mm thick	-	-	-	-	-	1000m^2	7495.50
1.00 mm thick	-	-	-	-	-	m^2	8.62
1.00 mm thick	-	-	-	-	-	1000m^2	8619.80

S PIPED SUPPLY SYSTEMS

Item Excluding site overheads and profit	PC £	Labour hours	Labour £	Plant £	Material £	Unit	Total rate £
S15 FOUNTAINS/WATER FEATURES - cont'd							
Lake liners; Landline Ltd; "Landflex" or "Alkorplan" geomembranes; to prepared surfaces (surfaces not included); all joints fully welded; installation by Landline employees							
"Landflex HC" polyethylene geomembranes							
0.50 mm thick	-	-	-	-	-	1000m²	3250.00
0.75 mm thick	-	-	-	-	-	1000m²	3750.00
1.00 mm thick	-	-	-	-	-	1000m²	4250.00
1.50 mm thick	-	-	-	-	-	1000m²	5000.00
2.00 mm thick	-	-	-	-	-	1000m²	5750.00
"Alkorplan PVC" geomembranes							
0.80 mm thick	-	-	-	-	-	1000 m	5950.00
1.20 mm thick	-	-	-	-	-	1000 m	8650.00
Lake liners; Monarflex Polyethylene lake and reservoir lining system; welding on site by Monarflex technicians (surface preparation and backfilling not included)							
"Blackline"; 500 micron	-	-	-	-	-	100 m²	341.00
"Blackline"; 750 micron	-	-	-	-	-	100 m²	432.00
"Blackline"; 1000 micron	-	-	-	-	-	100 m²	567.00
Operations over surfaces of lake liners							
Dug ballast; evenly spread over excavation already brought to grade							
150 mm thick	499.50	2.00	36.50	26.13	549.45	100 m²	612.08
200 mm thick	666.00	3.00	54.75	39.19	732.60	100 m²	826.54
300 mm thick	999.00	3.50	63.88	45.72	1098.90	100 m²	1208.49
Imported topsoil; evenly spread over excavation							
100 mm thick	260.00	1.50	27.38	19.59	312.00	100 m²	358.97
150 mm thick	390.00	2.00	36.50	26.13	468.00	100 m²	530.63
200 mm thick	520.00	3.00	54.75	39.19	624.00	100 m²	717.94
Blinding existing subsoil with 50 mm sand	143.60	1.00	18.25	26.13	157.96	100 m²	202.34
Topsoil from excavation; evenly spread over excavation							
100 mm thick	-	-	-	19.59	-	100 m²	19.59
200 mm thick	-	-	-	26.13	-	100 m²	26.13
300 mm thick	-	-	-	39.19	-	100 m²	39.19
Extra over for screening topsoil using a "Powergrid" screener; removing debris	-	-	-	5.17	0.35	m³	5.52
Lake construction; Fairwater Water Garden Design Consultants Ltd; lake construction at ground level; excavation; forming of lake; commissioning; excavated material spread on site Lined with existing site clay; natural edging with vegetation meeting the water							
500 m²	-	-	-	-	-	nr	8820.00
1000 m²	-	-	-	-	-	nr	15330.00
1500 m²	-	-	-	-	-	nr	21000.00
2000 m²	-	-	-	-	-	nr	26250.00

S PIPED SUPPLY SYSTEMS

Item Excluding site overheads and profit	PC £	Labour hours	Labour £	Plant £	Material £	Unit	Total rate £
Lined with 0.75 mm butyl on geotextile underlay; natural edging with vegetation meeting the water							
500 m²	-	-	-	-	-	nr	14175.00
1000 m²	-	-	-	-	-	nr	24570.00
1500 m²	-	-	-	-	-	nr	33600.00
2000 m²	-	-	-	-	-	nr	42000.00
Extra over to the above for hard block edging to secure and protect liner; measured at lake perimeter	-	-	-	-	-	m	44.10
Lined with imported puddling clay; natural edging with vegetation meeting the water							
500 m²	-	-	-	-	-	nr	23100.00
1000 m²	-	-	-	-	-	nr	39900.00
1500 m²	-	-	-	-	-	nr	54600.00
2000 m²	-	-	-	-	-	nr	68250.00
Waterfall construction; Fairwater Water Garden Design Consultants Ltd							
Stone placed on top of butyl liner; securing with concrete and dressing to form natural rock pools and edgings							
Portland stone; m³ rate	-	-	-	-	-	m³	630.89
Portland stone; tonne rate	-	-	-	-	-	tonne	349.34
Balancing tank; blockwork construction; inclusive of recirculation pump and pond level control; 110 mm balancing pipe to pond; waterproofed with butyl rubber membrane; pipework mains water top-up and overflow							
450 x 600 x 1000 mm	-	-	-	-	-	nr	1312.14
Extra for pump							
2000 gallons per hour; submersible	-	-	-	-	-	nr	200.33
Water features; Fairwater Water Garden Design Consultants Ltd							
Natural stream; butyl lined level changes of 1.00 m; water pumped from lower pond (not included) via balancing tank; level changes via Purbeck stone water falls							
20 m long stream	-	-	-	-	-	nr	18900.00
Bespoke freestanding water wall; steel or glass panel; self contained recirculation system and reservoir							
water wall; 2.00 m high x 850 mm wide	-	-	-	-	-	nr	8925.00
Natural swimming pools; Fairwater Water Garden Design Consultants Ltd							
Natural swimming pool of 50 m² area; shingle regeneration zone 50 m² planted with marginal and aquatic plants	-	-	-	-	-	nr	57750.00

S PIPED SUPPLY SYSTEMS

Item Excluding site overheads and profit	PC £	Labour hours	Labour £	Plant £	Material £	Unit	Total rate £
S15 FOUNTAINS/WATER FEATURES - cont'd							
Swimming pools; Thompson Landscapes Conventional swimming pool construction inclusive of excavation reinforced gunnite construction, heating, lighting, pumping, filtration and mosaic lined							
rectangular construction; 14 x 6 m	-	-	-	-	-	nr	58650.00
Duraglide cover; inclusive of all electrical works	-	-	-	-	-	nr	25300.00
automatic cleaning system with main drains	-	-	-	-	-	nr	14950.00
Ornamental pools - General Preamble: Small pools may be lined with one of the materials mentioned under lakes and ponds, or may be in rendered brickwork, puddled clay or, for the smaller sizes, fibreglass. Most of these tend to be cheaper than waterproof concrete. Basic prices for various sizes of concrete pools are given in the Approximate Estimates section. Prices for excavation, grading, mass concrete, and precast concrete retaining walls are given in the relevant sections. The manufacturers should be consulted before specifying the type and thickness of pool liner, as this depends on the size, shape and proposed use of the pool. The manufacturer's recommendation on foundations and construction should be followed.							
Ornamental pools Pool liners; to 50 mm sand blinding to excavation (excavating not included); all stones over 50 mm; removing debris from surfaces of excavation; including all welding and jointing of liner sheets							
black polythene; 1000 gauge	-	-	-	-	-	m^2	2.57
blue polythene; 1000 gauge	-	-	-	-	-	m^2	2.86
coloured PVC; 1500 gauge	-	-	-	-	-	m^2	3.47
black PVC; 1500 gauge	-	-	-	-	-	m^2	2.96
black butyl; 0.75 mm thick	-	-	-	-	-	m^2	7.19
black butyl; 1.00 mm thick	-	-	-	-	-	m^2	8.19
black butyl; 1.50 mm thick	-	-	-	-	-	m^2	28.04
Fine gravel; 100 mm; evenly spread over area of pool; by hand	3.63	0.13	2.28	-	3.63	m^2	5.91
Selected topsoil from excavation; 100 mm; evenly spread over area of pool; by hand	-	0.13	2.28	-	-	m^2	2.28
Extra over selected topsoil for spreading imported topsoil over area of pool; by hand	26.00	-	-	-	31.20	m^3	-

S PIPED SUPPLY SYSTEMS

Item Excluding site overheads and profit	PC £	Labour hours	Labour £	Plant £	Material £	Unit	Total rate £
Pool surrounds and ornament; Haddonstone Ltd; Portland Bath or terracotta cast stone							
Pool surrounds; installed to pools or water feature construction priced separately; surrounds and copings to 112.5 mm internal brickwork							
C4HSKVP half small pool surround; internal diameter 1780 mm; kerb features continuous moulding enriched with ovolvo and palmette designs; inclusive of plinth and integral conch shell vases flanked by dolphins	1076.00	16.00	292.00	-	1087.93	nr	**1379.93**
C4SKVP small pool surround as above but with full circular construction; internal diameter 1780 mm	2152.00	48.00	876.00	-	2166.31	nr	**3042.31**
C4MKVP medium pool surround; internal diameter 2705 mm; inclusive of plinth and integral vases	3228.00	48.00	876.00	-	3257.89	nr	**4133.89**
C4XLKVP extra large pool surround; internal diameter 5450 mm	4304.00	140.00	2555.00	-	4339.15	nr	**6894.15**
Pool centre pieces and fountains; inclusive of plumbing and pumps							
HC350 Lotus Bowl; 1830 mm wide with C1700 triple Dolphin Fountain; HD2900 Doric Pedestal	3326.00	8.00	146.00	-	3328.35	nr	**3474.35**
C251 Gothic Fountain and Gothic Upper Base A350; freestanding fountain	915.00	4.00	73.00	-	917.35	nr	**990.35**
HC521 Romanesque Fountain; freestanding bowl with self-circulating fountain; filled with cobbles; 815 mm diameter x 348 mm high	420.00	2.00	36.50	-	510.97	nr	**547.47**
C300 Lion Fountain; 610 mm high on fountain base C305; 280 mm high	245.00	2.00	36.50	-	247.35	nr	**283.85**
Wall fountains, watertanks and fountains; inclusive of installation drainage, automatic top up, pump and balancing tank							
Capital Garden Products Ltd							
Lion Wall Fountain F010; 970 x 940 x 520 mm	-	4.00	73.00	-	2016.62	nr	**2089.62**
Dolphin Wall Fountain F001; 740 x 510 mm	-	4.00	73.00	-	1846.37	nr	**1919.37**
Dutch Master Wall Fountain F012; 740 x 410 mm	-	4.00	73.00	-	1873.37	nr	**1946.37**
James II Watertank 2801; 730 x 730 x 760 mm high; 405 litres	-	4.00	73.00	-	2028.37	nr	**2101.37**
James II Fountain 4901 bp; 710 x 1300 x 1440 mm high; 655 litres	-	4.00	73.00	-	1869.37	nr	**1942.37**

S PIPED SUPPLY SYSTEMS

Item Excluding site overheads and profit	PC £	Labour hours	Labour £	Plant £	Material £	Unit	**Total rate £**
S15 FOUNTAINS/WATER FEATURES - cont'd							
Marble wall fountains; Architectural Heritage Ltd; inclusive of installation, drainage, automatic top up, pump and balancing tank							
"Breccia Pernice"; marble wall fountain supported by two dolphins, 1770 mm high x 1000 mm wide x 640 mm deep	16400.00	4.00	73.00	-	18154.62	nr	**18227.62**
"The River God"; Verona marble wall mask fountain; 890 mm high x 810 mm wide x 230 mm deep	5400.00	4.00	73.00	-	7154.62	nr	**7227.62**
Fountain kits; typical prices of submersible units comprising fountain pumps, fountain nozzles, underwater spotlights, nozzle extension armatures, underwater terminal boxes and electrical control panels							
Single aerated white foamy water columns; ascending jet 70 mm diameter; descending water up to four times larger; jet height adjustable between 1.00 - 1.70 m	3900.00	-	-	-	4470.00	nr	**4470.00**
Single aerated white foamy water columns; ascending jet 110 mm diameter; descending water up to four times larger; jet height adjustable between 1.50 - 3.00 m	6500.00	-	-	-	7380.02	nr	**7380.02**

V ELECTRICAL SUPPLY/POWER/LIGHTING SYSTEMS

Item Excluding site overheads and profit	PC £	Labour hours	Labour £	Plant £	Material £	Unit	Total rate £
V41 STREET/AREA/FLOODLIGHTING							
Street area floodlighting - General Preamble: There are an enormous number of luminaires available which are designed for small scale urban and garden projects. The designs are continually changing and the landscape designer is advised to consult the manufacturer's latest catalogue. Most manufacturers supply light fittings suitable for column, bracket, bulkhead, wall or soffit mounting. Highway lamps and columns for trafficked roads are not included in this section as the design of highway lighting is a very specialized subject outside the scope of most landscape contracts. The IP reference number refers to the waterproof properties of the fitting; the higher the number the more waterproof the fitting. Most items can be fitted with time clocks or PIR controls.							
Market prices of lamps Lamps							
70 w HQIT-S	-	-	-	-	9.22	nr	-
70 w HQIT	-	-	-	-	18.87	nr	-
70 w SON	-	-	-	-	4.22	nr	-
70 w SONT	-	-	-	-	4.22	nr	-
100 w SONT	-	-	-	-	6.34	nr	-
150 w SONT	-	-	-	-	7.71	nr	-
28 w 2D	-	-	-	-	22.00	nr	-
100 w GLS\E27	-	-	-	-	2.07	nr	-
Bulkhead and canopy fittings; including fixing to wall and light fitting (lamp, final painting, electric wiring, connections or related fixtures such as switch gear and time clock mechanisms not included unless otherwise indicated) Bulkhead and canopy fittings; Targetti Poulsen "Nyhavn Wall" small domed top conical shade with rings; copper wall lantern; finished untreated copper to achieve verdigris finish; also available in white aluminium; 310 mm diameter shade; with wall mounting arm; to IP 44	462.00	0.50	9.13	-	464.79	nr	473.92

V ELECTRICAL SUPPLY/POWER/LIGHTING SYSTEMS

Item Excluding site overheads and profit	PC £	Labour hours	Labour £	Plant £	Material £	Unit	Total rate £
V41 STREET/AREA/FLOODLIGHTING - cont'd							
Bulkhead and canopy fittings - cont'd							
Bulkhead and canopy fittings; Sugg Lighting							
"Princess" IP54 backlamp; 375 x 229 mm; copper frame; stove painted in black; with chimney and lampholder	338.08	0.50	9.13	-	340.87	nr	350.00
"Victoria" IP54 backlamp; 502 x 323 mm; copper frame; polished copper finish; with chimney, door and lampholder	445.64	0.50	9.13	-	448.43	nr	457.56
"Palace" IP54 backlamp; 457 x 321 mm; copper frame; stove painted black finish; with chimney, door and lampholder	444.68	0.50	9.13	-	447.47	nr	456.60
"Windsor" IP54 backlamp; 650 x 306 mm; copper frame; polished and lacquered finish; with door and lampholder	435.28	0.50	9.13	-	438.07	nr	447.19
"Windsor" IP54 gas backlamp; 650 x 306 mm; copper frame; polished and lacquered finish; hinged door; double inverted cluster mantle with permanent pilot and mains solenoid	779.73	0.50	9.13	-	782.52	nr	791.64
Floodlighting; ground, wall or pole mounted; including fixing (lamp, final painting, electric wiring, connections or related fixtures such as switch gear and time clock mechanisms not included unless otherwise indicated)							
Floodlighting; Targetti Poulsen - Landscape Division							
SPR-12; multi purpose ground/spike mounted wide angle floodlight; 70 w HIT or SON; to IP65	219.00	1.00	18.25	-	224.65	nr	242.90
Floodlight accessories; Targetti Poulsen - Landscape Division							
earth spike for SPR-12	27.00	-	-	-	27.00	nr	-
louvre for SPR-12	47.00	-	-	-	47.00	nr	-
cowl for SPR-12	42.00	-	-	-	42.00	nr	-
barn doors for SPR-12	101.00	-	-	-	101.00	nr	-
Large area/pitch floodlighting; CU Phosco							
ref FL444 1000 w SON-T; floodlights with lamp and loose gear; narrow asymmetric beam	248.85	1.00	18.25	-	248.85	nr	267.10
ref FL444 2.0 kw MBIOS; floodlight with lamp and loose gear; projector beam	474.85	1.00	18.25	-	474.85	nr	493.10
Large area floodlighting; CU Phosco							
ref FL345/G/250S; floodlight with lamp and integral gear	307.20	1.00	18.25	-	307.20	nr	325.45
ref FL345/G/400 MBI; floodlight with lamp and integral gear	236.60	1.00	18.25	-	236.60	nr	254.85
Small area floodlighting; Sugg Lighting; floodlight for feature lighting; clear or toughened glass							
"Scenario"; lamp; 150 w HPS-T	670.24	1.00	18.25	-	670.24	nr	688.49
"Scenario"; lamp; 150 w HQI-T	670.24	1.00	18.25	-	670.24	nr	688.49

V ELECTRICAL SUPPLY/POWER/LIGHTING SYSTEMS

Item Excluding site overheads and profit	PC £	Labour hours	Labour £	Plant £	Material £	Unit	Total rate £
Spotlights for uplighting and for illuminating signs and notice boards, statuary and other features) ground, wall or pole mounted; including fixing, light fitting and priming (lamp, final painting, electric wiring, connections or related fixtures such as switch gear and time clock mechanisms not included); all mains voltage (240 v) unless otherwise stated							
Spotlights; Targetti Poulsen - Landscape Division							
"Maxispotter" miniature solid copper spotlight designed to patternate and age naturally; with mounting bracket and internal anti glare louvre; low voltage halogen reflector; 20/35/50 w; to IP56	79.95	1.00	18.25	-	85.60	nr	**103.85**
"WeeBee Spot"; ref SP-05 for low voltage halogen reflector; 20/35/50 w; c/w integral transformer and wall mounting box; to IP65	152.00	1.00	18.25	-	157.65	nr	**175.90**
Spotlight accessories; Targetti Poulsen - Landscape Division							
"ES"; earth spike for WeeBee Spot	17.00	-	-	-	17.00	nr	-
Spotlighters, uplighters and cowl lighting; Havells Sylvania; TECHNO - SHORT ARM; head adjustable 130°; rotation 350°; projection 215 mm on 210 mm base plate; 355 mm high; integral gear; PG16 cable gland; black or aluminium							
ref S.3517.09/14; 70 w CDMT/HQIT	463.03	1.00	18.25	-	463.03	nr	**481.28**
ref S.3518.09/14; 150 w CDMT/HQIT	486.72	1.00	18.25	-	486.72	nr	**504.97**
Recessed uplighting; including walk/drive over fully recessed uplighting; excavating, ground fixing, concreting in and making good surfaces (electric wiring, connections or related fixtures such as switch gear and time-clock mechanisms not included unless otherwise stated) (Note: transformers will power multiple lights dependent on the distance between the light units); all mains voltage (240 v) unless otherwise stated							
Recessed uplighting; Targetti Poulsen - Landscape Division; "WeeBee Up"; ultra-compact recessed halogen uplight in diecast aluminium and stainless steel top plate and toughened safety glass; 20/35 w; complete with installation sleeve; low voltage requires transformer; to IP67							
ref BU-01	105.00	2.00	36.50	-	110.65	nr	**147.15**

V ELECTRICAL SUPPLY/POWER/LIGHTING SYSTEMS

Item Excluding site overheads and profit	PC £	Labour hours	Labour £	Plant £	Material £	Unit	Total rate £
V41 STREET/AREA/FLOODLIGHTING - cont'd							
Recessed uplighting - cont'd							
Recessed uplighting; Targetti Poulsen - Landscape Division; 266 mm diameter; diecast aluminium with stainless steel top plate 10 mm toughened safety glass; 2000 kg drive over; integral control gear; to IP67							
IPR-14; HIT metal halide; white light; spot or flood or wall wash distribution	384.00	1.50	27.38	-	389.65	nr	**417.03**
IRR-14 HIT compact fluorescent; white light; low power consumption; flood distribution	384.00	1.50	27.38	-	389.65	nr	**417.03**
Accessories for IPR-14 uplighters; Targetti Poulsen - Landscape Division							
rockguard	110.00	-	-	-	110.00	nr	-
stainless steel installation sleeve	47.00	-	-	-	47.00	nr	-
anti glare louvre (internal tilt)	76.00	-	-	-	76.00	nr	-
Recessed uplighting; Targetti Poulsen - Landscape Division; Nimbus 125-150 mm diameter uplighters; manufactured from diecast aluminium with stainless steel top plate 8 mm toughened safety glass; 2000 kg drive over; integral control gear; to IP67							
BU-02; 95 mm deep; 20/35/50 w low voltage halogen; requires remote transformer	112.11	2.00	36.50	-	117.76	nr	**154.26**
BU-03; 172 mm deep; 20/35/50 w low voltage halogen; with integral transformer	154.60	2.00	36.50	-	160.25	nr	**196.75**
Accessories for Nimbus uplighters; Targetti Poulsen - Landscape Division							
"IS" installation sleeve	41.00	-	-	-	41.00	nr	-
Wall recessed; Havells Sylvania; "EOS" range; integral gear; asymmetric reflector; toughened reeded glass							
"MINI EOS"; 145 x 90 mm; black or aluminium; 20 W QT9 lamp	63.87	1.25	22.81	-	63.87	nr	**86.68**
"RECTANGULAR EOS"; 270 x 145 mm; black or aluminium; 20 W QT9 lamp	123.24	1.25	22.81	-	139.34	nr	**162.15**
Underwater lighting							
Underwater lighting; Targetti Poulsen - Landscape Division							
UW-05 "Minipower" cast bronze underwater floodlight c\w mounting bracket; 50 w low voltage reflector lamp; requires remote transformer	206.24	1.00	18.25	-	206.24	nr	**224.49**

V ELECTRICAL SUPPLY/POWER/LIGHTING SYSTEMS

Item Excluding site overheads and profit	PC £	Labour hours	Labour £	Plant £	Material £	Unit	Total rate £
Low-level lighting; positioned to provide glare free light wash to pathways steps and terraces; including forming post holes, concreting in and making good to surfaces (final painting, electric wiring, connections or related fixtures such as switch gear and time clock mechanisms not included)							
Low-level lighting; Targetti Poulsen - Landscape Division							
"Sentry"; ref AM-08; single-sided bollard type sculptural pathway light; aluminium; 670 mm high; for 35 w HIT; to IP65	431.00	2.00	36.50	-	436.65	nr	473.15
"Footliter"; ref GL-05; low voltage pathlighter; 311 mm high; 6.00 m light distribution; in aluminium or solid copper; c\w earth spike; requires remote transformer; 20 w halogen; to IP44; copper finish	79.07	2.00	36.50	-	84.72	nr	121.22
Lighted bollards; including excavating, ground fixing, concreting in and making good surfaces (lamp, final painting, electric wiring, connections or related fixtures such as switch gear and time-clock mechanisms not included) (Note: all illuminated bollards must be earthed); heights given are from ground level to top of bollards							
Lighted bollards; Targetti Poulsen "Orbiter" vandal resistant; head of cast aluminium; domed top; anti-glare rings; pole extruded aluminium; diffuser clear UV stabilized polycarbonate; powder coated; 1040 mm high x 255 mm diameter; with root or base plate; IP44	499.00	2.50	45.63	-	504.95	nr	550.57
"Waterfront" solidly proportioned; head of cast silumin; domed top; symmetrical distribution; pole extruded aluminium sandblasted or painted white; internal diffuser clear UV stabilized polycarbonate; 865 mm high x diameter 260 mm; IP55	583.00	2.50	45.63	-	588.95	nr	634.57
"Bysted" concentric louvred bollard; head cast iron; post COR-TEN steel; externally untreated to provide natural aging effect of uniform oxidized red surface finish; internal painted white; lamp diffuser rings of clear polycarbonate; 1130 mm high x 280 mm diameter; to IP44	826.00	2.50	45.63	-	831.95	nr	877.57
Lighted bollards; Woodscape Ltd							
Illuminated bollard; in "very durable hardwood"; integral die cast aluminium lighting unit; 165 x 165 mm x 1.00 m high	226.69	2.50	45.63	-	238.31	nr	283.94

V ELECTRICAL SUPPLY/POWER/LIGHTING SYSTEMS

Item Excluding site overheads and profit	PC £	Labour hours	Labour £	Plant £	Material £	Unit	Total rate £
V41 STREET/AREA/FLOODLIGHTING - cont'd							
Reproduction street lanterns and columns; including excavating, concreting in, backfilling and disposing of spoil, or fixing to ground or wall, making good surfaces, light fitting and priming (final painting, electric wiring, connections or related fixtures such as switch gear and time-clock mechanisms not included) (Note: lanterns up to 14 inches and columns up to 7 ft are suitable for residential lighting)							
Reproduction street lanterns; Sugg Lighting; reproduction lanterns hand made to original designs; all with ES lamp holder							
"Westminster" IP54 hexagonal hinged top lantern with door; two piece folded polycarbonate glazing; lamp 100 w HQI T; integral photo electric cell							
"small"; 1016 mm high x 356 mm wide	759.26	0.50	9.13	-	759.26	nr	**768.38**
"large"; 1124 mm high x 760 mm wide	1193.19	0.50	9.13	-	1193.19	nr	**1202.32**
"Guildhall" IP54; handcrafted copper frame; clear polycarbonate glazing circular tapered lantern with hemispherical top; integral photo electric cell; with door							
"small"; 711 mm high x 330 mm wide	650.50	0.50	9.13	-	653.29	nr	**662.41**
"medium"; 1150 mm high x 432 mm wide	685.00	0.50	9.13	-	687.79	nr	**696.91**
"large"; 1370 mm high x 550 mm wide	950.50	0.50	9.13	-	953.29	nr	**962.41**
"Grosvenor" circular lantern with door; copper frame; polished copper finish; polycarbonate glazing							
"small"; 790 x 330 mm; IP54	660.50	0.50	9.13	-	663.29	nr	**672.41**
"medium"; 1080 x 435 mm; IP65	685.50	0.50	9.13	-	688.29	nr	**697.41**
"Classic Globe" IP54 lantern with hinged outer frame; cast aluminium frame; black polyester powder coating							
"medium"; 965 x 483 mm	605.50	0.50	9.13	-	608.29	nr	**617.41**
"Windsor" lantern; copper frame; polished copper finish							
"small"; 905 x 356 mm; IP54; with door	530.19	0.50	9.13	-	532.98	nr	**542.11**
"small"; 905 x 356 mm; IP65; without door	528.02	0.50	9.13	-	530.81	nr	**539.93**
"medium"; 1124 x 420 mm; IP54; with door	575.82	0.50	9.13	-	578.61	nr	**587.74**
"large"; 1124 x 470 mm; IP54; with door	654.06	0.50	9.13	-	656.85	nr	**665.97**
"medium"; gas lantern; 1124 x 420 mm; IP54; with door; integral solenoid and pilot	958.00	0.50	9.13	-	960.79	nr	**969.91**

V ELECTRICAL SUPPLY/POWER/LIGHTING SYSTEMS

Item Excluding site overheads and profit	PC £	Labour hours	Labour £	Plant £	Material £	Unit	Total rate £
Reproduction brackets and suspensions; Sugg Lighting							
Iron brackets							
"Bow" bracket; 6'0"	929.41	2.00	36.50	-	932.20	nr	**968.70**
"Ornate" iron bracket; large	415.10	2.00	36.50	-	417.89	nr	**454.39**
"Ornate" iron bracket; medium	399.11	2.00	36.50	-	401.90	nr	**438.40**
"Swan neck" iron bracket; large	390.92	2.00	36.50	-	393.71	nr	**430.21**
"Swan neck" iron bracket; medium	221.55	2.00	36.50	-	224.34	nr	**260.84**
Cast brackets							
"Universal" cast bracket	296.66	2.00	36.50	-	299.45	nr	**335.95**
"Abbey" bracket	142.49	2.00	36.50	-	145.28	nr	**181.78**
"Short Abbey" bracket	134.97	2.00	36.50	-	137.76	nr	**174.26**
"Plaza" cast bracket	141.85	2.00	36.50	-	144.64	nr	**181.14**
Base mountings							
"Universal" pedestal	294.25	2.00	36.50	-	297.04	nr	**333.54**
"Universal" plinth	166.23	2.00	36.50	-	169.02	nr	**205.52**
Reproduction lighting columns							
Reproduction lighting columns; Sugg Lighting							
"Harborne" C11; fabricated iron/steel heavy duty post with integral cast root; 3 - 5 m	1126.34	8.00	146.00	-	1133.78	nr	**1279.78**
"Aylesbury" C12; base fabricated heavy gauge 89 mm aluminium post; 3 - 5 m	1816.95	8.00	146.00	-	1824.39	nr	**1970.39**
"Cannonbury" C13; fabricated heavy gauge 89 mm aluminium post; 3 - 5 m	1132.05	8.00	146.00	-	1139.49	nr	**1285.49**
standard column C14; rooted British Steel 168/89 mm embellished post; 5 - 8 m	608.97	8.00	146.00	-	616.41	nr	**762.41**
"Large Constitution Hill" C22X; cast aluminium, steel cored post with extended spigot; 4.3 m	4935.00	6.00	109.50	-	4942.44	nr	**5051.94**
"Cardiff" C29; cast aluminium post; 3.5 m	2356.56	6.00	109.50	-	2364.00	nr	**2473.50**
"Seven Dials" C36; cast aluminium rooted post; 3.8 m	1768.15	8.00	146.00	-	1775.59	nr	**1921.59**
"Royal Exchange" C42; traditional cast aluminium post; welded multi arm construction; 2.1 m	2580.90	6.00	109.50	-	2588.34	nr	**2697.84**
Precinct lighting lanterns; ground, wall or pole mounted; including fixing and light fitting (lamps, poles, brackets, final painting, electric wiring, connections or related fixtures such as switch gear and time clock mechanisms not included)							
Precinct lighting lanterns; Targetti Poulsen							
"Nyhavn Park" side entry 90° curved arm mounted; steel rings over domed top; housing cast silumin sandblasted with integral gear; protected by UV stabilized clear polycarbonate diffuser; IP55	615.00	1.00	18.25	-	615.00	nr	**633.25**
"Kipp"; bottom entry pole-top; shot blasted or lacquered Hanover design award; hinged diffuser for simple maintenance; indirect lighting technique ensures low glare; IP55	402.00	1.00	18.25	-	402.00	nr	**420.25**
Precinct lighting lanterns; Sugg Lighting							
Juno Dome; IP66; molded GRP body; graphite finish	525.80	1.00	18.25	-	525.80	nr	**544.05**

V ELECTRICAL SUPPLY/POWER/LIGHTING SYSTEMS

Item Excluding site overheads and profit	PC £	Labour hours	Labour £	Plant £	Material £	Unit	Total rate £
V41 STREET/AREA/FLOODLIGHTING - cont'd							
Precinct lighting lanterns - cont'd							
Juno Cone; IP66; molded GRP body; graphite finish	525.80	1.00	18.25	-	525.80	nr	**544.05**
Sharkon; IP65; cast aluminium body; silver and blue finish	525.80	1.00	18.25	-	525.80	nr	**544.05**
Transformers for low voltage lighting; Targetti Poulsen - Landscape Division; distance from electrical supply 25 m; trenching and backfilling measured separately							
Boxed transformer							
50 Va for maximum 50 w of lamp	64.42	2.00	36.50	-	91.92	nr	**128.42**
100Va for maximum 100 w of lamp	68.15	2.00	36.50	-	95.65	nr	**132.15**
150 Va for maximum 150 w of lamp	68.16	2.00	36.50	-	95.66	nr	**132.16**
200 Va for maximum 200 w of lamp	84.95	2.00	36.50	-	112.45	nr	**148.95**
250 Va for maximum 50 w of lamp	99.36	2.00	36.50	-	126.86	nr	**163.36**
Buried transformers							
50 Va for maximum 50 w of lamp	64.37	2.00	36.50	-	91.87	nr	**128.37**
Installation; electric cable in trench 500 mm deep (trench not included); all in accordance with IEE regulations							
Light duty 600 volt grade armoured							
3 core 2.5 mm	1.10	0.01	0.09	-	1.10	m	**1.19**
4 core 2.5 mm	1.50	0.01	0.09	-	1.50	m	**1.59**
Twin and earth PVC cable in plastic conduit (conduit not included)							
2.50 mm^2	0.35	0.01	0.18	-	0.35	m	**0.53**
4.00 mm^2	0.72	0.01	0.18	-	0.72	m	**0.90**
16 mm heavy gauge high impact PVC conduit	3.60	0.03	0.61	-	3.60	m	**4.21**
Main switch and fuse unit; 30 A	70.23	-	-	-	70.23	nr	**-**
Weatherproof switched socket; 13 A	14.00	2.00	36.50	-	14.00	nr	**50.50**

ESSENTIAL READING FROM TAYLOR AND FRANCIS

Construction Delays
Extensions of Time and Prolongation Claims

Roger Gibson

Providing guidance on delay analysis, the author gives readers the information and practical details to be considered in formulating and resolving extension of time submissions and time-related prolongation claims. Useful guidance and recommended good practice is given on all the common delay analysis techniques. Worked examples of extension of time submissions and time-related prolongation claims are included.

Selected Contents:

1. Introduction
2. Programmes & Record Keeping
3. Contracts and Case Law
4. The 'Thorny Issues'
5. Extensions of Time
6. Prolongation Claims Summary

April 2008: 234x156: 374pp
Hb: 978-0-415-35486-6 **£75.00**

To Order: Tel: +44 (0) 1235 400524 **Fax:** +44 (0) 1235 400525
or Post: Taylor and Francis Customer Services,
Bookpoint Ltd, Unit T1, 200 Milton Park, Abingdon, Oxon, OX14 4TA UK
Email: book.orders@tandf.co.uk

For a complete listing of all our titles visit:
www.tandf.co.uk

Taylor & Francis
Taylor & Francis Group

ESSENTIAL READING FROM TAYLOR AND FRANCIS

Spon's Estimating Costs Guide to Minor Works,

Alterations and Repairs to Fire, Flood, Gale and Theft Damage

Fourth Edition

Bryan Spain

Specially written for contractors, quantity surveyors and clients carrying out small works, Spon's Estimating Costs Guide to Minor Works, Alterations and Repairs to Fire, Flood, Gale and Theft Damage contains accurate information on thousands of rates each broken down to labour, material overheads and profit.

Selected Contents: Introduction. Standard Method of Measurement/Trades Link. Part 1: Unit Rates. Part 2: Damage Repairs. Part 3: Approximate Estimating. Part 4: Plant and Tool Hire. Part 5: General Construction Data. Part 6: Business Matters

August 2008: 216x138: 320pp
Pb: 978-0-415-46906-7: **£29.99**

To Order: Tel: +44 (0) 1235 400524 **Fax:** +44 (0) 1235 400525
or Post: Taylor and Francis Customer Services,
Bookpoint Ltd, Unit T1, 200 Milton Park, Abingdon, Oxon, OX14 4TA UK
Email: book.orders@tandf.co.uk

For a complete listing of all our titles visit:
www.tandf.co.uk

Taylor & Francis
Taylor & Francis Group

PART 4

Approximate Estimates – Minor Works

APPROXIMATE ESTIMATES

Prices in this section are based upon the Prices for Measured Works, but allow for incidentals which would normally be measured separately in a Bill of Quantities. They do not include for Preliminaries which are priced elsewhere in this book.

Items shown as subcontract or specialist rates would normally include the specialist's overhead and profit. All other items which could fall within the scope of works of general landscape and external works contractors would not include profit.

Based on current commercial rates, profits of 15% to 35% may be added to these rates to indicate the likely "with profit" values of the tasks below. The variation quoted above is dependent on the sector in which the works are taking place - domestic, public or commercial.

DEMOLITION AND SITE CLEARANCE

Item Excluding site overheads and profit	Unit	Total rate £
DEMOLITION AND SITE CLEARANCE		
Demolish existing surfaces by hand held electric breaker; removal by grab		
Break up plain concrete slab; remove to licensed tip;		
150 mm thick	m^2	20.00
200 mm thick	m^2	20.00
Break up reinforced concrete slab and remove to licensed tip		
150 mm thick	m^2	25.00
200 mm thick	m^2	32.00
300 mm thick	m^2	60.00
Break out existing surface and associated 150 mm thick granular base load to remove off site by grab		
macadam 70 mm thick	m^2	14.76
block paving 50 mm thick	m^2	20.15
block paving 80 mm thick	m^2	22.78
Demolish existing free standing walls; grub out foundations; remove arisings to skip; backfill with imported topsoil; works by excavator, dumper and diesel breaker		
Brick wall; 112 mm thick		
300 mm high	m	23.09
500 mm high	m	31.34
Brick wall; 225 mm thick		
300 mm high	m	32.05
500 mm high	m	34.39
1.00 m high	m	42.98
1.20 m high	m	50.01
1.50 m high	m	52.92
1.80 m high	m	60.47
Demolish existing free standing walls; grub out foundations; remove arisings to skip; backfill with imported topsoil; works by hand and diesel breaker		
Brick wall; 112 mm thick		
300 mm high	m	24.58
500 mm high	m	27.07
Brick wall; 225 mm thick		
300 mm high	m	34.69
500 mm high	m	37.91
1.00 m high	m	49.72
1.20 m high	m	59.64
1.50 m high	m	63.06
1.80 m high	m	72.31
Break out existing free standing building; break out plain concrete base 150 mm thick and remove to skip distance 50 m; backfill with imported topsoil; all works by hand		
Timber buildings		
shed 6.0 m^2	nr	311.02
shed 10.0 m^2	nr	683.03
shed 15.0 m^2	nr	947.93
Site clearance - General		
Clear away light fencing and gates (chain link, chestnut paling, light boarded fence or similar) and remove to licensed tip	100 m	374.00
Strip turf; strip topsoil 250 mm thick move to stockpile 25 m		
all by machine; disposal of turf of to skip	100 m	711.86
by machine; disposal of turf by grab	100 m	729.05
strip and stack turf for preservation by hand; strip soil by machine	100 m	890.25
by hand; disposal to skip	100 m	2197.75

DEMOLITION AND SITE CLEARANCE

Item Excluding site overheads and profit	Unit	Total rate £
Clear mixed shrub area; dig out roots Groundcovers and small shrubs 20%; shrubs 1.00 -2.00 m high 40%; shrubs 2. 00 - 3.00 m high 20%; shrubs over 3.00 m high 20%		
clearance only	m²	**30.41**
disposal to skip; chipped	m²	**18.00**
disposal to skip; unchipped	m²	**35.65**
disposal on site; unchipped	m²	**4.67**

GROUNDWORK

Item Excluding site overheads and profit	Unit	Total rate £
EXCAVATION AND FILLING		
Cut and strip by machine turves 50 mm thick		
Load to barrows and stack on site not exceeding 25 m travel to stack	100 m²	530.00
Load to barrows and disposal off site by skip; distance 25 m	100 m²	1170.00
Excavate to reduce levels; mechanical		
Removal ro spoil heaps		
excavated directly to loading position	m³	7.80
transporting to loading position 25 m distance	m³	14.00
Excavate to reduce levels; mechanical		
Removal off site by grab		
excavated directly to loading position	m³	48.00
transporting to loading position 25 m distance	m³	53.00
Excavate to reduce levels; mechanical		
Removal off site by skip		
excavated directly to loading position	m³	57.00
transporting to loading position 25 m distance	m³	62.00
Excavate to reduce levels; hand		
Removal off site by skip		
excavated directly to loading position	m³	180.00
excavated directly to loading position	m³	230.00
filled to bags and transporting to loading position 25 m distant	m³	360.00
Excavation and filling; mechanical		
Excavate existing soil on proposed turf or planting area to reduce levels; grade to levels; fill excavated area with topsoil from spoil heaps; removal of and excavated material by skip		
100 mm deep	m²	7.20
200 mm deep	m²	15.00
300 mm deep	m²	22.00
Spread excavated material to levels in layers not exceeding 150 mm; grade to finished levels to receive surface treatments		
By machine		
average thickness 100 mm	m²	2.00
average thickness 100 mm but with imported topsoil	m²	5.00
average thickness 200 mm	m²	2.00
average thickness 200 mm but with imported topsoil	m²	9.00
average thickness 250 mm	m²	2.00
average thickness 250 mm but with imported topsoil	m²	20.00
extra for work to banks exceeding 30° slope	30%	-
Filling by hand		
Excavate existing soil on proposed turf or planting area to reduce levels; grade to levels; fill excavated area with topsoil from spoil heaps; removal of excavated material by skip		
100 mm deep	m²	8.80
200 mm deep	m²	18.00
300 mm deep	m²	27.00
500 mm deep	m²	44.00

GROUNDWORK

Item Excluding site overheads and profit	Unit	Total rate £
TRENCHES		
Excavate trenches; remove excavated material off site by grab 25 m distance; fill trench to ground level with site mixed concrete 1:3:6; allow for movement of concrete and excavated material		
By machine		
300 mm wide x 250 mm deep	m	**15.05**
500 mm wide x 250 mm deep	m	**22.06**
750 mm wide x 350 mm deep	m	**44.29**
1200 mm wide x 600 mm deep	m	**122.49**
By hand		
300 mm wide x 250 mm deep	m	**24.10**
500 mm wide x 250 mm deep	m	**38.66**
750 mm wide x 350 mm deep	m	**84.90**
1200 mm wide x 600 mm deep	m	**233.73**

IN SITU CONCRETE

Item Excluding site overheads and profit	Unit	Total rate £
IN SITU CONCRETE		
Mix concrete on site; aggregates delivered in 10 tonne loads; deliver mixed concrete to location by mechanical dumper distance 25 m		
1:3:6	m^3	111.00
1:2:4	m^3	124.00
As above but ready mixed concrete		
10 N/mm^2	m^3	125.00
15 N/mm^2	m^3	132.00
Mix concrete on site; aggregates delivered in 10 tonne loads; deliver mixed concrete to location by barrow distance 25 m		
1:3:6	m^3	154.00
1:2:4	m^3	168.00
As above but aggregates delivered in 850 kg bulk bags		
1:3:6	m^3	190.00
1:2:4	m^3	204.00
As above but concrete discharged directly from ready mix lorry to required location		
10 N/mm^2	m^3	120.00
15 N/mm^2	m^3	130.00
Excavate foundation trench mechanically; remove spoil offsite by grab; lay 1:3:6 site mixed concrete foundations; distance from mixer 25 m; depth of trench to be 225 mm deeper than foundation to allow for three underground brick courses priced separately		
Foundation size		
200 mm deep x 400 mm wide	m	21.00
300 mm deep x 500 mm wide	m	36.00
400 mm deep x 400 mm wide	m	37.00
400 mm deep x 600 mm wide	m	55.00
600 mm deep x 600 mm wide	m	78.00
As above but hand excavation and disposal to spoil heap 25 m by barrow; disposal off site by grab		
200 mm deep x 400 mm wide	m	44.00
300 mm deep x 500 mm wide	m	51.00
400 mm deep x 400 mm wide	m	49.00
400 mm deep x 600 mm wide	m	110.00
600 mm deep x 600 mm wide	m	150.00

BRICK/BLOCK WALLING

Item Excluding site overheads and profit	Unit	Total rate £
BRICK WALLING		
Excavate foundation trench 500 mm deep; remove spoil to dump off site; (all by machine) lay site mixed concrete foundations 1:3:6 350 x 150 mm thick; construct half brick wall with one brick piers at 2.0 m centres; laid in cement:lime:sand (1:1:6) mortar with flush joints; fair face one side; DPC two courses underground; engineering brick in cement:sand (1:3) mortar; coping of headers on end		
Wall 900 mm high above DPC		
in engineering brick (class B) PC £290.00/1000	m	236.00
in sandfaced facings PC £300.00/1000	m	245.00
in reclaimed bricks PC £1,000.00/1000	m	354.00
Excavate foundation trench 400 mm deep; remove spoil to dump off site; lay GEN 1 concrete foundations 450 mm wide x 250 mm thick; construct one brick wall with one and a half brick piers at 3.0 m centres; all in English Garden Wall bond; laid in cement:lime:sand (1:1:6) mortar with flush joints, fair face one side; DPC two courses engineering brick in cement:sand (1:3) mortar; engineering brick coping		
Wall 900 mm high above DPC		
in engineering brick (class B) PC £290.00/1000	m	359.00
in sandfaced facings PC £300.00/1000	m	362.00
in reclaimed bricks PC £1,000.00/1000	m	445.00
Wall 1200 mm high above DPC		
in engineering brick (class B) PC £290.00/1000	m	419.00
in sandfaced facings PC £300.00/1000	m	421.00
in reclaimed bricks PC £1,000.00/1000	m	521.00
Wall 1800 mm high above DPC		
in engineering brick (class B) PC £290.00/1000	m	614.00
in sandfaced facings PC £300.00/1000	m	619.00
in reclaimed bricks PC £1,000.00/1000	m	770.00
BLOCK WALLING		
Excavate foundation trench 450 mm deep; remove spoil to dump off site; lay GEN 1 concrete foundations 600 x 300 mm thick; construct wall of concrete block; two courses below ground		
Solid blocks 7 N/mm^2; wall 1.00 m high		
100 mm thick	m	100.00
Hollow blocks filled with concrete; wall 1.00 m high		
215 mm thick	m	160.00
Hollow blocks but with steel bar cast into the foundation; wall 1.00 m high		
215 mm thick	m	164.00
Solid blocks 7 N/mm^2; wall 1.80 m high		
100 mm thick	m	150.00
Hollow blocks filled with concrete; wall 1.80 m high		
215 mm thick	m	240.00
Hollow blocks but with steel bar cast into the foundation wall 1.80 m high		
215 mm thick	m	248.17

ROADS AND PAVINGS

Item Excluding site overheads and profit	Unit	Total rate £
BASES FOR PAVING		
Excavate ground and reduce levels to receive 38 mm thick slab and 25 mm mortar bed; dispose of excavated material off site; treat substrate with total herbicide Lay granular fill Type 1 150 mm thick laid to falls and compacted		
all by machine	m^2	19.00
all by hand except disposal by grab	m^2	37.00
Excavate ground and reduce levels to receive 65 mm thick surface and bed (not included); dispose of excavated material off site; treat substrate with total herbicide Lay 100 mm compacted hardcore; lay 1:2:4 concrete base 150 mm thick laid to falls		
all by machine	m^2	42.00
all by hand except disposal by grab	m^2	84.00
As above but inclusive of reinforcement fabric A142		
all by machine	m^2	48.00
all by hand except disposal by grab	m^2	90.00
Lay 100 mm compacted hardcore; lay 1:3:6 concrete base 150 mm thick laid to falls		
all by machine	m^2	36.00
all by hand except disposal by grab	m^2	78.00
As above but inclusive of reinforcement fabric A142		
all by machine	m^2	42.00
all by hand except disposal by grab	m^2	83.00
KERBS AND EDGINGS		
Note: excavation is by machine unless otherwise mentioned		
Excavate trench and construct concrete foundation 150 mm wide x 150 mm deep; lay precast concrete kerb units bedded in semi-dry concrete; slump 35 mm maximum; haunching one side; disposal of arisings off site Edgings laid straight		
50 x 150 mm	m	33.00
125 mm high x 150 mm wide; bullnosed	m	33.00
50 x 200 mm	m	34.00
Second hand granite setts		
100 x 100 mm	m	64.00
Single course; brick or block edgings laid to stretcher		
concrete blocks 200 x 100 mm	m	27.00
engineering bricks	m	32.00
paving bricks; PC £500.00/1000	m	33.00
Double course; brick or block edgings laid to stretcher		
concrete blocks 200 x 100 mm	m	30.00
engineering bricks	m	37.00
paving bricks; PC £500.00/1000	m	39.00
Single course brick or block edgings laid to header course (soldier course)		
blocks 200 x 100 x 60 mm; PC £8.03/m^2; butt jointed	m	30.00
bricks 200 x 100 x 50 mm; PC £300.00/1000; butt jointed	m	33.00
bricks 200 x 100 x 50 mm; PC £300.00/1000; mortar joints	m	37.00
Sawn yorkstone edgings; excavate for groundbeam; lay concrete 1:2:4 150 mm deep x 33.3% wider than the edging; on 35 mm thick mortar bed; inclusive of haunching one side Yorkstone 50 mm thick		
100 mm wide x random lengths	m	38.00
100 x 100 mm	m	42.00
200 mm wide x 100 mm long	m	50.00
250 mm wide x random lengths	m	47.00
500 mm wide x random lengths	m	84.00

ROADS AND PAVINGS

Item Excluding site overheads and profit	Unit	Total rate £
Timber edgings; softwood		
150 x 38 mm		
laid straight	m	4.59
laid to curves	m	6.64
INTERLOCKING BLOCK PAVING		
Excavate ground; supply and lay granular fill Type 1 150 mm thick laid to falls and compacted; supply and lay block pavers; laid on 50 mm compacted sharp sand; vibrated; joints filled with loose sand excluding edgings or kerbs measured separately		
Concrete blocks		
200 x 100 x 60 mm	m²	62.00
200 x 100 x 80 mm	m²	64.00
Reduce levels; lay 150 mm granular material Type 1; lay precast concrete edging 50 x 150 mm; on concrete foundation 1:2:4; lay 200 x 100 x 60 mm vehicular block paving to 90° herringbone pattern; on 50 mm compacted sand bed; vibrated; jointed in sand and vibrated		
1.0 m wide clear width between edgings	m	110.00
1.5 m wide clear width between edgings	m	140.00
2.0 m wide clear width between edgings	m	170.00
3.0 m wide clear width between edgings	m	230.00
Works by hand; reduce levels; lay 150 mm granular material Type 1; lay edge restraint of block paving 200 mm wide on 150 mm thick concrete foundation 1:2:4 haunched; lay 200 x 100 x 60 mm vehicular block paving to 90° herringbone pattern; on 50 mm compacted sand bed; vibrated; jointed in sand and vibrated;		
1.0 m wide clear width between edgings	m	96.00
1.5 m wide clear width between edgings	m	130.00
2.0 m wide clear width between edgings	m	160.00
3.0 m wide clear width between edgings	m	220.00
Works by hand; reduce levels; lay 150 mm granular material Type 1; lay edge restraint of block paving 200 mm wide on 150 mm thick concrete foundation 1:2:4 haunched; lay 200 x 100 x 60 mm vehicular block paving to 90° herringbone pattern; on 50 mm compacted sand bed; vibrated; jointed in sand and vibrated		
1.0 m wide clear width between edgings	m	120.00
1.5 m wide clear width between edgings	m	150.00
2.0 m wide clear width between edgings	m	190.00
3.0 m wide clear width between edgings	m	260.00
BRICK PAVING		
WORKS BY MACHINE		
Excavate and lay base Type 1 150 mm thick remove arisings; all by machine; lay clay brick paving		
200 x 100 x 50 mm thick; butt jointed on 50 mm sharp sand bed		
PC £300.00/1000	m²	72.00
PC £600.00/1000	m²	87.00
200 x 100 x 50 mm thick; 10 mm mortar joints on 35 mm mortar bed		
PC £300.00/1000	m²	86.00
PC £600.00/1000	m²	100.00

ROADS AND PAVINGS

Item Excluding site overheads and profit	Unit	Total rate £
WORKS BY MACHINE – cont'd		
Excavate and lay 150 mm site mixed concrete base 1:3:6 reinforced with A142 mesh; all by machine; remove arisings; lay clay brick paving		
200 x 100 x 50 mm thick; 10 mm mortar joints on 35 mm mortar bed; running or stretcher bond		
PC £300.00/1000	m^2	**120.00**
PC £600.00/1000	m^2	**130.00**
200 x 100 x 50 mm thick; 10 mm mortar joints on 35 mm mortar bed; butt jointed; herringbone bond		
PC £300.00/1000	m^2	**98.00**
PC £600.00/1000	m^2	**120.00**
Excavate and lay readymix concrete base 150 mm thick reinforced with A393 mesh; all by machine; remove arisings; lay clay brick paving		
215 x 102.5 x 50 mm thick; 10 mm mortar joints on 35 mm mortar bed		
PC £300.00/1000	m^2	**88.00**
PC £600.00/1000	m^2	**110.00**
WORKS BY HAND		
Excavate and lay base Type 1 150 mm thick by hand; arisings barrowed to spoil heap maximum distance 25 m and removal off site by grab; lay clay brick paving		
200 x 100 x 50 mm thick; butt jointed on 50 mm sharp sand bed		
PC £300.00/1000	m^2	**83.90**
PC £600.00/1000	m^2	**99.30**
Excavate and lay 150 mm concrete base; 1:3:6: site mixed concrete reinforced with A142 mesh; remove arisings to stockpile and then off site by grab; lay clay brick paving		
215 x 102.5 x 50 mm thick; 10 mm mortar joints on 35 mm mortar bed		
PC £300.00/1000	m^2	**107.00**
PC £600.00/1000	m^2	**122.00**
NATURAL STONE/SLAB PAVING		
WORKS BY MACHINE		
Yorkstone slabs; excavate ground by machine and reduce levels to receive 65 mm thick slab and 35 mm mortar bed; dispose of excavated material off site; treat substrate with total herbicide; lay granular fill Type 1 150 mm thick laid to falls and compacted; lay to random rectangular pattern on 35 mm mortar bed; Yorkstone		
New riven slabs		
laid random rectangular	m^2	**142.00**
New riven slabs; but to 150 mm plain concrete base		
laid random rectangular	m^2	**158.00**
Reclaimed Cathedral grade riven slabs		
laid random rectangular	m^2	**181.00**
Reclaimed Cathedral grade riven slabs; but to 150 mm plain concrete base		
laid random rectangular	m^2	**197.00**
New slabs sawn 6 sides		
laid random rectangular	m^2	**126.00**
3 sizes; laid to coursed pattern	m^2	**130.00**
New slabs sawn 6 sides; but to 150 mm plain concrete base		
laid random rectangular	m^2	**140.00**
3 sizes; laid to coursed pattern	m^2	**145.00**

ROADS AND PAVINGS

Item Excluding site overheads and profit	Unit	Total rate £
Indian sandstone slabs; excavate ground by machine and reduce levels to receive 65 mm thick slab and 35 mm mortar bed; dispose of excavated material off site; treat substrate with total herbicide; lay granular fill Type 1 150 mm thick laid to falls and compacted; lay to random rectangular pattern on 35 mm mortar bed;		
New riven slabs		
laid random rectangular	m²	98.60
WORKS BY HAND		
Yorkstone slabs; excavate ground by hand and reduce levels to receive 65 mm thick slab and 35 mm mortar bed; barrow all materials and arisings 25 m; dispose of excavated material off site by grab; treat substrate with total herbicide; lay granular fill Type 1 150 mm thick laid to falls and compacted; lay to random rectangular pattern on 35 mm mortar bed		
New riven slabs		
laid random rectangular	m²	158.00
New riven slabs laid random rectangular; but to 150 mm plain concrete base		
laid random rectangular	m²	173.00
Reclaimed Cathedral grade riven slabs		
laid random rectangular	m²	197.00
Reclaimed Cathedral grade riven slabs; but to 150 mm plain concrete base		
laid random rectangular	m²	211.00
New slabs sawn 6 sides		
laid random rectangular	m²	132.00
3 sizes; laid to coursed pattern	m²	141.00
New slabs sawn 6 sides; but to 150 mm plain concrete base		
laid random rectangular	m²	146.00
3 sizes; laid to coursed pattern	m²	154.00
Indian sandstone slabs; excavate ground by hand and reduce levels to receive 65 mm thick slab and 35 mm mortar bed; barrow all materials and arisings 25 m; dispose of excavated material off site by grab; treat substrate with total herbicide; lay granular fill Type 1 150 mm thick laid to falls and compacted; lay to random rectangular pattern on 35 mm mortar bed		
New riven slabs		
laid random rectangular	m²	115.00

PREPARATION FOR SEEDING/TURFING

Item Excluding site overheads and profit	Unit	Total rate £
SURFACE PREPARATIONS		
Cultivate existing ground by pedestrian operated rotavator; spread and lightly consolidate topsoil brought from spoil heap 25 m distance in layers not exceeding 150 mm; grade to specified levels; remove stones over 25 mm; rake and grade to a fine tilth		
By machine		
100 mm thick	100 m^2	163.00
200 mm thick	100 m^2	256.00
300 mm thick	100 m^2	349.00
500 mm thick	100 m^2	535.00
By hand		
100 mm thick	100 m^2	480.00
150 mm thick	100 m^2	686.00
300 mm thick	100 m^2	1300.00
500 mm thick	100 m^2	2350.00
Lift and remove existing turf to skip 25 m distance; cultivate surface to receive new turf; rake to a fine tilth		
Works by hand		
normal turfed area	100 m^2	606.00
compacted turfed area	100 m^2	623.00
Lift and remove existing turf to skip 25 m distance; cultivate surface to receive new turf; rake to a fine tilth; fill area to receive turf with 50 mm imported topsoil		
Works by hand		
normal turfed area	100 m^2	846.15
compacted turfed area	100 m^2	863.15
SEEDING AND TURFING		
Domestic lawn areas		
Cultivate recently filled topsoil area; grade to levels and falls and rake to remove stones and debris; add fertilizers; add surface treatment as specified		
Turf areas		
Rolawn medallion	m^2	5.89
Seeded areas		
grass seed PC £4.50/kg; application rate 35 g/m^2	m^2	1.17
grass seed PC £4.50/kg; application rate 50 g/m^2	m^2	1.24
grass seed PC £6.00/kg; application rate 35 g/m^2	m^2	1.22
grass seed PC £6.00/kg; application rate 50 g/m^2	m^2	1.36
Cultivate recently filled topsoil area; grade to levels and falls and rake to remove stones and debris; add fertilizers; add surface treatment as specified; maintain for one year watering and cutting 26 times during the summer; pedestrian mower with grass box; arisings removed off site		
Turf areas		
Rolawn medallion	m^2	8.09
Seeded areas		
grass seed PC £4.50/kg; application rate 35 g/m^2	m^2	3.37
grass seed PC £4.50/kg; application rate 50 g/m^2	m^2	3.44
grass seed PC £6.00/kg; application rate 35 g/m^2	m^2	3.46
grass seed PC £6.00/kg; application rate 50 g/m^2	m^2	3.56

PLANTING

Item Excluding site overheads and profit	Unit	Total rate £
TREE PLANTING		
Excavate tree pit by hand; fork over bottom of pit; plant tree with roots well spread out; backfill with excavated material, incorporating treeplanting compost at 1 m³ per 3 m³ of soil, one tree stake and two ties; tree pits square in sizes shown		
Light standard bare root tree in pit; PC £9.45		
600 x 600 mm deep	each	38.00
900 x 900 mm deep	each	49.00
Standard tree bare root tree in pit; PC £11.81		
600 x 600 mm deep	each	40.76
900 x 600 mm deep	each	53.00
Standard root balled tree in pit; PC £19.69		
600 x 600 mm deep	each	51.00
900 x 600 mm deep	each	66.00
Selected standard bare root tree in pit; PC £19.16		
900 x 900 mm deep	each	68.00
1.00 x 1.00 m x 600 mm deep	each	95.00
Selected standard root ball tree in pit; PC £29.66		
900 x 600 mm deep	each	78.00
1.00 x 1.00 m x 600 mm deep	each	105.00
Heavy standard bare root tree in pit; PC £35.17		
900 x 900 mm deep	each	89.00
1.00 x 1.00 m x 600 mm deep	each	97.00
Heavy standard root ball tree in pit; PC £50.92		
900 x 600 mm deep	each	110.00
1.00 x 1.00 m x 600 mm deep	each	131.00
Extra heavy standard bare root tree in pit; PC £44.10		
1.00 x 1.00 m deep	each	129.00
Extra heavy standard root ball tree in pit; PC £59.85		
1.00 x 1.00 m deep	each	149.00
1.50 m x 750 mm deep	each	178.00
SHRUB PLANTING		
Treat recently filled ground with systemic weedkiller; cultivate ground and clear arisings; add mushroom composts and fertilizers		
Cultivation by rotavator; all other works by hand		
compost 50 mm; general purpose fertilizer 35 g/m²	m²	4.30
compost 100 mm; general purpose fertilizer 35 g/m²	m²	7.00
compost 100 mm; Enmag 35 g/m²	m²	7.00
All works by hand		
compacted ground; compost 50 mm; general purpose fertilizer 35 g/m²	m²	5.00
ground previously planted but cleared of vegetable matter; compost 50 mm; general purpose fertilizer 35 g/m²	m²	4.80
ground previously planted but cleared of vegetable matter; compost 50 mm; general purpose fertilizer 35 g/m²	m²	4.70
Excavate planting holes on 300 x 300 x 300 mm deep to area previously prepared; plant shrubs PC £3.00 each in groups of 3 to 5 inclusive of transportfrom holding area setting out and final mulching 50 mm thick		
By hand		
300 mm centres (11.11 plants per m²)	m²	66.00
400 mm centres (6.26 plants per m²)	m²	41.00
500 mm centres (4 plants per m²)	m²	26.00
750 mm centres (1.78 plants per m²)	m²	13.00

PLANTING

Item Excluding site overheads and profit	Unit	Total rate £
SHRUB PLANTING – cont'd		
Cultivate and grade shrub bed; bring top 300 mm of topsoil to a fine tilth, incorporating mushroom compost at 50 mm and Enmag slow release fertilizer; rake and bring to given levels; remove all stones and debris over 50 mm; dig planting holes average 300 x 300x 300 mm deep; supply and plant specified shrubs in quantities as shown below; backfill with excavated material as above; water to field capacity and mulch 50 mm bark chips 20 - 40 mm size; water and weed regularly for 12 months; shrubs 2 litre PC £2.80; ground covers 9 cm PC £1.50		
Shrubs at centres shown below		
300 mm centres	m²	63.00
400 mm centres	m²	41.00
500 mm centres	m²	28.00
600 mm centres	m²	22.00
750 mm centres	m²	17.00
900 mm centres	m²	14.00
Groundcover 30% / shrubs 70% at the distances shown below		
200/300 mm	m²	67.00
300/400 mm	m²	57.00
300/500 mm	m²	34.00
Groundcover 50% / shrubs 50% at the distances shown below		
200/300 mm	m²	77.00
300/400 mm	m²	42.00
300/500 mm	m²	37.00
400/500 mm	m²	28.00
Cultivate ground by machine and rake to level; plant bulbs as shown; bulbs PC £25.00/100		
15 bulbs per m²	m²	6.90
25 bulbs per m²	m²	12.00
50 bulbs per m²	m²	23.00
BEDDING		
Spray surface with glyphosate; lift and dispose of turf when herbicide action is complete; cultivate new area for bedding plants to 400 mm deep; spread compost 100 mm deep and chemical fertilizer "Enmag" and rake to fine tilth to receive new bedding plants; remove all arisings to skip		
Existing turf area		
disposal to skip	100 m²	849.00
disposal to compost area on site; distance 25 m	100 m²	499.00
Plant bedding to existing planting area; bedding planting PC £0.25 each		
Clear existing bedding; cultivate soil to 230 mm deep; incorporate compost 75 mm and rake to fine tilth; collect bedding from nursery and plant at 100 mm ccs; irrigate on completion; maintain weekly for 12 weeks		
mass planted; 100 mm centres	m²	33.00
to patterns; 100 mm centres	m²	36.00
mass planted; 150 mm centres	m²	19.00
to patterns; 150 mm centres	m²	22.00
mass planted; 200 mm centres	m²	13.00
to patterns; 200 mm centres	m²	16.00
Extra for watering by hand held hose pipe		
Flow rate 25 litres/minute		
10 litres/m²	100 m²	16.00
15 litres/m²	100 m²	23.00
20 litres/m²	100 m²	30.00

PLANTING

Item Excluding site overheads and profit	Unit	Total rate £
25 litres/m^2	100 m^2	38.00
Flow rate 40 litres/minute		
10 litres/m^2	100 m^2	9.50
15 litres/m^2	100 m^2	15.00
20 litres/m^2	100 m^2	19.00
25 litres/m^2	100 m^2	24.00

MAINTENANCE

Item Excluding site overheads and profit	Unit	Total rate £
GRASS MAINTENANCE		
Cut grass with pedestrian mower; strim edges of shrub beds and edges to pavings; length of edges is 10% of the grass area; remove arisings to compost heap on site		
lawn cut 22 times per year; cost per occassion		
pedestrian mower 400 mm wide	100 m^2	4.46
pedestrian mower 450 mm wide	100 m^2	3.76
pedestrian mower 500 mm wide	100 m^2	3.57
lawn cut 18 times per year; cost per occassion		
pedestrian mower 400 mm wide	100 m^2	4.98
pedestrian mower 450 mm wide	100 m^2	4.28
pedestrian mower 500 mm wide	100 m^2	4.09
SHRUB BED MAINTENANCE		
Prune shrubs; cut back to soft growth; shrubs wich have previously been pruned in the last 3 years		
Shrubs 1.00 - 2.00 m high		
1.00 m centres	10 m^2	11.40
750 mm centres	10 m^2	20.26
Shrubs		
1.00 m centres	10 m^2	8.20
750 mm centres	10 m^2	14.57
500 mm centres	10 m^2	32.80
400 mm centres	10 m^2	51.25
Weed newly planted shrub bed; irrigate to field capacity by hand; fork over surface of plant bed every four visits		
Regular visits		
newly planted areas; per visit	m^2	0.79
established areas; per visit	m^2	0.48
HEDGE MAINTENANCE		
Cut hedge by hand operated machinery; remove arising to compost heap on site; 2 sides and top		
Hedge to 2.00 m high; depth of cut 300 mm		
hedge to 1.0 m tall	m	0.95
hedge to 1.5 m tall	m	1.70
hedge to 2.0 m tall	m	3.40
Reduce height of ornamental hedge using hand tools or mechanical hand tools; thickness of branches average 20-50 mm diameter; collect arisings and dispose on site; hedge measured 2 sides and top; assumed thickness 600 mm		
Hedge to 2.00 m high		
reduce by 600 mm	m	2.32
reduce by 900 mm	m	7.11
Hedge 2.00 - 4.00 m high		
reduce by 600 mm	m	3.39
reduce by 900 mm	m	8.57

DRAINAGE

Item Excluding site overheads and profit	Unit	Total rate £
DRAINAGE		
Pipe laying		
Excavate trench by excavator 600 mm deep; lay bedding and backfill as per material specification below; lay non woven geofabric and fill with topsoil to ground level		
100 mm vitrified clay; laid on earth with excavated backfill	m	15.00
110 mm PVC-u drainpipe; laid on sand bed with gravel backfill	m	18.00
Excavate trench by hand 600 mm deep; lay bedding and backfill as per material specification below; lay non woven geofabric and fill with topsoil to ground level		
100 mm vitrified clay; laid on earth with excavated backfill	m	20.00
110 mm PVC-u drainpipe; laid on sand bed with gravel backfill	m	22.00
LINEAR DRAINAGE		
Linear drainage to design sensitive areas		
Excavate trench by machine; lay Aco Brickslot channel drain on concrete base and surround to falls; all to manufacturers specifications		
paving surround to both sides of channel	m	130.00
Linear drainage to pedestrian area		
Excavate trench by machine; lay Aco MultiDrain MD Brickslot; offset galvanized slot drain grating for M100PPD; load class C250 channel drain on concrete base and surround to falls; all to manufacturers specifications; paving surround to channel with brick paving PC £300.00/1000		
brickslot galvanized grating; paving surround to one side of channel	m	97.00
slotted galvanized grating; paving surround to both sides of channel	m	110.00
stainless steel grating; paving surround to one side of channel	m	180.00
Linear drainage to light vehicular area		
Excavate trench by machine; lay Aco NK100 channel drain on concrete base and surround to falls; all to manufacturers specifications; paving surround to channel with brick paving PC £300.00/1000		
slotted galvanized grating; paving surround to one side of channel	m	130.00
slotted galvanized grating; paving surround to both sides of channel	m	139.00
"Heelguard" composite grating; paving surround to one side of channel	m	129.00
"Heelguard" ductile grating; paving surround to both sides of channel	m	131.00
Accessories for channel drain		
Sump unit with sediment bucket	nr	120.00
End cap; inlet/outlet	nr	19.00
MANHOLES		
Inspection chambers; brick manhole; excavate pit for inspection chamber including earthwork support and disposal of spoil to dump on site not exceeding 100 m; lay concrete (1:2:4) base 1500 mm diameter x 200 mm thick; 110 mm vitrified clay channels; benching in concrete (1:3:6) allowing one outlet and two inlets for 110 mm diameter pipe; construct inspection chamber one brick thick walls of engineering brick Class B; backfill with excavated material; complete with two cast iron step irons; cover 600 x 450 mm ductile iron; light vehicle loading		
1200 x 1200 x 1200 mm		
excavation by machine	each	1030.00
excavation by hand	each	2050.00
1200 x 1200 x 1200 mm; by machine but with recessed cover		
600 x 450 mm; 5 tonne loading; filled with block pavers	each	1150.00
1200 x 1200 x 1500 mm		
excavation by machine	each	1190.00
excavation by hand	each	2350.00

DRAINAGE

Item Excluding site overheads and profit	Unit	Total rate £
MANHOLES – cont'd		
Inspection chambers; polypropylene; excavate pit for inspection chamber including earthwork support and disposal of spoil to dump on site not exceeding 100 m; lay concrete (1:2:4) base 700 mm diameter x 200 mm thick		
Polypropylene 600 mm deep		
excavation by machine	each	460.00
excavation by hand	each	427.00
Polypropylene 1200 mm		
excavation by machine	each	459.00
excavation by hand	each	526.00
GULLIES		
Clay gully		
Excavate hole; supply and set in concrete vitrified clay trapped mud (dirt) gully complete with galvanized bucket and cast iron hinged locking grate and frame; lay kerb to gully		
1 tonne loading; RGP5; 100 mm outlet; 100 mm diameter; 225 mm internal width; 585 mm internal depth	each	330.00
Gullies PVC-u		
Excavate hole and lay 100 mm concrete (C20P) base 150 x 150 mm to suit given invert level of drain; connect to drainage system; backfill with Type 1 granular fill; install gully; complete with grate and frame; brick kerb to gully surround		
PVC-u universal gully with vertical hopper; plastic grid and frame included	each	190.00
PVC-u gully 100 mm with P trap; grid and frame included	each	200.00
yard gully trapped; 300 mm diameter; 600 mm deep; sediment bucket and ductile iron cover	each	350.00
SOAKAWAYS		
Soakaway "Aquacell" Wavin Plastics Ltd; preformed polypropylene soakaway infiltration crate units		
Excavate pits or trenches for soakaway units; place Aquacell polypropylene soakaway units in recommended configurations laid on 100 mm shingle; cover with terram and backfill to sides and top with 100 mm shingle; backfill with 150 mm topsoil		
4 crates; 760 litres	nr	300.00
8 crates; 1520 litres	nr	532.00
12 crates; 2280 litres	nr	874.00
16 crates; 3050 litres	nr	1060.00

ESSENTIAL READING FROM TAYLOR AND FRANCIS

Ethics for the Built Environment

Peter Fewings

Much closer relationships are being formed in the development of the built environment and cultural changes in procurement methods are taking place. A need has grown to re-examine the ethical frameworks required to sustain collaborative trust and transparency. Young professionals are moving around between companies on a frequent basis and take their personal values with them. What can companies do to support their employees?

The book looks at how people develop their personal values and tries to set up a model for making effective ethical decisions.

Selected Contents:

1 Business Ethics
2 Professional Practice
3 The Ethics of Employment
4 Environmental Sustainability
5 Health and Safety Ethics
6 Ethical Relationships
7 Corporate Social Responsibility
CASE STUDIES

August 2008: 234x156: 384pp
Hb: 978-0-415-42982-5 **£90.00**
Pb: 978-0-415-42983-2 **£29.99**

To Order: Tel: +44 (0) 1235 400524 **Fax:** +44 (0) 1235 400525
or Post: Taylor and Francis Customer Services,
Bookpoint Ltd, Unit T1, 200 Milton Park, Abingdon, Oxon, OX14 4TA UK
Email: book.orders@tandf.co.uk

For a complete listing of all our titles visit:
www.tandf.co.uk

Taylor & Francis
Taylor & Francis Group

ESSENTIAL READING FROM TAYLOR AND FRANCIS

Project Management Demystified

Third Edition

Geoff Reiss

Concise, practical and entertaining to read, this excellent introduction to project management is an indispensable book for both professionals and students working in or studying project management in business, engineering or the public sector.

Approachable and written in an easy-to-use style, it shows readers how, where and when to use the various project management techniques, demonstrating how to achieve efficient management of human, material and financial resources to make major contributions to projects and be an appreciated and successful project manager.

This new edition contains expanded sections on programme management, portfolio management, and the public sector. An entirely new chapter covers the evaluation, analysis and management of risks and issues. A much expanded section explores the rise and utilisation of methodologies like Prince2.

Contents: Introduction. Setting the Stage. Getting the Words in the Right Order. Nine Steps to a Successful Project. The Scope of the Project and its Objectives. Project Planning. A Fly on the Wall. Resource Management. Progress Monitoring and Control. Advanced Critical-Path Topics. The People Issues. Risk and Issue Management. Terminology

June 2007: 234x156mm: 224 pages
Pb: 978-0-415-42163-8: **£21.99**

To Order: Tel: +44 (0) 1235 400524 **Fax:** +44 (0) 1235 400525
or Post: Taylor and Francis Customer Services,
Bookpoint Ltd, Unit T1, 200 Milton Park, Abingdon, Oxon, OX14 4TA UK
Email: book.orders@tandf.co.uk

For a complete listing of all our titles visit:
www.tandf.co.uk

Taylor & Francis
Taylor & Francis Group

ESSENTIAL READING FROM TAYLOR AND FRANCIS

Environmental Noise Barriers
2nd edition
Benz Kotzen and Colin English

Environmental Noise Barriers is a unique one-stop reference for practitioners, whether acoustical engineers, landscape architects or manufacturers, and for highways departments in local and central authorities. This extensively revised new edition is updated in line with UK and EU legislation and international provision of barriers.

Contents
Introduction;
Defining the Need for Barriers; Acoustic and Visual Impact Assessment – Environmental Impact Assessment Techniques and Mitigation;
Acoustic Performance of Barriers; Barrier Morphology and Design;
Types of Barrier and Barrier Materials; Planting and Bio-barriers;
Noise Amelioration and Roads; Tunnels;
Airports;
Environmental Noise Barriers in Other Locations – Industrial Uses

April 2009 246 x 189mm: 280pp
Hb: 978-0-415-43708-0 **£90.00**

To Order: Tel: +44 (0) 1235 400524 **Fax:** +44 (0) 1235 400525
or Post: Taylor and Francis Customer Services,
Bookpoint Ltd, Unit T1, 200 Milton Park, Abingdon, Oxon, OX14 4TA UK
Email: book.orders@tandf.co.uk

For a complete listing of all our titles visit:
www.tandf.co.uk

Taylor & Francis
Taylor & Francis Group

ESSENTIAL READING FROM TAYLOR AND FRANCIS

Spon's Irish Construction Price Book

Third Edition

Franklin + Andrews

This new edition of *Spon's Irish Construction Price Book*, edited by Franklin + Andrews, is the only complete and up-to-date source of cost data for this important market.

- All the materials costs, labour rates, labour constants and cost per square metre are based on current conditions in Ireland
- Structured according to the new Agreed Rules of Measurement (second edition)
- 30 pages of Approximate Estimating Rates for quick pricing

This price book is an essential aid to profitable contracting for all those operating in Ireland's booming construction industry.

Franklin + Andrews, Construction Economists, have offices in 100 countries and in-depth experience and expertise in all sectors of the construction industry.

April 2008: 246x174 mm: 510 pages
Hb: 978-0-415-45637-1: **£145.00**

To Order: Tel: +44 (0) 1235 400524 **Fax:** +44 (0) 1235 400525
or Post: Taylor and Francis Customer Services,
Bookpoint Ltd, Unit T1, 200 Milton Park, Abingdon, Oxon, OX14 4TA UK
Email: book.orders@tandf.co.uk

For a complete listing of all our titles visit:
www.tandf.co.uk

Taylor & Francis
Taylor & Francis Group

PART 5

Prices for Measured Works – Minor Works

INTRODUCTION
Typical Project Profile

Contract value	£10,000.00 - £70,000.00
Labour rate (see page 7)	£20.50 per hour
Labour rate for maintenance contracts	-
Number of site staff	6 - 9
Project area	1200 m²
Project location	Outer London
Project components	20% hard landscape 80% soft landscape and planting
Access to works areas	Very good
Contract	Main contract
Delivery of materials	Part loads
Profit and site overheads	Excluded

NEW ITEMS FOR THIS EDITION

Item Excluding site overheads and profit	PC £	Labour hours	Labour £	Plant £	Material £	Unit	Total rate £
Q10 EDGINGS - EXCELEDGE							
Exceledge "SuperEdge"; recycled vinyl edging; black; 3 x 150 mm high with 5 mm diameter bead on each edge; for edging to soft landscape turf or bedding areas; formed by heat to required angle, or fixed to form straight or curvilinear lines using 350 mm galvanized fixing stakes							
150 mm deep	5.49	0.02	0.37	-	5.49	m	**5.86**
Exceledge "BrickEdge"; recycled vinyl hard landscape edging former; black; 50 mm high with 76 mm return; 3 mm thick wall for the retention of bricks, blocks and loose or bonded gravels; fixed with steel fixing spikes at 300 mm centres							
Flexible; 50 mm deep	6.05	0.02	0.37	-	6.05	m	**6.42**
Rigid; 50 mm deep	6.05	0.02	0.37	-	6.05	m	**6.42**
Exceledge "EdgeKing"; polyethylene flexible edging; black; used for turfed areas and retention of loose fill materials; with 25 mm diameter rounded hose top for bold demarcation and safety in play areas; fixed with steel vee-stakes at 45° angle into adjacent material or to prefixed timber supports using screws and washers							
125 mm deep	3.98	0.02	0.37	-	3.98	m	**4.35**
Corner pieces	5.71	-	-	-	5.71	ea	**-**
Joiner plugs	0.73	-	-	-	0.73	ea	**-**
Exceledge "Borderline"; tree rings; preformed unit from 100 mm high rolled steel in galvanized or powder coated (black or brown) finish; for circular traditional lawn edging around trees, bushes, flag poles or part of a feature; fixed with flat steel matching stakes							
700 mm	14.37	0.50	10.25	-	14.37	ea	**24.62**
900 mm	17.25	0.55	11.28	-	17.25	ea	**28.52**
1200 mm	24.97	0.60	12.30	-	24.97	ea	**37.27**
1500 mm	28.32	0.65	13.32	-	28.32	ea	**41.64**
1800 mm	31.65	0.70	14.35	-	31.65	ea	**46.00**

NEW ITEMS FOR THIS EDITION

Item Excluding site overheads and profit	PC £	Labour hours	Labour £	Plant £	Material £	Unit	Total rate £
Q10 EDGINGS - GRANITE SETT EDGES							
New granite sett edgings; CED Ltd; 100 x 100 x 100 mm; bedding in cement mortar (1:4); including haunching with in situ concrete; 11.50 N/mm^2 - 40 mm aggregate one side; concrete foundations (not included)							
Edgings 100 x 100 x 100 mm; ref 1R "better quality" with 20 mm pointing gaps							
100 mm wide	2.71	1.33	27.27	-	5.20	m	**32.47**
2 rows; 220 mm wide	5.42	2.22	45.51	-	8.65	m	**54.16**
3 rows; 340 mm wide	8.13	3.00	61.50	-	10.67	m	**72.17**
Edgings 100 x 100 x 200 mm; ref 1R "better quality" with 20 mm pointing gaps							
100 mm wide	2.40	0.75	15.38	-	4.89	m	**20.27**
2 rows; 220 mm wide	4.79	1.50	30.75	-	8.02	m	**38.77**
3 rows; 340 mm wide	7.19	2.25	46.13	-	9.73	m	**55.85**
Edgings 100 x 100 x 100 mm; ref 8R "standard quality" with 20 mm pointing gaps							
100 mm wide	2.69	1.33	27.27	-	5.18	m	**32.44**
2 rows; 220 mm wide	5.37	2.22	45.51	-	8.60	m	**54.11**
3 rows; 340 mm wide	8.06	3.00	61.50	-	10.59	m	**72.09**
Edgings 100 x 100 x 200 mm; ref 8R "standard quality" with 20 mm pointing gaps							
100 mm wide	3.76	0.75	15.38	-	6.26	m	**21.63**
2 rows; 220 mm wide	7.52	1.50	30.75	-	10.75	m	**41.50**
3 rows; 340 mm wide	11.29	2.25	46.13	-	13.82	m	**59.95**
Reclaimed granite setts edgings; CED Ltd; 100 x 100 mm; bedding in cement mortar (1:4); including haunching with in situ concrete; 11.50 N/mm^2 - 40 mm aggregate one side; on concrete foundations (not included)							
Edgings 100 x 100 x 100 mm; with 20 mm pointing gaps							
100 mm wide	3.94	1.33	27.27	-	6.44	m	**33.70**
2 rows; 220 mm wide	7.88	2.22	45.51	-	11.11	m	**56.62**
3 rows; 340 mm wide	11.83	3.00	61.50	-	14.36	m	**75.86**
Q22 BITUCHEM							
Resin bound macadam pavings; hand laid; Bituchem							
"Natratex" clear resin bound macadam to pedestrian or vehicular hard landscape areas; laid to base course (not included)							
25 mm thick to pedestrian areas	22.19	-	-	-	22.19	m^2	-
30 mm thick to vehicular areas	26.63	-	-	-	26.63	m^2	-
"Colourtex" coloured resin bound macadam to pedestrian or vehicular hard landscape areas							
25 mm thick	22.19	-	-	-	22.19	m^2	-

NEW ITEMS FOR THIS EDITION

Item Excluding site overheads and profit	PC £	Labour hours	Labour £	Plant £	Material £	Unit	Total rate £
Q22 RESIN PAVINGS - ADDAGRIP							
Bound aggregates; Addagrip Surface Treatments UK Ltd; natural decorative resin bonded surface dressing laid to concrete, macadam or to plywood panels priced separately							
"Cobalt Blue Glass"; 6 mm							
15 mm depth	-	-	-	-	-	m²	88.00
"Midnight Grey", "Chocolate", "Derbyshire Cream", "All Gold"; 6 mm							
15 mm depth	-	-	-	-	-	m²	60.50
18 mm depth	-	-	-	-	-	m²	71.50

C EXISTING SITE/BUILDINGS/SERVICES

Item Excluding site overheads and profit	PC £	Labour hours	Labour £	Plant £	Material £	Unit	Total rate £
C20 DEMOLITION							
Demolish existing structures; removal to skip maximum distance 50 m; mechanical demolition; with 3 tonne excavator and dumper							
Brick wall							
112.5 mm thick	-	0.13	2.73	6.40	-	m^2	9.13
225 mm thick	-	0.17	3.42	12.21	-	m^2	15.62
337.5 mm thick	-	0.20	4.10	19.31	-	m^2	23.41
450 mm thick	-	0.27	5.47	23.19	-	m^2	28.65
Demolish existing structures; removal to skip maximum distance 50 m; all works by hand							
Brick wall							
112.5 mm thick	-	0.33	6.83	5.61	-	m^2	12.45
225 mm thick	-	0.50	10.25	11.23	-	m^2	21.48
337.5 mm thick	-	0.67	13.67	18.14	-	m^2	31.80
450 mm thick	-	1.00	20.50	21.62	-	m^2	42.12
Demolish existing structures; removal to skip maximum distance 50 m; by diesel or electric breaker; all other works by hand							
Brick wall							
112.5 mm thick	-	0.17	3.42	6.90	-	m^2	10.32
225 mm thick	-	0.20	4.10	12.78	-	m^2	16.88
337.5 mm thick	-	0.25	5.13	20.07	-	m^2	25.20
450 mm thick	-	0.33	6.83	24.20	-	m^2	31.03
Break out concrete footings associated with free standing walls; inclusive of all excavation, removal to skip and backfilling with excavated material							
By mechanical breaker; diesel or electric							
plain concrete	-	1.50	30.75	69.95	-	m^3	100.70
as above but loading to skip by hand; backfilling by hand	-	5.50	112.75	82.53	-	m^3	195.28
reinforced concrete	-	2.50	51.25	85.51	-	m^3	136.76
Remove existing free standing buildings; demolition by hand							
Timber buildings with suspended timber floor; hardstanding or concrete base not included; loading to skip							
shed 6.0 m^2	-	2.00	41.00	79.81	-	nr	120.81
shed 10.0 m^2	-	3.00	61.50	112.24	-	nr	173.74
shed 15.0 m^2	-	3.50	71.75	112.24	-	nr	183.99
Timber building; insulated; with timber or concrete posts set in concrete, felt covered timber or tiled roof; internal walls cladding with timber or plasterboard; load arisings to skip							
timber structure 6.0 m^2	-	2.50	51.25	269.37	-	nr	320.62
timber structure 12.0 m^2	-	3.50	71.75	377.11	-	nr	448.86
timber structure 20.0 m^2	-	8.00	164.00	510.30	-	nr	674.30

C EXISTING SITE/BUILDINGS/SERVICES

Item Excluding site overheads and profit	PC £	Labour hours	Labour £	Plant £	Material £	Unit	Total rate £
C20 DEMOLITION – cont'd							
Demolition of free standing brick buildings with tiled or sheet roof; concrete foundations measured separately; mechanical demolition maximum distance to stockpile 25 m; including for all access scaffolding and the like; maximum height of roof 4.00 m; including all doors, windows, guttering and down pipes; including disposal by grab							
Half brick thick							
10 m^2	-	8.00	164.00	323.46	265.20	nr	**752.66**
20 m^2	-	16.00	328.00	567.42	341.12	nr	**1236.54**
1 brick thick							
10 m^2	-	10.00	205.00	487.92	530.40	nr	**1223.32**
20 m^2	-	16.00	328.00	731.88	682.24	nr	**1742.12**
Cavity wall with blockwork inner skin and brick outer skin; insulated							
10 m^2	-	12.00	246.00	652.38	737.10	nr	**1635.48**
20 m^2	-	20.00	410.00	896.34	947.70	nr	**2254.04**
Extra over to the above for disconnection of services							
Electrical							
disconnection	-	-	-	-	-	nr	**100.00**
grub out cables and dispose; backfilling; by machine	-	-	-	4.18	-	m	**4.18**
grub out cables and dispose; backfilling; by hand	-	0.50	10.25	-	-	m	**10.25**
Water supply; foul or surface water drainage							
disconnection; capping off	-	1.00	20.50	-	100.00	nr	**120.50**
grub out pipes and dispose; backfilling; by machine	-	-	-	4.92	-	m	**4.92**
grub out pipes and dispose; backfilling; by hand	-	0.50	10.25	0.74	-	m	**10.99**

D GROUNDWORK

Item Excluding site overheads and profit	PC £	Labour hours	Labour £	Plant £	Material £	Unit	Total rate £
D11 SOIL STABILIZATION							
Timber log retaining walls; Timber fencing; AVS Fencing Supplies Ltd; tanalized softwood posts; all timber posts are "kiln dried redwood" with 15 year guarantee; walls below which have cut posts would not be subject to the suppliers guarantee							
Machine rounded softwood logs to trenches priced separately; disposal of excavated material priced separately; inclusive of 75 mm hardcore blinding to trench and backfilling trench with site mixed concrete 1:3:6; geofabric pinned to rear of logs; heights of logs above ground							
500 mm (constructed from 1.80 m lengths)	31.15	1.50	30.75	-	42.92	m	**73.67**
1.20 m (constructed from 1.80 m lengths)	62.30	1.30	26.65	-	83.12	m	**109.77**
1.60 m (constructed from 2.40 m lengths)	83.20	2.50	51.25	-	110.83	m	**162.08**
2.00 m (constructed from 3.00 m lengths)	104.00	3.50	71.75	-	138.47	m	**210.22**
As above but with 150 mm machine rounded timbers							
500 mm	42.70	2.50	51.25	-	54.47	m	**105.72**
1.20 m	142.19	1.75	35.88	-	163.01	m	**198.88**
1.60 m	142.33	3.00	61.50	-	170.01	m	**231.51**
As above but with 200 mm machine rounded timbers							
1.80 m (constructed from 2.40 m lengths)	189.75	4.00	82.00	-	224.28	m	**306.28**
Railway sleeper walls; AVS Fencing Supplies Ltd							
Construct retaining wall from railway sleepers; fixed with steel galvanized pins 12 mm driven into the ground; sleepers laid flat							
Grade 1 softwood; 2590 x 250 x 125 mm							
150 mm; 1 sleeper high	7.39	0.50	10.25	-	11.79	m	**22.04**
300 mm; 2 sleepers high	14.79	1.00	20.50	-	15.67	m	**36.17**
450 mm; 3 sleepers high	22.00	1.50	30.75	-	23.16	m	**53.91**
600 mm; 4 sleepers high	29.64	2.00	41.00	-	30.80	m	**71.80**
Grade 1 softwood as above but with 2 nr galvanized angle iron stakes set into concrete internally and screwed to the inside face of the sleepers							
750 mm; 5 sleepers high	36.67	2.50	51.25	-	64.18	m^2	**115.43**
900 mm; 6 sleepers high	43.89	2.75	56.38	-	74.12	m^2	**130.49**
Grade 1 hardwood; 2590 x 250 x 150 mm							
150 mm; 1 sleeper high	7.72	0.50	10.25	-	8.16	m	**18.41**
300 mm; 2 sleepers high	15.43	1.00	20.50	-	16.31	m	**36.81**
450 mm; 3 sleepers high	22.96	1.50	30.75	-	24.12	m	**54.87**
600 mm; 4 sleepers high	30.93	2.00	41.00	-	32.09	m	**73.09**
Grade 1 hardwood as above but with 2 nr galvanized angle iron stakes set into concrete internally and screwed to the inside face of the sleepers							
750 mm; 5 sleepers high	38.27	2.50	51.25	-	65.79	m^2	**117.04**
900 mm; 6 sleepers high	45.81	2.75	56.38	-	76.04	m^2	**132.41**

D GROUNDWORK

Item Excluding site overheads and profit	PC £	Labour hours	Labour £	Plant £	Material £	Unit	Total rate £
D11 SOIL STABILIZATION – cont'd							
Construct retaining wall from railway sleepers; fixed with steel galvanized pins 12 mm driven into the ground; sleepers laid flat – cont'd							
New pine softwood; 2400 x 250 x 125 mm							
120 mm; 1 sleeper high	9.57	0.50	10.25	-	10.01	m	**20.26**
240 mm; 2 sleepers high	19.14	1.00	20.50	-	20.02	m	**40.52**
360 mm; 3 sleepers high	28.72	1.50	30.75	-	29.87	m	**60.62**
480 mm; 4 sleepers high	38.29	2.00	41.00	-	39.44	m	**80.44**
New pine softwood as above but with 2 nr galvanized angle iron stakes set into concrete internally and screwed to the inside face of the sleepers							
600 mm; 5 sleepers high	47.86	2.50	51.25	-	75.37	m^2	**126.62**
720 mm; 6 sleepers high	57.43	2.75	56.38	-	87.66	m^2	**144.04**
New oak hardwood; 2600 x 220 x 130 mm							
130 mm; 1 sleeper high	11.42	0.50	10.25	-	11.86	m	**22.11**
260 mm; 2 sleepers high	22.85	1.00	20.50	-	23.73	m	**44.23**
390 mm; 3 sleepers high	34.27	1.50	30.75	-	35.43	m	**66.18**
520 mm; 4 sleepers high	45.70	2.00	41.00	-	46.85	m	**87.85**
New oak hardwood as above but with 2 nr galvanized angle iron stakes set into concrete internally and screwed to the inside face of the sleepers							
640 mm; 5 sleepers high	57.12	2.50	51.25	-	84.63	m^2	**135.88**
760 mm; 6 sleepers high	68.54	2.75	56.38	-	98.77	m^2	**155.15**
Excavate foundation trench; set railway sleepers vertically on end in concrete 1:3:6 continuous foundation to 33.3% of their length to form retaining wall							
Grade 1 hardwood; finished height above ground level							
300 mm	13.68	3.00	61.50	2.05	26.60	m	**90.16**
500 mm	19.83	3.00	61.50	2.05	35.82	m	**99.38**
600 mm	27.56	3.00	61.50	2.05	52.60	m	**116.16**
750 mm	29.75	3.50	71.75	2.05	55.88	m	**129.68**
1.00 m	40.65	3.75	76.88	2.05	63.11	m	**142.04**
Excavate and place vertical steel universal beams 165 wide in concrete base at 2.590 centres; fix railway sleepers set horizontally between beams to form horizontal fence or retaining wall							
Grade 1 hardwood; bay length 2.590 m							
0.50 m high; (2 sleepers)	16.50	2.50	51.25	-	59.93	bay	**111.18**
750 m high; (3 sleepers)	24.25	2.60	53.30	-	88.97	bay	**142.27**
1.00 m high; (4 sleepers)	32.50	3.00	61.50	-	114.64	bay	**176.14**
1.25 m high; (5 sleepers)	41.50	3.50	71.75	-	147.97	bay	**219.72**
1.50 m high; (6 sleepers)	48.75	3.00	61.50	-	176.10	bay	**237.60**
1.75 m high; (7 sleepers)	56.75	3.00	61.50	-	184.10	bay	**245.60**

D GROUNDWORK

Item Excluding site overheads and profit	PC £	Labour hours	Labour £	Plant £	Material £	Unit	Total rate £
Retaining walls; Maccaferri Ltd							
Wire mesh gabions; galvanized mesh							
80 mm x 100 mm; filling with broken stones							
125 mm - 200 mm size; wire down securely							
to manufacturer's instructions; filling front							
face by hand							
2 x 1 x 0.50 m	22.16	2.00	41.00	6.88	91.46	nr	**139.33**
2 x 1 x 1.00 m	31.04	4.00	82.00	13.75	169.64	nr	**265.39**
PVC coated gabions							
2 x 1 x 0.5m	28.67	2.00	41.00	6.88	97.97	nr	**145.84**
2 x 1 x 1.00 m	40.36	4.00	82.00	13.75	178.96	nr	**274.71**
D20 EXCAVATING AND FILLING							
MACHINE SELECTION TABLE							
Road Equipment Ltd; machine volumes							
for excavating/filling only and placing							
excavated material alongside or to a							
dumper; no bulkages are allowed for in							
the material volumes; these rates should							
be increased by user-preferred							
percentages to suit prevailing site							
conditions; the figures in the next section							
for "Excavation mechanical" and							
filling allow for the use of banksmen							
within the rates shown below							
1.5 tonne excavators; digging volume							
1 cycle/minute; 0.04 m^3	-	0.42	8.54	2.70	-	m^3	**11.24**
2 cycles/minute; 0.08 m^3	-	0.21	4.27	1.68	-	m^3	**5.95**
3 cycles/minute; 0.12 m^3	-	0.14	2.85	1.35	-	m^3	**4.20**
3 tonne excavators; digging volume							
1 cycle/minute; 0.13 m^3	-	0.13	2.63	1.52	-	m^3	**4.15**
2 cycles/minute; 0.26 m^3	-	0.06	1.31	1.04	-	m^3	**2.35**
3 cycles/minute; 0.39 m^3	-	0.04	0.88	1.01	-	m^3	**1.88**
5 tonne excavators; digging volume							
1 cycle/minute; 0.28 m^3	-	0.06	1.22	1.01	-	m^3	**2.23**
2 cycles/minute; 0.56 m^3	-	0.03	0.61	0.54	-	m^3	**1.15**
3 cycles/minute; 0.84 m^3	-	0.02	0.41	0.49	-	m^3	**0.89**
Dumpers; Road Equipment Ltd							
1 tonne high tip skip loader; volume							
0.485 m^3 (775 kg)							
5 loads per hour	-	0.41	8.45	1.45	-	m^3	**9.91**
7 loads per hour	-	0.29	6.04	1.08	-	m^3	**7.12**
10 loads per hour	-	0.21	4.23	0.81	-	m^3	**5.04**
3 tonne dumper; maximum volume 2.40 m^3							
(3.38 t); available volume 1.9 m^3							
4 loads per hour	-	0.14	2.85	0.53	-	m^3	**3.38**
5 loads per hour	-	0.11	2.28	0.43	-	m^3	**2.71**
7 loads per hour	-	0.08	1.63	0.32	-	m^3	**1.95**
10 loads per hour	-	0.06	1.14	0.24	-	m^3	**1.38**
Lifting turf for preservation							
hand lift and stack	-	0.08	1.71	-	-	m^2	**1.71**
Load to wheelbarrows and dispose offsite by							
skip							
25 m distance	-	0.02	0.43	3.08	-	m^2	**3.50**

D GROUNDWORK

Item Excluding site overheads and profit	PC £	Labour hours	Labour £	Plant £	Material £	Unit	Total rate £
D20 EXCAVATING AND FILLING – cont'd							
Lifting turf for preservation – cont'd Note: The figures in this section relate to the machine capacities shown in the Major Works section of this book. The figures below allow for dig efficiency based on depth, bulkages of 25% on loamy soils (adjustments should be made for different soil types) and for a banksman.							
Site clearance; by machine; clear site of mature shrubs from existing cultivated beds; dig out roots by machine Mixed shrubs in beds; planting centres 500 mm average							
height less than 1 m	-	0.03	0.51	0.73	-	m²	1.24
1.00 - 1.50 m	-	0.04	0.82	1.17	-	m²	1.99
1.50 - 2.00 m; pruning to ground level by hand	-	0.10	2.05	2.92	-	m²	4.97
2.00 - 3.00 m; pruning to ground level by hand	-	0.10	2.05	5.85	-	m²	7.90
3.00 - 4.00 m; pruning to ground level by hand	-	0.20	4.10	9.74	-	m²	13.84
Site clearance; by hand; clear site of mature shrubs from existing cultivated beds; dig out roots Mixed shrubs in beds; planting centres 500 mm average							
height less than 1 m	-	0.33	6.83	-	-	m²	6.83
1.00 - 1.50 m	-	0.50	10.25	-	-	m²	10.25
1.50 - 2.00 m	-	1.00	20.50	-	-	m²	20.50
2.00 - 3.00 m	-	2.00	41.00	-	-	m²	41.00
3.00 - 4.00 m	-	3.00	61.50	-	-	m²	61.50
Disposal of material from site clearance operations Shrubs and groundcovers less than 1.00 m height; disposal to skip							
deciduous shrubs; not chipped; winter	-	0.05	1.02	12.32	-	m²	13.34
deciduous shrubs; chipped; winter	-	0.05	1.02	2.78	-	m²	3.80
evergreen or deciduous shrubs; chipped; summer	-	0.08	1.54	8.32	-	m²	9.86
Shrubs 1.00 m - 2.00 m height; disposal to skip							
deciduous shrubs; not chipped; winter	-	0.17	3.42	36.95	-	m²	40.37
deciduous shrubs; chipped; winter	-	0.25	5.13	8.32	-	m²	13.45
evergreen or deciduous shrubs; chipped; summer	-	0.30	6.15	16.02	-	m²	22.17
Shrubs or hedges 2.00 m - 3.00 m height; disposal to skip							
deciduous plants non woody growth; not chipped; winter	-	0.25	5.13	36.95	-	m²	42.08
deciduous shrubs; chipped; winter	-	0.50	10.25	19.41	-	m²	29.66
evergreen or deciduous shrubs non woody growth; chipped; summer	-	0.67	13.67	24.03	-	m²	37.70
evergreen or deciduous shrubs woody growth; chipped; summer	-	1.50	30.75	25.89	-	m²	56.64

D GROUNDWORK

Item Excluding site overheads and profit	PC £	Labour hours	Labour £	Plant £	Material £	Unit	Total rate £
Shrubs and groundcovers less than 1.00 m height; disposal to spoil heap							
deciduous shrubs; not chipped; winter	-	0.05	1.02	-	-	m²	**1.02**
deciduous shrubs; chipped; winter	-	0.05	1.02	0.93	-	m²	**1.96**
evergreen or deciduous shrubs; chipped; summer	-	0.08	1.54	0.93	-	m²	**2.47**
Shrubs 1.00 m - 2.00 m height; disposal to spoil heap							
deciduous shrubs; chipped; winter	-	0.25	5.13	0.93	-	m²	**6.06**
deciduous shrubs; not chipped; winter	-	0.17	3.42	-	-	m²	**3.42**
evergreen or deciduous shrubs; chipped; summer	-	0.30	6.15	1.24	-	m²	**7.39**
Shrubs or hedges 2.00 m - 3.00 m height; disposal to spoil heaps							
deciduous plants non woody growth; not chipped; winter	-	0.25	5.13	-	-	m²	**5.13**
deciduous shrubs; chipped; winter	-	0.50	10.25	0.93	-	m²	**11.18**
evergreen or deciduous shrubs non woody growth; chipped; summer	-	0.67	13.67	1.86	-	m²	**15.53**
evergreen or deciduous shrubs woody growth; chipped; summer	-	1.50	30.75	3.72	-	m²	**34.47**
Excavating; mechanical; topsoil for preservation; depositing alongside 3 tonne tracked excavator (bucket volume 0.13 m³)							
average depth 100 mm	-	0.02	0.49	0.76	-	m²	**1.25**
average depth 150 mm	-	0.03	0.69	1.07	-	m²	**1.76**
average depth 200 mm	-	0.04	0.82	1.27	-	m²	**2.09**
average depth 250 mm	-	0.04	0.90	1.40	-	m²	**2.30**
average depth 300 mm	-	0.05	0.98	1.52	-	m²	**2.51**
Excavating; mechanical; topsoil for preservation; depositing to spoil heaps by wheeled dumper; spoil heaps maximum 50 m distance 3 tonne tracked excavator (bucket volume 0.13 m³)							
average depth 100 mm	-	0.02	0.49	1.18	-	m²	**1.68**
average depth 150 mm	-	0.03	0.69	1.70	-	m²	**2.39**
average depth 200 mm	-	0.04	0.82	2.11	-	m²	**2.93**
average depth 250 mm	-	0.04	0.90	2.45	-	m²	**3.35**
average depth 300 mm	-	0.05	0.98	2.79	-	m²	**3.77**
Excavating; mechanical; to reduce levels 3 tonne excavator (bucket volume 0.13 m³)							
maximum depth not exceeding 0.25 m	-	0.13	2.63	5.08	-	m³	**7.71**
maximum depth not exceeding 1.00 m	-	0.13	2.63	4.82	-	m³	**7.45**
Pits; 3 tonne tracked excavator							
maximum depth not exceeding 0.25 m	-	0.58	11.89	1.02	-	m³	**12.91**
maximum depth not exceeding 1.00 m	-	0.50	10.25	4.93	-	m³	**15.18**
maximum depth not exceeding 2.00 m	-	0.60	12.30	5.91	-	m³	**18.21**
Trenches; width not exceeding 0.30 m; 3 tonne excavator							
maximum depth not exceeding 0.25 m	-	1.33	27.33	5.48	-	m³	**32.81**
maximum depth not exceeding 1.00 m	-	0.69	14.04	2.81	-	m³	**16.86**
maximum depth not exceeding 2.00 m	-	0.60	12.30	2.46	-	m³	**14.76**

D GROUNDWORK

Item Excluding site overheads and profit	PC £	Labour hours	Labour £	Plant £	Material £	Unit	Total rate £
D20 EXCAVATING AND FILLING – cont'd							
Excavating; mechanical; to reduce levels – cont'd							
Trenches; width exceeding 0.30 m; 3 tonne excavator							
maximum depth not exceeding 0.25 m	-	0.60	12.30	2.46	-	m^3	14.76
maximum depth not exceeding 1.00 m	-	0.50	10.25	2.05	-	m^3	12.30
maximum depth not exceeding 2.00 m	-	0.38	7.88	1.58	-	m^3	9.46
Extra over any types of excavating irrespective of depth for breaking out existing materials; heavy duty 110 volt breaker tool							
hard rock	-	5.00	102.50	57.50	-	m^3	160.00
concrete	-	3.00	61.50	34.50	-	m^3	96.00
reinforced concrete	-	4.00	82.00	58.75	-	m^3	140.75
brickwork, blockwork or stonework	-	1.50	30.75	17.25	-	m^3	48.00
Extra over any types of excavating irrespective of depth for breaking out existing hard pavings; 1600 watt, 110 volt breaker							
concrete; 100 mm thick	-	0.20	4.10	1.55	-	m^2	5.65
concrete; 150 mm thick	-	0.25	5.13	1.94	-	m^2	7.06
concrete; 200 mm thick	-	0.29	5.86	2.21	-	m^2	8.07
concrete; 300 mm thick	-	0.40	8.20	3.10	-	m^2	11.30
reinforced concrete; 100 mm thick	-	0.33	6.83	8.96	-	m^2	15.79
reinforced concrete; 150 mm thick	-	0.40	8.20	10.75	-	m^2	18.95
reinforced concrete; 200 mm thick	-	0.50	10.25	13.44	-	m^2	23.69
reinforced concrete; 300 mm thick	-	1.00	20.50	26.88	-	m^2	47.38
tarmacadam; 75 mm thick	-	0.14	2.93	1.11	-	m^2	4.04
tarmacadam and hardcore; 150 mm thick	-	0.20	4.10	1.55	-	m^2	5.65
Extra over any types of excavating irrespective of depth for taking up							
precast concrete paving slabs	-	0.08	1.71	0.65	-	m^2	2.35
natural stone paving	-	0.13	2.56	0.97	-	m^2	3.53
cobbles	-	0.17	3.42	1.29	-	m^2	4.71
brick paviors	-	0.17	3.42	1.29	-	m^2	4.71
Excavating; hand							
Topsoil for preservation; loading to barrows							
average depth 100 mm	-	0.30	6.15	-	-	m^2	6.15
average depth 200 mm	-	0.60	12.30	-	-	m^2	12.30
average depth 300 mm	-	0.90	18.45	-	-	m^2	18.45
over 300 mm	-	3.00	61.50	-	-	m^3	61.50
Excavating; hand							
Topsoil to reduce levels							
maximum depth not exceeding 0.25 m	-	2.52	51.66	-	-	m^3	51.66
maximum depth not exceeding 1.00 m	-	3.12	63.96	-	-	m^3	63.96
Pits							
maximum depth not exceeding 0.25 m	-	2.67	54.67	-	-	m^3	54.67
maximum depth not exceeding 1.00 m	-	3.47	71.07	-	-	m^3	71.07
maximum depth not exceeding 2.00 m (includes earthwork support)	-	6.93	142.13	-	-	m^3	204.67
Trenches; width not exceeding 0.30 m							
maximum depth not exceeding 0.25 m	-	2.86	58.57	-	-	m^3	58.57
maximum depth not exceeding 1.00 m	-	3.72	76.28	-	-	m^3	76.28
maximum depth not exceeding 1.00 m	-	3.72	76.28	-	-	m^3	107.55

D GROUNDWORK

Item Excluding site overheads and profit	PC £	Labour hours	Labour £	Plant £	Material £	Unit	Total rate £
Trenches; width exceeding 0.30 m wide							
maximum depth not exceeding 0.25 m	-	2.86	58.57	-	-	m^3	58.57
maximum depth not exceeding 1.00 m	-	4.00	82.00	-	-	m^3	82.00
maximum depth not exceeding 2.00 m (includes earthwork support)	-	6.00	123.00	-	-	m^3	185.53
Filling to make up levels; mechanical (3 tonne tracked excavator); depositing in layers 150 mm maximum thickness							
Arising from the excavations							
maximum thickness less than 0.25 m	-	0.10	2.13	3.30	-	m^3	5.43
average thickness exceeding 0.25 m	-	0.08	1.64	2.54	-	m^3	4.18
Obtained from on site spoil heaps; average 25 m distance; multiple handling							
maximum thickness less than 0.25 m	-	0.16	3.29	6.00	-	m^3	9.28
average thickness exceeding 0.25 m	-	0.13	2.67	4.71	-	m^3	7.38
Obtained off site; planting quality topsoil PC £26.00/m^3							
maximum thickness less than 0.25 m	26.00	0.14	2.94	5.45	32.50	m^3	40.89
average thickness exceeding 0.25 m	26.00	0.13	2.67	5.04	32.50	m^3	40.21
Obtained off site; crushed concrete hardcore; PC £20.70/tonne							
maximum thickness less than 0.25 m	20.70	0.14	2.87	12.52	25.25	m^3	40.64
maximum thickness exceeding 0.25 m	20.70	0.13	2.61	12.01	25.25	m^3	39.87
Filling to make up levels; hand; depositing in layers 150 mm maximum thickness							
Arising from the excavations							
average thickness exceeding 0.25 m	-	0.75	15.38	-	-	m^3	15.38
Obtained from on site spoil heaps; average 25 m distance; multiple handling							
average thickness exceeding 0.25 m thick	-	1.25	25.62	-	-	m^3	25.62
Obtained off site; planting quality topsoil PC £43.50/m^3 (10 tonne loads)							
average thickness exceeding 0.25 m thick	71.78	1.25	25.62	-	89.72	m^3	115.34
Disposal; mechanical							
Light soils and loams (bulking factor - 1.25) Excavated material; off site; to tip; mechanically loaded by grab; capacity of load 7.25 m^3							
inert material	-	-	-	-	-	m^3	39.60
soil (sandy and loam) dry	-	-	-	-	-	m^3	39.60
soil (sandy and loam) wet	-	-	-	-	-	m^3	42.00
broken out compacted materials such as road bases and the like	-	-	-	-	-	m^3	44.00
soil (clay) dry	-	-	-	-	-	m^3	51.40
soil (clay) wet	-	-	-	-	-	m^3	53.40
Disposal by skip; 6 yd^3 (4.6 m^3)							
Excavated material; off site; to tip							
by machine	-	-	-	48.65	-	m^3	48.65
by hand	-	4.00	82.00	46.19	-	m^3	128.19
Disposal on site							
Excavated material to spoil heaps							
average 25 m distance	-	-	-	5.57	-	m^3	5.57
average 50 m distance	-	-	-	6.13	-	m^3	6.13

D GROUNDWORK

Item Excluding site overheads and profit	PC £	Labour hours	Labour £	Plant £	Material £	Unit	Total rate £
D20 EXCAVATING AND FILLING – cont'd							
Disposal on site – cont'd							
average 100 m distance	-	-	-	6.69	-	m³	6.69
average 200 m distance	-	-	-	8.36	-	m³	8.36
Excavated material; spreading on site							
average 25 m distance	-	-	-	6.53	-	m³	6.53
average 50 m distance	-	-	-	7.17	-	m³	7.17
average 100 m distance	-	-	-	8.24	-	m³	8.24
average 200 m distance	-	-	-	9.09	-	m³	9.09
Disposal; hand							
Excavated material; on site; in spoil heaps							
average 25 m distance	-	2.40	49.20	-	-	m³	49.20
average 50 m distance	-	2.64	54.12	-	-	m³	54.12
average 100 m distance	-	3.00	61.50	-	-	m³	61.50
average 200 m distance	-	3.60	73.80	-	-	m³	73.80
Disposal hand; excavated material in bags							
Filling bags 25 kg and barrowing through buildings							
average 25 m distance	-	8.40	172.20	-	-	m³	172.20
average 50 m distance	-	9.20	188.60	-	-	m³	188.60
average 100 m distance	-	10.10	207.05	-	-	m³	207.05
average 200 m distance	-	11.60	237.80	-	-	m³	237.80
Excavated material; spreading on site							
average 25 m distance	-	2.64	54.12	-	-	m³	54.12
average 50 m distance	-	3.00	61.50	-	-	m³	61.50
average 100 m distance	-	3.60	73.80	-	-	m³	73.80
average 200 m distance	-	4.20	86.10	-	-	m³	86.10
Surface treatments							
Compacting							
bottoms of excavations	-	0.01	0.10	0.10	-	m²	0.20
Grading operations; surface previously excavated to reduce levels to prepare to receive subsequent treatments; grading to accurate levels and falls 20 mm tolerances							
Clay or heavy soils or hardcore							
3 tonne excavator	-	0.02	0.41	0.58	-	m²	0.99
5 tonne excavator	-	0.04	0.82	0.25	-	m²	1.07
by hand	-	0.10	2.05	-	-	m²	2.05
Loamy topsoils							
3 tonne excavator	-	0.01	0.27	0.39	-	m²	0.66
5 tonne excavator	-	0.03	0.55	0.17	-	m²	0.71
by hand	-	0.05	1.02	-	-	m²	1.02
Sand or graded granular materials							
3 tonne excavator	-	0.01	0.20	0.29	-	m²	0.50
5 tonne excavator	-	0.02	0.41	0.13	-	m²	0.53
by hand	-	0.03	0.68	-	-	m²	0.68
Excavating; by hand							
Topsoil for preservation; loading to barrows							
average depth 100 mm	-	0.24	4.92	-	-	m²	4.92
average depth 150 mm	-	0.36	7.38	-	-	m²	7.38
average depth 200 mm	-	0.58	11.81	-	-	m²	11.81

D GROUNDWORK

Item Excluding site overheads and profit	PC £	Labour hours	Labour £	Plant £	Material £	Unit	Total rate £
average depth 250 mm	-	0.72	14.76	-	-	m²	**14.76**
average depth 300 mm	-	0.86	17.71	-	-	m²	**17.71**
Grading operations; surface recently filled to raise levels to prepare to receive subsequent treatments; grading to accurate levels and falls 20 mm tolerances							
Clay or heavy soils or hardcore							
3 tonne excavator	-	0.02	0.31	0.44	-	m²	**0.76**
by hand	-	0.08	1.71	-	-	m²	**1.71**
Loamy topsoils							
3 tonne excavator	-	0.01	0.21	0.30	-	m²	**0.51**
by hand	-	0.04	0.82	-	-	m²	**0.82**
Sand or graded granular materials							
3 tonne excavator	-	0.01	0.15	0.22	-	m²	**0.37**
by hand	-	0.03	0.59	-	-	m²	**0.59**
Surface preparation trimming surfaces of cultivated ground to final levels, removing roots stones and debris exceeding 50 mm in any direction to tip offsite; slopes less than 15°							
clean ground with minimal stone content	-	0.25	5.13	-	-	100 m²	**5.13**
slightly stony - 0.5 kg stones per m²	-	0.33	6.83	0.02	-	100 m²	**6.85**
very stony - 1.0 - 3.0 kg stones per m²	-	0.50	10.25	0.04	-	100 m²	**10.29**
clearing mixed slightly contaminated rubble inclusive of roots and vegetation	-	0.50	10.25	9.24	-	100 m²	**19.49**
clearing brick-bats, stones and clean rubble	-	0.60	12.30	0.17	-	100 m²	**12.47**
Hand cultivation; preparation for turfing/ seeding areas							
Cultivate surface to receive landscape surface treatments using hand fork or implement; 1 spit (300 mm deep), break up ground and bring to a fine tilth by rake							
compacted ground	-	0.07	1.37	-	-	m²	**1.37**
ground previously supporting turf; lifting of turf not included	-	0.06	1.28	-	-	m²	**1.28**
ground previously planted with small to medium shrubs; removal of planting not included	-	0.06	1.14	-	-	m²	**1.14**
soft ground	-	0.04	0.82	-	-	m²	**0.82**
Hand cultivation; preparation for planting							
Cultivate surface to receive landscape surface treatments using hand fork or implement; 1 spit (300 mm deep), break up ground and bring to a fine tilth by rake							
compacted ground	-	0.05	1.02	-	-	m²	**1.02**
ground previously supporting turf; lifting of turf not included	-	0.04	0.85	-	-	m²	**0.85**
ground previously planted with small to medium shrubs; removal of planting not included	-	0.04	0.73	-	-	m²	**0.73**
soft ground	-	0.03	0.59	-	-	m²	**0.59**

E IN SITU CONCRETE/LARGE PRECAST CONCRETE

Item Excluding site overheads and profit	PC £	Labour hours	Labour £	Plant £	Material £	Unit	Total rate £
E10 MIXING/CASTING/CURING IN SITU CONCRETE							
General Please see the notes on concrete mix designations in the Major Works section of this book under E10 In Situ Concrete.							
Concrete mixes; mixed on site; costs for producing concrete; prices for commonly used mixes for various types of work; based on an 85 litre concrete mixer Roughest type mass concrete such as footings, road haunchings 300 mm thick; aggregates delivered in 10 tonne loads							
1:3:6	40.50	1.51	31.02	-	73.61	m³	**104.63**
1:3:6 sulphate resisting	92.53	1.51	31.02	-	92.53	m³	**123.55**
As above but aggregates delivered in 850 kg bulk bags							
1:3:6	33.11	1.51	31.02	-	109.70	m³	**140.72**
1:3:6 sulphate resisting	52.03	1.51	31.02	-	128.62	m³	**159.64**
Most ordinary use of concrete such as mass walls above ground, road slabs, etc. and general reinforced concrete work; aggregates delivered in 10 tonne loads							
1:2:4	87.32	1.51	31.02	-	87.32	m³	**118.34**
1:2:4 sulphate resisting	114.07	1.51	31.02	-	114.07	m³	**145.09**
As above but aggregates delivered in 850 kg bulk bags							
1:2:4	46.82	1.51	31.02	-	123.41	m³	**154.43**
1:2:4 sulphate resisting	73.57	1.51	31.02	-	150.16	m³	**181.18**
Watertight floors, pavements and walls, tanks, pits, steps, paths, surface of two course roads; reinforced concrete where extra strength is required (aggregates delivered in 10 tonne loads)							
1:1.5:3	100.41	1.51	31.02	-	100.41	m³	**131.43**
As above but aggregates delivered in 850 kg bulk bags							
1:1.5:3	59.91	1.51	31.02	-	136.50	m³	**167.52**
Plain in situ concrete; site mixed; 10 N/mm² - 40 mm aggregate (1:3:6) (aggregate delivery indicated) Foundations							
ordinary Portland cement; 10 tonne ballast loads	104.63	1.00	20.50	-	104.63	m³	**125.13**
ordinary Portland cement; 850 kg bulk bags	140.72	1.00	20.50	-	140.72	m³	**161.22**
sulphate resistant cement; 10 tonne ballast loads	123.55	1.00	20.50	-	123.55	m³	**144.05**
sulphate resistant cement; 850 kg bulk bags	159.64	1.00	20.50	-	159.64	m³	**180.14**
Foundations; poured on or against earth or unblinded hardcore							
ordinary Portland cement; 10 tonne ballast loads	104.63	1.00	20.50	-	109.86	m³	**130.36**
ordinary Portland cement; 850 kg bulk bags	140.72	1.00	20.50	-	147.76	m³	**168.26**

E IN SITU CONCRETE/LARGE PRECAST CONCRETE

Item Excluding site overheads and profit	PC £	Labour hours	Labour £	Plant £	Material £	Unit	Total rate £
sulphate resistant cement; 10 tonne ballast loads	123.55	1.00	20.50	-	129.73	m³	**150.23**
sulphate resistant cement; 850 kg bulk bags	159.64	1.00	20.50	-	167.62	m³	**188.12**
Isolated foundations							
ordinary Portland cement; 10 tonne ballast loads	104.63	2.00	41.00	-	107.25	m³	**148.25**
ordinary Portland cement; 850 kg bulk bags	140.72	2.00	41.00	-	144.24	m³	**185.24**
sulphate resistant cement; 10 tonne ballast loads	123.55	2.00	41.00	-	126.64	m³	**167.64**
sulphate resistant cement; 850 kg bulk bags	159.64	2.00	41.00	-	163.63	m³	**204.63**
Plain in situ concrete; site mixed; 21 N/mm² - 20 mm aggregate (1:2:4)							
Foundations							
ordinary Portland cement; 10 tonne ballast loads	118.34	1.00	20.50	-	121.30	m³	**141.80**
ordinary Portland cement; 850 kg bulk bags	-	1.00	20.50	-	158.29	m³	**178.79**
sulphate resistant cement; 10 tonne ballast loads	145.09	1.00	20.50	-	148.72	m³	**169.22**
sulphate resistant cement; 850 kg bulk bags	181.18	1.00	20.50	-	185.71	m³	**206.21**
Foundations; poured on or against earth or unblinded hardcore							
ordinary Portland cement; 10 tonne ballast loads	118.34	1.00	20.50	-	124.26	m³	**144.76**
ordinary Portland cement; 850 kg bulk bags	118.34	1.50	30.75	-	124.26	m³	**155.01**
sulphate resistant cement; 10 tonne ballast loads	145.09	1.10	22.55	-	152.34	m³	**174.89**
sulphate resistant cement; 850 kg bulk bags	181.18	1.50	30.75	-	190.24	m³	**220.99**
Isolated foundations							
ordinary Portland cement	145.09	2.00	41.00	-	145.09	m³	**186.09**
sulphate resistant cement	145.09	2.00	41.00	-	148.72	m³	**189.72**
Reinforced in situ concrete; site mixed; 21 N/mm² - 20 mm aggregate (1:2:4); aggregates delivered in 10 tonne loads							
Foundations							
ordinary Portland cement	118.34	1.10	22.55	-	121.30	m³	**143.85**
sulphate resistant cement	145.09	1.10	22.55	-	148.72	m³	**171.27**
Foundations; poured on or against earth or unblinded hardcore							
ordinary Portland cement	145.09	1.10	22.55	-	152.34	m³	**174.89**
sulphate resistant cement	145.09	1.10	22.55	-	152.34	m³	**174.89**
Isolated foundations							
ordinary Portland cement	118.34	2.20	45.10	-	121.30	m³	**166.40**
sulphate resistant cement	145.09	2.20	45.10	-	152.34	m³	**197.44**
Plain in situ concrete; ready mixed; Tarmac Southern; 10 N/mm mixes; suitable for mass concrete fill and blinding							
Foundations							
GEN1; designated mix	85.95	1.50	30.75	-	88.10	m³	**118.85**

E IN SITU CONCRETE/LARGE PRECAST CONCRETE

Item Excluding site overheads and profit	PC £	Labour hours	Labour £	Plant £	Material £	Unit	Total rate £
E10 MIXING/CASTING/CURING IN SITU CONCRETE – cont'd							
Plain in situ concrete; ready mixed; Tarmac Southern; 10 N/mm mixes; suitable for mass concrete fill and blinding – cont'd							
ST2; standard mix	92.45	1.50	30.75	-	94.76	m^3	125.51
Foundations; poured on or against earth or unblinded hardcore							
GEN1; designated mix	85.95	1.57	32.29	-	90.25	m^3	122.54
ST2; standard mix	92.45	1.57	32.29	-	97.07	m^3	129.36
Isolated foundations							
GEN1; designated mix	85.95	2.00	41.00	-	88.10	m^3	129.10
ST2; standard mix	92.45	2.00	41.00	-	94.76	m^3	135.76
Plain in situ concrete; ready mixed; Tarmac Southern; 15 N/mm mixes; suitable for oversite below suspended slabs and strip footings in non aggressive soils							
Foundations							
GEN 2; designated mix	89.55	1.50	30.75	-	91.79	m^3	122.54
ST3; standard mix	84.49	1.50	30.75	-	88.71	m^3	119.46
Foundations; poured on or against earth or unblinded hardcore							
GEN2; designated mix	89.55	1.57	32.29	-	91.79	m^3	124.08
ST3; standard mix	92.68	1.57	32.29	-	95.00	m^3	127.28
Isolated foundations							
GEN2; designated mix	89.55	2.00	41.00	-	91.79	m^3	132.79
ST3; standard mix	92.68	2.00	41.00	-	95.00	m^3	136.00
Concrete mixed on site; Easymix Concrete Ltd							
Concrete mixer on lorry; maximum standing time 3 m^3/hr							
C20 concrete	95.00	1.00	20.50	-	95.00	m^3	115.50
C25 concrete	95.00	1.00	20.50	-	95.00	m^3	115.50
C30 concrete	105.00	1.00	20.50	-	105.00	m^3	125.50
Concrete mixed on site; Easymix Concrete Ltd; concrete delivered by pump							
Concrete mixer on lorry; maximum standing time 3 m^3/hr							
C20 concrete	95.00	0.05	1.02	-	112.50	m^3	113.53
C25 concrete	95.00	1.00	20.50	-	112.50	m^3	133.00
C30 concrete	105.00	0.05	1.02	-	122.50	m^3	123.53
E20 FORMWORK FOR IN SITU CONCRETE							
Plain vertical formwork; basic finish							
Sides of foundations							
height exceeding 1.00 m	-	2.00	41.00	-	5.19	m^2	46.19
height not exceeding 250 mm	-	1.00	20.50	-	1.45	m	21.95
height 250 - 500 mm	-	1.00	20.50	-	2.59	m	23.10
height 500 mm - 1.00 m	-	1.50	30.75	-	5.19	m	35.94

E IN SITU CONCRETE/LARGE PRECAST CONCRETE

Item Excluding site overheads and profit	PC £	Labour hours	Labour £	Plant £	Material £	Unit	Total rate £
Sides of foundations; left in							
height over 1.00 m	-	2.00	41.00	-	10.92	m²	**51.92**
height not exceeding 250 mm	-	1.00	20.50	-	5.63	m	**26.13**
height 250 - 500 mm	-	1.00	20.50	-	10.94	m	**31.44**
height 500 mm - 1.00 m	-	1.50	30.75	-	21.89	m	**52.64**
E30 REINFORCEMENT FOR IN SITU CONCRETE							
For reinforcement bar prices please see the Major Works section of this book.							
Reinforcement fabric; lapped; in beds or suspended slabs							
Fabric 3.6 x 2.0 m²							
ref A142 (2.22 kg/m²)	3.77	0.22	4.51	-	4.14	m²	**8.65**
ref A393 (6.16 kg/m²)	10.45	0.28	5.74	-	11.50	m²	**17.24**
Fabric 2.4 x 4.8 m²							
ref A98 (1.54 kg/m²)	0.87	0.22	4.51	-	0.96	m²	**5.47**
ref A142 (2.22 kg/m²)	1.25	0.22	4.51	-	1.38	m²	**5.88**
ref A193 (3.02 kg/m²)	1.70	0.22	4.51	-	1.87	m²	**6.38**
ref A252 (3.95 kg/m²)	2.22	0.24	4.92	-	2.44	m²	**7.36**
ref A393 (6.16 kg/m²)	3.47	0.28	5.74	-	3.82	m²	**9.56**

F MASONRY

Item Excluding site overheads and profit	PC £	Labour hours	Labour £	Plant £	Material £	Unit	Total rate £
F10 BRICK/BLOCK WALLING							
Batching quantities for these mortar mixes may be found in the Memoranda section of this book.							
Mortar mixes; common mixes for various types of work; mortar mixed on site; prices based on builders merchant rates for cement; mechanically mixed							
Aggregates delivered in 850 kg bulk bags							
1:3	-	0.75	15.38	-	163.02	m^3	**178.39**
1:4	-	0.75	15.38	-	128.39	m^3	**143.77**
1:1:6	-	0.75	15.38	-	151.04	m^3	**166.42**
1:1:6 sulphate resisting	-	0.75	15.38	-	174.80	m^3	**190.18**
Aggregates delivered in 10 tonne loads							
1:2	-	0.75	15.38	-	147.13	m^3	**162.50**
1:3	-	0.75	15.38	-	119.79	m^3	**135.16**
1:4	-	0.75	15.38	-	92.13	m^3	**107.50**
1:1:6	-	0.75	15.38	-	114.78	m^3	**130.15**
1:1:6 sulphate resisting	-	0.75	15.38	-	138.54	m^3	**153.91**
Mortar mixes; common mixes for various types of work; mortar mixed on site; prices based on builders merchant rates for cement; hand mixed							
Aggregates delivered in 850 kg bulk bags							
1:3	-	2.00	41.00	-	165.02	m^3	**206.02**
1:4	-	2.00	41.00	-	130.06	m^3	**171.06**
1:1:6	-	2.00	41.00	-	152.71	m^3	**193.71**
1:1:6 sulphate resisting	-	2.00	41.00	-	176.47	m^3	**217.47**
Variation in brick prices							
Add or subtract the following amounts for every £1.00/1000 difference in the PC price of the measured items below							
half brick thick	-	-	-	-	0.06	m^2	-
one brick thick	-	-	-	-	0.13	m^2	-
one and a half brick thick	-	-	-	-	0.19	m^2	-
two brick thick	-	-	-	-	0.25	m^2	-
Movement of materials							
Loading to wheelbarrows and transporting to location; per 215 mm thick walls; maximum 25 m distance	-	0.42	8.54	-	-	m^2	**8.54**
Class B engineering bricks; PC £290.00/ 1000; double Flemish bond in cement mortar (1:3)							
Mechanically offloading; maximum 25 m distance; loading to wheelbarrows; transporting to location; per 215 mm thick walls	-	0.42	8.54	-	-	m^2	**8.54**
Walls							
half brick thick	-	3.19	65.44	-	19.43	m^2	**84.86**
one brick thick	-	6.38	130.87	-	38.85	m^2	**169.73**
one and a half brick thick	-	9.58	196.34	-	58.28	m^2	**254.62**
two brick thick	-	12.77	261.74	-	77.71	m^2	**339.45**

F MASONRY

Item Excluding site overheads and profit	PC £	Labour hours	Labour £	Plant £	Material £	Unit	Total rate £
Walls; curved; mean radius 6 m							
half brick thick	-	4.81	98.51	-	20.10	m²	118.61
one brick thick	-	9.61	196.93	-	38.85	m²	235.78
Walls; curved; mean radius 1.50 m							
half brick thick	-	6.40	131.16	-	20.10	m²	151.26
one brick thick	-	12.80	262.48	-	38.85	m²	301.34
Walls; tapering; one face battering; average							
one and a half brick thick	-	10.34	212.07	-	58.28	m²	270.35
two brick thick	-	15.52	318.10	-	77.71	m²	395.81
Walls; battering (retaining)							
one and a half brick thick	-	11.49	235.63	-	58.28	m²	293.92
two brick thick	-	17.24	353.45	-	77.71	m²	431.16
Isolated piers							
one brick thick	-	9.31	190.85	-	42.91	m²	233.76
one and a half brick thick	-	11.97	245.38	-	64.36	m²	309.75
two brick thick	-	13.30	272.65	-	87.17	m²	359.82
three brick thick	-	16.49	338.09	-	128.73	m²	466.81
Projections; vertical							
one brick x half brick	-	0.70	14.35	-	4.33	m	18.68
one brick x one brick	-	1.40	28.70	-	8.68	m	37.38
one and a half brick x one brick	-	2.10	43.05	-	13.02	m	56.07
two brick x one brick	-	2.30	47.15	-	17.36	m	64.51
Walls; half brick thick							
in honeycomb bond	-	1.80	36.90	-	14.30	m²	51.20
in quarter bond	-	1.67	34.17	-	18.75	m²	52.92
Facing bricks; PC £300.00/1000; English garden wall bond; in gauged mortar (1:1:6); facework one side							
Mechanically offloading; maximum 25 m distance; loading to wheelbarrows and transporting to location; per 215 mm thick walls		0.42	8.54	-	-	m²	8.54
Walls							
half brick thick	-	3.19	65.44	-	20.85	m²	86.29
half brick thick half brick thick (using site cut snap headers to form bond)	-	4.99	102.24	-	20.85	m²	123.10
one brick thick	-	6.38	130.87	-	39.75	m²	170.62
one and a half brick thick	-	9.58	196.31	-	62.56	m²	258.86
two brick thick	-	12.77	261.74	-	83.41	m²	345.15
Walls; curved; mean radius 6 m							
half brick thick	-	4.81	98.51	-	22.40	m²	120.91
one brick thick	-	9.61	196.93	-	43.50	m²	240.43
Walls; curved; mean radius 1.50 m							
half brick thick	-	6.40	131.16	-	21.95	m²	153.11
one brick thick	-	12.80	262.48	-	42.60	m²	305.09
Walls; tapering; one face battering; average							
one and a half brick thick	-	11.49	235.59	-	65.26	m²	300.84
two brick thick	-	17.24	353.45	-	87.01	m²	440.46
Walls; battering (retaining)							
one and a half brick thick	-	10.34	212.03	-	65.26	m²	277.29
two brick thick	-	15.52	318.10	-	87.01	m²	405.11
Isolated piers; English bond; facework all round							
one brick thick	-	9.31	190.85	-	45.61	m²	236.46
one and a half brick thick	-	11.97	245.38	-	68.41	m²	313.80
two brick thick	-	13.30	272.65	-	92.52	m²	365.17
three brick thick	-	16.49	338.09	-	136.83	m²	474.91

F MASONRY

Item Excluding site overheads and profit	PC £	Labour hours	Labour £	Plant £	Material £	Unit	Total rate £
F10 BRICK/BLOCK WALLING – cont'd							
Facing bricks; PC £300.00/1000; English garden wall bond; in gauged mortar (1:1:6); facework one side – cont'd							
Projections; vertical							
one brick x half brick	-	0.93	19.09	-	4.65	m	**23.74**
one brick x one brick	-	1.86	38.17	-	9.31	m	**47.48**
one and a half brick x one brick	-	2.79	57.26	-	13.97	m	**71.22**
two brick x one brick	-	3.06	62.71	-	18.62	m	**81.33**
Brickwork fair faced both sides; facing bricks in gauged mortar (1:1:6)							
Extra for fair face both sides; flush, struck, weathered or bucket-handle pointing	-	0.89	18.18	-	-	m²	**18.18**
Reclaimed bricks; PC £1000.00/1000; English garden wall bond; in gauged mortar (1:1:6)							
Walls							
half brick thick (stretcher bond)	-	3.19	65.44	-	64.95	m²	**130.39**
half brick thick (using site cut snap headers to form bond)	-	4.99	102.24	-	64.95	m²	**167.20**
one brick thick	-	6.38	130.87	-	127.50	m²	**258.38**
one and a half brick thick	-	9.58	196.34	-	191.26	m²	**387.60**
two brick thick	-	12.77	261.74	-	255.01	m²	**516.75**
Walls; curved; mean radius 6 m							
half brick thick	-	4.81	98.51	-	68.60	m²	**167.11**
one brick thick	-	9.61	196.93	-	135.90	m²	**332.83**
Walls; curved; mean radius 1.50 m							
half brick thick	-	6.40	131.16	-	71.60	m²	**202.76**
one brick thick	-	12.77	261.74	-	141.90	m²	**403.65**
Walls; stretcher bond; wall ties at 450 centres vertically and horizontally							
one brick thick	-	6.96	142.77	-	128.02	m²	**270.79**
two brick thick	-	13.93	285.50	-	256.56	m²	**542.06**
Brickwork fair faced both sides; facing bricks in gauged mortar (1:1:6)							
extra for fair face both sides; flush, struck, weathered or bucket-handle pointing	-	0.67	13.67	-	-	m²	**13.67**
Brick copings							
Copings; all brick headers-on-edge; two angles rounded 53 mm radius; flush pointing top and both sides as work proceeds; one brick wide; horizontal							
machine-made specials	25.65	0.71	14.57	-	27.23	m	**41.80**
hand-made specials	25.65	0.71	14.57	-	27.23	m	**41.81**
Extra over copings for two courses machine-made tile creasings, projecting 25 mm each side; 260 mm wide copings; horizontal	6.16	0.50	10.25	-	6.87	m	**17.12**
Copings; all brick headers-on-edge; flush pointing top and both sides as work proceeds; one brick wide; horizontal							
bricks PC £300.00/1000	4.00	0.71	14.57	-	4.57	m	**19.14**
bricks PC £500.00/1000	6.67	0.71	14.57	-	7.12	m	**21.69**

F MASONRY

Item Excluding site overheads and profit	PC £	Labour hours	Labour £	Plant £	Material £	Unit	Total rate £
Dense aggregate concrete blocks; "Tarmac Topblock" or other equal and approved; in gauged mortar (1:2:9)							
Solid blocks 7 N/mm^2							
440 x 215 x 100 mm thick	7.64	1.20	24.60	-	8.73	m^2	**33.33**
440 x 215 x 140 mm thick	12.30	1.30	26.65	-	13.51	m^2	**40.16**
Solid blocks 7 N/mm^2 laid flat							
440 x 100 x 215 mm thick	15.60	3.22	66.01	-	18.50	m^2	**84.51**
Hollow concrete blocks							
440 x 215 x 215 mm thick	18.90	1.30	26.65	-	20.11	m^2	**46.76**
Filling of hollow concrete blocks with concrete as work proceeds; tamping and compacting							
440 x 215 x 215 mm thick	21.63	0.20	4.10	-	21.63	m^2	**25.73**

G STRUCTURAL/CASCASSING METAL/TIMBER

Item Excluding site overheads and profit	PC £	Labour hours	Labour £	Plant £	Material £	Unit	Total rate £
G31 TIMBER DECKING							
Timber decking							
Supports for timber decking; softwood joists to receive decking boards; joists at 400 mm centres; Southern yellow pine							
38 x 88 mm	16.55	1.00	20.50	-	20.19	m^2	**40.69**
50 x 150 mm	11.40	1.00	20.50	-	14.53	m^2	**35.03**
50 x 125 mm	9.50	1.00	20.50	-	12.44	m^2	**32.94**
Hardwood decking; yellow Balau; grooved or smooth; 6 mm joints							
deck boards; 90 mm wide x 19 mm thick	19.99	1.00	20.50	-	22.26	m^2	**42.76**
deck boards; 145 mm wide x 21 mm thick	29.33	1.00	20.50	-	34.54	m^2	**55.04**
deck boards; 145 mm wide x 28 mm thick	30.19	1.00	20.50	-	35.48	m^2	**55.98**
Hardwood decking; Ipe; smooth; 6 mm joints							
deck boards; 90 mm wide x 19 mm thick	38.10	1.00	20.50	-	40.38	m^2	**60.88**
deck boards; 145 mm wide x 19 mm thick	28.00	1.00	20.50	-	30.28	m^2	**50.78**
Western Red Cedar; 6 mm joints							
Prime deck grade; 90 mm wide x 40 mm thick	34.67	1.00	20.50	-	40.41	m^2	**60.91**
Prime deck grade; 142 mm wide x 40 mm thick	36.85	1.00	20.50	-	42.82	m^2	**63.32**
Handrails and base rail; fixed to posts at 2.00 m centres							
Posts 100 x 100 x 1370 high	8.48	1.00	20.50	-	9.73	m	**30.23**
Posts turned 1220 high	17.12	1.00	20.50	-	18.36	m	**38.86**
Handrails; balusters							
Square balusters at 100 mm centres	39.90	0.50	10.25	-	40.43	m	**50.69**
Square balusters at 300 mm centres	13.29	0.33	6.83	-	13.71	m	**20.55**
Turned balusters at 100 mm centres	63.00	0.50	10.25	-	63.53	m	**73.78**
Turned balusters at 300 mm centres	20.98	0.33	6.83	-	21.41	m	**28.24**

J WATERPROOFING

Item Excluding site overheads and profit	PC £	Labour hours	Labour £	Plant £	Material £	Unit	Total rate £
J30 LIQUID APPLIED TANKING/DAMP PROOFING							
Tanking and damp proofing; Ruberoid Building Products; "Synthaprufe" cold applied bituminous emulsion waterproof coating							
"Synthaprufe"; to smooth finished concrete or screeded slabs; flat; blinding with sand							
two coats	5.24	0.22	4.56	-	5.94	m^2	**10.50**
three coats	7.86	0.31	6.36	-	8.82	m^2	**15.18**
"Synthaprufe"; to fair faced brickwork with flush joints, rendered brickwork or smooth finished concrete walls; vertical							
two coats	5.89	0.29	5.86	-	6.66	m^2	**12.52**
three coats	8.64	0.40	8.20	-	9.69	m^2	**17.89**
Tanking and damp proofing; RIW Ltd							
Liquid asphaltic composition; to smooth finished concrete screeded slabs or screeded slabs; flat							
two coats	10.10	0.33	6.83	-	11.11	m^2	**17.94**
Liquid asphaltic composition; fair faced brickwork with flush joints, rendered brickwork or smooth finished concrete walls; vertical							
two coats	10.10	0.50	10.25	-	11.11	m^2	**21.36**
"Heviseal"; to smooth finished concrete or screeded slabs; to surfaces of ponds, tanks, planters; flat							
two coats	6.67	0.33	6.83	-	7.34	m^2	**14.17**
"Heviseal"; to fair faced brickwork with flush joints, rendered brickwork or smooth finished concrete walls; to surfaces of retaining walls, ponds, tanks, planters; vertical							
two coats	6.67	0.50	10.25	-	7.34	m^2	**17.59**

M SURFACE FINISHES

Item Excluding site overheads and profit	PC £	Labour hours	Labour £	Plant £	Material £	Unit	Total rate £
M60 PAINTING/CLEAR FINISHING							
Prepare; touch up primer; two undercoats and one finishing coat of gloss oil paint; on metal surfaces							
General surfaces							
girth exceeding 300 mm	0.96	0.33	6.83	-	0.96	m^2	7.79
isolated surfaces; girth not exceeding 300 mm	0.28	0.13	2.73	-	0.28	m	3.01
isolated areas not exceeding 0.50 m^2 irrespective of girth	0.51	0.13	2.73	-	0.51	nr	3.24
Ornamental railings; each side measured overall							
girth exceeding 300 mm	0.96	0.75	15.38	-	0.96	m^2	16.33
Prepare; one coat primer; two undercoats and one finishing coat of gloss oil paint; on wood surfaces							
General surfaces							
girth exceeding 300 mm	1.29	0.40	8.20	-	1.29	m^2	9.49
isolated surfaces; girth not exceeding 300 mm	0.42	0.20	4.10	-	0.42	m	4.52
isolated areas not exceeding 0.50 m^2 irrespective of girth	0.50	0.36	7.45	-	0.50	nr	7.95
Prepare; two coats of creosote; on wood surfaces							
General surfaces							
girth exceeding 300 mm	0.21	0.21	4.26	-	0.21	m^2	4.47
isolated surfaces; girth not exceeding 300 mm	0.07	0.12	2.36	-	0.07	m	2.43
Prepare; proprietary solution primer; two coats of dark stain; on wood surfaces							
General surfaces							
girth exceeding 300 mm	3.01	0.10	2.05	-	3.10	m^2	5.15
isolated surfaces; girth not exceeding 300 mm	0.39	0.05	1.02	-	0.40	m	1.42
Three coats "Dimex Shield"; to clean, dry surfaces; in accordance with manufacturer's instructions							
Brick or block walls							
girth exceeding 300 mm	1.28	0.28	5.74	-	1.28	m^2	7.02
Cement render or concrete walls							
girth exceeding 300 mm	1.01	0.25	5.13	-	1.01	m^2	6.13
Two coats resin based paint "Sandtex Matt"; in accordance with manufacturer's instructions							
Brick or block walls							
girth exceeding 300 mm	7.83	0.20	4.10	-	7.83	m^2	11.93
Cement render or concrete walls							
girth exceeding 300 mm	5.32	0.17	3.42	-	5.32	m^2	8.74

Q PAVING/PLANTING/FENCING/SITE FURNITURE

Item Excluding site overheads and profit	PC £	Labour hours	Labour £	Plant £	Material £	Unit	Total rate £
Q10 KERBS/EDGINGS/CHANNELS/ PAVING ACCESSORIES							
Foundations to kerbs							
Excavating trenches; width 300 mm; 3 tonne excavator							
depth 300 mm	-	0.10	2.05	0.41	2.96	m	**5.42**
depth 400 mm	-	0.11	2.28	0.46	3.96	m	**6.70**
By hand	-	0.60	12.35	-	4.93	m	**17.28**
Excavating trenches; width 450 mm; 3 tonne excavator; disposal off site							
depth 300 mm	-	0.13	2.56	0.62	4.44	m	**7.62**
depth 400 mm	-	0.14	2.93	0.70	5.94	m	**9.57**
Disposal							
Excavated material; off site; to tip							
grab loaded 7.25 m^3	-	-	-	-	-	m^3	**2.96**
Disposal by skip; 6 yd3 (4.6 m^3)							
by machine	-	-	-	41.26	-	m^3	**41.26**
by hand	-	4.00	82.00	44.34	-	m^3	**126.34**
Foundations; to kerbs, edgings or channels; in situ concrete; 21 N/mm^2 - 20 mm aggregate (1:2:4) site mixed; one side against earth face, other against formwork (not included); site mixed concrete							
Site mixed concrete							
150 wide x 100 mm deep	-	0.13	2.73	-	1.78	m	**4.51**
150 wide x 150 mm deep	-	0.17	3.42	-	2.66	m	**6.08**
200 wide x 150 mm deep	-	0.20	4.10	-	3.55	m	**7.65**
300 wide x 150 mm deep	-	0.23	4.78	-	5.33	m	**10.11**
600 wide x 200 mm deep	-	0.29	5.86	-	14.20	m	**20.06**
Ready mixed concrete							
150 wide x 100 mm deep	-	0.13	2.73	-	1.36	m	**4.09**
150 wide x 150 mm deep	-	0.17	3.42	-	2.03	m	**5.45**
200 wide x 150 mm deep	-	0.20	4.10	-	2.71	m	**6.81**
300 wide x 150 mm deep	-	0.23	4.78	-	4.07	m	**8.85**
600 wide x 200 mm deep	-	0.29	5.86	-	10.84	m	**16.70**
Formwork; sides of foundations (this will usually be required to one side of each kerb foundation adjacent to road sub-bases)							
100 mm deep	-	0.04	0.85	-	0.18	m	**1.03**
150 mm deep	-	0.04	0.85	-	0.26	m	**1.12**
Precast concrete edging units; including haunching with in situ concrete; 11.50 N/mm^2 - 40 mm aggregate both sides							
Edgings; rectangular, bullnosed or chamfered							
50 x 150 mm	1.54	0.33	6.83	-	6.77	m	**13.61**
125 x 150 mm bullnosed	1.68	0.33	6.83	-	6.91	m	**13.74**
50 x 200 mm	2.29	0.33	6.83	-	7.53	m	**14.36**
50 x 250 mm	2.66	0.33	6.83	-	7.90	m	**14.73**
50 x 250 mm flat top	3.29	0.33	6.83	-	8.52	m	**15.35**

Q PAVING/PLANTING/FENCING/SITE FURNITURE

Item Excluding site overheads and profit	PC £	Labour hours	Labour £	Plant £	Material £	Unit	Total rate £
Q10 KERBS/EDGINGS/CHANNELS/ PAVING ACCESSORIES – cont'd							
Marshalls Mono; small element precast concrete kerb system; Keykerb Large (KL) upstand of 100 - 125 mm; on 150 mm concrete foundation including haunching with in situ concrete 1:3:6 1 side							
Bullnosed or half battered; 100 x 127 x 200 mm							
laid straight	12.54	0.80	16.40	-	13.64	m	30.04
radial blocks laid to curve; 8 blocks/1/4 circle - 500 mm radius	9.65	0.80	16.40	-	10.75	m	27.15
radial blocks laid to curve; 8 radial blocks; alternating 8 standard blocks/1/4 circle - 1000 mm radius	15.92	1.25	25.63	-	17.02	m	42.65
radial blocks; alternating 16 standard blocks/1/4 circle - 1500 mm radius	14.65	1.50	30.75	-	15.75	m	46.50
internal angle 90°	5.29	0.20	4.10	-	5.29	nr	9.39
external angle	5.29	0.20	4.10	-	5.29	nr	9.39
drop crossing kerbs; KL half battered to KL splay; LH and RH	11.08	1.00	20.50	-	12.18	pair	32.68
drop crossing kerbs; KL half battered to KS bullnosed	10.78	1.00	20.50	-	11.88	pair	32.38
Marshalls Mono; small element precast concrete kerb system; Keykerb Small (KS) upstand of 25 - 50 mm; on 150 mm concrete foundation including haunching with in situ concrete 1:3:6 1 side							
Half battered							
laid straight	8.81	0.80	16.40	-	9.91	m	26.31
radial blocks laid to curve; 8 blocks/1/4 circle - 500 mm radius	7.25	1.00	20.50	-	8.34	m	28.84
radial blocks laid to curve; 8 radial blocks; alternating 8 standard blocks/1/4 circle - 1000 mm radius	11.65	1.25	25.63	-	12.75	m	38.37
radial blocks; alternating 16 standard blocks/1/4 circle - 1500 mm radius	12.47	1.50	30.75	-	13.57	m	44.32
internal angle 90°	5.29	0.20	4.10	-	5.29	nr	9.39
external angle	5.29	0.20	4.10	-	5.29	nr	9.39
Second-hand granite setts; 100 x 100 mm; bedding in cement mortar (1:4); on 150 mm deep concrete foundations; including haunching with in situ concrete; 11.50 N/mm^2 - 40 mm aggregate one side							
Edgings							
300 mm wide	13.24	1.33	27.33	-	17.50	m	44.84
Brick or block stretchers; bedding in cement mortar (1:4); on 150 mm deep concrete foundations, including haunching with in situ concrete; 11.50 N/mm^2 - 40 mm aggregate one side							
Single course							
concrete paving blocks; PC £8.72/m^2; 200 x 100 x 60 mm	0.87	0.31	6.31	-	3.84	m	10.15

Q PAVING/PLANTING/FENCING/SITE FURNITURE

Item Excluding site overheads and profit	PC £	Labour hours	Labour £	Plant £	Material £	Unit	Total rate £
engineering bricks; PC £300.00/1000; 215 x 102.5 x 65 mm	1.29	0.40	8.20	-	4.32	m	12.52
paving bricks; PC £500.00/1000; 215 x 102.5 x 65 mm	2.22	0.40	8.20	-	5.19	m	13.39
Two courses							
concrete paving blocks; PC £8.72/m²; 200 x 100 x 60 mm	1.74	0.40	8.20	-	5.20	m	13.40
engineering bricks; PC £300.00/1000; 215 x 102.5 x 65 mm	2.58	0.57	11.71	-	6.16	m	17.87
paving bricks; PC £500.00/1000; 215 x 102.5 x 65 mm	4.44	0.57	11.71	-	7.90	m	19.61
Three courses							
concrete paving blocks; PC £8.72/m²; 200 x 100 x 60 mm	2.62	0.44	9.11	-	6.07	m	15.18
engineering bricks; PC £300.00/1000; 215 x 102.5 x 65 mm	3.87	0.67	13.67	-	7.51	m	21.18
paving bricks; PC £500.00/1000; 215 x 102.5 x 65 mm	6.67	0.67	13.67	-	10.12	m	23.78
Brick or block headers; brick on flat; bedding in cement mortar (1:4); on 150 mm deep concrete foundations; including haunching with in situ concrete; 11.50 N/mm² - 40 mm aggregate one side							
Single course							
concrete paving blocks; PC £8.72/m²; 200 x 100 x 60 mm	1.74	0.62	12.61	-	4.71	m	17.32
engineering bricks; PC £300.00/1000; 215 x 102.5 x 65 mm	2.58	0.80	16.40	-	5.68	m	22.08
paving bricks; PC £500.00/1000; 215 x 102.5 x 65 mm	4.44	0.80	16.40	-	7.41	m	23.81
Bricks on edge; bedding in cement mortar (1:4); on 150 mm deep concrete foundations; including haunching with in situ concrete; 11.50 N/mm² - 40 mm aggregate one side							
One brick wide							
engineering bricks; 215 x 102.5 x 65 mm	3.87	0.57	11.71	-	7.63	m	19.34
paving bricks; 215 x 102.5 x 65 mm	6.67	0.57	11.71	-	10.24	m	21.95
Two courses; stretchers laid on edge; 225 mm wide							
engineering bricks; 215 x 102.5 x 65 mm	7.73	1.20	24.60	-	12.77	m	37.37
paving bricks; 215 x 102.5 x 65 mm	13.33	1.20	24.60	-	17.98	m	42.58
Extra over bricks on edge for standard kerbs to one side; haunching in concrete							
125 x 255 mm; ref HB2; SP	3.93	0.44	9.11	-	7.07	m	16.18
New granite sett edgings; CED Ltd; 100 x 100 x 100 mm; bedding in cement mortar (1:4); including haunching with in situ concrete; 11.50 N/mm² - 40 mm aggregate one side; concrete foundations (not included)							
Edgings 100 x 100 x 100 mm; ref 1R "better quality" with 20 mm pointing gaps							
100 mm wide	2.71	1.33	27.27	-	5.20	m	32.47
2 rows; 220 mm wide	5.42	2.22	45.51	-	8.65	m	54.16

Q PAVING/PLANTING/FENCING/SITE FURNITURE

Item Excluding site overheads and profit	PC £	Labour hours	Labour £	Plant £	Material £	Unit	Total rate £
Q10 KERBS/EDGINGS/CHANNELS/ **PAVING ACCESSORIES** – cont'd							
New granite sett edgings; CED Ltd; 100 x 100 x 100 mm; bedding in cement mortar (1:4); including haunching with in situ concrete; 11.50 N/mm^2 - 40 mm aggregate one side; concrete foundations (not included) – cont'd							
3 rows; 340 mm wide	8.13	3.00	61.50	-	10.67	m	72.17
Edgings 100 x 100 x 200 mm; ref 1R "better quality" with 20 mm pointing gaps							
100 mm wide	2.40	0.75	15.38	-	4.89	m	20.27
2 rows; 220 mm wide	4.79	1.50	30.75	-	8.02	m	38.77
3 rows; 340 mm wide	7.19	2.25	46.13	-	9.73	m	55.85
Edgings 100 x 100 x 100 mm; ref 8R "standard quality" with 20 mm pointing gaps							
100 mm wide	2.69	1.33	27.27	-	5.18	m	32.44
2 rows; 220 mm wide	5.37	2.22	45.51	-	8.60	m	54.11
3 rows; 340 mm wide	8.06	3.00	61.50	-	10.59	m	72.09
Edgings 100 x 100 x 200 mm; ref 8R "standard quality" with 20 mm pointing gaps							
100 mm wide	3.76	0.75	15.38	-	6.26	m	21.63
2 rows; 220 mm wide	7.52	1.50	30.75	-	10.75	m	41.50
3 rows; 340 mm wide	11.29	2.25	46.13	-	13.82	m	59.95
Reclaimed granite setts edgings; CED Ltd; 100 x 100 mm; bedding in cement mortar (1:4); including haunching with in situ concrete; 11.50 N/mm^2 - 40 mm aggregate one side; on concrete foundations (not included)							
Edgings 100 x 100 x 100 mm; with 20 mm pointing gaps							
100 mm wide	3.94	1.33	27.27	-	6.44	m	33.70
2 rows; 220 mm wide	7.88	2.22	45.51	-	11.11	m	56.62
3 rows; 340 mm wide	11.83	3.00	61.50	-	14.36	m	75.86
Channels; bedding in cement mortar (1:3); joints pointed flush; on concrete foundations (not included)							
Three courses stretchers; 350 mm wide; quarter bond to form dished channels							
engineering bricks; PC £300.00/1000; 215 x 102.5 x 65 mm	3.87	1.00	20.50	-	5.15	m	25.65
paving bricks; PC £500.00/1000; 215 x 102.5 x 65 mm	6.67	1.00	20.50	-	7.75	m	28.25
Three courses granite setts; 340 mm wide; to form dished channels							
340 mm wide	13.24	2.00	41.00	-	15.73	m	56.73

Q PAVING/PLANTING/FENCING/SITE FURNITURE

Item Excluding site overheads and profit	PC £	Labour hours	Labour £	Plant £	Material £	Unit	Total rate £
Permaloc "AshphaltEdge"; RTS Ltd; extruded aluminium alloy L shaped edging with 5.33 mm exposed upper lip; edging fixed to roadway base and edge profile with 250 mm steel fixing spike; laid to straight or curvilinear road edge; subsequently filled with macadam (not included)							
Depth of macadam							
38 mm	7.34	0.02	0.34	-	7.52	m	**7.87**
51 mm	8.27	0.02	0.35	-	8.48	m	**8.83**
64 mm	9.14	0.02	0.37	-	9.37	m	**9.74**
76 mm	11.05	0.02	0.41	-	11.33	m	**11.74**
102 mm	12.86	0.02	0.43	-	13.18	m	**13.61**
Permaloc "Cleanline"; RTS Ltd; heavy duty straight profile edging; for edgings to soft landscape beds or turf areas; 3.2 mm x 102 high; 3.2 mm thick with 4.75 mm exposed upper lip; fixed to form straight or curvilinear edge with 305 mm fixing spike							
Milled aluminium							
100 deep	8.41	0.02	0.37	-	8.62	m	**8.99**
Black							
100 deep	8.41	0.02	0.37	-	8.62	m	**8.99**
Permaloc "Permastrip"; RTS Ltd; heavy duty L shaped profile maintenance strip; 3.2 mm x 89 mm high with 5.2 mm exposed top lip; for straight or gentle curves on paths or bed turf interfaces; fixed to form straight or curvilinear edge with standard 305 mm stake; other stake lengths available							
Milled aluminium							
89 deep	8.41	0.02	0.37	-	8.62	m	**8.99**
Black							
89 deep	8.41	0.02	0.37	-	8.62	m	**8.99**
Permaloc "Proline"; RTS Ltd; medium duty straight profiled maintenance strip; 3.2 mm x 102 mm high with 3.18 mm exposed top lip; for straight or gentle curves on paths or bed turf interfaces; fixed to form straight or curvilinear edge with standard 305 mm stake; other stake lengths available							
Milled aluminium							
89 deep	7.32	0.02	0.37	-	7.50	m	**7.88**
Black							
89 deep	7.32	0.02	0.37	-	7.50	m	**7.88**

Q PAVING/PLANTING/FENCING/SITE FURNITURE

Item Excluding site overheads and profit	PC £	Labour hours	Labour £	Plant £	Material £	Unit	Total rate £
Q10 KERBS/EDGINGS/CHANNELS/ PAVING ACCESSORIES – cont'd							
Exceledge "SuperEdge"; recycled vinyl edging; black; 3 x 150 mm high with 5 mm diameter bead on each edge; for edging to soft landscape turf or bedding areas; formed by heat to required angle, or fixed to form straight or curvilinear lines using 350 mm galvanized fixing stakes							
150 mm deep	5.49	0.02	0.37	-	5.49	m	**5.86**
Exceledge "BrickEdge"; recycled vinyl hard landscape edging former; black; 50 mm high with 76 mm return; 3 mm thick wall for the retention of bricks, blocks and loose or bonded gravels; fixed with steel fixing spikes at 300 mm centres							
Flexible; 50 mm deep	6.05	0.02	0.37	-	6.05	m	**6.42**
Rigid; 50 mm deep	6.05	0.02	0.37	-	6.05	m	**6.42**
Exceledge "EdgeKing"; polyethylene flexible edging; black; used for turfed areas and retention of loose fill materials; with 25 mm diameter rounded hose top for bold demarcation and safety in play areas; fixed with steel vee-stakes at 45° angle into adjacent material or to prefixed timber supports using screws and washers							
125 mm deep	3.98	0.02	0.37	-	3.98	m	**4.35**
Corner pieces	5.71	-	-	-	5.71	ea	**-**
Joiner plugs	0.73	-	-	-	0.73	ea	**-**
Exceledge "Borderline"; tree rings; preformed unit from 100 mm high rolled steel in galvanized or powder coated (black or brown) finish; for circular traditional lawn edging around trees, bushes, flag poles or part of a feature; fixed with flat steel matching stakes							
700 mm	14.37	0.50	10.25	-	14.37	ea	**24.62**
900 mm	17.25	0.55	11.28	-	17.25	ea	**28.52**
1200 mm	24.97	0.60	12.30	-	24.97	ea	**37.27**
1500 mm	28.32	0.65	13.32	-	28.32	ea	**41.64**
1800 mm	31.65	0.70	14.35	-	31.65	ea	**46.00**
Q20 GRANULAR SUB-BASES TO ROADS/PAVING							
Hardcore bases; obtained off site; PC £8.00/m^3							
By machine							
100 mm thick	2.07	0.05	1.02	2.09	2.07	m^2	**5.19**
150 mm thick	3.10	0.07	1.37	2.62	3.10	m^2	**7.09**
over 150 thick	20.70	0.50	10.25	20.90	20.70	m^3	**51.85**

Q PAVING/PLANTING/FENCING/SITE FURNITURE

Item Excluding site overheads and profit	PC £	Labour hours	Labour £	Plant £	Material £	Unit	Total rate £
By hand							
100 mm thick	2.07	0.20	4.10	0.50	2.07	m^2	**6.67**
150 mm thick	3.10	0.30	6.16	0.50	3.73	m^2	**10.38**
Hardcore; difference for each £1.00 increase/decrease in PC price per m^3; price will vary with type and source of hardcore							
average 75 mm thick	-	-	-	-	0.08	m^2	-
average 100 mm thick	-	-	-	-	0.10	m^2	-
average 150 mm thick	-	-	-	-	0.15	m^2	-
average 200 mm thick	-	-	-	-	0.20	m^2	-
average 250 mm thick	-	-	-	-	0.25	m^2	-
average 300 mm thick	-	-	-	-	0.30	m^2	-
exceeding 300 mm thick	-	-	-	-	1.10	m^3	-
Type 1 granular fill base; PC £19.20/tonne (£42.24/m^3 compacted)							
By machine							
100 mm thick	4.22	0.03	0.57	0.56	4.22	m^2	**5.36**
150 mm thick	6.34	0.03	0.51	0.78	6.34	m^2	**7.63**
By hand (mechanical compaction)							
100 mm thick	4.22	0.17	3.42	0.17	4.22	m^2	**7.81**
150 mm thick	6.34	0.25	5.13	0.25	6.34	m^2	**11.71**
Surface treatments							
Sand blinding; to hardcore base (not included); 25 mm thick	1.01	0.03	0.68	-	1.01	m^2	**1.70**
Sand blinding; to hardcore base (not included); 50 mm thick	2.02	0.05	1.02	-	2.02	m^2	**3.05**
Filter fabrics; to hardcore base (not included)	0.50	0.01	0.20	-	0.52	m^2	**0.73**
Q21 IN SITU CONCRETE ROADS/ PAVINGS							
Unreinforced concrete; on prepared sub-base (not included)							
Paths and roads; 21.00 N/mm^2 - 20 mm aggregate (1:2:4) mechanically mixed on site							
100 mm thick	11.83	0.13	2.56	-	12.43	m^2	**14.99**
150 mm thick	17.75	0.17	3.42	-	18.64	m^2	**22.06**
Formwork for in situ concrete							
Sides of foundations							
height not exceeding 250 mm	0.44	0.05	1.02	-	0.73	m	**1.76**
Extra over formwork for curved work 6 m radius	-	0.25	5.13	-	-	m	**5.13**
Steel road forms; to edges of beds or faces of foundations							
150 mm wide	-	0.20	4.10	1.40	-	m	**5.50**
Reinforcement; fabric; side laps 150 mm; head laps 300 mm; mesh 200 x 200 mm; in roads, footpaths or pavings							
Fabric							
ref A142 (2.22 kg/m^2)	3.77	0.08	1.70	-	4.14	m^2	**5.85**
ref A193 (3.02 kg/m^2)	5.13	0.08	1.70	-	5.64	m^2	**7.34**

Q PAVING/PLANTING/FENCING/SITE FURNITURE

Item Excluding site overheads and profit	PC £	Labour hours	Labour £	Plant £	Material £	Unit	Total rate £
Q21 IN SITU CONCRETE ROADS/ PAVINGS – cont'd							
Reinforced in situ concrete; mechanically mixed on site; normal Portland cement; on hardcore base (not included); reinforcement (not included)							
Roads; 11.50 N/mm^2 - 40 mm aggregate (1:3:6)							
100 mm thick	10.46	0.40	8.20	-	10.72	m^2	**18.92**
150 mm thick	15.69	0.60	12.30	-	16.09	m^2	**28.39**
200 mm thick	20.93	0.80	16.40	-	21.45	m^2	**37.85**
Roads; 21.00 N/mm^2 - 20 mm aggregate (1:2:4)							
100 mm thick	11.83	0.40	8.20	-	12.13	m^2	**20.33**
150 mm thick	17.75	0.60	12.30	-	18.20	m^2	**30.50**
200 mm thick	23.67	0.80	16.40	-	24.26	m^2	**40.66**
Roads; 25.00 N/mm^2 - 20 mm aggregate GEN 4 ready mixed							
100 mm thick	9.37	0.40	8.20	-	9.60	m^2	**17.80**
150 mm thick	14.05	0.60	12.30	-	14.41	m^2	**26.71**
200 mm thick	18.73	0.80	16.40	-	19.20	m^2	**35.60**
250 mm thick	23.42	1.00	20.50	0.68	24.00	m^2	**45.18**
Reinforced in situ concrete; ready mixed; discharged directly into location from supply lorry; normal Portland cement; on hardcore base (not included); reinforcement (not included)							
Roads; 11.50 N/mm^2 - 40 mm aggregate (1:3:6)							
100 mm thick	9.24	0.16	3.28	-	9.24	m^2	**12.52**
150 mm thick	13.87	0.24	4.92	-	13.87	m^2	**18.79**
200 mm thick	18.49	0.36	7.38	-	18.49	m^2	**25.87**
250 mm thick	23.11	0.54	11.07	0.68	23.11	m^2	**34.86**
300 mm thick	27.73	0.66	13.53	0.68	27.73	m^2	**41.95**
Roads; 21.00 N/mm^2 - 20 mm aggregate (1:2:4)							
100 mm thick	9.29	0.16	3.28	-	9.29	m^2	**12.57**
150 mm thick	13.94	0.24	4.92	-	13.94	m^2	**18.86**
200 mm thick	18.59	0.36	7.38	-	18.59	m^2	**25.97**
250 mm thick	23.23	0.54	11.07	0.68	23.23	m^2	**34.99**
300 mm thick	27.88	0.66	13.53	0.68	27.88	m^2	**42.09**
Roads; 26.00 N/mm^2 - 20 mm aggregate (1:1:5:3)							
100 mm thick	9.73	0.16	3.28	-	9.73	m^2	**13.01**
150 mm thick	14.60	0.24	4.92	-	14.60	m^2	**19.52**
200 mm thick	19.46	0.36	7.38	-	19.46	m^2	**26.84**
250 mm thick	24.33	0.54	11.07	0.68	24.33	m^2	**36.08**
300 mm thick	29.20	0.66	13.53	0.68	29.20	m^2	**43.41**
Concrete sundries							
Treating surfaces of unset concrete; grading to cambers; tamping with 75 mm thick steel shod tamper or similar	-	0.13	2.73	-	-	m^2	**2.73**
Expansion joints							
13 mm thick joint filler; formwork							
width or depth not exceeding 150 mm	1.81	0.20	4.10	-	2.49	m	**6.59**

Q PAVING/PLANTING/FENCING/SITE FURNITURE

Item Excluding site overheads and profit	PC £	Labour hours	Labour £	Plant £	Material £	Unit	Total rate £
width or depth 150 - 300 mm	1.55	0.25	5.13	-	2.92	m	8.04
width or depth 300 - 450 mm	0.70	0.30	6.15	-	2.75	m	8.90
25 mm thick joint filler; formwork							
width or depth not exceeding 150 mm	2.65	0.20	4.10	-	3.33	m	7.43
width or depth 150 - 300 mm	2.65	0.25	5.13	-	4.02	m	9.14
width or depth 300 - 450 mm	2.65	0.30	6.15	-	4.70	m	10.85
Sealants; sealing top 25 mm of joint with rubberized bituminous compound	1.28	0.25	5.13	-	1.28	m	6.40
Q22 COATED MACADAM/ASPHALT ROADS/PAVINGS							
Coated macadam/asphalt roads/pavings - General							
Preamble: The prices for all in situ finishings to roads and footpaths include for work to falls, crossfalls or slopes not exceeding 15° from horizontal; for laying on prepared bases (not included) and for rolling with an appropriate roller. Users should note the new terminology for the surfaces described below which is to European standard descriptions. The now redundant descriptions for each course are shown in brackets. Note: The rates below are for smaller areas up to 200 m².							
Macadam surfacing; Spadeoak Construction Co Ltd; surface (wearing) course; 20 mm of 6 mm dense bitumen macadam							
Hand lay							
limestone aggregate	-	-	-	-	-	m²	9.04
granite aggregate	-	-	-	-	-	m²	9.02
red	-	-	-	-	-	m²	11.12
Macadam surfacing; Spadeoak Construction Co Ltd; surface (wearing) course; 30 mm of 10 mm dense bitumen macadam							
Hand lay							
limestone aggregate	-	-	-	-	-	m²	10.47
granite aggregate	-	-	-	-	-	m²	10.60
red	-	-	-	-	-	m²	-
Macadam surfacing; Spadeoak Construction Co Ltd; surface (wearing) course; 40 mm of 10 mm dense bitumen macadam							
Hand lay							
limestone aggregate	-	-	-	-	-	m²	12.95
granite aggregate	-	-	-	-	-	m²	13.11
red	-	-	-	-	-	m²	22.89

Q PAVING/PLANTING/FENCING/SITE FURNITURE

Item Excluding site overheads and profit	PC £	Labour hours	Labour £	Plant £	Material £	Unit	Total rate £
Q22 COATED MACADAM/ASPHALT ROADS/PAVINGS – cont'd							
Macadam surfacing; Spadeoak Construction Co Ltd; binder (base) course; 50 mm of 20 mm dense bitumen macadam							
Hand lay							
limestone aggregate	-	-	-	-	-	m²	12.98
granite aggregate	-	-	-	-	-	m²	13.16
Macadam surfacing; Spadeoak Construction Co Ltd; binder (base) course; 60 mm of 20 mm dense bitumen macadam							
Hand lay							
limestone aggregate	-	-	-	-	-	m²	14.12
granite aggregate	-	-	-	-	-	m²	14.33
Macadam surfacing; Spadeoak Construction Co Ltd; base (roadbase) course; 75 mm of 28 mm dense bitumen macadam							
Hand lay							
limestone aggregate	-	-	-	-	-	m²	17.01
granite aggregate	-	-	-	-	-	m²	17.29
Areas up to 300 m²							
Macadam surfacing; Spadeoak Construction Co Ltd; base (roadbase) course; 100 mm of 28 mm dense bitumen macadam							
Hand lay							
limestone aggregate	-	-	-	-	-	m²	72.46
granite aggregate	-	-	-	-	-	m²	21.10
Macadam surfacing; Spadeoak Construction Co Ltd; base (roadbase) course; 150 mm of 28 mm dense bitumen macadam in two layers							
Hand lay							
limestone aggregate	-	-	-	-	-	m²	33.29
granite aggregate	-	-	-	-	-	m²	33.84
Base (roadbase) course; 200 mm of 28 mm dense bitumen macadam in two layers							
Hand lay							
limestone aggregate	-	-	-	-	-	m²	28.16
granite aggregate	-	-	-	-	-	m²	42.25
Resin bound macadam pavings; hand laid; Bituchem "Natratex" clear resin bound macadam to pedestrian or vehicular hard landscape areas; laid to base course (not included)							
25 mm thick to pedestrian areas	22.19	-	-	-	22.19	m²	-
30 mm thick to vehicular areas	26.63	-	-	-	26.63	m²	-

Q PAVING/PLANTING/FENCING/SITE FURNITURE

Item Excluding site overheads and profit	PC £	Labour hours	Labour £	Plant £	Material £	Unit	Total rate £
"Colourtex" coloured resin bound macadam to pedestrian or vehicular hard landscape areas							
25 mm thick	22.19	-	-	-	22.19	m²	-
Bound aggregates; Addagrip Surface Treatments UK Ltd; natural decorative resin bonded surface dressing laid to concrete, macadam or to plywood panels priced separately							
Primer coat to macadam or concrete base	-	-	-	-	-	m²	5.06
Golden pea gravel 1-3 mm							
buff adhesive	-	-	-	-	-	m²	25.30
red adhesive	-	-	-	-	-	m²	25.30
green adhesive	-	-	-	-	-	m²	25.30
Golden pea gravel 2-5 mm							
buff adhesive	-	-	-	-	-	m²	28.75
Chinese bauxite 1-3 mm							
buff adhesive	-	-	-	-	-	m²	25.30
"Cobalt Blue Glass"; 6 mm							
15 mm depth	-	-	-	-	-	m²	88.00
"Midnight Grey", "Chocolate", "Derbyshire Cream", "All Gold"; 6 mm							
15 mm depth	-	-	-	-	-	m²	60.50
18 mm depth	-	-	-	-	-	m²	71.50
Q23 GRAVEL/HOGGIN/WOODCHIP ROADS/PAVINGS							
Excavation and path preparation; excavating; 300 mm deep; to width of path; depositing excavated material at sides of excavation							
By machine							
width 1.00 m	-	-	-	4.23	-	m²	4.23
width 1.50 m	-	-	-	3.53	-	m²	3.53
width 2.00 m	-	-	-	3.02	-	m²	3.02
width 3.00 m	-	-	-	2.54	-	m²	2.54
By hand							
width 1.00 m	-	0.70	14.44	-	-	m²	14.44
width 1.50 m	-	1.05	21.53	-	-	m²	21.53
width 2.00 m	-	1.40	28.70	-	-	m²	28.70
width 3.00 m	-	2.10	43.05	-	-	m²	43.05
Excavating trenches; in centre of pathways; 100 flexible drain pipes; filling with clean broken stone or gravel rejects							
300 x 450 mm deep							
by machine	5.10	0.10	2.05	1.59	5.20	m	8.84
by hand	5.10	0.40	8.20	-	5.20	m	13.40
Hand trimming and compacting reduced surface of pathway; by vibrating roller							
width 1.00 m	-	0.05	1.02	0.50	-	m	1.53
width 1.50 m	-	0.04	0.91	0.45	-	m	1.36
width 2.00 m	-	0.04	0.82	0.40	-	m	1.22
width 3.00 m	-	0.04	0.82	0.40	-	m	1.22
Permeable membranes; to trimmed and compacted surface of pathway							
"Terram 1000"	0.50	0.02	0.41	-	0.52	m²	0.93

Q PAVING/PLANTING/FENCING/SITE FURNITURE

Item Excluding site overheads and profit	PC £	Labour hours	Labour £	Plant £	Material £	Unit	Total rate £
Q23 GRAVEL/HOGGIN/WOODCHIP ROADS/PAVINGS – cont'd							
Permaloc "AshphaltEdge"; RTS Ltd; extruded aluminium alloy L shaped edging with 5.33 mm exposed upper lip; edging fixed to roadway base and edge profile with 250 mm steel fixing spike; laid to straight or curvilinear road edge; subsequently filled with macadam (not included)							
Depth of macadam							
38 mm	7.34	0.02	0.34	-	7.52	m	**7.87**
51 mm	8.27	0.02	0.35	-	8.48	m	**8.83**
64 mm	9.14	0.02	0.37	-	9.37	m	**9.74**
76 mm	11.05	0.02	0.41	-	11.33	m	**11.74**
102 mm	12.86	0.02	0.43	-	13.18	m	**13.61**
Permaloc "Cleanline"; RTS Ltd; heavy duty straight profile edging; for edgings to soft landscape beds or turf areas; 3.2 mm x 102 high; 3.2 mm thick with 4.75 mm exposed upper lip; fixed to form straight or curvilinear edge with 305 mm fixing spike							
Milled aluminium							
100 deep	8.41	0.02	0.37	-	8.62	m	**8.99**
Black							
100 deep	8.41	0.02	0.37	-	8.62	m	**8.99**
Permaloc "Permastrip"; RTS Ltd; heavy duty L shaped profile maintenance strip; 3.2 mm x 89 mm high with 5.2 mm exposed top lip; for straight or gentle curves on paths or bed turf interfaces; fixed to form straight or curvilinear edge with standard 305 mm stake; other stake lengths available							
Milled aluminium							
89 deep	8.41	0.02	0.37	-	8.62	m	**8.99**
Black							
89 deep	8.41	0.02	0.37	-	8.62	m	**8.99**
Permaloc "Proline"; RTS Ltd; medium duty straight profiled maintenance strip; 3.2 mm x 102 mm high with 3.18 mm exposed top lip; for straight or gentle curves on paths or bed turf interfaces; fixed to form straight or curvilinear edge with standard 305 mm stake; other stake lengths available							
Milled aluminium							
89 deep	7.32	0.02	0.37	-	7.50	m	**7.88**
Black							
89 deep	7.32	0.02	0.37	-	7.50	m	**7.88**

Q PAVING/PLANTING/FENCING/SITE FURNITURE

Item Excluding site overheads and profit	PC £	Labour hours	Labour £	Plant £	Material £	Unit	Total rate £
Timber edging boards							
Boards; 50 x 50 x 750 mm timber pegs at 1000 mm centres (excavations and hardcore under edgings not included)							
Straight							
38 x 150 mm treated softwood edge boards	2.54	0.10	2.05	-	2.54	m	4.59
50 x 150 mm treated softwood edge boards	2.75	0.10	2.05	-	2.75	m	4.80
38 x 150 mm hardwood (iroko) edge boards	9.26	0.10	2.05	-	9.26	m	11.31
50 x 150 mm hardwood (iroko) edge boards	10.69	0.10	2.05	-	10.69	m	12.74
Curved							
38 x 150 mm treated softwood edge boards	2.54	0.20	4.10	-	2.54	m	6.64
50 x 150 mm treated softwood edge boards	2.75	0.25	5.13	-	2.75	m	7.88
38 x 150 mm hardwood (iroko) edge boards	9.26	0.20	4.10	-	9.26	m	13.36
50 x 150 mm hardwood (iroko) edge boards	10.69	0.25	5.13	-	10.69	m	15.81
Filling to make up levels							
Obtained off site; hardcore; PC £20.70/m^3							
150 mm thick	3.10	0.07	1.37	2.62	3.10	m^2	7.09
Obtained off site; granular fill type 1; PC £19.20/tonne (£42.24/m^3 compacted)							
100 mm thick	4.22	0.03	0.57	0.56	4.22	m^2	5.36
150 mm thick	6.34	0.03	0.51	0.78	6.34	m^2	7.63
Surface treatments							
Sand blinding; to hardcore (not included)							
50 mm thick	2.02	0.04	0.82	-	2.23	m^2	3.05
Filter fabric; to hardcore (not included)	0.50	0.01	0.20	-	0.52	m^2	0.73
Granular Pavings							
Footpath gravels; porous self binding gravel							
Breedon Special Aggregates; "Golden Gravel" or equivalent; rolling wet; on hardcore base (not included); for pavements; to falls and crossfalls and to slopes not exceeding 15° from horizontal; over 300 mm wide							
50 mm thick	11.82	0.05	1.02	1.15	11.82	m^2	13.99
by hand; 50 mm thick	11.82	0.20	4.10	1.00	11.82	m^2	16.92
Breedon Special Aggregates; "Wayfarer" specially formulated fine gravel for use on golf course pathways							
50 mm thick	10.68	0.03	0.59	0.72	10.68	m^2	11.98
by hand; 50 mm thick	10.68	0.10	2.05	0.57	10.68	m^2	13.30
Hoggin (stabilized); PC £54.58/m^3 on hardcore base (not included); to falls and crossfalls and to slopes not exceeding 15° from horizontal; over 300 mm wide							
100 mm thick	5.46	0.02	0.41	0.57	8.19	m^2	9.16
by hand; 100 mm thick	5.46	0.10	2.05	0.40	8.19	m^2	10.64

Q PAVING/PLANTING/FENCING/SITE FURNITURE

Item Excluding site overheads and profit	PC £	Labour hours	Labour £	Plant £	Material £	Unit	Total rate £
Q23 GRAVEL/HOGGIN/WOODCHIP ROADS/PAVINGS – cont'd							
Footpath gravels; porous self binding gravel – cont'd							
150 mm thick	8.19	0.03	0.51	0.71	12.28	m^2	**13.50**
Ballast; as dug; watering; rolling; on hardcore base (not included)							
100 mm thick	3.33	0.13	2.73	1.89	4.16	m^2	**8.79**
150 mm thick	5.00	0.03	0.51	0.61	6.24	m^2	**7.36**
CED Ltd; Cedec gravel; self-binding; laid to inert (non-limestone) base measured separately; compacting							
red, gold or silver; 50 mm thick	13.54	0.03	0.59	0.44	13.54	m^2	**14.56**
by hand; red, gold or silver; 50 mm thick	13.54	0.07	1.37	0.20	13.54	m^2	**15.11**
Grundon Ltd; Coxwell self-binding path gravels laid and compacted to excavation or base measured separately							
50 mm thick	3.27	0.08	1.71	0.40	3.27	m^2	**5.38**
Footpath gravels; porous loose gravels							
Washed shingle; on prepared base (not included)							
25 - 50 size; 25 mm thick	1.01	0.02	0.36	0.07	1.07	m^2	**1.50**
25 - 50 size; 75 mm thick	3.04	0.05	1.08	0.22	3.20	m^2	**4.49**
50 - 75 size; 25 mm thick	1.01	0.02	0.41	0.08	1.07	m^2	**1.56**
50 - 75 size; 75 mm thick	3.04	0.07	1.37	0.33	3.20	m^2	**4.89**
Pea shingle; on prepared base (not included)							
10 - 15 size; 25 mm thick	1.01	0.02	0.36	0.07	1.07	m^2	**1.50**
5 - 10 size; 75 mm thick	3.04	0.05	1.08	0.22	3.20	m^2	**4.49**
Breedon Special Aggregates; Breedon Buff decorative limestone chippings							
50 mm thick	5.21	0.01	0.20	0.08	5.47	m^2	**5.76**
by hand; 50 mm thick	5.21	0.03	0.68	-	5.47	m^2	**6.15**
Breedon Special Aggregates; Brindle or Moorland Black chippings							
50 mm thick	6.74	0.01	0.20	0.08	7.08	m^2	**7.37**
by hand; 50 mm thick	6.74	0.03	0.68	-	7.08	m^2	**7.76**
Breedon Special Aggregates; slate chippings; plum/blue/green							
50 mm thick	9.78	0.01	0.20	0.08	10.27	m^2	**10.55**
by hand; 50 mm thick	9.78	0.03	0.68	-	10.27	m^2	**10.95**
Wood chip surfaces; Melcourt Industries Ltd							
Wood chips; to surface of pathways by hand; levelling and spreading by hand (excavation and preparation not included) (items labelled FSC are Forest Stewardship Council certified)							
"Walk Chips"; FSC; 100 mm thick (25 m^3 loads)	3.42	0.02	0.34	-	3.59	m^2	**3.93**
"Woodfibre"; FSC; 100 mm thick	2.00	0.05	1.03	-	2.09	m^2	**3.12**

Q PAVING/PLANTING/FENCING/SITE FURNITURE

Item Excluding site overheads and profit	PC £	Labour hours	Labour £	Plant £	Material £	Unit	Total rate £
Q24 INTERLOCKING BRICK/BLOCK ROADS/PAVINGS							
Precast concrete block edgings; PC £8.03/m^2; 200 x 100 x 60 mm; on prepared base (not included); haunching one side Edgings; butt joints							
stretcher course	0.80	0.17	3.42	-	3.73	m	7.14
header course	1.61	0.27	5.47	-	4.74	m	10.20
Precast concrete vehicular paving blocks; Marshalls Plc; on prepared base (not included); on 50 mm compacted sharp sand bed; blocks laid in 7 mm loose sand and vibrated; joints filled with sharp sand and vibrated; level and to falls only "Trafica" paving blocks; 450 x 450 x 70 mm							
"Perfecta" finish; colour natural	28.64	0.50	10.25	0.32	31.59	m^2	42.16
"Perfecta" finish; colour buff	33.13	0.50	10.25	0.32	36.19	m^2	46.76
"Saxon" finish; colour natural	25.14	0.50	10.25	0.32	28.00	m^2	38.57
"Saxon" finish; colour buff	28.94	0.50	10.25	0.32	31.89	m^2	42.46
Precast concrete vehicular paving blocks; "Keyblock" Marshalls Plc; on prepared base (not included); on 50 mm compacted sharp sand bed; blocks laid in 7 mm loose sand and vibrated; joints filled with sharp sand and vibrated; level and to falls only Herringbone bond							
200 x 100 x 60 mm; natural grey	8.03	1.50	30.75	0.32	10.70	m^2	41.77
200 x 100 x 60 mm; colours	8.72	1.50	30.75	0.32	11.42	m^2	42.49
200 x 100 x 80 mm; natural grey	8.94	1.50	30.75	0.32	11.65	m^2	42.72
200 x 100 x 80 mm; colours	10.10	1.50	30.75	0.32	12.87	m^2	43.94
Basketweave bond							
200 x 100 x 60 mm; natural grey	8.03	1.20	24.60	0.32	10.70	m^2	35.62
200 x 100 x 60 mm; colours	8.72	1.20	24.60	0.32	11.42	m^2	36.34
200 x 100 x 80 mm; natural grey	8.94	1.20	24.60	0.32	11.65	m^2	36.57
200 x 100 x 80 mm; colours	10.10	1.20	24.60	0.32	12.87	m^2	37.79
Precast concrete vehicular paving blocks; Charcon Hard Landscaping; on prepared base (not included); on 50 mm compacted sharp sand bed; blocks laid in 7 mm loose sand and vibrated; joints filled with sharp sand and vibrated; level and to falls only "Europa" concrete blocks							
200 x 100 x 60 mm; natural grey	12.11	1.50	30.75	0.32	14.98	m^2	46.05
200 x 100 x 60 mm; colours	12.96	1.50	30.75	0.32	15.87	m^2	46.94
"Parliament" concrete blocks							
200 x 100 x 65 mm; natural grey	29.30	1.50	30.75	0.32	32.30	m^2	63.37
200 x 100 x 65 mm; colours	29.30	1.50	30.75	0.32	32.30	m^2	63.37

Q PAVING/PLANTING/FENCING/SITE FURNITURE

Item Excluding site overheads and profit	PC £	Labour hours	Labour £	Plant £	Material £	Unit	Total rate £
Q24 INTERLOCKING BRICK/BLOCK ROADS/PAVINGS – cont'd							
Recycled polyethylene grassblocks; Boddingtons Grass Reinforcement Solutions; Boddingtons Ltd; interlocking units laid to prepared base or rootzone (not included)							
"Bodpave"; load bearing >300 tonnes per m^2; 500 x 500 x 50 mm deep							
1 - 50 m^2	16.50	0.20	4.10	0.20	19.77	m^2	24.07
51 - 500 m^2	13.75	0.20	4.10	0.20	17.02	m^2	21.32
"GrassProtecta"; extruded expanded polyethylene flexible mesh laid to existing grass surface or newly seeded areas to provide heavy surface protection from traffic and pedestrians							
"Standard"; 1.2 kg/m^2; 20 x 2 m; up to 320 m^2	7.50	0.01	0.17	-	7.50	m^2	7.67
"Standard"; 1.2 kg/m^2; 20 x 2 m; 321 m^2 to 3000 m^2	6.50	0.01	0.17	-	6.50	m^2	6.67
"Heavy"; 2 kg/m^2; 20 x 2 m; up to 320 m^2	8.50	0.01	0.17	-	8.50	m^2	8.67
"Heavy"; 2 kg/m^2; 20 x 2 m; 321 m^2 to 3000 m^2	8.00	0.01	0.17	-	8.00	m^2	8.17
"TurfProtecta"; extruded polyethylene flexible mesh laid to existing grass surface or newly seeded areas to provide surface protection from traffic including vehicle or animal wear and tear							
"Standard"; 30 m x 2 m; up to 300 m^2	2.90	0.01	0.17	-	2.90	m^2	3.07
"Standard"; 30 m x 2 m; 301 m^2 to 600 m^2	2.41	0.01	0.17	-	2.41	m^2	2.58
"Premium"; 30 m x 2 m; up to 300 m^2	2.99	0.01	0.17	-	2.99	m^2	3.16
"Premium"; 30 m x 2 m; 301 m^2 to 600 m^2	2.57	0.01	0.17	-	2.57	m^2	2.74
Q25 SLAB/BRICK/SETT/COBBLE PAVINGS							
Bricks - General Preamble: Bricks shall be hard, well burnt, non-dusting, resistant to frost and sulphate attack and true to shape, size and sample.							
Movement of materials Mechanically offloading bricks; loading wheelbarrows; transporting maximum 25 m distance	-	0.20	4.10	-	-	m^2	4.10
Edge restraints; to brick paving; on prepared base (not included); 65 mm thick bricks; PC £300.00/1000; haunching one side							
Header course							
200 x 100 mm; butt joints	3.00	0.27	5.47	-	6.13	m	11.59
210 x 105 mm; mortar joints	2.67	0.50	10.25	-	5.97	m	16.22
Stretcher course							
200 x 100 mm; butt joints	1.50	0.17	3.42	-	4.42	m	7.84
210 x 105 mm; mortar joints	1.36	0.33	6.83	-	4.36	m	11.20

Q PAVING/PLANTING/FENCING/SITE FURNITURE

Item Excluding site overheads and profit	PC £	Labour hours	Labour £	Plant £	Material £	Unit	Total rate £
Variation in brick prices; add or subtract the following amounts for every £1.00/ 1000 difference in the PC price							
Edgings							
100 mm wide	-	-	-	-	0.05	10 m	-
200 mm wide	-	-	-	-	0.11	10 m	-
102.5 mm wide	-	-	-	-	0.05	10 m	-
215 mm wide	-	-	-	-	0.09	10 m	-
Clay brick pavings; on prepared base (not included); bedding on 50 mm sharp sand; kiln dried sand joints							
Pavings; 200 x 100 x 65 mm wirecut chamfered paviors							
brick; PC £500.00/1000	25.00	1.97	40.37	0.25	28.10	m²	68.72
Clay brick pavings; 200 x 100 x 50 mm; laid to running stretcher or stack bond only; on prepared base (not included); bedding on cement:sand (1:4) pointing mortar as work proceeds							
PC £600.00/1000							
laid on edge	47.62	4.76	97.62	-	56.06	m²	153.68
laid on edge but pavior 65 mm thick	40.00	3.81	78.09	-	48.25	m²	126.35
laid flat	25.97	2.20	45.10	-	30.95	m²	76.05
PC £500.00/1000							
laid on edge	39.68	4.76	97.62	-	47.93	m²	145.55
laid on edge but pavior 65 mm thick	33.33	3.81	78.09	-	41.42	m²	119.52
laid flat	21.64	2.20	45.10	-	26.51	m²	71.61
PC £400.00/1000							
laid on edge	31.74	4.76	97.62	-	39.80	m²	137.41
laid on edge but pavior 65 mm thick	26.66	3.81	78.09	-	34.59	m²	112.68
laid flat	17.32	2.20	45.10	-	22.07	m²	67.17
PC £300.00/1000							
laid on edge	23.81	4.76	97.62	-	31.66	m²	129.28
laid on edge but pavior 65 mm thick	20.00	3.81	78.09	-	27.76	m²	105.85
laid flat	12.99	2.20	45.10	-	17.64	m²	62.74
Clay brick pavings; 200 x 100 x 50 mm; butt jointed laid herringbone or basketweave pattern only; on prepared base (not included); bedding on 50 mm sharp sand							
PC £600.00/1000							
laid flat	30.00	1.44	29.59	0.46	33.24	m²	63.30
PC £500.00/1000							
laid flat	25.00	1.44	29.59	0.46	28.12	m²	58.17
PC £400.00/1000							
laid flat	20.00	1.44	29.59	0.46	22.99	m²	53.05
PC £300.00/1000							
laid flat	15.00	1.44	29.59	0.46	17.87	m²	47.92
Clay brick pavings; 215 x 102.5 x 65 mm; on prepared base (not included); bedding on cement:sand (1:4) pointing mortar as work proceeds							
Paving bricks; PC £600.00/1000; herringbone bond							
laid on edge	35.55	3.55	72.88	-	42.17	m²	115.04

Q PAVING/PLANTING/FENCING/SITE FURNITURE

Item Excluding site overheads and profit	PC £	Labour hours	Labour £	Plant £	Material £	Unit	Total rate £
Q25 SLAB/BRICK/SETT/COBBLE PAVINGS – cont'd							
Clay brick pavings; 215 x 102.5 x 65 mm; on prepared base (not included); bedding on cement:sand (1:4) pointing mortar as work proceeds – cont'd							
laid flat	23.70	2.37	48.59	-	29.43	m²	**78.02**
Paving bricks; PC £600.00/1000; basketweave bond							
laid on edge	35.55	2.37	48.59	-	42.17	m²	**90.76**
laid flat	23.70	1.58	32.40	-	29.43	m²	**61.83**
Paving bricks; PC £600.00/1000; running or stack bond							
laid on edge	35.55	1.90	38.88	-	42.17	m²	**81.05**
laid flat	23.70	1.26	25.92	-	29.43	m²	**55.35**
Paving bricks; PC £500.00/1000; herringbone bond							
laid on edge	29.63	3.55	72.88	-	39.38	m²	**112.26**
laid flat	19.75	2.37	48.59	-	25.48	m²	**74.07**
Paving bricks; PC £500.00/1000; basketweave bond							
laid on edge	29.63	2.37	48.59	-	33.66	m²	**82.24**
laid flat	19.75	1.58	32.40	-	25.48	m²	**57.88**
Paving bricks; PC £500.00/1000; running or stack bond							
laid on edge	29.63	1.90	38.88	-	33.66	m²	**72.53**
laid flat	19.75	1.26	25.92	-	25.48	m²	**51.40**
Paving bricks; PC £400.00/1000; herringbone bond							
laid on edge	23.70	3.55	72.88	-	27.58	m²	**100.46**
laid flat	15.80	2.37	48.59	-	21.42	m²	**70.01**
Paving bricks; PC £400.00/1000; basketweave bond							
laid on edge	23.70	2.37	48.59	-	27.58	m²	**76.17**
laid flat	15.80	1.58	32.40	-	21.42	m²	**53.82**
Paving bricks; PC £400.00/1000; running or stack bond							
laid on edge	23.70	1.90	38.88	-	27.58	m²	**66.46**
laid flat	15.80	1.26	25.92	-	21.42	m²	**47.34**
Paving bricks; PC £300.00/1000; herringbone bond							
laid on edge	17.77	3.55	72.88	-	21.06	m²	**93.94**
laid flat	11.85	2.37	48.59	-	17.58	m²	**66.17**
Paving bricks; PC £300.00/1000; basketweave bond							
laid on edge	17.77	2.37	48.59	-	21.06	m²	**69.65**
laid flat	11.85	1.58	32.40	-	17.58	m²	**49.98**
Paving bricks; PC £300.00/1000; running or stack bond							
laid on edge	17.77	1.90	38.87	-	21.06	m²	**59.93**
laid flat	11.85	1.26	25.92	-	17.58	m²	**43.50**
Cutting							
curved cutting	-	0.44	9.11	14.46	-	m	**23.57**
raking cutting	-	0.33	6.83	11.01	-	m	**17.84**

Q PAVING/PLANTING/FENCING/SITE FURNITURE

Item Excluding site overheads and profit	PC £	Labour hours	Labour £	Plant £	Material £	Unit	Total rate £
Add or subtract the following amounts for every £10.00/1000 difference in the prime cost of bricks							
Butt joints							
200 x 100	-	-	-	-	0.50	m²	-
215 x 102.5	-	-	-	-	0.45	m²	-
10 mm mortar joints							
200 x 100	-	-	-	-	0.43	m²	-
215 x 102.5	-	-	-	-	0.40	m²	-
Precast concrete pavings; Charcon Hard Landscaping; on prepared sub-base (not included); bedding on 25 mm thick cement:sand mortar (1:4); butt joints; straight both ways; jointing in cement: sand (1:3) brushed in; on 50 mm thick sharp sand base							
Pavings; natural grey							
400 x 400 x 65 mm chamfered	27.31	0.40	8.20	-	32.02	m²	**40.23**
450 x 450 x 50 mm chamfered	14.72	0.44	9.11	-	19.43	m²	**28.54**
450 x 450 x 70 mm chamfered	16.94	0.44	9.11	-	21.65	m²	**30.76**
600 x 300 x 50 mm	13.33	0.44	9.11	-	18.05	m²	**27.16**
450 x 600 x 50 mm	11.78	0.44	9.11	-	16.49	m²	**25.60**
600 x 600 x 50 mm	9.47	0.40	8.20	-	14.18	m²	**22.38**
750 x 600 x 50 mm	8.96	0.40	8.20	-	13.67	m²	**21.87**
900 x 600 x 50 mm	8.19	0.40	8.20	-	12.90	m²	**21.10**
Pavings; coloured							
450 x 450 x 70 mm chamfered	19.90	0.44	9.11	-	24.61	m²	**33.72**
450 x 600 x 50 mm	17.78	0.44	9.11	-	22.49	m²	**31.60**
400 x 400 x 65 mm chamfered	27.31	0.40	8.20	-	32.02	m²	**40.23**
600 x 600 x 50 mm	14.81	0.40	8.20	-	19.52	m²	**27.72**
750 x 600 x 50 mm	13.67	0.40	8.20	-	18.38	m²	**26.58**
900 x 600 x 50 mm	12.14	0.40	8.20	-	16.85	m²	**25.05**
Precast concrete pavings; Charcon Hard Landscaping; on prepared sub-base (not included); bedding on 25 mm thick cement:sand mortar (1:4); butt joints; straight both ways; jointing in cement: sand (1:3) brushed in; on 50 mm thick sharp sand base							
"Appalacian" rough textured exposed aggregate pebble paving							
600 x 600 x 65 mm	35.27	0.50	10.25	-	38.84	m²	**49.09**

Q PAVING/PLANTING/FENCING/SITE FURNITURE

Item Excluding site overheads and profit	PC £	Labour hours	Labour £	Plant £	Material £	Unit	Total rate £
Q25 SLAB/BRICK/SETT/COBBLE PAVINGS – cont'd							
Precast concrete pavings; Marshalls Plc; "Heritage" imitation riven yorkstone paving; on prepared sub-base measured separately; bedding on 25 mm thick cement:sand mortar (1:4); pointed straight both ways cement:sand (1:3)							
Square and rectangular paving							
450 x 300 x 38 mm	39.94	1.00	20.50	-	43.77	m^2	**64.27**
450 x 450 x 38 mm	24.06	0.75	15.38	-	27.90	m^2	**43.27**
600 x 300 x 38 mm	26.29	0.80	16.40	-	30.13	m^2	**46.53**
600 x 450 x 38 mm	26.68	0.75	15.38	-	30.52	m^2	**45.89**
600 x 600 x 38 mm	24.33	0.50	10.25	-	28.24	m^2	**38.49**
Extra labours for laying the a selection of the above sizes to random rectangular pattern	-	0.33	6.83	-	-	m^2	**6.83**
Radial paving for circles							
circle with centre stone and first ring (8 slabs); 450 x 230/560 x 38 mm; diameter 1.54 m (total area 1.86 m^2)	65.70	1.50	30.75	-	70.67	nr	**101.42**
circle with second ring (16 slabs); 450 x 300/460 x 38 mm; diameter 2.48 m (total area 4.83 m^2)	169.38	4.00	82.00	-	182.14	nr	**264.14**
circle with third ring (16 slabs); 450 x 470/625 x 38 mm; diameter 3.42 m (total area 9.18 m^2)	308.26	8.00	164.00	-	332.30	nr	**496.30**
Stepping stones							
380 diameter x 38 mm	4.95	0.20	4.10	-	10.32	nr	**14.43**
asymmetrical 560 x 420 x 38 mm	7.38	0.20	4.10	-	12.75	nr	**16.85**
Precast concrete pavings; Marshalls Plc; "Chancery" imitation reclaimed riven yorkstone paving; on prepared sub-base measured separately; bedding on 25 mm thick cement:sand mortar (1:4); pointed straight both ways cement:sand (1:3)							
Square and rectangular paving							
300 x 300 x 45 mm	27.77	1.00	20.50	-	31.61	m^2	**52.11**
450 x 300 x 45 mm	26.14	0.90	18.45	-	29.97	m^2	**48.42**
600 x 300 x 45 mm	25.39	0.80	16.40	-	29.23	m^2	**45.63**
600 x 450 x 45 mm	25.49	0.75	15.38	-	29.33	m^2	**44.71**
450 x 450 x 45 mm	22.98	0.75	15.38	-	26.82	m^2	**42.19**
600 x 600 x 45 mm	24.50	0.50	10.25	-	28.34	m^2	**38.59**
Extra labours for laying the a selection of the above sizes to random rectangular pattern	-	0.33	6.83	-	-	m^2	**6.83**
Radial paving for circles							
circle with centre stone and first ring (8 slabs); 450 x 230/560 x 38 mm; diameter 1.54 m (total area 1.86 m^2)	83.94	1.50	30.75	-	88.91	nr	**119.66**
circle with second ring (16 slabs); 450 x 300/460 x 38 mm; diameter 2.48 m (total area 4.83 m^2)	216.58	4.00	82.00	-	229.34	nr	**311.34**
circle with third ring (16 slabs); 450 x 470/625 x 38 mm; diameter 3.42 m (total area 9.18 m^2)	394.34	8.00	164.00	-	418.38	nr	**582.38**
Squaring off set for 2 ring circle							
16 slabs; 2.72 m^2	230.67	1.00	20.50	-	230.67	nr	**251.17**

Q PAVING/PLANTING/FENCING/SITE FURNITURE

Item Excluding site overheads and profit	PC £	Labour hours	Labour £	Plant £	Material £	Unit	Total rate £
Edge restraints; to block paving; on prepared base (not included); 200 x 100 x 80 mm; PC £8.94/m^2; haunching one side							
Header course							
200 x 100 mm; butt joints	1.79	0.27	5.47	-	4.92	m	**10.38**
Stretcher course							
200 x 100 mm; butt joints	0.89	0.17	3.42	-	3.82	m	**7.23**
Concrete paviors; Marshalls Plc; on prepared base (not included); bedding on 50 mm sand; kiln dried sand joints swept in							
"Keyblock" paviors							
200 x 100 x 60 mm; grey	8.03	0.40	8.20	0.32	10.29	m^2	**18.82**
200 x 100 x 60 mm; colours	8.72	0.40	8.20	0.32	10.98	m^2	**19.51**
200 x 100 x 80 mm; grey	8.94	0.44	9.11	0.32	11.20	m^2	**20.64**
200 x 100 x 80 mm; colours	10.10	0.44	9.11	0.32	12.36	m^2	**21.80**
Concrete cobble paviors; Charcon Hard Landscaping; Concrete Products; on prepared base (not included); bedding on 50 mm sand; kiln dried sand joints swept in							
Paviors							
"Woburn" blocks; 100 - 201 x 134 x 80 mm; random sizes	30.79	0.67	13.67	0.32	33.82	m^2	**47.81**
"Woburn" blocks; 100 - 201 x 134 x 80 mm; single size	30.79	0.50	10.25	0.32	33.82	m^2	**44.40**
"Woburn" blocks; 100 - 201 x 134 x 60 mm; random sizes	24.93	0.67	13.67	0.32	27.82	m^2	**41.81**
"Woburn" blocks; 100 - 201 x 134 x 60 mm; single size	24.93	0.50	10.25	0.32	27.79	m^2	**38.37**
Concrete setts; on 25 mm sand; compacted; vibrated; joints filled with sand; natural or coloured; well rammed hardcore base (not included)							
Marshalls Plc; Tegula cobble paving							
60 mm thick; random sizes	21.61	0.57	11.71	0.32	24.95	m^2	**36.99**
60 mm thick; single size	21.61	0.45	9.32	0.32	25.16	m^2	**34.80**
80 mm thick; random sizes	24.89	0.57	11.71	0.32	28.60	m^2	**40.64**
80 mm thick; single size	24.89	0.45	9.32	0.32	28.60	m^2	**38.24**
cobbles 80 x 80 x 60 mm thick; traditional	37.73	0.56	11.39	0.32	41.14	m^2	**52.85**
Cobbles							
Charcon Hard Landscaping; country setts							
100 mm thick; random sizes	38.03	1.00	20.50	0.32	40.50	m^2	**61.32**
100 mm thick; single size	38.03	0.67	13.67	0.32	41.45	m^2	**55.44**

Q PAVING/PLANTING/FENCING/SITE FURNITURE

Item Excluding site overheads and profit	PC £	Labour hours	Labour £	Plant £	Material £	Unit	Total rate £
Q25 SLAB/BRICK/SETT/COBBLE PAVINGS – cont'd							
Natural stone, slab or granite paving - General							
Preamble: Provide paving slabs of the specified thickness in random sizes but not less than 25 slabs per 10 m² of surface area, to be laid in parallel courses with joints alternately broken and laid to falls.							
Reconstituted yorkstone aggregate pavings; Marshalls Plc; "Saxon" on prepared sub-base measured separately; bedding on 25 mm thick cement:sand mortar (1:4); on 50 mm thick sharp sand base							
Square and rectangular paving in various colours; butt joints straight both ways							
300 x 300 x 35 mm	36.74	0.80	16.40	-	42.13	m²	**58.53**
600 x 300 x 35 mm	22.79	0.65	13.32	-	28.18	m²	**41.50**
450 x 450 x 50 mm	24.89	0.75	15.38	-	30.28	m²	**45.65**
600 x 600 x 35 mm	17.08	0.50	10.25	-	22.47	m²	**32.72**
600 x 600 x 50 mm	17.08	0.55	11.28	-	22.47	m²	**33.74**
Square and rectangular paving in natural; butt joints straight both ways							
300 x 300 x 35 mm	30.64	0.80	16.40	-	36.02	m²	**52.42**
450 x 450 x 50 mm	21.22	0.70	14.35	-	26.60	m²	**40.95**
600 x 300 x 35 mm	20.05	0.75	15.38	-	25.43	m²	**40.81**
600 x 600 x 35 mm	14.73	0.50	10.25	-	20.11	m²	**30.36**
600 x 600 x 50 mm	17.86	0.60	12.30	-	23.25	m²	**35.55**
Radial paving for circles; 20 mm joints							
circle with centre stone and first ring (8 slabs); 450 x 230/560 x 35 mm; diameter 1.54 m (total area 1.86 m²)	67.92	1.50	30.75	-	73.17	nr	**103.92**
circle with second ring (16 slabs); 450 x 300/460 x 35 mm; diameter 2.48 m (total area 4.83 m²)	165.52	4.00	82.00	-	179.06	nr	**261.06**
circle with third ring (24 slabs); 450 x 310/430 x 35 mm; diameter 3.42 m (total area 9.18 m²)	311.92	8.00	164.00	-	337.83	nr	**501.83**
Granite setts; bedding on 25 mm cement:sand (1:3)							
Natural granite setts; 100 x 100 mm to 125 x 150 mm; x 150 to 250 mm length; riven surface; silver grey							
new; standard grade	37.32	2.00	41.00	-	45.43	m²	**86.43**
new; high grade	26.89	2.00	41.00	-	35.00	m²	**76.00**
reclaimed; cleaned	39.11	2.00	41.00	-	47.99	m²	**88.99**
Natural stone, slate or granite flag pavings; CED Ltd; on prepared base (not included); bedding on 25 mm cement:sand (1:3); cement:sand (1:3) joints							
Yorkstone; riven laid random rectangular							
new slabs; 40 - 60 mm thick	71.79	1.71	35.14	-	82.52	m²	**117.66**
reclaimed slabs; Cathedral grade; 50 - 75 mm thick	87.54	2.80	57.40	-	99.85	m²	**157.25**

Q PAVING/PLANTING/FENCING/SITE FURNITURE

Item Excluding site overheads and profit	PC £	Labour hours	Labour £	Plant £	Material £	Unit	Total rate £
Donegal quartzite slabs; standard tiles							
200 x random lengths x 15-25 mm	79.70	4.20	86.13	-	83.25	m²	**169.39**
250 x random lengths x 15-25 mm	79.70	3.82	78.28	-	83.25	m²	**161.53**
300 x random lengths x 15-25 mm	79.70	3.47	71.18	-	83.25	m²	**154.43**
350 x random lengths x 15-25 mm	79.70	3.16	64.69	-	83.25	m²	**147.94**
400 x random lengths x 15-25 mm	79.70	2.87	58.82	-	83.25	m²	**142.08**
450 x random lengths x 15-25 mm	79.70	2.61	53.47	-	87.24	m²	**140.71**
Natural yorkstone, pavings or edgings; Johnsons Wellfield Quarries; sawn 6 sides; 50 mm thick; on prepared base measured separately; bedding on 25 mm cement:sand (1:3); cement:sand (1:3) joints							
Paving							
laid to random rectangular pattern	59.52	1.71	35.14	-	66.47	m²	**101.61**
laid to coursed laying pattern; 3 sizes	64.55	1.72	35.34	-	71.16	m²	**106.50**
Paving; single size							
600 x 600 mm	71.47	0.85	17.43	-	78.42	m²	**95.85**
600 x 400 mm	70.32	1.00	20.50	-	77.22	m²	**97.72**
300 x 200 mm	81.51	2.00	41.00	-	88.96	m²	**129.96**
215 x 102.5 mm	85.28	2.50	51.25	-	92.92	m²	**144.17**
Paving; cut to template off site; 600 x 600 mm; radius							
1.00 m	204.00	3.33	68.33	-	207.38	m²	**275.71**
2.50 m	196.67	2.00	41.00	-	200.05	m²	**241.05**
5.00 m	191.43	2.00	41.00	-	194.81	m²	**235.81**
Edgings							
100 mm wide x random lengths	7.85	0.50	10.25	-	8.58	m	**18.83**
100 mm x 100 mm	8.15	0.50	10.25	-	11.94	m	**22.19**
100 mm x 200 mm	16.30	0.50	10.25	-	20.50	m	**30.75**
250 mm wide x random lengths	15.39	0.40	8.20	-	19.54	m	**27.74**
500 mm wide x random lengths	30.79	0.33	6.83	-	35.71	m	**42.54**
Yorkstone edgings; 600 mm long x 250 mm wide; cut to radius							
1.00 m to 3.00	53.35	0.50	10.25	-	56.36	m	**66.61**
3.00 m to 5.00 m	51.78	0.44	9.11	-	54.71	m	**63.82**
exceeding 5.00 m	50.21	0.40	8.20	-	53.06	m	**61.26**
Natural yorkstone, pavings or edgings; Johnsons Wellfield Quarries; sawn 6 sides; 75 mm thick; on prepared base measured separately; bedding on 25 mm cement:sand (1:3); cement:sand (1:3) joints							
Paving							
laid to random rectangular pattern	65.96	0.95	19.48	-	72.64	m²	**92.11**
laid to coursed laying pattern; 3 sizes	74.12	0.95	19.48	-	81.20	m²	**100.68**
Paving; single size							
600 x 600 mm	82.60	0.95	19.48	-	90.11	m²	**109.58**
600 x 400 mm	82.60	0.95	19.48	-	90.11	m²	**109.58**
300 x 200 mm	94.85	0.75	15.38	-	102.97	m²	**118.35**
215 x 102.5 mm	99.56	2.50	51.25	-	107.92	m²	**159.17**
Paving; cut to template off site; 600 x 600 mm; radius							
1.00 m	238.68	4.00	82.00	-	242.06	m²	**324.06**
2.50 m	232.40	2.50	51.25	-	235.78	m²	**287.03**
5.00 m	226.14	2.50	51.25	-	229.52	m²	**280.77**

Q PAVING/PLANTING/FENCING/SITE FURNITURE

Item Excluding site overheads and profit	PC £	Labour hours	Labour £	Plant £	Material £	Unit	Total rate £
Q25 SLAB/BRICK/SETT/COBBLE PAVINGS – cont'd							
Natural yorkstone, pavings or edgings; Johnsons Wellfield Quarries; sawn 6 sides; 75 mm thick – cont'd							
Edgings							
100 mm wide x random lengths	10.05	0.60	12.30	-	10.89	m	23.19
100 mm x 100 mm	99.56	0.60	12.30	-	107.92	m	120.22
100 mm x 200 mm	19.93	0.50	10.25	-	24.31	m	34.56
250 mm wide x random lengths	18.53	0.50	10.25	-	22.84	m	33.09
500 mm wide x random lengths	37.06	0.40	8.20	-	42.29	m	50.49
Edgings; 600 mm long x 250 mm wide; cut to radius							
1.00 m to 3.00 m	62.45	0.60	12.30	-	65.91	m	78.21
3.00 m to 5.00 m	60.71	0.50	10.25	-	64.09	m	74.34
exceeding 5.00 m	59.91	0.44	9.11	-	63.24	m	72.35
CED Ltd; Indian sandstone, riven pavings or edgings; 25-35 mm thick; on prepared base measured separately; bedding on 25 mm cement:sand (1:3); cement:sand (1:3) joints							
Paving							
laid to random rectangular pattern	21.45	2.40	49.21	-	25.90	m^2	75.11
laid to coursed laying pattern; 3 sizes	21.45	2.00	41.00	-	25.90	m^2	66.90
Paving; single size							
600 x 600 mm	21.45	1.00	20.50	-	25.90	m^2	46.40
600 x 400 mm	21.45	1.25	25.63	-	25.90	m^2	51.53
400 x 400 mm	21.45	1.67	34.17	-	25.90	m^2	60.07
Natural stone, slate or granite flag pavings; CED Ltd; on prepared base (not included); bedding on 25 mm cement:sand (1:3); cement:sand (1:3) joints							
Granite paving; sawn 6 sides; textured top							
new slabs; silver grey, blue grey or yellow; 50 mm thick	49.09	1.71	35.14	-	55.10	m^2	90.24
new slabs; black; 50 mm thick	69.11	1.71	35.14	-	76.12	m^2	111.26
Cobble pavings - General							
Cobbles should be embedded by hand, tight-butted, endwise to a depth of 60% of their length. A dry grout of rapid-hardening cement:sand (1:2) shall be brushed over the cobbles until the interstices are filled to the level of the adjoining paving. Surplus grout shall then be brushed off and a light, fine spray of water applied over the area.							
Cobble pavings							
Cobbles; to present a uniform colour in panels or varied in colour as required							
Scottish beach cobbles; 200 - 100 mm	57.42	2.00	41.00	-	64.94	m^2	105.94
Scottish beach cobbles; 100 - 75 mm	40.62	2.50	51.25	-	48.14	m^2	99.39
Scottish beach cobbles; 75 - 50 mm	23.65	3.33	68.33	-	31.17	m^2	99.50

Q PAVING/PLANTING/FENCING/SITE FURNITURE

Item Excluding site overheads and profit	PC £	Labour hours	Labour £	Plant £	Material £	Unit	Total rate £
Q26 SPECIAL SURFACINGS/PAVINGS FOR SPORT/GENERAL AMENITY							
Market prices of surfacing materials							
Surfacings; Melcourt Industries Ltd (items labelled FSC are Forest Stewardship Council certified)							
"Playbark 10/50®"; per 25 m³ load	-	-	-	-	57.10	m³	-
"Playbark 8/25®"; per 25 m³ load	-	-	-	-	55.60	m³	-
"Playchips®"; per 25 m³ load; FSC	-	-	-	-	39.75	m³	-
"Kushyfall"; per 25 m³ load; FSC	-	-	-	-	35.20	m³	-
"Softfall"; per 25 m³ load	-	-	-	-	28.90	m³	-
"Playsand"; per 10 t load	-	-	-	-	94.86	m³	-
"Woodfibre"; per 25 m³ load; FSC	-	-	-	-	32.95	m³	-
Playgrounds; Wicksteed Leisure Ltd							
Safety tiles; on prepared base (not included)							
1000 x 1000 x 60 mm; red or green	57.00	0.20	4.10	-	57.00	m²	**61.10**
1000 x 1000 x 60 mm; black	54.00	0.20	4.10	-	54.00	m²	**58.10**
1000 x 1000 x 43 mm; red or green	54.00	0.20	4.10	-	54.00	m²	**58.10**
1000 x 1000 x 43 mm; black	49.00	0.20	4.10	-	49.00	m²	**53.10**
Playgrounds; SMP Playgrounds Ltd							
Tiles; on prepared base (not included)							
"Premier 25"; 1000 x 1000 x 25 mm; black; for general use	46.00	0.20	4.10	-	54.25	m²	**58.35**
"Premier 70"; 1000 x 1000 x 70 mm; black; for higher equipment	80.00	0.25	5.13	-	88.25	m²	**93.38**
Playgrounds; Melcourt Industries Ltd							
"Playbark®"; on drainage layer (not included); minimum 300 mm settled depth							
"Playbark® 8/25", 8 - 25 mm particles; red/brown	16.68	0.35	7.19	-	16.68	m²	**23.87**
"Playbark® 10/50"; 10 - 50 mm particles; red/brown	19.03	0.55	11.29	-	19.03	m²	**30.32**
Playgrounds; timber edgings							
Timber edging boards; 50 x 50 x 750 mm timber pegs at 1000 mm centres; excavations and hardcore under edgings (not included)							
50 x 150 mm; hardwood (iroko) edge boards	9.97	0.10	2.05	-	10.69	m	**12.74**
38 x 150 mm; hardwood (iroko) edge boards	8.55	0.10	2.05	-	9.26	m	**11.31**
50 x 150 mm; treated softwood edge boards	2.75	0.10	2.05	-	2.75	m	**4.80**
38 x 150 mm; treated softwood edge boards	1.82	0.10	2.05	-	2.54	m	**4.59**

Q PAVING/PLANTING/FENCING/SITE FURNITURE

Item Excluding site overheads and profit	PC £	Labour hours	Labour £	Plant £	Material £	Unit	Total rate £
Q30 SEEDING/TURFING							
Seeding/turfing - General							
Market prices of seeding materials Please see market prices in the Major Works section of this book.							
Market prices of chemicals and application rates Please see market prices in the Major Works section of this book.							
Cultivation							
Breaking up existing ground; using pedestrian operated tine cultivator or rotavator							
100 mm deep	-	0.50	10.25	16.58	-	100 m²	**26.83**
150 mm deep	-	0.57	11.71	18.95	-	100 m²	**30.66**
200 mm deep	-	0.67	13.67	22.11	-	100 m²	**35.78**
As above but in heavy clay or wet soils							
100 mm deep	-	0.67	13.67	22.11	-	100 m²	**35.78**
150 mm deep	-	0.80	16.40	26.53	-	100 m²	**42.93**
200 mm deep	-	1.00	20.50	33.17	-	100 m²	**53.67**
Rolling cultivated ground lightly; using self-propelled agricultural roller	-	0.06	1.14	0.84	-	100 m²	**1.98**
Importing and storing selected and approved topsoil; inclusive of settlement							
small quantities; less than 15 m³	26.00	-	-	-	31.20	m³	**-**
over 15 m³	26.00	-	-	-	31.20	m³	**-**
Spreading and lightly consolidating approved topsoil (imported or from spoil heaps); in layers not exceeding 150 mm; travel distance from spoil heaps not exceeding 25 m							
By machine (imported topsoil not included)							
minimum depth 100 mm	-	-	-	0.93	-	m²	**0.93**
minimum depth 200 mm	-	-	-	1.86	-	m²	**1.86**
minimum depth 300 mm	-	-	-	2.79	-	m²	**2.79**
minimum depth 500 mm	-	-	-	4.65	-	m²	**4.65**
over 500 mm	-	-	-	9.30	-	m³	**9.30**
By hand (imported topsoil not included)							
minimum depth 100 mm	-	0.20	4.10	-	-	m²	**4.10**
minimum depth 150 mm	-	0.30	6.16	-	-	m²	**6.16**
minimum depth 300 mm	-	0.60	12.28	-	-	m²	**12.28**
minimum depth 450 mm	-	0.90	18.47	-	-	m²	**18.47**
over 450 mm deep	-	2.22	45.51	-	-	m³	**45.51**
Extra over for spreading topsoil to slopes 15 - 30°; by machine or hand	-	-	-	-	-	10%	**-**
Extra over for spreading topsoil to slopes over 30°; by machine or hand	-	-	-	-	-	25%	**-**
Extra over for spreading topsoil from spoil heaps; travel exceeding 100 m; by machine							
100 - 150 m	-	0.01	0.25	0.07	-	m³	**0.32**
150 - 200 m	-	0.02	0.38	0.10	-	m³	**0.48**
200 - 300 m	-	0.03	0.57	0.16	-	m³	**0.72**

Q PAVING/PLANTING/FENCING/SITE FURNITURE

Item Excluding site overheads and profit	PC £	Labour hours	Labour £	Plant £	Material £	Unit	Total rate £
Extra over spreading topsoil for travel exceeding 100 m; by hand							
100 m	-	0.83	17.08	-	-	m³	17.08
200 m	-	1.67	34.17	-	-	m³	34.17
300 m	-	2.50	51.25	-	-	m³	51.25
Evenly grading; to general surfaces to bring to finished levels							
by pedestrian operated rotavator	-	-	0.08	0.13	-	m²	0.21
by hand	-	0.01	0.20	-	-	m²	0.20
Extra over grading for slopes 15 - 30°; by machine or hand	-	-	-	-	-	10%	-
Extra over grading for slopes over 30°; by machine or hand	-	-	-	-	-	25%	-
Apply screened topdressing to grass surfaces; spread using Tru-Lute							
sand soil mixes 90/10 to 50/50	0.12	-	0.04	0.03	0.12	m²	0.20
Spread only existing cultivated soil to final levels using Tru-Lute							
cultivated soil	-	-	0.04	0.03	-	m²	0.08
Clearing stones; disposing off site; to distance not exceeding 13 km							
by hand; stones not exceeding 50 mm in any direction; loading to skip 4.6 m³	-	0.01	0.20	0.04	-	m²	0.24
Lightly cultivating; weeding; to fallow areas; disposing debris off site							
by hand	-	0.01	0.29	0.30	-	m²	0.59
Surface applications and soil additives; pre-seeding; material delivered to a maximum of 25 m from area of application; applied by hand							
Soil conditioners; to cultivated ground; medium bark; based on deliveries of 25 m³ loads; PC £35.65/m³; including turning in							
1 m³ per 40 m² = 25 mm thick	0.89	0.02	0.46	-	0.89	m²	1.35
1 m³ per 20 m² = 50 mm thick	1.78	0.04	0.91	-	1.78	m²	2.69
1 m³ per 13.33 m² = 75 mm thick	2.67	0.07	1.37	-	2.67	m²	4.04
1 m³ per 10 m² = 100 mm thick	3.56	0.08	1.64	-	3.56	m²	5.21
Soil conditioners; to cultivated ground; mushroom compost; delivered in 25 m³ loads; PC £21.20/m³; including turning in							
1 m³ per 40 m² = 25 mm thick	0.53	0.02	0.46	-	0.53	m²	0.99
1 m³ per 20 m² = 50 mm thick	1.06	0.04	0.91	-	1.06	m²	1.97
1 m³ per 13.33 m² = 75 mm thick	1.59	0.07	1.37	-	1.59	m²	2.96
1 m³ per 10 m² = 100 mm thick	2.12	0.08	1.64	-	2.12	m²	3.76
Soil conditioners; to cultivated ground; mushroom compost; delivered in 60 m³ loads; PC £9.50/m³; including turning in							
1 m³ per 40 m² = 25 mm thick	0.24	0.02	0.46	-	0.24	m²	0.69
1 m³ per 20 m² = 50 mm thick	0.42	0.04	0.91	-	0.42	m²	1.33
1 m³ per 13.33 m² = 75 mm thick	0.71	0.07	1.37	-	0.71	m²	2.08
1 m³ per 10 m² = 100 mm thick	0.95	0.08	1.64	-	0.95	m²	2.59

Q PAVING/PLANTING/FENCING/SITE FURNITURE

Item Excluding site overheads and profit	PC £	Labour hours	Labour £	Plant £	Material £	Unit	Total rate £
Q30 SEEDING/TURFING – cont'd							
Preparation of seedbeds - General Preamble: for preliminary operations see "Cultivation" section.							
Preparation of seedbeds; soil preparation Lifting selected and approved topsoil from spoil heaps; passing through 6 mm screen; removing debris	-	0.08	1.71	4.57	0.02	m^3	6.30
Topsoil; supply only; PC £26.00/m^3; allowing for 20% settlement; 20 tonne loads							
25 mm	0.65	-	-	-	0.78	m^2	-
50 mm	1.30	-	-	-	1.56	m^2	-
100 mm	2.60	-	-	-	3.12	m^2	-
150 mm	3.90	-	-	-	4.68	m^2	-
200 mm	5.20	-	-	-	6.24	m^2	-
250 mm	6.50	-	-	-	7.80	m^2	-
300 mm	7.80	-	-	-	9.36	m^2	-
400 mm	10.40	-	-	-	12.48	m^2	-
450 mm	11.70	-	-	-	14.04	m^2	-
Topsoil; supply only; PC £43.50/m^3; allowing for 20% settlement; 10 tonne loads							
25 mm	1.09	-	-	-	1.30	m^2	-
50 mm	2.17	-	-	-	2.61	m^2	-
100 mm	4.35	-	-	-	5.22	m^2	-
150 mm	6.53	-	-	-	7.83	m^2	-
200 mm	8.70	-	-	-	10.44	m^2	-
250 mm	10.88	-	-	-	13.05	m^2	-
300 mm	13.05	-	-	-	15.66	m^2	-
400 mm	17.40	-	-	-	20.88	m^2	-
450 mm	19.57	-	-	-	23.49	m^2	-
Spreading topsoil to form seedbeds average thickness not exceeding 200 mm (topsoil not included); by machine; grading and cultivation not included							
Excavated material; from spoil heaps							
average 25 m distance	-	0.10	2.05	5.98	-	m^3	8.03
average 50 m distance	-	0.10	2.05	6.60	-	m^3	8.65
average 100 m distance	-	0.10	2.05	7.20	-	m^3	9.25
average 200 m distance	-	0.10	2.05	9.00	-	m^3	11.05
Excavated material; spreading on site							
25 mm deep	-	-	-	0.22	-	m^2	0.22
50 mm deep	-	-	-	0.44	-	m^2	0.44
75 mm deep	-	-	-	0.65	-	m^2	0.65
100 mm deep	-	-	-	0.87	-	m^2	0.87
150 mm deep	-	-	-	1.04	-	m^2	1.04
Spreading only topsoil to form seedbeds (topsoil not included); by hand							
25 mm deep	-	0.03	0.51	-	-	m^2	0.51
50 mm deep	-	0.03	0.68	-	-	m^2	0.68
75 mm deep	-	0.04	0.88	-	-	m^2	0.88
100 mm deep	-	0.05	1.02	-	-	m^2	1.02
150 mm deep	-	0.08	1.54	-	-	m^2	1.54
Bringing existing topsoil to a fine tilth for seeding; by raking or harrowing; stones not to exceed 6 mm; by machine	-	-	0.08	0.04	-	m^2	0.12

Q PAVING/PLANTING/FENCING/SITE FURNITURE

Item Excluding site overheads and profit	PC £	Labour hours	Labour £	Plant £	Material £	Unit	Total rate £
Bringing existing topsoil to a fine tilth for seeding; by raking or harrowing; stones not to exceed 6 mm; by hand	-	0.01	0.18	-	-	m^2	0.18
Preparation of seedbeds; soil treatments							
For the following topsoil improvement and seeding operations add or subtract the following amounts for every £0.10 difference in the material cost price							
35 g/m^2	-	-	-	-	0.35	100 m^2	-
50 g/m^2	-	-	-	-	0.50	100 m^2	-
70 g/m^2	-	-	-	-	0.70	100 m^2	-
100 g/m^2	-	-	-	-	1.00	100 m^2	-
Pre-seeding fertilizers (12+00+09); PC £0.85/kg; to seedbeds; by hand							
35 g/m^2	2.68	0.17	3.42	-	2.68	100 m^2	6.09
50 g/m^2	3.82	0.17	3.42	-	3.82	100 m^2	7.24
70 g/m^2	5.35	0.17	3.42	-	5.35	100 m^2	8.77
100 g/m^2	7.65	0.20	4.10	-	7.65	100 m^2	11.75
125 g/m^2	9.56	0.20	4.10	-	9.56	100 m^2	13.66
Seeding							
Grass seed; spreading in two operations; PC £4.50/kg (for changes in material prices please refer to table above); by hand							
35 g/m^2	15.75	0.17	3.42	-	15.75	100 m^2	19.17
50 g/m^2	22.50	0.17	3.42	-	22.50	100 m^2	25.92
70 g/m^2	31.50	0.17	3.42	-	31.50	100 m^2	34.92
100 g/m^2	45.00	0.17	3.42	-	45.00	100 m^2	48.42
125 g/m^2	56.25	0.20	4.10	-	56.25	100 m^2	60.35
Grass seed; spreading in two operations; PC £6.00/kg (for changes in material prices please refer to table above); by hand							
35 g/m^2	21.00	0.17	3.42	-	21.00	100 m^2	24.42
50 g/m^2	30.00	0.17	3.42	-	30.00	100 m^2	33.42
70 g/m^2	42.00	0.17	3.42	-	42.00	100 m^2	45.42
100 g/m^2	60.00	0.17	3.42	-	60.00	100 m^2	63.42
125 g/m^2	75.00	0.20	4.10	-	75.00	100 m^2	79.10
Extra over seeding by hand for slopes over 30° (allowing for the actual area but measured in plan)							
35 g/m^2	2.34	-	0.08	-	2.34	100 m^2	2.42
50 g/m^2	3.38	-	0.08	-	3.38	100 m^2	3.45
70 g/m^2	4.72	-	0.08	-	4.72	100 m^2	4.80
100 g/m^2	6.75	-	0.09	-	6.75	100 m^2	6.84
125 g/m^2	8.41	-	0.09	-	8.41	100 m^2	8.50
Harrowing seeded areas; light chain harrow	-	-	-	0.12	-	100 m^2	0.12
Raking over seeded areas							
by hand	-	0.80	16.40	-	-	100 m^2	16.40
Rolling seeded areas; light roller							
by pedestrian operated mechanical roller	-	0.08	1.71	1.68	-	100 m^2	3.38
by hand drawn roller	-	0.17	3.42	-	-	100 m^2	3.42
Extra over harrowing, raking or rolling seeded areas for slopes over 30°; by machine or hand	-	-	-	-	-	25%	-
Turf edging; to seeded areas; 300 mm wide	2.19	0.13	2.73	-	2.19	m^2	4.92

Q PAVING/PLANTING/FENCING/SITE FURNITURE

Item Excluding site overheads and profit	PC £	Labour hours	Labour £	Plant £	Material £	Unit	Total rate £
Q30 SEEDING/TURFING – cont'd							
Preparation of turf beds							
Rolling turf to be lifted; lifting by hand or mechanical turf stripper; stacks to be not more than 1 m high							
cutting only preparing to lift; pedestrian turf cutter	-	0.75	15.38	9.01	-	100 m²	24.38
lifting and stacking; by hand	-	8.33	170.83	-	-	100 m²	170.83
Rolling up; moving to stacks							
distance not exceeding 100 m	-	2.50	51.25	-	-	100 m²	51.25
extra over rolling and moving turf to stacks to transport per additional 100 m	-	0.83	17.08	-	-	100 m²	17.08
Lifting selected and approved topsoil from spoil heaps							
passing through 6 mm screen; removing debris	-	0.17	3.42	9.14	-	m³	12.56
Extra over lifting topsoil and passing through screen for imported topsoil; plus 20% allowance for settlement	26.00	-	-	-	31.20	m³	-
Spreading topsoil to form turfbeds (topsoil not included); by machine							
25 mm deep	-	-	-	0.23	-	m²	0.23
50 mm deep	-	-	-	0.46	-	m²	0.46
75 mm deep	-	-	-	0.70	-	m²	0.70
100 mm deep	-	-	-	0.93	-	m²	0.93
150 mm deep	-	-	-	1.39	-	m²	1.39
Bringing existing topsoil to a fine tilth for turfing by raking or harrowing; stones not to exceed 6 mm; by hand	-	0.01	0.27	-	-	m²	0.27
Spreading topsoil to form turfbeds (topsoil not included); by hand							
25 mm deep	-	0.04	0.82	-	-	m²	0.82
50 mm deep	-	0.08	1.64	-	-	m²	1.64
75 mm deep	-	0.12	2.46	-	-	m²	2.46
100 mm deep	-	0.15	3.08	-	-	m²	3.08
150 mm deep	-	0.23	4.62	-	-	m²	4.62
Operations after spreading of topsoil; by hand							
Bringing existing topsoil to a fine tilth for turfing by raking or harrowing; stones not to exceed 6 mm							
topsoil spread by machine (soil compacted)	-	0.03	0.51	-	-	m²	0.51
topsoil spread by hand	-	0.02	0.34	-	-	m²	0.34
Turfing							
Turfing; laying only; to stretcher bond; butt joints; including providing and working from barrow plank runs where necessary to surfaces not exceeding 30° from horizontal							
specially selected lawn turves from previously lifted stockpile	-	0.20	4.10	-	-	m²	4.10
cultivated lawn turves; to larger open areas	-	0.11	2.19	-	-	m²	2.19
cultivated lawn turves; to domestic or garden areas	-	0.13	2.73	-	-	m²	2.73

Q PAVING/PLANTING/FENCING/SITE FURNITURE

Item Excluding site overheads and profit	PC £	Labour hours	Labour £	Plant £	Material £	Unit	Total rate £
Industrially grown turf; PC prices listed represent the general range of industrial turf prices for sportsfields and amenity purposes; prices will vary with quantity and site location							
"Rolawn"							
ref RB Medallion; sports fields, domestic lawns, general landscape	2.19	0.13	2.73	-	2.19	m²	4.92
"Inturf"							
ref Inturf 1; fine lawns, golf greens, bowling greens	2.72	0.13	2.73	-	2.72	m²	5.45
ref Inturf 2; football grounds, parks, hardwearing areas	2.15	0.13	2.73	-	2.15	m²	4.88
ref Inturf 2 Bargold; fine turf; football grounds, parks, hardwearing areas	1.86	0.13	2.73	-	1.86	m²	4.59
ref Inturf 3; hockey grounds, polo, medium wearing areas	2.15	0.13	2.73	-	2.15	m²	4.88
Firming turves with wooden beater	-	0.01	0.20	-	-	m²	0.20
Rolling turfed areas; light roller							
by pedestrian operated mechanical roller	-	0.08	1.71	1.68	-	100 m²	3.38
by hand drawn roller	-	0.17	3.42	-	-	100 m²	3.42
Dressing with finely sifted topsoil; brushing into joints	0.03	0.05	1.02	-	0.03	m²	1.06
Turfing; laying only							
to slopes over 30°; to diagonal bond (measured as plan area - add 15% to these rates for the incline area of 30° slopes)	-	0.12	2.46	-	-	m²	2.46
Extra over laying turfing for pegging down turves							
wooden or galvanized wire pegs; 200 mm long; 2 pegs per 0.50 m²	1.58	0.01	0.27	-	1.58	m²	1.85
Artificial grass; Artificial Lawn Company; laid to sharp sand bed priced separately 15 kg kiln sand brushed in per m²							
"Leisure Lawn"; 24 mm thick artificial sports turf; sand filled	-	-	-	-	-	m²	27.80
"Budget Grass"; for general use; budget surface; sand filled	-	-	-	-	-	m²	21.95
"Multi Grass"; patios conservatories and pool surrounds; sand filled	-	-	-	-	-	m²	23.90
"Premier"; lawns and patios	-	-	-	-	-	m²	30.00
"Play Lawn"; grass/sand and rubber filled	-	-	-	-	-	m²	33.85
"Grassflex"; safety surfacing for play areas	-	-	-	-	-	m²	49.50
Maintenance operations (Note: the following rates apply to aftercare maintenance executed as part of a landscaping contract only)							
Initial cutting; to turfed areas							
20 mm high; using pedestrian guided power driven cylinder mower; including boxing off cuttings (stone picking and rolling not included)	-	0.18	3.69	0.29	-	100 m²	3.98
Repairing damaged grass areas							
scraping out; removing slurry; from ruts and holes; average 100 mm deep	-	0.13	2.73	-	-	m²	2.73
100 mm topsoil	-	0.13	2.73	-	3.12	m²	5.85

Q PAVING/PLANTING/FENCING/SITE FURNITURE

Item Excluding site overheads and profit	PC £	Labour hours	Labour £	Plant £	Material £	Unit	Total rate £
Q30 SEEDING/TURFING – cont'd							
Maintenance operations – cont'd							
Repairing damaged grass areas; sowing grass seed to match existing or as specified; to individually prepared worn patches							
35 g/m^2	0.16	0.01	0.20	-	0.18	m^2	**0.39**
50 g/m^2	0.23	0.01	0.20	-	0.26	m^2	**0.46**
Using pedestrian operated mechanical equipment and blowers							
grassed areas with perimeters of mature trees such as sports fields and amenity areas	-	0.04	0.82	0.05	-	100 m^2	**0.87**
grassed areas containing ornamental trees and shrub beds	-	0.10	2.05	0.12	-	100 m^2	**2.17**
verges	-	0.07	1.37	0.08	-	100 m^2	**1.45**
By hand							
grassed areas with perimeters of mature trees such as sports fields and amenity areas	-	0.05	1.02	0.06	-	100 m^2	**1.08**
grassed areas containing ornamental trees and shrub beds	-	0.08	1.71	0.10	-	100 m^2	**1.81**
verges	-	1.00	20.50	1.20	-	100 m^2	**21.70**
Removal of arisings							
areas with perimeters of mature trees	-	0.01	0.14	0.10	1.28	100 m^2	**1.51**
areas containing ornamental trees and shrub beds	-	0.02	0.41	0.37	3.20	100 m^2	**3.98**
Cutting grass to specified height; per cut							
ride-on triple cylinder mower	-	0.01	0.29	0.14	-	100 m^2	**0.42**
ride-on triple rotary mower	-	0.01	0.29	-	-	100 m^2	**0.29**
pedestrian mower (open areas)	-	0.18	3.69	0.37	-	100 m^2	**4.06**
pedestrian mower (small areas or areas with obstacles)	-	0.33	6.83	0.44	-	100 m^2	**7.28**
Cutting rough grass; per cut							
power flail or scythe cutter	-	0.04	0.72	-	-	100 m^2	**0.72**
Extra over cutting grass for slopes not exceeding 30°	-	-	-	-	-	10%	**-**
Extra over cutting grass for slopes exceeding 30°	-	-	-	-	-	40%	**-**
Cutting fine sward							
pedestrian operated seven-blade cylinder lawn mower	-	0.14	2.87	0.22	-	100 m^2	**3.09**
Extra over cutting fine sward for boxing off cuttings							
pedestrian mower	-	0.03	0.57	0.04	-	100 m^2	**0.62**
Cutting areas of rough grass							
scythe	-	1.00	20.50	-	-	100 m^2	**20.50**
sickle	-	2.00	41.00	-	-	100 m^2	**41.00**
petrol operated strimmer	-	0.30	6.16	0.44	-	100 m^2	**6.60**
Cutting areas of rough grass which contain trees or whips							
petrol operated strimmer	-	0.40	8.20	0.59	-	100 m^2	**8.79**
Extra over cutting rough grass for on site raking up and dumping	-	0.33	6.83	-	-	100 m^2	**6.83**

Q PAVING/PLANTING/FENCING/SITE FURNITURE

Item Excluding site overheads and profit	PC £	Labour hours	Labour £	Plant £	Material £	Unit	Total rate £
Trimming edge of grass areas; edging tool							
with petrol powered strimmer	-	0.13	2.73	0.20	-	100 m	2.93
by hand	-	0.67	13.67	-	-	100 m	13.67
Rolling grass areas; light roller							
by pedestrian operated mechanical roller	-	0.08	1.71	1.68	-	100 m^2	3.38
by hand drawn roller	-	0.17	3.42	-	-	100 m^2	3.42
Aerating grass areas; to a depth of 100 mm							
using pedestrian-guided motor powered solid or slitting tine turf aerator	-	0.18	3.59	2.68	-	100 m^2	6.27
using hollow tine aerator; including sweeping up and dumping corings	-	0.50	10.25	5.36	-	100 m^2	15.61
using hand aerator or fork	-	1.67	34.17	-	-	100 m^2	34.17
Extra over aerating grass areas for on site sweeping up and dumping corings	-	0.17	3.42	-	-	100 m^2	3.42
Switching off dew; from fine turf areas	-	0.20	4.10	-	-	100 m^2	4.10
Scarifying grass areas to break up thatch; removing dead grass							
using self-propelled scarifier; including removing and disposing of grass on site	-	0.33	6.83	0.13	-	100 m^2	6.97
by hand	-	3.03	62.12	-	-	100 m^2	62.12
For the following topsoil improvement and seeding operations add or subtract the following amounts for every £0.10 difference in the material cost price							
35 g/m^2	-	-	-	-	0.35	100 m^2	-
50 g/m^2	-	-	-	-	0.50	100 m^2	-
70 g/m^2	-	-	-	-	0.70	100 m^2	-
100 g/m^2	-	-	-	-	1.00	100 m^2	-
Top dressing fertilizers (7+7+7); PC £1.13/kg; to seedbeds; by hand							
35 g/m^2	3.95	0.17	3.42	-	3.95	100 m^2	7.37
50 g/m^2	5.65	0.17	3.42	-	5.65	100 m^2	9.06
70 g/m^2	7.91	0.17	3.42	-	7.91	100 m^2	11.32
Watering turf; evenly; at a rate of 5 litre/m^2							
using sprinkler equipment and with sufficient water pressure to run 1 nr 15 m radius sprinkler	-	0.02	0.41	-	-	100 m^2	0.41
using hand-held watering equipment	-	0.25	5.13	-	-	100 m^2	5.13

Q PAVING/PLANTING/FENCING/SITE FURNITURE

Item Excluding site overheads and profit	PC £	Labour hours	Labour £	Plant £	Material £	Unit	Total rate £
Q31 PLANTING							
Site protection; temporary protective fencing							
Cleft chestnut rolled fencing; to 100 mm diameter chestnut posts; driving into firm ground at 3 m centres; pales at 50 mm centres							
900 mm high	3.88	0.11	2.19	-	8.95	m	**11.14**
1100 mm high	4.61	0.11	2.19	-	9.48	m	**11.67**
1500 mm high	7.07	0.11	2.19	-	11.95	m	**14.13**
Extra over temporary protective fencing for removing and making good (no allowance for re-use of material)	-	0.07	1.37	0.22	-	m	**1.59**
Cultivation							
Breaking up existing ground; using pedestrian operated tine cultivator or rotavator							
100 mm deep	-	0.55	11.28	16.58	-	100 m²	**27.86**
150 mm deep	-	0.63	12.88	18.95	-	100 m²	**31.83**
200 mm deep	-	0.73	15.03	22.11	-	100 m²	**37.15**
As above but in heavy clay or wet soils							
100 mm deep	-	0.73	15.03	22.11	-	100 m²	**37.15**
150 mm deep	-	0.88	18.04	26.53	-	100 m²	**44.57**
200 mm deep	-	1.10	22.55	33.17	-	100 m²	**55.72**
Importing only selected and approved topsoil							
1 - 14 m³	26.00	-	-	-	78.00	m³	**-**
over 15 m³	26.00	-	-	-	26.00	m³	**-**
Imported topsoil; spreading and lightly consolidating approved topsoil in layers not exceeding 150 mm; travel distance from offloading point not exceeding 25 m							
By machine							
minimum depth 100 mm	-	-	-	0.93	3.25	m²	**4.18**
minimum depth 200 mm	-	-	-	1.86	6.50	m²	**8.36**
minimum depth 300 mm	-	-	-	2.79	9.75	m²	**12.54**
minimum depth 500 mm	-	-	-	4.65	16.25	m²	**20.90**
over 500 mm	-	-	-	9.30	32.50	m³	**41.80**
By hand							
minimum depth 100 mm	-	0.20	4.10	-	3.25	m²	**7.35**
minimum depth 150 mm	-	0.30	6.16	-	4.88	m²	**11.03**
minimum depth 300 mm	-	0.60	12.28	-	9.75	m²	**22.03**
minimum depth 500 mm	-	0.90	18.47	-	16.25	m²	**34.72**
over 500 mm deep	-	2.22	45.51	-	32.50	m³	**78.01**
Spreading and lightly consolidating approved topsoil (imported or from spoil heaps); in layers not exceeding 150 mm; travel distance from spoil heaps not exceeding 25 m							
By machine (imported topsoil not included)							
minimum depth 100 mm	-	-	-	0.93	-	m²	**0.93**
minimum depth 200 mm	-	-	-	1.86	-	m²	**1.86**
minimum depth 300 mm	-	-	-	2.79	-	m²	**2.79**
minimum depth 500 mm	-	-	-	4.65	-	m²	**4.65**
over 500 mm	-	-	-	9.30	-	m³	**9.30**

Q PAVING/PLANTING/FENCING/SITE FURNITURE

Item Excluding site overheads and profit	PC £	Labour hours	Labour £	Plant £	Material £	Unit	Total rate £
By hand (imported topsoil not included)							
minimum depth 100 mm	-	0.20	4.10	-	-	m^2	4.10
minimum depth 150 mm	-	0.30	6.16	-	-	m^2	6.16
minimum depth 300 mm	-	0.60	12.28	-	-	m^2	12.28
minimum depth 500 mm	-	0.90	18.47	-	-	m^2	18.47
over 500 mm deep	-	2.22	45.51	-	-	m^3	45.51
Extra over for spreading topsoil to slopes 15 - 30° by machine or hand	-	-	-	-	-	10%	-
Extra over for spreading topsoil to slopes over 30° by machine or hand	-	-	-	-	-	25%	-
Extra over spreading topsoil for travel exceeding 100 m; by machine							
100 - 150 m	-	-	-	0.61	-	m^3	0.61
150 - 200 m	-	-	-	0.82	-	m^3	0.82
200 - 300 m	-	-	-	1.09	-	m^3	1.09
Extra over spreading topsoil for travel exceeding 100 m; by hand							
100 m	-	2.50	51.25	-	-	m^3	51.25
200 m	-	3.50	71.75	-	-	m^3	71.75
300 m	-	4.50	92.25	-	-	m^3	92.25
Evenly grading; to general surfaces to bring to finished levels							
by pedestrian operated rotavator	-	-	0.08	0.13	-	m^2	0.21
by hand	-	0.01	0.20	-	-	m^2	0.20
Extra over grading for slopes 15 - 30° by machine or hand	-	-	-	-	-	10%	-
Extra over grading for slopes over 30° by machine or hand	-	-	-	-	-	25%	-
Clearing stones; disposing off site							
by hand; stones not exceeding 50 mm in any direction; loading to skip 4.6 m^3	-	0.01	0.20	0.04	-	m^2	0.24
Lightly cultivating; weeding; to fallow areas; disposing debris off site							
by hand	-	0.01	0.29	0.30	-	m^2	0.59
Preparation of planting operations; herbicides and granular additives							
For the following topsoil improvement and planting operations add or subtract the following amounts for every £0.10 difference in the material cost price							
35 g/m^2	-	-	-	-	0.35	100 m^2	-
50 g/m^2	-	-	-	-	0.50	100 m^2	-
70 g/m^2	-	-	-	-	0.70	100 m^2	-
100 g/m^2	-	-	-	-	1.00	100 m^2	-
General herbicides; in accordance with manufacturer's instructions; "Knapsack" spray application							
"Roundup Pro Biactive 360" at 50 ml/ 100 m^2	0.55	0.33	6.83	-	0.55	100 m^2	7.38
Fertilizers; in top 150 mm of topsoil; at 35 g/m^2							
"Mascot Microfine"; 8+0+6 + 2% Mg + 4% Fe	6.25	0.12	2.51	-	6.25	100 m^2	8.76
"Enmag"; controlled release fertilizer (8-9 months); 11+22+09	8.27	0.12	2.51	-	8.69	100 m^2	11.20
"Mascot Sports Amenity"; 7+7+7	3.95	0.12	2.51	-	3.95	100 m^2	6.46
"Super Phosphate Powder"	4.27	0.12	2.51	-	4.48	100 m^2	6.99
"Mascot Sports Amenity"; 15+5+10	4.72	0.12	2.51	-	4.72	100 m^2	7.23
Bone meal	3.88	0.12	2.51	-	4.07	100 m^2	6.58

Q PAVING/PLANTING/FENCING/SITE FURNITURE

Item Excluding site overheads and profit	PC £	Labour hours	Labour £	Plant £	Material £	Unit	Total rate £
Q31 PLANTING – cont'd							
Preparation of planting operations; **herbicides and granular additives** – cont'd							
Fertilizers; in top 150 mm of topsoil at 70 g/m^2							
"Mascot Microfine"; 8+0+6 + 2% Mg + 4% Fe	12.50	0.12	2.51	-	12.50	100 m^2	15.01
"Enmag"; controlled release fertilizer (8-9 months); 11+22+09	16.55	0.12	2.51	-	17.37	100 m^2	19.89
"Mascot Sports Amenity"; 7+7+7	7.91	0.12	2.51	-	7.91	100 m^2	10.42
"Super Phosphate Powder"	8.54	0.12	2.51	-	8.97	100 m^2	11.48
"Mascot Sports Amenity"; 15+5+10	9.44	0.12	2.51	-	9.44	100 m^2	11.95
Bone meal	7.76	0.12	2.51	-	8.14	100 m^2	10.66
Preparation of planting areas; movement **of materials to location maximum 25 m** **from offload location**							
By machine	-	-	-	6.96	-	m^3	6.96
By hand	-	1.00	20.50	-	-	m^3	20.50
Preparation of planting operations; **spreading only; movement of material to** **planting beds not included**							
Composted bark and manure soil conditioner (20 m^3 loads); from not further than 25 m from location; cultivating into topsoil by pedestrian operated cultivator							
50 mm thick	151.50	2.86	58.57	5.53	159.07	100 m^2	223.18
100 mm thick	303.00	6.05	124.00	5.53	318.15	100 m^2	447.68
150 mm thick	454.50	8.90	182.55	5.53	477.23	100 m^2	665.30
200 mm thick	606.00	12.90	264.55	5.53	636.30	100 m^2	906.38
Mushroom compost (20 m^3 loads); from not further than 25 m from location; cultivating into topsoil by pedestrian operated cultivator							
50 mm thick	106.00	2.86	58.57	5.53	106.00	100 m^2	170.10
100 mm thick	212.00	6.05	124.00	5.53	212.00	100 m^2	341.53
150 mm thick	318.00	8.90	182.55	5.53	318.00	100 m^2	506.08
200 mm thick	424.00	12.90	264.55	5.53	424.00	100 m^2	694.08
Tree planting; pre-planting operations							
Excavating tree pits; depositing soil alongside pits; by machine							
600 mm x 600 mm x 600 mm deep	-	0.15	3.02	0.60	-	nr	3.62
900 mm x 900 mm x 600 mm deep	-	0.33	6.76	1.36	-	nr	8.12
1.00 m x 1.00 m x 600 mm deep	-	0.61	12.58	1.68	-	nr	14.26
1.25 m x 1.25 m x 600 mm deep	-	0.96	19.69	2.63	-	nr	22.32
1.00 m x 1.00 m x 1.00 m deep	-	1.02	20.97	2.80	-	nr	23.77
1.50 m x 1.50 m x 750 mm deep	-	1.73	35.38	4.73	-	nr	40.11
1.50 m x 1.50 m x 1.00 m deep	-	2.30	47.05	6.28	-	nr	53.34
1.75 m x 1.75 m x 1.00 mm deep	-	3.13	64.21	8.58	-	nr	72.79
2.00 m x 2.00 m x 1.00 mm deep	-	4.09	83.86	11.20	-	nr	95.06

Q PAVING/PLANTING/FENCING/SITE FURNITURE

Item Excluding site overheads and profit	PC £	Labour hours	Labour £	Plant £	Material £	Unit	Total rate £
Excavating tree pits; depositing soil alongside pits; by hand							
600 mm x 600 mm x 600 mm deep	-	0.44	9.02	-	-	nr	9.02
900 mm x 900 mm x 600 mm deep	-	1.00	20.50	-	-	nr	20.50
1.00 m x 1.00 m x 600 mm deep	-	1.13	23.06	-	-	nr	23.06
1.25 m x 1.25 m x 600 mm deep	-	1.93	39.56	-	-	nr	39.56
1.00 m x 1.00 m x 1.00 m deep	-	2.06	42.23	-	-	nr	42.23
1.50 m x 1.50 m x 750 mm deep	-	3.47	71.14	-	-	nr	71.14
1.75 m x 1.50 m x 750 mm deep	-	4.05	83.03	-	-	nr	83.03
1.50 m x 1.50 m x 1.00 m deep	-	4.63	94.92	-	-	nr	94.92
2.00 m x 2.00 m x 750 mm deep	-	6.17	126.48	-	-	nr	126.48
2.00 m x 2.00 m x 1.00 m deep	-	8.23	168.73	-	-	nr	168.73
Breaking up subsoil in tree pits; to a depth of 200 mm	-	0.03	0.68	-	-	m^2	0.68
Spreading and lightly consolidating approved topsoil (imported or from spoil heaps); in layers not exceeding 150 mm; distance from spoil heaps not exceeding 50 m (imported topsoil not included); by machine							
minimum depth 100 mm	-	-	-	1.32	-	m^2	1.32
minimum depth 150 mm	-	-	-	1.98	-	m^2	1.98
minimum depth 300 mm	-	-	-	3.63	-	m^2	3.63
minimum depth 450 mm	-	-	-	5.22	-	m^2	5.22
Spreading and lightly consolidating approved topsoil (imported or from spoil heaps); in layers not exceeding 150 mm; distance from spoil heaps not exceeding 100 m (imported topsoil not included); by hand							
minimum depth 100 mm	-	0.13	2.56	-	-	m^2	2.56
minimum depth 150 mm	-	0.19	3.85	-	-	m^2	3.85
minimum depth 300 mm	-	0.38	7.71	-	-	m^2	7.71
minimum depth 450 mm	-	0.56	11.58	-	-	m^2	11.58
Extra for filling tree pits with imported topsoil; PC £26.00/m^3; plus allowance for 20% settlement							
depth 100 mm	-	-	-	-	3.12	m^2	-
depth 150 mm	-	-	-	-	4.68	m^2	-
depth 200 mm	-	-	-	-	6.24	m^2	-
depth 300 mm	-	-	-	-	9.36	m^2	-
depth 400 mm	-	-	-	-	12.48	m^2	-
depth 450 mm	-	-	-	-	14.04	m^2	-
depth 500 mm	-	-	-	-	15.60	m^2	-
depth 600 mm	-	-	-	-	18.72	m^2	-
Add or deduct the following amounts for every £0.50 change in the material price of topsoil							
depth 100 mm	-	-	-	-	0.06	m^2	-
depth 150 mm	-	-	-	-	0.09	m^2	-
depth 200 mm	-	-	-	-	0.12	m^2	-
depth 300 mm	-	-	-	-	0.18	m^2	-
depth 400 mm	-	-	-	-	0.24	m^2	-
depth 450 mm	-	-	-	-	0.27	m^2	-
depth 500 mm	-	-	-	-	0.30	m^2	-
depth 600 mm	-	-	-	-	0.36	m^2	-

Q PAVING/PLANTING/FENCING/SITE FURNITURE

Item Excluding site overheads and profit	PC £	Labour hours	Labour £	Plant £	Material £	Unit	Total rate £
Q31 PLANTING – cont'd							
Tree staking							
J Toms Ltd; extra over trees for tree stake(s); driving 500 mm into firm ground; trimming to approved height; including two tree ties to approved pattern							
one stake; 1.52 m long x 32 mm x 32 mm	1.47	0.20	4.10	-	1.47	nr	**5.57**
two stakes; 1.21 m long x 25 mm x 25 mm	1.60	0.30	6.15	-	1.60	nr	**7.75**
two stakes; 1.52 m long x 32 mm x 32 mm	2.16	0.30	6.15	-	2.16	nr	**8.31**
three stakes; 1.52 m long x 32 mm x 32 mm	3.24	0.36	7.38	-	3.24	nr	**10.62**
Tree anchors							
Platipus Anchors Ltd; extra over trees for tree anchors							
ref RF1 rootball kit; for 75 to 220 mm girth; 2 to 4.5 m high; inclusive of "Plati-Mat" PM1	33.32	1.00	20.50	-	33.32	nr	**53.82**
ref RF2; rootball kit; for 220 to 450 mm girth; 4.5 to 7.5 m high; inclusive of "Plati-Mat" PM2	55.70	1.33	27.27	-	55.70	nr	**82.97**
ref RF3; rootball kit; for 450 to 750 mm girth; 7.5 to 12 m high; inclusive of "Plati-Mat" PM3	111.33	1.50	30.75	-	111.33	nr	**142.08**
ref CG1; guy fixing kit; 75 to 220 mm girth; 2 to 4.5 m high	18.17	1.67	34.17	-	18.17	nr	**52.34**
ref CG2; guy fixing kit; 220 to 450 mm girth; 4.5 to 7.5 m high	31.96	2.00	41.00	-	31.96	nr	**72.96**
installation tools; drive rod for RF1/CG1 kits	-	-	-	-	58.90	nr	**-**
installation tools; drive rod for RF2/CG2 kits	-	-	-	-	88.76	nr	**-**
Extra over trees for land drain to tree pits; 100 mm diameter perforated flexible agricultural drain; including excavating drain trench; laying pipe; backfilling	1.10	1.00	20.50	-	1.21	m	**21.71**
Tree planting; tree pit additives							
Melcourt Industries Ltd; "Topgrow"; incorporating into topsoil at 1 part "Topgrow" to 3 parts excavated topsoil; supplied in 75 L bags; pit size							
600 x 600 x 600 mm	1.44	0.02	0.41	-	1.44	nr	**1.85**
900 x 900 x 900 mm	4.86	0.06	1.23	-	4.86	nr	**6.09**
1.00 x 1.00 x 1.00 m	6.66	0.24	4.92	-	6.66	nr	**11.58**
1.25 x 1.25 x 1.25 m	13.02	0.40	8.20	-	13.02	nr	**21.22**
1.50 x 1.50 x 1.50 m	22.50	0.90	18.45	-	22.50	nr	**40.95**
Melcourt Industries Ltd; "Topgrow"; incorporating into topsoil at 1 part "Topgrow" to 3 parts excavated topsoil; supplied in 65 m³ loose loads; pit size							
600 x 600 x 600 mm	1.11	0.02	0.34	-	1.11	nr	**1.45**
900 x 900 x 900 mm	3.75	0.05	1.02	-	3.75	nr	**4.77**
1.00 x 1.00 x 1.00 m	5.14	0.20	4.10	-	5.14	nr	**9.24**
1.25 x 1.25 x 1.25 m	10.03	0.33	6.83	-	10.03	nr	**16.87**
1.50 x 1.50 x 1.50 m	17.34	0.75	15.38	-	17.34	nr	**32.72**

Q PAVING/PLANTING/FENCING/SITE FURNITURE

Item Excluding site overheads and profit	PC £	Labour hours	Labour £	Plant £	Material £	Unit	Total rate £
Mulching of tree pits; Melcourt Industries Ltd (items labelled FSC are Forest Stewardship Council certified)							
Spreading mulch; to individual trees; maximum distance 25 m (mulch not included)							
50 mm thick	-	0.05	1.00	-	-	m²	1.00
75 mm thick	-	0.07	1.49	-	-	m²	1.49
100 mm thick	-	0.10	1.99	-	-	m²	1.99
Mulch; "Bark Nuggets®"; to individual trees; delivered in 25 m³ loads; maximum distance 25 m							
50 mm thick	2.51	0.05	1.00	-	2.64	m²	3.63
75 mm thick	3.77	0.05	1.02	-	3.96	m²	4.98
100 mm thick	5.02	0.07	1.37	-	5.27	m²	6.64
Mulch; "Amenity Bark Mulch"; FSC; to individual trees; delivered in 25 m³ loads; maximum distance 25 m							
50 mm thick	1.78	0.05	1.00	-	1.87	m²	2.87
75 mm thick	2.67	0.07	1.49	-	2.81	m²	4.30
100 mm thick	3.56	0.07	1.37	-	3.74	m²	5.11
Trees; planting labours only							
Bare root trees; including backfilling with previously excavated material (all other operations and materials not included)							
light standard; 6 - 8 cm girth	-	0.35	7.17	-	-	nr	7.17
standard; 8 - 10 cm girth	-	0.40	8.20	-	-	nr	8.20
selected standard; 10 - 12 cm girth	-	0.58	11.89	-	-	nr	11.89
heavy standard; 12 - 14 cm girth	-	0.83	17.08	-	-	nr	17.08
extra heavy standard; 14 - 16 cm girth	-	1.00	20.50	-	-	nr	20.50
Root balled trees; including backfilling with previously excavated material (all other operations and materials not included)							
standard; 8 - 10 cm girth	-	0.50	10.25	-	-	nr	10.25
selected standard; 10 - 12 cm girth	-	0.60	12.30	-	-	nr	12.30
heavy standard; 12 - 14 cm girth	-	0.80	16.40	-	-	nr	16.40
extra heavy standard;14 - 16 cm girth	-	1.50	30.75	-	-	nr	30.75
16 - 18 cm girth	-	1.30	26.57	24.44	-	nr	51.01
18 - 20 cm girth	-	1.60	32.80	30.17	-	nr	62.97
20 - 25 cm girth	-	4.50	92.25	97.59	-	nr	189.84
25 - 30 cm girth	-	6.00	123.00	128.43	-	nr	251.43
30 - 35 cm girth	-	11.00	225.50	226.53	-	nr	452.03
Tree planting; root balled trees; advanced nursery stock and semi-mature - General							
Preamble: the cost of planting semi-mature trees will depend on the size and species, and on the access to the site for tree handling machines. Prices should be obtained for individual trees and planting.							

Q PAVING/PLANTING/FENCING/SITE FURNITURE

Item Excluding site overheads and profit	PC £	Labour hours	Labour £	Plant £	Material £	Unit	Total rate £
Q31 PLANTING – cont'd							
Tree planting; bare root trees; nursery stock; James Coles & Sons (Nurseries) Ltd							
"Acer platanoides"; including backfillling with excavated material (other operations not included)							
light standard; 6 - 8 cm girth	9.45	0.35	7.17	-	9.45	nr	**16.63**
standard; 8 - 10 cm girth	11.81	0.40	8.20	-	11.81	nr	**20.01**
selected standard; 10 - 12 cm girth	19.16	0.58	11.89	-	19.16	nr	**31.05**
heavy standard; 12 - 14 cm girth	35.17	0.83	17.08	-	35.17	nr	**52.25**
extra heavy standard; 14 - 16 cm girth	44.10	1.00	20.50	-	44.10	nr	**64.60**
"Carpinus betulus"; including backfillling with excavated material (other operations not included)							
light standard; 6 - 8 cm girth	13.91	0.35	7.17	-	13.91	nr	**21.09**
standard; 8 - 10 cm girth	27.82	0.40	8.20	-	27.82	nr	**36.02**
selected standard; 10 - 12 cm girth	39.64	0.58	11.89	-	39.64	nr	**51.53**
heavy standard; 12 - 14 cm girth	41.21	0.83	17.09	-	41.21	nr	**58.30**
extra heavy standard; 14 - 16 cm girth	48.56	1.00	20.50	-	48.56	nr	**69.06**
"Fraxinus excelsior"; including backfillling with excavated material (other operations not included)							
light standard; 6 - 8 cm girth	10.24	0.35	7.17	-	10.24	nr	**17.41**
standard; 8 - 10 cm girth	16.80	0.40	8.20	-	16.80	nr	**25.00**
selected standard; 10 - 12 cm girth	23.63	0.58	11.89	-	23.63	nr	**35.52**
heavy standard; 12 - 14 cm girth	41.21	0.83	17.02	-	41.21	nr	**58.23**
extra heavy standard; 14 - 16 cm girth	49.88	1.00	20.50	-	49.88	nr	**70.38**
"Prunus avium Plena"; including backfillling with excavated material (other operations not included)							
light standard; 6 - 8 cm girth	10.24	0.36	7.46	-	10.24	nr	**17.70**
standard; 8 - 10 cm girth	16.80	0.40	8.20	-	16.80	nr	**25.00**
selected standard; 10 - 12 cm girth	30.98	0.58	11.89	-	30.98	nr	**42.87**
heavy standard; 12 - 14 cm girth	49.88	0.83	17.08	-	49.88	nr	**66.96**
extra heavy standard; 14 - 16 cm girth	46.99	1.00	20.50	-	46.99	nr	**67.49**
"Quercus robur"; including backfillling with excavated material (other operations not included)							
light standard; 6 - 8 cm girth	22.05	0.35	7.17	-	22.05	nr	**29.23**
standard; 8 - 10 cm girth	32.29	0.40	8.20	-	32.29	nr	**40.49**
selected standard; 10 - 12 cm girth	44.10	0.58	11.89	-	44.10	nr	**55.99**
heavy standard; 12 - 14 cm girth	61.69	0.83	17.08	-	61.69	nr	**78.77**
"Robinia pseudoacacia Frisia"; including backfillling with excavated material (other operations not included)							
light standard; 6 - 8 cm girth	22.05	0.35	7.17	-	22.05	nr	**29.23**
standard; 8 - 10 cm girth	29.40	0.40	8.20	-	29.40	nr	**37.60**
selected standard; 10 - 12 cm girth	46.99	0.58	11.89	-	46.99	nr	**58.88**

Q PAVING/PLANTING/FENCING/SITE FURNITURE

Item Excluding site overheads and profit	PC £	Labour hours	Labour £	Plant £	Material £	Unit	Total rate £
Tree planting; root balled trees; nursery stock; James Coles & Sons (Nurseries) Ltd							
"Acer platanoides"; including backfilling with excavated material (other operations not included)							
standard; 8 - 10 cm girth	19.69	0.48	9.84	-	19.69	nr	**29.53**
selected standard; 10 - 12 cm girth	29.66	0.56	11.49	-	29.66	nr	**41.15**
heavy standard; 12 - 14 cm girth	50.92	0.76	15.67	-	50.92	nr	**66.59**
extra heavy standard; 14 - 16 cm girth	59.85	1.20	24.60	-	59.85	nr	**84.45**
"Carpinus betulus"; including backfilling with excavated material (other operations not included)							
standard; 8 - 10 cm girth	35.70	0.48	9.84	-	35.70	nr	**45.54**
selected standard; 10 - 12 cm girth	50.14	0.56	11.49	-	50.14	nr	**61.63**
heavy standard; 12 - 14 cm girth	83.47	0.76	15.67	-	83.47	nr	**99.14**
extra heavy standard; 14 - 16 cm girth	111.30	1.20	24.60	-	111.30	nr	**135.90**
Tree planting; containerized trees; nursery stock; James Coles & Sons (Nurseries) Ltd							
"Acer platanoides"; including backfilling with excavated material (other operations not included)							
standard; 8 - 10 cm girth	36.75	0.48	9.84	-	36.75	ea	**46.59**
selected standard; 10 - 12 cm girth	57.75	0.56	11.49	-	57.75	ea	**69.24**
heavy standard; 12 - 14 cm girth	68.25	0.76	15.67	-	68.25	ea	**83.92**
extra heavy standard; 14 - 16 cm girth	78.75	1.20	24.60	-	78.75	ea	**103.35**
"Carpinus betulus"; including backfilling with excavated material (other operations not included)							
standard; 8 - 10 cm girth	42.00	0.48	9.84	-	42.00	ea	**51.84**
selected standard; 10 - 12 cm girth	63.00	0.56	11.49	-	63.00	ea	**74.49**
heavy standard; 12 - 14 cm girth	78.75	0.76	15.67	-	78.75	ea	**94.42**
extra heavy standard; 14 - 16 cm girth	89.25	1.20	24.60	-	89.25	ea	**113.85**
"Fraxinus excelsior"; including backfilling with excavated material (other operations not included)							
standard; 8 - 10 cm girth	36.75	0.48	9.84	-	36.75	ea	**46.59**
selected standard; 10 - 12 cm girth	57.75	0.56	11.49	-	57.75	ea	**69.24**
heavy standard; 12 - 14 cm girth	68.25	0.76	15.67	-	68.25	ea	**83.92**
extra heavy standard; 14 - 16 cm girth	89.25	1.20	24.60	-	89.25	ea	**113.85**
"Prunus avium Plena"; including backfilling with excavated material (other operations not included)							
selected standard; 10 - 12 cm girth	63.00	0.56	11.49	-	63.00	ea	**74.49**
heavy standard; 12 - 14 cm girth	73.50	0.76	15.67	-	73.50	ea	**89.17**
extra heavy standard; 14 - 16 cm girth	84.00	1.20	24.60	-	84.00	ea	**108.60**
"Quercus robur"; including backfilling with excavated material (other operations not included)							
standard; 8 - 10 cm girth	42.00	0.48	9.84	-	42.00	ea	**51.84**
selected standard; 10 - 12 cm girth	63.00	0.56	11.49	-	63.00	ea	**74.49**
heavy standard; 12 - 14 cm girth	78.75	0.76	15.67	-	78.75	ea	**94.42**
extra heavy standard; 14 - 16 cm girth	89.25	1.20	24.60	-	89.25	ea	**113.85**

Q PAVING/PLANTING/FENCING/SITE FURNITURE

Item Excluding site overheads and profit	PC £	Labour hours	Labour £	Plant £	Material £	Unit	Total rate £
Q31 PLANTING – cont'd							
Tree planting; containerized trees; nursery stock; James Coles & Sons (Nurseries) Ltd – cont'd							
"Betula utilis jaquemontii"; multistemmed; including backfillling with excavated material (other operations not included)							
175/200 mm high	24.15	0.48	9.84	-	24.15	ea	33.99
200/250 mm high	31.50	0.56	11.49	-	31.50	ea	42.99
250/300 mm high	36.75	0.76	15.67	-	36.75	ea	52.42
300/350 mm high	57.75	1.20	24.60	-	57.75	ea	82.35
Tree planting; "Airpot" container grown trees; advanced nursery stock and semi-mature; Deepdale Trees Ltd							
"Acer platanoides Emerald Queen"; including backfilling with excavated material (other operations not included)							
16 - 18 cm girth	95.00	1.98	40.59	31.44	95.00	nr	167.03
18 - 20 cm girth	130.00	2.18	44.65	34.58	130.00	nr	209.23
20 - 25 cm girth	190.00	2.38	48.71	37.72	190.00	nr	276.43
25 - 30 cm girth	250.00	2.97	60.88	62.87	250.00	nr	373.76
30 - 35 cm girth	450.00	3.96	81.18	88.35	450.00	nr	619.53
"Aesculus briotti"; including backfilling with excavated material (other operations not included)							
16 - 18 cm girth	90.00	1.98	40.59	31.44	90.00	nr	162.03
18 - 20 cm girth	130.00	1.60	32.80	34.58	130.00	nr	197.38
20 - 25 cm girth	180.00	2.38	48.71	37.72	180.00	nr	266.43
25 - 30 cm girth	240.00	2.97	60.88	80.24	240.00	nr	381.12
30 - 35 cm girth	330.00	3.96	81.18	88.35	330.00	nr	499.53
Tree planting; "Airpot" container grown trees; semi-mature and mature trees; Deepdale Trees Ltd; planting and back filling; planted by telehandler or by crane; delivery included; all other operations priced separately							
Semi mature trees indicative prices							
40 - 45 cm girth	550.00	4.00	82.00	51.75	550.00	nr	683.75
45 - 50 cm girth	750.00	4.00	82.00	51.75	750.00	nr	883.75
55 - 60 cm girth	1350.00	6.00	123.00	51.75	1350.00	nr	1524.75
60 - 70 cm girth	2500.00	7.00	143.50	69.69	2500.00	nr	2713.19
70 - 80 cm girth	3500.00	7.50	153.75	87.63	3500.00	nr	3741.38
80 - 90 cm girth	4500.00	8.00	164.00	103.51	4500.00	nr	4767.51
Tree planting; root balled trees; advanced nursery stock and semi-mature; Lorenz von Ehren							
"Acer platanoides Emerald Queen"; including backfilling with excavated material (other operations not included)							
16 - 18 cm girth	93.50	1.30	26.57	4.11	93.50	nr	124.18
18 - 20 cm girth	110.00	1.60	32.80	4.11	110.00	nr	146.91
20 - 25 cm girth	137.50	4.50	92.25	18.90	137.50	nr	248.65
25 - 30 cm girth	181.50	6.00	123.00	23.50	181.50	nr	328.00
30 - 35 cm girth	330.00	11.00	225.50	31.43	330.00	nr	586.93

Q PAVING/PLANTING/FENCING/SITE FURNITURE

Item Excluding site overheads and profit	PC £	Labour hours	Labour £	Plant £	Material £	Unit	Total rate £
"Quercus palustris"; including backfilling with excavated material (other operations not included)							
16 - 18 cm girth	121.00	1.30	26.57	4.11	121.00	nr	**151.68**
18 - 20 cm girth	143.00	1.60	32.80	4.11	143.00	nr	**179.91**
20 - 25 cm girth	154.00	4.50	92.25	18.90	154.00	nr	**265.15**
25 - 30 cm girth	242.00	6.00	123.00	23.50	242.00	nr	**388.50**
30 - 35 cm girth	374.00	11.00	225.50	31.43	374.00	nr	**630.93**
"Tilia cordata Greenspire"; including backfilling with excavated material (other operations not included)							
16 - 18 cm girth	104.50	1.30	26.57	4.11	104.50	nr	**135.18**
18 - 20 cm girth	110.00	1.60	32.80	4.11	110.00	nr	**146.91**
20 - 25 cm girth	132.00	4.50	92.25	18.90	132.00	nr	**243.15**
25 - 30 cm girth; 5 x transplanted 4.0 - 5.0 m tall	170.50	6.00	123.00	23.50	170.50	nr	**317.00**
30 - 35 cm girth; 5 x transplanted 5.0 - 7.0 m tall	286.00	11.00	225.50	31.43	286.00	nr	**542.93**
"Betula pendula (3 stems)"; including backfilling with excavated material (other operations not included)							
3.0 - 3.5 m high	77.00	1.98	40.59	31.44	77.00	nr	**149.03**
3.5 - 4.0 m high	99.00	1.60	32.80	34.58	99.00	nr	**166.38**
4.0 - 4.5 m high	154.00	2.38	48.71	53.01	154.00	nr	**255.72**
4.5 - 5.0 m high	187.00	2.97	60.88	80.24	187.00	nr	**328.12**
5.0 - 6.0 m high	231.00	3.96	81.18	88.35	231.00	nr	**400.53**
6.0 - 7.0 m high	440.00	4.50	92.25	104.89	440.00	nr	**637.14**
"Pinus sylvestris"; including backfilling with excavated material (other operations not included)							
3.0 - 3.5 m high	440.00	1.98	40.59	31.44	440.00	nr	**512.03**
3.5 - 4.0 m high	550.00	1.60	32.80	34.58	550.00	nr	**617.38**
4.0 - 4.5 m high	880.00	2.38	48.71	37.72	880.00	nr	**966.43**
4.5 - 5.0 m high	1210.00	2.97	60.88	80.24	1210.00	nr	**1351.12**
5.0 - 6.0 m high	1650.00	3.96	81.18	88.35	1650.00	nr	**1819.53**
6.0 - 7.0 m high	2750.00	4.50	92.25	108.07	2750.00	nr	**2950.32**
Tree planting; containerized trees; Lorenz von Ehren; to the tree prices above add for Airpot containerization only							
Tree size							
20 - 25 cm	-	-	-	-	66.00	nr	-
25 - 30 cm	-	-	-	-	110.00	nr	-
30 - 35 cm	-	-	-	-	165.00	nr	-
35 - 40 cm	-	-	-	-	220.00	nr	-
40 - 50 cm	-	-	-	-	275.00	nr	-
50 - 70 cm	-	-	-	-	330.00	nr	-
70 - 80 cm	-	-	-	-	440.00	nr	-
80 - 90 cm	-	-	-	-	550.00	nr	-

Q PAVING/PLANTING/FENCING/SITE FURNITURE

Item Excluding site overheads and profit	PC £	Labour hours	Labour £	Plant £	Material £	Unit	Total rate £
Q31 PLANTING – cont'd							
Hedges							
Excavating trench for hedges; depositing soil alongside trench; by machine							
300 mm deep x 300 mm wide	-	0.03	0.61	0.30	-	m	**0.91**
300 mm deep x 450 mm wide	-	0.05	0.92	0.44	-	m	**1.37**
Excavating trench for hedges; depositing soil alongside trench; by hand							
300 mm deep x 300 mm wide	-	0.12	2.46	-	-	m	**2.46**
300 mm deep x 450 mm wide	-	0.23	4.62	-	-	m	**4.62**
Setting out; notching out; excavating trench; breaking up subsoil to minimum depth 300 mm							
minimum 400 mm deep	-	0.25	5.13	-	-	m	**5.13**
Hedge planting; including backfill with excavated topsoil; PC £0.36/nr							
single row; 200 mm centres	1.80	0.06	1.28	-	1.80	m	**3.08**
single row; 300 mm centres	1.20	0.06	1.14	-	1.20	m	**2.34**
single row; 400 mm centres	0.90	0.04	0.85	-	0.90	m	**1.75**
single row; 500 mm centres	0.72	0.03	0.68	-	0.72	m	**1.40**
double row; 200 mm centres	3.60	0.17	3.42	-	3.60	m	**7.02**
double row; 300 mm centres	2.40	0.13	2.73	-	2.40	m	**5.13**
double row; 400 mm centres	1.80	0.08	1.71	-	1.80	m	**3.51**
double row; 500 mm centres	1.44	0.07	1.37	-	1.44	m	**2.81**
Extra over hedges for incorporating manure; at 1 m^3 per 30 m	0.80	0.03	0.51	-	0.80	m	**1.32**
Shrub planting - General							
Preamble: for preparation of planting areas see "Cultivation" at the beginning of the planting section.							
Shrub planting							
Setting out; selecting planting from holding area; loading to wheelbarrows; planting as plan or as directed; distance from holding area maximum 50 m; plants 2 - 3 litre containers							
plants in groups of 100 nr minimum	-	0.01	0.24	-	-	nr	**0.24**
plants in groups of 10 - 100 nr	-	0.02	0.34	-	-	nr	**0.34**
plants in groups of 3 - 5 nr	-	0.03	0.51	-	-	nr	**0.51**
single plants not grouped	-	0.04	0.82	-	-	nr	**0.82**
Forming planting holes; in cultivated ground (cultivating not included); by mechanical auger; trimming holes by hand; depositing excavated material alongside holes							
250 diameter	-	0.03	0.68	0.13	-	nr	**0.81**
250 x 250 mm	-	0.04	0.82	0.24	-	nr	**1.06**
300 x 300 mm	-	0.08	1.54	0.30	-	nr	**1.83**

Q PAVING/PLANTING/FENCING/SITE FURNITURE

Item Excluding site overheads and profit	PC £	Labour hours	Labour £	Plant £	Material £	Unit	Total rate £
Hand excavation; forming planting holes; in cultivated ground (cultivating not included); depositing excavated material alongside holes							
100 x 100 x 100 mm deep; with mattock or hoe	-	0.01	0.14	-	-	nr	0.14
250 x 250 x 300 mm deep	-	0.04	0.82	-	-	nr	0.82
300 x 300 x 300 mm deep	-	0.06	1.14	-	-	nr	1.14
400 x 400 x 400 mm deep	-	0.13	2.56	-	-	nr	2.56
500 x 500 x 500 mm deep	-	0.25	5.13	-	-	nr	5.13
600 x 600 x 600 mm deep	-	0.43	8.87	-	-	nr	8.87
900 x 900 x 600 mm deep	-	1.00	20.50	-	-	nr	20.50
1.00 m x 1.00 m x 600 mm deep	-	1.23	25.21	-	-	nr	25.21
1.25 m x 1.25 m x 600 mm deep	-	1.93	39.56	-	-	nr	39.56
Hand excavation; forming planting holes; in uncultivated ground; depositing excavated material alongside holes							
100 x 100 x 100 mm deep; with mattock or hoe	-	0.03	0.51	-	-	nr	0.51
250 x 250 x 300 mm deep	-	0.06	1.14	-	-	nr	1.14
300 x 300 x 300 mm deep	-	0.06	1.28	-	-	nr	1.28
400 x 400 x 400 mm deep	-	0.25	5.13	-	-	nr	5.13
500 x 500 x 500 mm deep	-	0.33	6.67	-	-	nr	6.67
600 x 600 x 600 mm deep	-	0.55	11.28	-	-	nr	11.28
900 x 900 x 600 mm deep	-	1.25	25.63	-	-	nr	25.63
1.00 m x 1.00 m x 600 mm deep	-	1.54	31.52	-	-	nr	31.52
1.25 m x 1.25 m x 600 mm deep	-	2.41	49.46	-	-	nr	49.46
Bare root planting; to planting holes (forming holes not included); including backfilling with excavated material (bare root plants not included)							
bare root 1+1; 30 - 90 mm high	-	0.02	0.34	-	-	nr	0.34
bare root 1+2; 90 - 120 mm high	-	0.02	0.34	-	-	nr	0.34
Containerized planting; to planting holes (forming holes not included); including backfilling with excavated material (shrub or ground cover not included)							
9 cm pot	-	0.01	0.20	-	-	nr	0.20
2 litre container	-	0.02	0.41	-	-	nr	0.41
3 litre container	-	0.02	0.46	-	-	nr	0.46
5 litre container	-	0.03	0.68	-	-	nr	0.68
10 litre container	-	0.05	1.02	-	-	nr	1.02
15 litre container	-	0.07	1.37	-	-	nr	1.37
20 litre container	-	0.08	1.71	-	-	nr	1.71
Shrub planting; 2 litre containerized plants; in cultivated ground (cultivating not included); PC £3.00/nr							
average 2 plants per m^2	-	0.06	1.15	-	6.00	m^2	7.15
average 3 plants per m^2	-	0.08	1.72	-	9.00	m^2	10.72
average 4 plants per m^2	-	0.11	2.30	-	12.00	m^2	14.30
average 6 plants per m^2	-	0.17	3.45	-	18.00	m^2	21.45
Extra over shrubs for stakes	0.64	0.02	0.34	-	0.64	nr	0.98

Q PAVING/PLANTING/FENCING/SITE FURNITURE

Item Excluding site overheads and profit	PC £	Labour hours	Labour £	Plant £	Material £	Unit	Total rate £
Q31 PLANTING – cont'd							
Shrub planting – cont'd							
Composted bark soil conditioners; 20 m³ loads; on beds by mechanical loader; spreading and rotavating into topsoil; by machine							
50 mm thick	151.50	-	-	6.64	151.50	100 m²	158.14
100 mm thick	303.00	-	-	11.50	318.15	100 m²	329.65
150 mm thick	454.50	-	-	16.55	477.23	100 m²	493.77
200 mm thick	606.00	-	-	21.50	636.30	100 m²	657.80
Mushroom compost; 25 m³ loads; delivered not further than 25 m from location; cultivating into topsoil by pedestrian operated machine							
50 mm thick	106.00	2.86	58.57	5.53	106.00	100 m²	170.10
100 mm thick	212.00	6.05	124.00	5.53	212.00	100 m²	341.53
150 mm thick	318.00	8.90	182.55	5.53	318.00	100 m²	506.08
200 mm thick	424.00	12.90	264.55	5.53	424.00	100 m²	694.08
Manure; 20 m³ loads; delivered not further than 25 m from location; cultivating into topsoil by pedestrian operated machine							
50 mm thick	236.25	2.86	58.57	5.53	236.25	100 m²	300.35
100 mm thick	472.50	6.05	124.00	5.53	496.13	100 m²	625.66
150 mm thick	708.75	8.90	182.55	5.53	744.19	100 m²	932.27
200 mm thick	945.00	12.90	264.55	5.53	992.25	100 m²	1262.33
Fertilizers (7+7+7); PC £1.13/kg; to beds; by hand							
35 g/m²	3.95	0.17	3.42	-	3.95	100 m²	7.37
50 g/m²	5.65	0.17	3.42	-	5.65	100 m²	9.06
70 g/m²	7.91	0.17	3.42	-	7.91	100 m²	11.32
Fertilizers; "Enmag"; PC £2.63/kg; controlled release fertilizer; to beds; by hand							
35 g/m²	8.27	0.17	3.42	-	8.27	100 m²	11.69
50 g/m²	11.82	0.17	3.42	-	11.82	100 m²	15.24
70 g/m²	17.73	0.17	3.42	-	17.73	100 m²	21.15
Note: for machine incorporation of fertilizers and soil conditioners see "Cultivation".							
Herbaceous and groundcover planting							
Herbaceous plants; PC £1.50/nr; including forming planting holes in cultivated ground (cultivating not included); backfilling with excavated material; 1 litre containers							
average 4 plants per m² - 500 mm centres	6.00	0.09	1.91	-	6.00	m²	7.91
average 6 plants per m² - 408 mm centres	-	0.14	2.87	-	9.00	m²	11.87
average 8 plants per m² - 354 mm centres	-	0.19	3.83	-	12.00	m²	15.83
Note: for machine incorporation of fertilizers and soil conditioners see "Cultivation".							
Bulb planting							
Bulbs; including forming planting holes in cultivated area (cultivating not included); backfilling with excavated material							
small	13.00	0.83	17.08	-	13.00	100 nr	30.08
medium	22.00	0.83	17.08	-	22.00	100 nr	39.08
large	25.00	0.91	18.64	-	25.00	100 nr	43.64

Q PAVING/PLANTING/FENCING/SITE FURNITURE

Item Excluding site overheads and profit	PC £	Labour hours	Labour £	Plant £	Material £	Unit	Total rate £
Bulbs; in grassed area; using bulb planter; including backfilling with screened topsoil or peat and cut turf plug							
small	13.00	1.67	34.17	-	13.00	100 nr	47.17
medium	22.00	1.67	34.17	-	22.00	100 nr	56.17
large	25.00	2.00	41.00	-	25.00	100 nr	66.00
Aquatic planting							
Aquatic plants; in prepared growing medium in pool; plant size 2 - 3 litre containerized (plants not included)	-	0.04	0.82	-	-	nr	0.82
Operations after planting							
Initial cutting back to shrubs and hedge plants; including disposal of all cuttings	-	1.00	20.50	-	-	100 m^2	20.50
Mulch; Melcourt Industries Ltd; "Bark Nuggets®"; to plant beds; delivered in 25 m^3 loads; maximum distance 25 m							
50 mm thick	2.23	0.03	0.60	-	2.34	m^2	2.94
75 mm thick	3.77	0.07	1.37	-	3.96	m^2	5.32
100 mm thick	5.02	0.09	1.82	-	5.27	m^2	7.09
Mulch; Melcourt Industries Ltd; "Amenity Bark Mulch" (FSC); to plant beds; delivered in 25 m^3 loads; maximum distance 25 m							
50 mm thick	1.78	0.04	0.91	-	1.87	m^2	2.78
75 mm thick	2.67	0.07	1.37	-	2.81	m^2	4.18
100 mm thick	3.56	0.09	1.82	-	3.74	m^2	5.57
Maintenance operations (Note: the following rates apply to aftercare maintenance executed as part of a landscaping contract only)							
Weeding and hand forking planted areas; including disposing weeds and debris on site; areas maintained weekly	-	-	0.08	-	-	m^2	0.08
Weeding and hand forking planted areas; including disposing weeds and debris on site; areas maintained monthly	-	0.01	0.20	-	-	m^2	0.20
Mulch; Melcourt Industries Ltd; "Bark Nuggets®"; to plant beds; delivered in 25 m^3 loads; maximum distance 25 m							
50 mm thick	2.51	0.04	0.91	-	2.64	m^2	3.55
75 mm thick	3.77	0.07	1.37	-	3.96	m^2	5.32
Mulch; Melcourt Industries Ltd; "Amenity Bark Mulch" (FSC); to plant beds; delivered in 25 m^3 loads; maximum distance 25 m							
50 mm thick	1.78	0.04	0.91	-	1.87	m^2	2.78
75 mm thick	2.67	0.07	1.37	-	2.81	m^2	4.18
Watering planting; evenly; at a rate of 5 litre/m^2							
using hand-held watering equipment	-	0.25	5.13	-	-	100 m^2	5.13
using sprinkler equipment and with sufficient water pressure to run 1 nr 15 m radius sprinkler	-	0.14	2.85	-	-	100 m^2	2.85

Q PAVING/PLANTING/FENCING/SITE FURNITURE

Item Excluding site overheads and profit	PC £	Labour hours	Labour £	Plant £	Material £	Unit	Total rate £
Q31 PLANTING – cont'd							
Work to existing planting							
Cutting and trimming ornamental hedges; to specified profiles; including cleaning out hedge bottoms; hedge cut 2 occasions per annum; by hand							
up to 2.00 m high	-	0.03	0.68	-	0.50	m	**1.18**
2.00 - 4.00 m high	-	0.05	1.02	4.54	1.00	m	**6.56**
Q35 LANDSCAPE MAINTENANCE							
Grass cutting - pedestrian operated equipment							
Using cylinder lawn mower fitted with not less than five cutting blades, front and rear rollers; on surface not exceeding 30° from horizontal; arisings let fly; width of cut							
51 cm	-	0.06	1.25	0.20	-	100 m	**1.45**
61 cm	-	0.05	1.04	0.18	-	100 m	**1.22**
Using rotary self-propelled mower; width of cut							
45 cm	-	0.08	1.58	0.06	-	100 m^2	**1.64**
40 cm	-	0.11	2.28	0.06	-	100 m^2	**2.34**
Add for using grass box for collecting and depositing arisings							
removing and depositing arisings	-	0.05	1.02	-	-	100 m^2	**1.02**
Add for 30 to 50° from horizontal	-	-	-	-	-	33%	**-**
Add for slopes exceeding 50°	-	-	-	-	-	100%	**-**
Cutting grass or light woody undergrowth; using trimmer with nylon cord or metal disc cutter; on surface							
not exceeding 30° from horizontal	-	0.20	4.10	0.29	-	100 m^2	**4.39**
30 - 50° from horizontal	-	0.40	8.20	0.59	-	100 m^2	**8.79**
exceeding 50° from horizontal	-	0.50	10.25	0.73	-	100 m^2	**10.98**
Grass cutting - collecting arisings							
Extra over for tractor drawn and self-propelled machinery using attached grass boxes; depositing arisings							
22 cuts per year	-	0.05	1.02	-	-	100 m^2	**1.02**
18 cuts per year	-	0.08	1.54	-	-	100 m^2	**1.54**
12 cuts per year	-	0.10	2.05	-	-	100 m^2	**2.05**
4 cuts per year	-	0.25	5.13	-	-	100 m^2	**5.13**
Disposing arisings							
22 cuts per year	-	0.01	0.16	0.03	0.04	100 m^2	**0.23**
18 cuts per year	-	0.01	0.20	0.03	0.05	100 m^2	**0.28**
12 cuts per year	-	0.01	0.30	0.04	0.08	100 m^2	**0.42**
4 cuts per year	-	0.04	0.89	0.14	0.26	100 m^2	**1.29**
Scarifying by hand							
hand implement	-	0.50	10.25	-	-	100 m^2	**10.25**
add for disposal of arisings	-	0.03	0.51	1.10	16.00	100 m^2	**17.61**
Rolling							
Rolling grassed area; equipment towed by tractor; once over; using							
smooth roller	-	0.01	0.29	0.22	-	100 m^2	**0.51**

Q PAVING/PLANTING/FENCING/SITE FURNITURE

Item Excluding site overheads and profit	PC £	Labour hours	Labour £	Plant £	Material £	Unit	Total rate £
Turf aeration							
By hand							
hand fork; to effect a minimum penetration of 100 mm and spaced 150 mm apart	-	1.33	27.33	-	-	100 m²	27.33
hollow tine hand implement; to effect a minimum penetration of 100 mm and spaced 150 mm apart	-	2.00	41.00	-	-	100 m²	41.00
collection of arisings by hand	-	3.00	61.50	-	-	100 m²	61.50
Turf areas; surface treatments and top dressing; Boughton Loam Ltd							
Apply screened topdressing to grass surfaces; spread using Tru-Lute							
sand soil mixes 90/10 to 50/50	0.12	-	0.04	0.03	0.12	m²	0.20
Leaf clearance							
Using pedestrian operated mechanical equipment and blowers							
grassed areas with perimeters of mature trees such as sports fields and amenity areas	-	0.04	0.82	0.05	-	100 m²	0.87
grassed areas containing ornamental trees and shrub beds	-	0.10	2.05	0.12	-	100 m²	2.17
verges	-	0.07	1.37	0.08	-	100 m²	1.45
By hand							
grassed areas with perimeters of mature trees such as sports fields and amenity areas	-	0.05	1.02	0.06	-	100 m²	1.08
grassed areas containing ornamental trees and shrub beds	-	0.08	1.71	0.10	-	100 m²	1.81
verges	-	1.00	20.50	1.20	-	100 m²	21.70
Removal of arisings							
areas with perimeters of mature trees	-	0.01	0.14	0.10	1.28	100 m²	1.51
areas containing ornamental trees and shrub beds	-	0.02	0.41	0.37	3.20	100 m²	3.98
Litter clearance							
Collection and disposal of litter from grassed area							
areas exceeding 1000 m²	-	0.01	0.20	0.30	-	100 m²	0.50
area not exceeding 1000 m²	-	0.04	0.82	0.30	-	100 m²	1.12
Edge maintenance							
Maintain edges where lawn abuts pathway or hard surface using							
strimmer	-	0.01	0.10	0.01	-	m	0.11
shears	-	0.02	0.34	-	-	m	0.34
Maintain edges where lawn abuts plant bed using							
mechanical edging tool	-	0.01	0.14	0.03	-	m	0.17
shears	-	0.01	0.23	-	-	m	0.23
half moon edging tool	-	0.02	0.41	-	-	m	0.41
Tree guards, stakes and ties							
Adjusting existing tree tie	-	0.03	0.68	-	-	nr	0.68
Taking up single or double tree stake and ties; removing and disposing	-	0.05	1.02	-	-	nr	1.02

Q PAVING/PLANTING/FENCING/SITE FURNITURE

Item Excluding site overheads and profit	PC £	Labour hours	Labour £	Plant £	Material £	Unit	Total rate £
Q35 LANDSCAPE MAINTENANCE – cont'd							
Pruning shrubs							
Trimming ground cover planting							
soft groundcover; vinca ivy and the like	-	1.00	20.50	-	-	100 m^2	20.50
woody groundcover; cotoneaster and the like	-	1.50	30.75	-	-	100 m^2	30.75
Pruning massed shrub border (measure ground area)							
shrub beds pruned annually	-	0.01	0.20	-	-	m^2	0.20
shrub beds pruned hard every 3 years	-	0.03	0.57	-	-	m^2	0.57
Cutting off dead heads							
bush or standard rose	-	0.05	1.02	-	-	nr	1.02
climbing rose	-	0.08	1.71	-	-	nr	1.71
Pruning roses							
bush or standard rose	-	0.05	1.02	-	-	nr	1.02
climbing or rambling rose; tying in as required	-	0.07	1.37	-	-	nr	1.37
Pruning ornamental shrubs; height before pruning (increase these rates by 50% if pruning work has not been executed during the previous two years)							
not exceeding 1m	-	0.04	0.82	-	-	nr	0.82
1 to 2 m	-	0.06	1.14	-	-	nr	1.14
exceeding 2 m	-	0.13	2.56	-	-	nr	2.56
Removing excess growth etc. from face of building etc.; height before pruning							
not exceeding 2 m	-	0.03	0.59	-	-	nr	0.59
2 to 4 m	-	0.05	1.02	-	-	nr	1.02
4 to 6 m	-	0.08	1.71	-	-	nr	1.71
6 to 8 m	-	0.13	2.56	-	-	nr	2.56
8 to 10 m	-	0.14	2.93	-	-	nr	2.93
Removing epicormic growth from base of shrub or trunk and base of tree; any height; any diameter; number of growths							
not exceeding 10	-	0.05	1.02	-	-	nr	1.02
10 to 20	-	0.07	1.37	-	-	nr	1.37
Beds, borders and planters							
Lifting							
bulbs	-	0.50	10.25	-	-	100 nr	10.25
tubers or corms	-	0.40	8.20	-	-	100 nr	8.20
established herbaceous plants; hoeing and depositing for replanting	-	2.00	41.00	-	-	100 nr	41.00
Temporary staking and tying in herbaceous plant	-	0.03	0.68	-	0.11	nr	0.79
Cutting down spent growth of herbaceous plant; clearing arisings							
unstaked	-	0.02	0.41	-	-	nr	0.41
staked; not exceeding 4 stakes per plant; removing stakes and putting into store	-	0.03	0.51	-	-	nr	0.51
Hand weeding							
newly planted areas	-	2.00	41.00	-	-	100 m^2	41.00
established areas	-	0.50	10.25	-	-	100 m^2	10.25
Removing grasses from groundcover areas	-	3.00	61.56	-	-	100 m^2	61.56
Hand digging with fork; not exceeding 150 mm deep; breaking down lumps; leaving surface with a medium tilth	-	1.33	27.33	-	-	100 m^2	27.33

Q PAVING/PLANTING/FENCING/SITE FURNITURE

Item Excluding site overheads and profit	PC £	Labour hours	Labour £	Plant £	Material £	Unit	Total rate £
Hand digging with fork or spade to an average depth of 230 mm; breaking down lumps; leaving surface with a medium tilth	-	2.00	41.00	-	-	100 m^2	**41.00**
Hand hoeing; not exceeding 50 mm deep; leaving surface with a medium tilth	-	0.40	8.20	-	-	100 m^2	**8.20**
Hand raking to remove stones etc.; breaking down lumps; leaving surface with a fine tilth prior to planting	-	0.67	13.67	-	-	100 m^2	**13.67**
Hand weeding; planter or window box; not exceeding 1.00 m^2							
ground level box	-	0.05	1.02	-	-	nr	**1.02**
box accessed by stepladder	-	0.08	1.71	-	-	nr	**1.71**
Spreading only compost, mulch or processed bark to a depth of 75 mm							
on shrub bed with existing mature planting	-	0.09	1.86	-	-	m^2	**1.86**
recently planted areas	-	0.07	1.37	-	-	m^2	**1.37**
groundcover and herbaceous areas	-	0.08	1.54	-	-	m^2	**1.54**
Clearing cultivated area of leaves, litter and other extraneous debris; using hand implement							
weekly maintenance	-	0.13	2.56	-	-	100 m^2	**2.56**
daily maintenance	-	0.02	0.34	-	-	100 m^2	**0.34**
Bedding							
Lifting							
bedding plants; hoeing and depositing for disposal	-	3.00	61.50	-	-	100 m^2	**61.50**
Hand digging with fork; not exceeding 150 mm deep; breaking down lumps; leaving surface with a medium tilth	-	0.75	15.38	-	-	100 m^2	**15.38**
Hand weeding							
newly planted areas	-	2.00	41.00	-	-	100 m^2	**41.00**
established areas	-	0.50	10.25	-	-	100 m^2	**10.25**
Hand digging with fork or spade to an average depth of 230 mm; breaking down lumps; leaving surface with a medium tilth	-	0.50	10.25	-	-	100 m^2	**10.25**
Hand hoeing: not exceeding 50 mm deep; leaving surface with a medium tilth	-	0.40	8.20	-	-	100 m^2	**8.20**
Hand raking to remove stones etc.; breaking down lumps; leaving surface with a fine tilth prior to planting	-	0.67	13.67	-	-	100 m^2	**13.67**
Hand weeding; planter or window box; not exceeding 1.00 m^2							
ground level box	-	0.05	1.02	-	-	nr	**1.02**
box accessed by stepladder	-	0.08	1.71	-	-	nr	**1.71**
Spreading only; compost, mulch or processed bark to a depth of 75 mm							
on shrub bed with existing mature planting	-	0.09	1.86	-	-	m^2	**1.86**
recently planted areas	-	0.07	1.37	-	-	m^2	**1.37**
groundcover and herbaceous areas	-	0.08	1.54	-	-	m^2	**1.54**
Collecting bedding from nursery	-	3.00	61.50	14.67	-	100 m^2	**76.17**
Setting out							
mass planting single variety	-	0.13	2.56	-	-	m^2	**2.56**
pattern	-	0.33	6.83	-	-	m^2	**6.83**
Planting only							
massed bedding plants	-	0.20	4.10	-	-	m^2	**4.10**

Q PAVING/PLANTING/FENCING/SITE FURNITURE

Item Excluding site overheads and profit	PC £	Labour hours	Labour £	Plant £	Material £	Unit	Total rate £
Q35 LANDSCAPE MAINTENANCE – cont'd							
Bedding – cont'd							
Clearing cultivated area of leaves, litter and other extraneous debris; using hand implement							
weekly maintenance	-	0.13	2.56	-	-	100 m^2	**2.56**
daily maintenance	-	0.02	0.34	-	-	100 m^2	**0.34**
Irrigation and watering							
Hand held hosepipe; flow rate 25 litres per minute; irrigation requirement							
10 litres/m^2	-	0.74	15.11	-	-	100 m^2	**15.11**
15 litres/m^2	-	1.10	22.55	-	-	100 m^2	**22.55**
20 litres/m^2	-	1.46	29.99	-	-	100 m^2	**29.99**
25 litres/m^2	-	1.84	37.66	-	-	100 m^2	**37.66**
Hand held hosepipe; flow rate 40 litres per minute; irrigation requirement							
10 litres/m^2	-	0.46	9.47	-	-	100 m^2	**9.47**
15 litres/m^2	-	0.69	14.21	-	-	100 m^2	**14.21**
20 litres/m^2	-	0.91	18.72	-	-	100 m^2	**18.72**
25 litres/m^2	-	1.15	23.50	-	-	100 m^2	**23.50**
Hedge cutting; field hedges cut once or twice annually							
Trimming sides and top using hand tool or hand held mechanical tools							
not exceeding 2 m high	-	0.10	2.05	0.15	-	10 m^2	**2.20**
2 to 4 m high	-	0.33	6.83	0.49	-	10 m^2	**7.32**
Hedge cutting; ornamental							
Trimming sides and top using hand tool or hand held mechanical tools							
not exceeding 2 m high	-	0.13	2.56	0.18	-	10 m^2	**2.75**
2 to 4 m high	-	0.50	10.25	0.73	-	10 m^2	**10.98**
Hedge cutting; reducing width; hand tool or hand held mechanical tools							
Not exceeding 2 m high							
average depth of cut not exceeding 300 mm	-	0.10	2.05	0.15	-	10 m^2	**2.20**
average depth of cut 300 to 600 mm	-	0.83	17.08	1.23	-	10 m^2	**18.31**
average depth of cut 600 to 900 mm	-	1.25	25.63	1.84	-	10 m^2	**27.46**
2 to 4 m high							
average depth of cut not exceeding 300 mm	-	0.03	0.51	0.04	-	10 m^2	**0.55**
average depth of cut 300 to 600 mm	-	0.13	2.56	0.18	-	10 m^2	**2.75**
average depth of cut 600 to 900 mm	-	2.50	51.25	3.67	-	10 m^2	**54.92**
4 to 6 m high							
average depth of cut not exceeding 300 mm	-	0.10	2.05	0.15	-	m^2	**2.20**
average depth of cut 300 to 600 mm	-	0.17	3.42	0.25	-	m^2	**3.66**
average depth of cut 600 to 900 mm	-	0.50	10.25	0.73	-	m^2	**10.98**

Q PAVING/PLANTING/FENCING/SITE FURNITURE

Item Excluding site overheads and profit	PC £	Labour hours	Labour £	Plant £	Material £	Unit	Total rate £
Hedge cutting; reducing height; hand tool or hand held mechanical tools							
Not exceeding 2 m high							
average depth of cut not exceeding 300 mm	-	0.03	0.51	0.04	-	m^2	**0.55**
average depth of cut 300 to 600 mm	-	0.04	0.82	0.06	-	m^2	**0.88**
average depth of cut 600 to 900 mm	-	0.10	2.05	0.15	-	m^2	**2.20**
2 to 4 m high							
average depth of cut not exceeding 300 mm	-	0.05	1.02	0.07	-	m^2	**1.10**
average depth of cut 300 to 600 mm	-	0.07	1.37	0.10	-	m^2	**1.47**
average depth of cut 600 to 900 mm	-	0.13	2.56	0.18	-	m^2	**2.75**
4 to 6 m high							
average depth of cut not exceeding 300 mm	-	0.11	2.28	0.16	-	m^2	**2.44**
average depth of cut 300 to 600 mm	-	0.20	4.10	0.29	-	m^2	**4.39**
average depth of cut 600 to 900 mm	-	0.50	10.25	0.73	-	m^2	**10.98**
Hedge cutting; removal and disposal of arisings							
Sweeping up and depositing arisings							
300 mm cut	-	0.05	1.02	-	-	$10\ m^2$	**1.02**
600 mm cut	-	0.20	4.10	-	-	$10\ m^2$	**4.10**
900 mm cut	-	0.40	8.20	-	-	$10\ m^2$	**8.20**
Chipping arisings							
300 mm cut	-	0.02	0.41	0.37	-	$10\ m^2$	**0.78**
600 mm cut	-	0.08	1.71	1.55	-	$10\ m^2$	**3.26**
900 mm cut	-	0.20	4.10	3.72	-	$10\ m^2$	**7.82**
Disposal of unchipped arisings							
300 mm cut	-	0.02	0.34	0.55	2.13	$10\ m^2$	**3.03**
600 mm cut	-	0.03	0.68	1.10	3.20	$10\ m^2$	**4.98**
900 mm cut	-	0.08	1.71	2.75	8.00	$10\ m^2$	**12.46**
Disposal of chipped arisings							
300 mm cut	-	-	0.07	0.18	3.20	$10\ m^2$	**3.45**
600 mm cut	-	0.02	0.34	0.18	6.40	$10\ m^2$	**6.93**
900 mm cut	-	0.03	0.68	0.37	16.00	$10\ m^2$	**17.05**
Spraying; labour rates only; for chemical rates please use the tables at the beginning of the section Q35 Landscape Maintenance in the Major Works section of this Book							
Herbicide applications; standard backpack spray applicators; application to maintain 1.00 m diameter clear circles (0.79 m^2) around new planting							
plants at 1.50 m centres; 4444 nr/ha	-	0.25	5.06	-	-	$100\ m^2$	**5.06**
plants at 1.75 m centres; 3265 nr/ha	-	0.18	3.72	-	-	$100\ m^2$	**3.72**
plants at 2.00 m centres; 2500 nr/ha	-	0.14	2.85	-	-	$100\ m^2$	**2.85**
understory of mature planting or trees	-	0.13	2.56	-	-	$100\ m^2$	**2.56**
mass spraying low vegetation or hard surfaces	-	0.10	2.05	-	-	$100\ m^2$	**2.05**

Q PAVING/PLANTING/FENCING/SITE FURNITURE

Item Excluding site overheads and profit	PC £	Labour hours	Labour £	Plant £	Material £	Unit	Total rate £
Q40 FENCING							
Temporary fencing; HSS Hire; mesh framed mesh unclimbable fencing; including precast concrete supports and couplings							
Weekly hire; 2.85 x 2 m high							
weekly hire rate	-	-	-	2.10	-	m	2.10
erection of fencing; labour only	-	0.13	2.56	-	-	m	2.56
removal of fencing; loading to collection vehicle	-	0.08	1.71	-	-	m	1.71
delivery charge	-	-	-	0.80	-	m	0.80
return haulage charge	-	-	-	0.60	-	m	0.60
Protective fencing							
Cleft chestnut rolled fencing; fixing to 100 mm diameter chestnut posts; driving into firm ground at 3 m centres							
900 mm high	3.88	0.11	2.19	-	8.95	m	11.14
1200 mm high	4.61	0.16	3.28	-	9.71	m	12.99
1500 mm high; 3 strand	7.07	0.21	4.38	-	11.95	m	16.32
Timber fencing; AVS Fencing Supplies Ltd; tanalized softwood fencing							
Timber lap panels; pressure treated; fixed to timber posts 75 x 75 mm in 1:3:6 concrete; at 1.90 centres							
900 mm high	7.87	0.67	13.67	-	18.34	m	32.01
1200 mm high	7.92	0.75	15.38	-	18.68	m	34.05
1500 mm high	8.01	0.80	16.40	-	18.59	m	34.99
1800 mm high	8.03	0.90	18.45	-	19.98	m	38.43
Timber lap panels; fixed to slotted concrete posts 100 x 100 mm in 1:3:6 concrete at 1.88 m centres							
900 mm high	7.95	0.75	15.38	-	23.41	m	38.79
1200 mm high	8.01	0.80	16.40	-	23.47	m	39.87
1500 mm high	14.29	0.85	17.43	-	29.35	m	46.77
1800 mm high	19.04	0.90	18.45	-	34.09	m	52.54
extra for corner posts	20.53	0.90	18.45	-	32.88	nr	51.33
Extra over panel fencing for 300 mm high trellis tops; slats at 100 mm centres; including additional length of posts	4.46	0.10	2.05	-	4.46	m	6.51
Close boarded fencing; to concrete posts 100 x 100 mm; 2 nr softwood Arris rails; 100 x 22 mm softwood pales lapped 13 mm; including excavating and backfilling into firm ground at 3.00 m centres; setting in concrete 1:3:6; timber gravel board 150 x 50 mm							
900 mm high	14.98	1.00	20.50	-	20.64	m	41.14
1050 mm high	16.36	1.10	22.55	-	21.00	m	43.55
1500 mm high	18.16	1.15	23.57	-	22.80	m	46.37

Q PAVING/PLANTING/FENCING/SITE FURNITURE

Item Excluding site overheads and profit	PC £	Labour hours	Labour £	Plant £	Material £	Unit	Total rate £
Close boarded fencing; to concrete posts 100 x 100 mm; 3 nr softwood Arris rails; 100 x 22 mm softwood pales lapped 13 mm; including excavating and backfilling into firm ground at 3.00 m centres; setting in concrete; timber gravel board 150 x 50 mm							
1650 mm high	22.63	1.25	25.63	-	27.26	m	**52.89**
1800 mm high	22.79	1.30	26.65	-	27.42	m	**54.07**
extra over to the above for concrete gravel board 150 x 50 mm in lieu of timber gravel board	4.14	0.11	2.28	-	4.20	m	**6.47**
Close boarded fencing; 2 nr Arris rails; to timber posts 100 x 100 mm; 100 x 22 mm softwood pales lapped 13 mm; including excavating and backfilling into firm ground at 3.00 m centres; setting in concrete 1:3:6							
900 mm high	11.14	1.00	20.50	-	20.04	m	**40.54**
1050 mm high	15.77	1.10	22.55	-	20.40	m	**42.95**
1200 mm high	17.57	1.15	23.57	-	22.20	m	**45.78**
Close boarded fencing; 3 nr softwood Arris rails; to timber posts 100 x 100 mm; 100 x 22 mm softwood pales lapped 13 mm; including excavating and backfilling into firm ground at 3.00 m centres; setting in concrete 1:3:6							
1350 mm high	17.49	0.95	19.52	-	25.37	m	**44.90**
1650 mm high	21.58	1.30	26.65	-	26.21	m	**52.86**
1800 mm high	21.70	1.35	27.68	-	26.33	m	**54.01**
extra over for post 125 x 100 mm; for 1800 mm high fencing	8.05	-	-	-	8.05	nr	**-**
extra over to the above for counter rail	0.62	0.10	2.05	-	0.62	m	**2.67**
extra over to the above for capping rail	1.17	0.10	2.05	-	1.17	m	**3.22**
Close boarded fencing; 2 nr cant rails; nailed to timber posts 100 x 100 mm; 125 x 22 mm softwood pales lapped 13 mm; including excavating and backfilling into firm ground at 3.00 m centres; setting in concrete 1:3:6							
900 mm high	10.48	1.00	20.50	-	17.13	m	**37.63**
1050 mm high	12.85	1.10	22.55	-	17.49	m	**40.04**
1200 mm high	14.05	1.10	22.55	-	18.69	m	**41.24**
Close boarded fencing; 3 nr cant rails; nailed to timber posts 100 x 100 mm; 125 x 22 mm softwood pales lapped 13 mm; including excavating and backfilling into firm ground at 3.00 m centres; setting in concrete 1:3:6							
1350 mm high	14.97	1.15	23.57	-	21.62	m	**45.19**
1800 mm high	18.61	1.20	24.60	-	23.25	m	**47.85**
extra over for post 125 x 100 mm; for 1800 mm high fencing	-	-	-	-	-	nr	**-**

Q PAVING/PLANTING/FENCING/SITE FURNITURE

Item Excluding site overheads and profit	PC £	Labour hours	Labour £	Plant £	Material £	Unit	Total rate £
Q40 FENCING – cont'd							
Timber fencing; AVS Fencing Supplies Ltd; tanalized softwood fencing – cont'd							
Palisade fencing; 19 x 75 mm softwood vertical palings with pointed tops at 150 mm centres; nailing to 2 nr horizontal softwood Arris rails; morticed to 100 x 100 mm softwood posts with weathered tops at 3.00 m centres; setting in concrete							
900 mm high	16.12	0.90	18.45	-	20.83	m	**39.28**
1050 mm high	17.12	0.90	18.45	-	21.76	m	**40.21**
1200 mm high	17.46	0.95	19.48	-	22.17	m	**41.64**
extra over for rounded tops	0.73	-	-	-	0.73	m	**-**
Post-and-rail fencing; 90 x 38 mm softwood horizontal rails; fixing with galvanized nails to 150 x 75 mm softwood posts; including excavating and backfilling into firm ground at 1.80 m centres; all treated timber							
1200 mm high; 3 horizontal rails	7.89	0.35	7.18	-	7.97	m	**15.14**
1200 mm high; 4 horizontal rails	9.05	0.35	7.18	-	9.13	m	**16.30**
Cleft rail fencing; oak or chestnut tapered rails 2.80 m long; morticed into joints; to 125 x 100 mm softwood posts 1.95 m long; including excavating and backfilling into firm ground at 2.80 m centres							
two rails	17.71	0.25	5.13	-	17.71	m	**22.84**
three rails	20.18	0.28	5.74	-	20.18	m	**25.92**
four rails	22.65	0.35	7.18	-	22.65	m	**29.83**
"Hit and Miss" horizontal rail fencing; treated softwood; 87 x 38 mm top and bottom rails; 100 x 22 mm vertical boards arranged alternately on opposite side of rails; to 100 x 100 mm posts; including excavating and backfilling into firm ground; setting in concrete at 1.8 m centres							
1600 mm high	26.28	1.20	24.60	-	29.55	m	**54.15**
1800 mm high	28.93	1.33	27.33	-	32.20	m	**59.53**
2000 mm high	31.59	1.40	28.70	-	34.86	m	**63.56**
Palisade fencing; 100 x 22 mm softwood vertical palings with flat tops; nailing to 3 nr 50 x 100 mm horizontal softwood rails; housing into 100 x 100 mm softwood posts with weathered tops at 3.00 m centres; setting in concrete; all treated timber							
1800 mm high	20.56	0.47	9.58	-	25.43	m	**35.01**
1800 mm high	27.91	0.47	9.57	-	32.79	m	**42.35**
Post-and-rail fencing; 2 nr 90 x 38 mm softwood horizontal rails; fixing with galvanized nails to 150 x 75 mm softwood posts; including excavating and backfilling into firm ground at 1.80 m centres; all treated timber							
1200 mm high	6.73	0.30	6.15	-	6.81	m	**12.96**

Q PAVING/PLANTING/FENCING/SITE FURNITURE

Item Excluding site overheads and profit	PC £	Labour hours	Labour £	Plant £	Material £	Unit	Total rate £
Post-and-rail fencing; 3 nr 90 x 38 mm softwood horizontal rails; fixing with galvanized nails to 150 x 75 mm softwood posts; including excavating and backfilling into firm ground at 1.80 m centres; all treated timber							
1200 mm high	7.89	0.35	7.18	-	7.97	m	**15.14**
Morticed post-and-rail fencing; 3 nr horizontal 90 x 38 mm softwood rails; fixing with galvanized nails; 90 x 38 mm softwood centre prick posts; to 150 x 75 mm softwood posts; including excavating and backfilling into firm ground at 2.85 m centres; all treated timber							
1200 mm high	7.42	0.35	7.18	-	7.50	m	**14.68**
1350 mm high five rails	9.46	0.40	8.20	-	9.54	m	**17.74**
Cleft rail fencing; chestnut tapered rails 2.85 m long; morticed into joints; to 125 x 100 mm softwood posts 2.1 m long; including excavating and backfilling into firm ground at 2.75 m centres							
two rails	26.85	0.25	5.13	-	26.85	m	**31.97**
three rails	33.15	0.28	5.74	-	33.15	m	**38.89**
four rails	39.94	0.35	7.18	-	39.94	m	**47.12**
Boundary fencing; strained wire and wire mesh; AVS Fencing Supplies Ltd Strained wire fencing; concrete posts only at 2750 mm centres; 610 mm below ground; excavating holes; filling with concrete; replacing topsoil; disposing surplus soil off site							
900 mm high	3.03	0.56	11.39	1.85	4.63	m	**17.87**
1200 mm high	3.80	0.56	11.39	1.85	5.40	m	**18.64**
1800 mm high	5.20	0.78	15.94	4.62	6.79	m	**27.36**
Extra over strained wire fencing for concrete straining posts with one strut; posts and struts 610 mm below ground; struts, cleats, stretchers, winders, bolts and eye bolts; excavating holes; filling to within 150 mm of ground level with concrete (1:12) - 40 mm aggregate; replacing topsoil; disposing surplus soil off site							
900 mm high	31.85	0.67	13.73	4.16	70.15	nr	**88.04**
1200 mm high	23.55	0.67	13.67	4.17	55.94	nr	**73.78**
1800 mm high	33.13	0.67	13.67	4.17	67.98	nr	**85.82**
Extra over strained wire fencing for concrete straining posts with two struts; posts and struts 610 mm below ground; excavating holes; filling to within 150 mm of ground level with concrete (1:12) - 40 mm aggregate; replacing topsoil; disposing surplus soil off site							
900 mm high	42.70	0.91	18.66	5.34	103.43	nr	**127.43**
1200 mm high	35.70	0.85	17.43	4.17	88.68	nr	**110.28**
1800 mm high	50.67	0.85	17.43	4.17	107.45	nr	**129.05**

Q PAVING/PLANTING/FENCING/SITE FURNITURE

Item Excluding site overheads and profit	PC £	Labour hours	Labour £	Plant £	Material £	Unit	Total rate £
Q40 FENCING – cont'd							
Boundary fencing; strained wire and wire mesh; AVS Fencing Supplies Ltd – cont'd							
Strained wire fencing; painted steel angle posts only at 2750 mm centres; 610 mm below ground; driving in							
900 mm high; 40 x 40 x 5 mm	2.78	0.03	0.62	-	2.78	m	**3.40**
1200 mm high; 40 x 40 x 5 mm	3.16	0.03	0.68	-	3.16	m	**3.85**
1400 mm high; 40 x 40 x 5 mm	4.16	0.06	1.24	-	4.16	m	**5.40**
1800 mm high; 40 x 40 x 5 mm	4.53	0.07	1.49	-	4.53	m	**6.02**
1800 mm high; 45 x 45 x 5 mm; with extension for three rows barbed wire	5.81	0.12	2.48	-	5.81	m	**8.30**
Painted steel straining posts with two struts for strained wire fencing; setting in concrete							
900 mm high; 50 x 50 x 6 mm	49.56	1.00	20.50	-	49.56	nr	**70.06**
1200 mm high; 50 x 50 x 6 mm	59.86	1.00	20.50	-	59.86	nr	**80.36**
1500 mm high; 50 x 50 x 6 mm	74.79	1.00	20.50	-	74.79	nr	**95.29**
1800 mm high; 50 x 50 x 6 mm	77.52	1.00	20.50	-	77.52	nr	**98.02**
Strained wire; to posts (posts not included); 3 mm galvanized wire; fixing with galvanized stirrups							
900 mm high; 2 wire	0.17	0.03	0.68	-	0.33	m	**1.01**
1200 mm high; 3 wire	0.25	0.05	0.96	-	0.41	m	**1.37**
1400 mm high; 3 wire	0.25	0.05	0.96	-	0.41	m	**1.37**
1800 mm high; 3 wire	0.25	0.05	0.96	-	0.41	m	**1.37**
Barbed wire; to posts (posts not included); 3 mm galvanized wire; fixing with galvanized stirrups							
900 mm high; 2 wire	0.26	0.07	1.37	-	0.42	m	**1.79**
1200 mm high; 3 wire	0.39	0.09	1.91	-	0.55	m	**2.46**
1400 mm high; 3 wire	0.39	0.09	1.91	-	0.55	m	**2.46**
1800 mm high; 3 wire	0.39	0.09	1.91	-	0.55	m	**2.46**
Chain link fencing; AVS Fencing Supplies Ltd; to strained wire and posts priced separately; 3 mm galvanized wire; 50 mm mesh; galvanized steel components; fixing to line wires threaded through posts and strained with eye-bolts; posts (not included)							
900 mm high	4.08	0.07	1.37	-	4.24	m	**5.61**
1200 mm high	5.71	0.07	1.37	-	5.87	m	**7.24**
1800 mm high	7.91	0.10	2.05	-	8.07	m	**10.12**
2400 mm high	10.71	0.10	2.05	-	10.87	m	**12.92**
Chain link fencing; to strained wire and posts priced separately; 3.15 mm plastic coated galvanized wire (wire only 2.50 mm); 50 mm mesh; galvanized steel components; fencing with line wires threaded through posts and strained with eye-bolts; posts (not included) (Note: plastic coated fencing can be cheaper than galvanized finish as wire of a smaller cross-sectional area can be used)							
900 mm high	2.89	0.07	1.37	-	3.21	m	**4.57**
1200 mm high	3.98	0.07	1.37	-	4.30	m	**5.67**
1800 mm high	5.67	0.13	2.56	-	6.95	m	**9.51**

Q PAVING/PLANTING/FENCING/SITE FURNITURE

Item Excluding site overheads and profit	PC £	Labour hours	Labour £	Plant £	Material £	Unit	Total rate £
Extra over strained wire fencing for cranked arms and galvanized barbed wire							
1 row	3.65	0.02	0.34	-	3.65	m	3.99
2 row	3.78	0.05	1.02	-	3.78	m	4.80
3 row	3.91	0.05	1.02	-	3.91	m	4.93
Field fencing; Jacksons Fencing; welded wire mesh; fixed to posts and straining wires measured separately							
cattle fence; 1100 m high; 114 x 300 mm at bottom to 230 x 300 mm at top	1.36	0.10	2.05	-	1.61	m	3.66
sheep fence; 900 mm high; 140 x 300 mm at bottom to 230 x 300 mm at top	1.12	0.10	2.05	-	1.37	m	3.42
deer fence; 1900 mm high; 89 x 150 mm at bottom to 267 x 300 mm at top	2.86	0.13	2.56	-	3.11	m	5.68
Extra for concreting in posts	-	0.50	10.25	-	5.02	nr	15.27
Extra for straining post	10.35	0.75	15.38	-	10.35	nr	25.73
Rabbit netting; AVS Fencing Supplies Ltd; timber stakes; peeled kiln dried pressure treated; pointed; 1.8 m posts driven 900 mm into ground at 3 m centres (line wires and netting priced separately)							
75 -100 mm stakes	1.94	0.25	5.13	-	1.94	m	7.06
Corner posts or straining posts 150 mm diameter 2.3 m high set in concrete; centres to suit local conditions or changes of direction							
1 strut	12.69	1.00	20.50	11.88	20.26	each	52.64
2 strut	15.50	1.00	20.50	11.88	23.07	each	55.45
Strained wire; to posts (posts not included); 3 mm galvanized wire; fixing with galvanized stirrups							
900 mm high; 2 wire	0.17	0.03	0.68	-	0.33	m	1.01
1200 mm high; 3 wire	0.25	0.05	0.96	-	0.41	m	1.37
Rabbit netting; 31 mm; 19 gauge; 1050 mm high netting fixed to posts; line wires and straining posts or corner posts all priced separately							
900 high turned in	0.98	0.04	0.82	-	0.99	m	1.81
900 high buried 150 mm in trench	0.98	0.08	1.71	-	0.99	m	2.70
Boundary fencing; strained wire and wire mesh; Jacksons Fencing Tubular chain link fencing; galvanized; plastic coated; 60.3 mm diameter posts at 3.0 m centres; setting 700 mm into ground; choice of ten mesh colours; including excavating holes; backfilling and removing surplus soil; with top rail only							
900 mm high	-	-	-	-	-	m	36.91
1200 mm high	-	-	-	-	-	m	38.61
1800 mm high	-	-	-	-	-	m	41.92
2000 mm high	-	-	-	-	-	m	42.81

Q PAVING/PLANTING/FENCING/SITE FURNITURE

Item Excluding site overheads and profit	PC £	Labour hours	Labour £	Plant £	Material £	Unit	Total rate £
Q40 FENCING – cont'd							
Boundary fencing; strained wire and wire mesh; Jacksons Fencing – cont'd							
Tubular chain link fencing; galvanized; plastic coated; 60.3 mm diameter posts at 3.0 m centres; cranked arms and 3 lines barbed wire; setting 700 mm into ground; including excavating holes; backfilling and removing surplus soil; with top rail only							
1800 mm high	-	-	-	-	-	m	**43.23**
2000 mm high	-	-	-	-	-	m	**44.12**
Traditional trellis panels; The Garden Trellis Company; bespoke trellis panels for decorative, screening or security applications; timber planed all round; height of trellis 1800 mm							
Freestanding panels; posts 70 x 70 mm set in concrete; timber frames mitred and grooved 45 x 34 mm; heights and widths to suit; slats 32 x 10 mm joinery quality tanalized timber at 100 mm ccs; elements fixed by galvanized staples; capping rail 70 x 34 mm							
"HV68"; horizontal and vertical slats; softwood	72.00	1.00	20.50	-	78.70	m	**99.20**
"D68"; diagonal slats; softwood	86.40	1.00	20.50	-	93.10	m	**113.60**
"HV68"; horizontal and vertical slats; hardwood iroko	176.40	1.00	20.50	-	183.10	m	**203.60**
"D68"; diagonal slats; hardwood iroko or Western Red Cedar	201.60	1.00	20.50	-	208.30	m	**228.80**
Trellis panels fixed to face of existing wall or railings; timber frames mitred and grooved 45 x 34 mm; heights and widths to suit; slats 32 x 10 mm joinery quality tanalized timber at 100 mm ccs; elements fixed by galvanized staples; capping rail 70 x 34 mm							
"HV68"; horizontal and vertical slats; softwood	68.40	0.50	10.25	-	70.67	m	**80.92**
"D68"; diagonal slats; softwood	81.00	0.50	10.25	-	83.27	m	**93.52**
"HV68"; horizontal and vertical slats; iroko or Western Red Cedar	171.00	0.50	10.25	-	173.27	m	**183.52**
"D68"; diagonal slats; iroko or Western Red Cedar	189.00	0.50	10.25	-	191.27	m	**201.52**
Contemporary style trellis panels; The Garden Trellis Company; bespoke trellis panels for decorative, screening or security applications; timber planed all round							
Freestanding panels 30/15; posts 70 x 70 mm set in concrete with 90 x 30 mm top capping; slats 30 x 14 mm with 15 mm gaps; vertical support at 450 mm ccs							
joinery treated softwood	117.00	1.00	20.50	-	123.70	m	**144.20**
hardwood iroko or Western Red Cedar	198.00	1.00	20.50	-	204.70	m	**225.20**

Q PAVING/PLANTING/FENCING/SITE FURNITURE

Item Excluding site overheads and profit	PC £	Labour hours	Labour £	Plant £	Material £	Unit	Total rate £
Panels 30/15 face fixed to existing wall or fence; posts 70 x 70 mm set in concrete with 90 x 30 mm top capping; slats 30 x 14 mm with 15 mm gaps; vertical support at 450 mm ccs							
joinery treated softwood	108.00	0.50	10.25	-	110.27	m	**120.52**
hardwood iroko or Western Red Cedar	180.00	0.50	10.25	-	182.27	m	**192.52**
Integral arches to ornamental trellis panels; The Garden Trellis Company Arches to trellis panels in 45 x 34 mm grooved timbers to match framing; fixed to posts							
"R450"; 1/4 circle; joinery treated softwood	35.00	1.50	30.75	-	41.70	nr	**72.45**
"R450"; 1/4 circle; hardwood iroko or Western Red Cedar	45.00	1.50	30.75	-	51.70	nr	**82.45**
Arches to span 1800 mm wide							
joinery treated softwood	45.00	1.50	30.75	-	51.70	nr	**82.45**
hardwood iroko or Western Red Cedar	60.00	1.50	30.75	-	66.70	nr	**97.45**
Painting or staining of trellis panels; high quality coatings							
microporous opaque paint or spirit based stain	-	-	-	-	24.00	m^2	-
Concrete fencing Panel fencing; to precast concrete posts; in 2 m bays; setting posts 600 mm into ground; sandfaced finish							
900 mm high	11.74	0.25	5.13	-	13.72	m	**18.85**
1200 mm high	15.68	0.25	5.13	-	17.66	m	**22.78**
Panel fencing; to precast concrete posts; in 2 m bays; setting posts 750 mm into ground; sandfaced finish							
1500 mm high	20.46	0.33	6.83	-	22.44	m	**29.27**
1800 mm high	24.52	0.36	7.45	-	26.50	m	**33.95**
2100 mm high	28.39	0.40	8.20	-	30.37	m	**38.57**
2400 mm high	32.63	0.40	8.20	-	34.61	m	**42.81**
extra over capping panel to all of the above	6.72	-	-	-	6.72	m	-
Windbreak fencing Fencing; English Woodlands; "Shade and Shelter Netting" windbreak fencing; green; to 100 mm diameter treated softwood posts; setting 450 mm into ground; fixing with 50 x 25 mm treated softwood battens nailed to posts; including excavating and backfilling into firm ground; setting in concrete at 3 m centres							
1200 mm high	1.24	0.16	3.19	-	5.20	m	**8.39**
1800 mm high	1.76	0.16	3.19	-	5.72	m	**8.91**

Q PAVING/PLANTING/FENCING/SITE FURNITURE

Item Excluding site overheads and profit	PC £	Labour hours	Labour £	Plant £	Material £	Unit	Total rate £
Q40 FENCING – cont'd							
Gates - General Preamble: gates in fences; see specification for fencing, as gates in traditional or proprietary fencing systems are usually constructed of the same materials and finished as the fencing itself.							
Gates; hardwood; AVS Fencing Supplies Ltd Hardwood entrance gate; five bar diamond braced; curved hanging stile; planed iroko; fixed to 150 x 150 mm softwood posts; inclusive of hinges and furniture							
ref 1100 040; 0.9 m wide	211.08	5.00	102.50	-	267.96	nr	370.46
ref 1100 041; 1.2 m wide	225.06	5.00	102.50	-	281.94	nr	384.44
ref 1100 042; 1.5 m wide	293.35	5.00	102.50	-	350.23	nr	452.73
ref 1100 043; 1.8 m wide	310.63	5.00	102.50	-	367.51	nr	470.01
ref 1100 044; 2.1 m wide	340.12	5.00	102.50	-	397.00	nr	499.50
ref 1100 045; 2.4 m wide	320.10	5.00	102.50	-	376.98	nr	479.48
ref 1100 047; 3.0 m wide	390.95	5.00	102.50	-	447.83	nr	550.33
ref 1100 048; 3.3 m wide	408.34	5.00	102.50	-	465.22	nr	567.72
ref 1100 049; 3.6 m wide	422.28	5.00	102.50	-	479.16	nr	581.66
Hardwood field gate; five bar diamond braced; planed iroko; fixed to 150 x 150 mm softwood posts; inclusive of hinges and furniture							
ref 1100 100; 0.9 m wide	143.44	4.00	82.00	-	200.32	nr	282.32
ref 1100 101; 1.2 m wide	155.84	4.00	82.00	-	212.72	nr	294.72
ref 1100 102; 1.5 m wide	185.80	4.00	82.00	-	242.68	nr	324.68
ref 1100 103; 1.8 m wide	199.71	4.00	82.00	-	256.59	nr	338.59
ref 1100 104; 2.1 m wide	245.91	4.00	82.00	-	302.79	nr	384.79
ref 1100 105; 2.4 m wide	260.54	4.00	82.00	-	317.42	nr	399.42
ref 1100 106; 2.7 m wide	275.21	4.00	82.00	-	332.09	nr	414.09
ref 1100 108; 3.3 m wide	304.51	4.00	82.00	-	361.39	nr	443.39
ref 1100 109; 3.6 m wide	319.91	4.00	82.00	-	376.79	nr	458.79
Gates; softwood; Jacksons Fencing Timber field gates; treated softwood; including wrought iron ironmongery; five bar type; diamond braced; 1.80 m high; to 200 x 200 mm posts; setting 750 mm into firm ground							
width 2400 mm	81.63	10.00	205.00	-	202.92	nr	407.92
width 2700 mm	90.63	10.00	205.00	-	211.92	nr	416.92
width 3000 mm	97.56	10.00	205.00	-	218.85	nr	423.85
width 3300 mm	103.32	10.00	205.00	-	224.61	nr	429.61
Featherboard garden gates; treated softwood; including ironmongery; to 100 x 120 mm posts; 1 nr diagonal brace							
1.0 x 1.2 m high	25.11	3.00	61.50	-	108.05	nr	169.55
1.0 x 1.5 m high	32.94	3.00	61.50	-	118.08	nr	179.58
1.0 x 1.8 m high	36.90	3.00	61.50	-	126.41	nr	187.91

Q PAVING/PLANTING/FENCING/SITE FURNITURE

Item Excluding site overheads and profit	PC £	Labour hours	Labour £	Plant £	Material £	Unit	Total rate £
Picket garden gates; treated softwood; including ironmongery; to match picket fence; width 1000 mm; to 100 x 120 mm posts; 1 nr diagonal brace							
950 mm high	68.22	3.00	61.50	-	102.92	nr	**164.42**
1200 mm high	71.91	3.00	61.50	-	106.61	nr	**168.11**
1800 mm high	84.38	3.00	61.50	-	130.86	nr	**192.36**
Gates; tubular steel; Jacksons Fencing Tubular mild steel field gates; galvanized; including ironmongery; diamond braced; 1.80 m high; to tubular steel posts; setting in concrete							
width 3000 mm	125.37	5.00	102.50	-	228.76	nr	**331.26**
width 3300 mm	132.57	5.00	102.50	-	235.96	nr	**338.46**
width 3600 mm	139.91	5.00	102.50	-	243.30	nr	**345.80**
width 4200 mm	154.57	5.00	102.50	-	257.97	nr	**360.47**
Kissing gates Kissing gates; Jacksons Fencing; in galvanized metal bar; fixing to fencing posts (posts not included); 1.65 x 1.30 x 1.00 m high	221.40	5.00	102.50	-	221.40	nr	**323.90**
Stiles Stiles; 2 nr posts; setting into firm ground; 3 nr rails; 2 nr treads	72.00	3.00	61.50	-	85.39	nr	**146.89**

R DISPOSAL SYSTEMS

Item Excluding site overheads and profit	PC £	Labour hours	Labour £	Plant £	Material £	Unit	Total rate £
R12 DRAINAGE BELOW GROUND							
Silt pits and inspection chambers							
Excavating pits; starting from ground level; by machine							
maximum depth not exceeding 1.00 m	-	-	-	3.59	-	m^3	3.59
maximum depth not exceeding 2.00 m	-	-	-	4.06	-	m^3	4.06
Disposal of excavated material; depositing on site in permanent spoil heaps; average 50 m	-	-	-	6.96	-	m^3	6.96
Surface treatments; compacting; bottoms of excavations	-	0.05	1.02	-	-	m^2	1.02
Earthwork support; distance between opposing faces not exceeding 2.00 m							
maximum depth not exceeding 1.00 m	-	0.20	4.10	-	5.13	m^3	9.23
maximum depth not exceeding 2.00 m	-	0.30	6.15	-	5.13	m^3	11.28
maximum depth not exceeding 4.00 m	-	0.67	13.65	-	2.03	m^3	15.68
Silt pits and inspection chambers; in situ concrete							
Beds; plain in situ concrete; 11.50 N/mm^2 - 40 mm aggregate							
thickness not exceeding 150 mm	-	1.00	20.50	-	118.34	m^3	138.84
Benchings in bottoms; plain in situ concrete; 25.50 N/mm^2 - 20 mm aggregate							
thickness 150 - 450 mm	-	2.00	41.00	-	118.34	m^3	159.34
Isolated cover slabs; reinforced in situ concrete; 21.00 N/mm^2 - 20 mm aggregate							
thickness not exceeding 150 mm	-	4.00	82.00	-	118.34	m^3	200.34
Fabric reinforcement; A193 (3.02 kg/m^2) in cover slabs	5.13	0.06	1.28	-	5.64	m^2	6.92
Formwork to reinforced in situ concrete; isolated cover slabs							
soffits; horizontal	-	3.28	67.24	-	5.27	m^2	72.51
height not exceeding 250 mm	-	0.97	19.89	-	2.25	m	22.14
Brickwork							
Walls to manholes; bricks; PC £300.00/1000; in cement mortar (1:3)							
one brick thick	36.00	3.00	61.50	-	45.91	m^2	107.41
one and a half brick thick	54.00	4.00	82.00	-	68.86	m^2	150.86
two brick thick projection of footing or the like	72.00	4.80	98.40	-	91.82	m^2	190.22
Walls to manholes; engineering bricks; PC £260.00/1000; in cement mortar (1:3)							
one brick thick	34.80	3.00	61.50	-	44.65	m^2	106.15
one and a half brick thick	52.20	3.00	61.50	-	62.92	m^2	124.42
two brick thick projection of footing or the like	69.60	3.00	61.50	-	81.19	m^2	142.69
Extra over common or engineering bricks in any mortar for fair face; flush pointing as work proceeds; English bond walls or the like	-	0.13	2.73	-	-	m^2	2.73
In situ finishings; cement:sand mortar (1:3); steel trowelled; 13 mm one coat work to manhole walls; to brickwork or blockwork base; over 300 mm wide	-	0.80	16.40	-	2.70	m^2	19.10

R DISPOSAL SYSTEMS

Item Excluding site overheads and profit	PC £	Labour hours	Labour £	Plant £	Material £	Unit	Total rate £
Building into brickwork; ends of pipes; making good facings or renderings							
small	-	0.20	4.10	-	-	nr	4.10
large	-	0.30	6.15	-	-	nr	6.15
extra large	-	0.40	8.20	-	-	nr	8.20
extra large; including forming ring arch cover	-	0.50	10.25	-	-	nr	10.25
Inspection chambers; Hepworth Plc							
Polypropylene; up to 600 mm deep							
300 mm diameter; 960 mm deep; heavy duty round covers and frames; double seal recessed with 6 nr 110 mm outlets/inlets	277.02	3.00	61.50	-	280.16	nr	**341.66**
Polypropylene; up to 1200 mm deep							
475 mm diameter; 1030 mm deep; heavy duty round covers and frames; double seal recessed with 5 nr 110 mm outlets/inlets	112.69	4.00	82.00	-	358.79	nr	**440.79**
Step irons; Ashworth Ltd; drainage systems; malleable cast iron; galvanized; building into joints							
General purpose pattern; for one brick walls	4.39	0.17	3.48	-	4.39	nr	**7.87**
Best quality vitrified clay half section channels; Hepworth Plc; bedding and jointing in cement:mortar (1:2)							
Channels; straight							
100 mm	7.32	0.80	16.40	-	10.37	m	**26.77**
150 mm	12.19	1.00	20.50	-	15.24	m	**35.74**
225 mm	27.37	1.35	27.68	-	31.95	m	**59.62**
300 mm	56.19	1.80	36.90	-	59.24	m	**96.14**
Bends; 15, 30, 45 or 90°							
100 mm bends	6.59	0.75	15.38	-	8.11	nr	**23.49**
150 mm bends	11.38	0.90	18.45	-	14.44	nr	**32.89**
225 mm bends	44.17	1.20	24.60	-	47.22	nr	**71.82**
300 mm bends	90.06	1.10	22.55	-	93.88	nr	**116.43**
Intercepting traps							
Vitrified clay; inspection arms; brass stoppers; iron levers; chains and staples; galvanized; staples cut and pinned to brickwork; cement:mortar (1:2) joints to vitrified clay pipes and channels; bedding and surrounding in concrete; 11.50 N/mm^2 - 40 mm aggregate; cutting and fitting brickwork; making good facings							
100 mm inlet; 100 mm outlet	79.85	3.00	61.50	-	96.57	nr	**158.07**
150 mm inlet; 150 mm outlet	115.13	2.00	41.00	-	133.80	nr	**174.80**

R DISPOSAL SYSTEMS

Item Excluding site overheads and profit	PC £	Labour hours	Labour £	Plant £	Material £	Unit	Total rate £
R12 DRAINAGE BELOW GROUND – cont'd							
Excavating trenches; using 3 tonne tracked excavator; to receive pipes; grading bottoms; earthwork support; filling with excavated material to within 150 mm of finished surfaces and compacting; completing fill with topsoil; disposal of surplus soil							
Services not exceeding 200 mm nominal size							
average depth of run not exceeding 0.50 m	-	0.02	0.41	3.32	2.35	m	**6.08**
average depth of run not exceeding 0.75 m	-	0.03	0.68	4.94	2.35	m	**7.97**
average depth of run not exceeding 1.00 m	-	0.05	1.02	8.34	2.35	m	**11.71**
average depth of run not exceeding 1.25 m	-	0.10	2.05	10.99	2.35	m	**15.38**
Granular beds to trenches; lay granular material; to trenches excavated separately; to receive pipes (not included)							
300 mm wide x 100 mm thick							
reject sand	-	0.03	0.51	0.73	0.89	m	**2.13**
shingle 20 mm aggregate	-	0.03	0.51	0.73	1.22	m	**2.46**
sharp sand	-	0.03	0.51	0.73	1.22	m	**2.46**
300 mm wide x 150 mm thick							
reject sand	-	0.04	0.77	1.10	0.45	m	**2.31**
shingle 20 mm aggregate	-	0.04	0.77	1.10	0.61	m	**2.47**
sharp sand	-	0.04	0.77	1.10	0.61	m	**2.47**
Excavating trenches; using 3 tonne tracked excavator; to receive pipes; grading bottoms; earthwork support; filling with imported granular material type 2 and compacting; disposal of surplus soil							
Services not exceeding 200 mm nominal size							
average depth of run not exceeding 0.50 m	5.80	0.02	0.41	2.35	9.40	m	**12.16**
average depth of run not exceeding 0.75 m	8.70	0.03	0.51	2.74	14.10	m	**17.35**
average depth of run not exceeding 1.00 m	11.60	0.03	0.59	3.82	18.80	m	**23.21**
average depth of run not exceeding 1.25 m	14.50	0.03	0.59	6.42	23.50	m	**30.51**

R DISPOSAL SYSTEMS

Item Excluding site overheads and profit	PC £	Labour hours	Labour £	Plant £	Material £	Unit	Total rate £
Excavating trenches; using 3 tonne tracked excavator; to receive pipes; grading bottoms; earthwork support; filling with concrete ready mixed ST2; disposal of surplus soil							
Services not exceeding 200 mm nominal size							
average depth of run not exceeding 0.50 m	13.87	0.02	0.41	1.95	17.47	m	19.83
average depth of run not exceeding 0.75 m	20.80	0.03	0.51	2.74	26.20	m	29.45
average depth of run not exceeding 1.00 m	27.73	0.03	0.59	3.82	34.93	m	39.34
average depth of run not exceeding 1.25 m	34.67	0.03	0.59	6.42	43.67	m	50.68
Clay pipes and fittings; Hepworth Plc; "Supersleeve"							
100 mm clay pipes; polypropylene slip coupling; in trenches (trenches not included)							
laid straight	3.21	0.25	5.13	-	5.58	m	10.71
short runs under 3.00 m	3.21	0.31	6.41	-	5.58	m	11.99
Extra over 100 mm clay pipes for							
bends; 15-90°; single socket	10.70	0.25	5.13	-	10.70	nr	15.83
junction; 45 or 90°; double socket	22.57	0.25	5.13	-	22.57	nr	27.69
slip couplings polypropylene	3.79	0.08	1.71	-	3.79	nr	5.50
gully with "P" trap; 100 mm; 154 x 154 mm plastic grating	64.62	1.00	20.50	-	66.71	nr	87.21
PVC-u pipes and fittings; Wavin Plastics Ltd; "OsmaDrain" system							
110 mm PVC-u pipes; in trenches (trenches not included)							
laid straight	7.31	0.08	1.64	-	7.31	m	8.95
short runs under 3.00 m	7.31	0.12	2.46	-	7.31	m	9.77
Extra over 110 mm PVC-u pipes for							
bends; short radius	14.31	0.25	5.13	-	14.31	nr	19.44
bends; long radius	26.82	0.25	5.13	-	26.82	nr	31.95
junctions; equal; double socket	17.07	0.25	5.13	-	17.07	nr	22.20
slip couplings	-	0.10	2.05	-	8.25	nr	10.30
adaptors to clay	16.13	0.50	10.25	-	16.28	nr	26.53
Kerbs to gullies							
Kerbs to gullies; in one course Class B engineering bricks; to 4 nr sides; rendering in cement:mortar (1:3); dished to gully gratings	2.32	1.00	20.50	-	3.13	nr	23.63
Gullies; vitrified clay; Hepworth Plc; bedding in concrete; 11.50 N/mm^2 - 40 mm aggregate							
Vitrified clay yard gullies (mud); trapped; domestic duty (up to 1 tonne)							
"RGP5"; 100 mm outlet; 100 mm diameter; 225 mm internal width; 585 mm internal depth	124.61	3.50	71.75	-	125.28	nr	197.03

R DISPOSAL SYSTEMS

Item Excluding site overheads and profit	PC £	Labour hours	Labour £	Plant £	Material £	Unit	Total rate £
R12 DRAINAGE BELOW GROUND – cont'd							
Gullies; vitrified clay – cont'd							
Vitrified clay yard gullies (mud); trapped; medium duty (up to 5 tonnes)							
100 mm outlet; 100 mm diameter; 225 mm internal width; 585 mm internal depth	124.61	3.50	71.75	-	125.28	nr	**197.03**
150 mm outlet; 100 mm diameter; 225 mm internal width; 585 mm internal depth	193.01	3.50	71.75	-	193.69	nr	**265.44**
Combined filter and silt bucket for yard gullies							
225 mm wide	45.05	-	-	-	45.05	nr	-
Vitrified clay road gullies; trapped with rodding eye							
100 mm outlet; 300 mm internal diameter; 600 mm internal depth	119.85	3.50	71.75	-	120.53	nr	**192.28**
150 mm outlet; 300 mm internal diameter; 600 mm internal depth	122.73	3.50	71.75	-	123.41	nr	**195.16**
150 mm outlet; 400 mm internal dia, 750 mm internal depth	142.34	3.50	71.75	-	143.01	nr	**214.76**
150 mm outlet; 450 mm internal diameter; 900 mm internal depth	192.58	3.50	71.75	-	193.26	nr	**265.01**
Hinged gratings and frames for gullies; alloy							
120 x 120 mm	-	-	-	-	10.61	nr	-
150 x 150 mm	-	-	-	-	19.21	nr	-
193 mm for 150 mm diameter gully	-	-	-	-	29.85	nr	-
230 x 230 mm	-	-	-	-	35.12	nr	-
316 x 316 mm	-	-	-	-	93.11	nr	-
Hinged gratings and frames for gullies; cast iron							
265 mm for 225 mm diameter gully	-	-	-	-	59.66	nr	-
150 x 150 mm	-	-	-	-	19.21	nr	-
230 x 230 mm	-	-	-	-	35.12	nr	-
316 x 316 mm	-	-	-	-	93.11	nr	-
Universal gully trap; PVC-u; Wavin Plastics Ltd; "OsmaDrain" system; bedding in concrete; 11.50 N/mm^2 - 40 mm aggregate							
Universal gully fitting; comprising gully trap only							
110 mm outlet; 110 mm diameter; 205 mm internal depth	12.14	3.50	71.75	-	12.82	nr	**84.57**
Vertical inlet hopper c\w plastic grate							
272 x 183 mm	16.31	0.25	5.13	-	16.31	nr	**21.43**
Sealed access hopper							
110 x 110 mm	38.35	0.25	5.13	-	38.35	nr	**43.48**

R DISPOSAL SYSTEMS

Item Excluding site overheads and profit	PC £	Labour hours	Labour £	Plant £	Material £	Unit	Total rate £
Universal gully; PVC-u; Wavin Plastics Ltd; "OsmaDrain" system; accessories to universal gully trap							
Hoppers; backfilling with clean granular material; tamping; surrounding in lean mix concrete							
plain hopper; with 110 spigot 150 mm long	13.33	0.40	8.20	-	14.89	nr	**23.09**
vertical inlet hopper; with 110 spigot 150 mm long	16.31	0.40	8.20	-	17.87	nr	**26.07**
sealed access hopper; with 110 spigot 150 mm long	38.35	0.40	8.20	-	39.91	nr	**48.11**
plain hopper; solvent weld to trap	9.10	0.40	8.20	-	10.66	nr	**18.86**
vertical inlet hopper; solvent wed to trap	15.64	0.40	8.20	-	17.20	nr	**25.40**
sealed access cover; PVC-u	19.53	0.10	2.05	-	19.53	nr	**21.58**
Gullies PVC-u; Wavin Plastics Ltd; "OsmaDrain" system; bedding in concrete; 11.50 N/mm² - 40 mm aggregate							
Bottle gully; providing access to the drainage system for cleaning							
bottle gully; 228 x 228 x 317 mm deep	32.38	0.50	10.25	-	33.01	nr	**43.26**
sealed access cover; PVC-u; 217 x 217 mm	25.17	0.10	2.05	-	25.17	nr	**27.22**
grating; ductile iron; 215 x 215 mm	19.40	0.10	2.05	-	19.40	nr	**21.45**
bottle gully riser; 325 mm	4.17	0.50	10.25	-	5.53	nr	**15.78**
Yard gully; trapped; 300 mm diameter; 600 mm deep; including catchment bucket and ductile iron cover and frame; medium duty loading							
305 mm diameter; 600 mm deep	196.34	2.50	51.25	-	200.19	nr	**251.44**
Linear Drainage							
Loadings for slot drains A15 - 1.5 tonne; pedestrian use B125 - 12.5 tonne; domestic use C250 - 25 tonne; car parks, supermarkets, industrial units D400 - 40 tonne; highways E600 - 60 tonne; forklifts							
Linear drainage; channels; Aco Building Products; laid to concrete bed C25 on compacted granular base on 200 mm deep concrete bed; haunched with 200 mm concrete surround; all in 750 x 430 mm wide trench with compacted 200 mm granular base surround (excavation and sub-base not included)							
"Aco MultiDrain M100PPD"; recycled polypropolene drainage channel; range of gratings to complement installations which require discreet slot drainage							
142 mm wide x 150 mm deep	46.47	1.20	24.60	-	57.85	m	**82.45**
Accessories for "M100PPD"							
connectors; vertical outlet 110 mm	10.68	-	-	-	10.68	nr	**-**
connectors; vertical outlet 160 mm	10.68	-	-	-	10.68	nr	**-**
sump units; black plastic with plastic silt bucket; 110/160 mm inlet/outlets	84.47	1.50	30.75	-	105.40	nr	**136.15**
universal gully and bucket; 440 x 440 x 1315 mm deep	405.90	3.00	61.50	-	445.14	nr	**506.64**

R DISPOSAL SYSTEMS

Item Excluding site overheads and profit	PC £	Labour hours	Labour £	Plant £	Material £	Unit	Total rate £
R12 DRAINAGE BELOW GROUND – cont'd							
Linear drainage; channels – cont'd							
Aco drainlock gratings for "M100PPD" system to loading A15							
slotted galvanized steel	13.64	-	-	-	13.64	m	-
perforated galvanized steel	25.04	-	-	-	25.04	m	-
Aco drainlock gratings for "M100PPD" system to loading C250							
"Heelguard"; composite black; 500 mm long with security locking	34.03	-	-	-	34.03	m	-
intercept; ductile iron; 500 mm long	40.94	-	-	-	40.94	m	-
slotted galvanized steel; 1.00 m long	35.31	-	-	-	35.31	m	-
perforated galvanized steel; 1.00 m long	51.90	-	-	-	51.90	m	-
mesh galvanized steel; 1.00 m long	35.26	-	-	-	35.26	m	-
Aco MultiDrain MD brickslot; offset galvanized slot drain grating for "M100PPD"; load class C250							
brickslot galvanized steel; 1.00 m	67.20	-	-	-	67.20	m	-
Aco MultiDrain MD brickslot; offset galvanized slot drain grating for "M100PPD"; load class C250 - 400 Aco gratings for "M100Ppd" brickslot; galvanized steel; load class C250/D400; 1.00 m							
brickslot; galvanized steel; 1.00 m	89.13	-	-	-	89.13	m	-
brickslot; stainless steel; 1.00 m	134.77	-	-	-	134.77	m	-
Aco ProFile; one piece pre-galvanized steel slot drainage system							
Aco ProFile; load class C250; discreet slot drainage system ideal for visually sensitive paved areas	83.46	1.20	24.60	-	94.84	m	**119.44**
Extra over for laying of paving (PC £300.00/1000) adjacent to channel on epoxy mortar bed; stretcher bond	-	0.33	6.83	-	3.48	m	**10.31**
Channel footpath drain							
Aco channel drain to pedestrian footpaths; connection to existing kerbdrain; laying on 150 mm concrete base and surround; excavation sub-base not included	27.09	1.20	24.60	-	38.48	m	**63.08**

R DISPOSAL SYSTEMS

Item Excluding site overheads and profit	PC £	Labour hours	Labour £	Plant £	Material £	Unit	Total rate £
Channel drains; Aco Technologies Ltd; Aco MultiDrain MD polymer concrete channel drainage system; traditional channel and grate drainage solution							
Shallow depth channels "M100D" system; 135 mm wide							
100 mm deep	46.47	1.20	24.60	-	57.85	m	82.45
Accessories for MultiDrain MD							
sump unit; complete with sediment bucket	83.67	1.50	30.75	-	83.67	nr	114.42
end cap; closing piece	16.92	-	-	-	16.92	nr	-
extra over for laying of paving (PC £300.00/1000) adjacent to channel on epoxy mortar bed; stretcher bond	-	0.33	6.83	-	2.24	m	9.07
Channel footpath drain							
Aco channel drain to pedestrian footpaths; connection to existing kerbdrain; laying on 150 mm concrete base and surround; excavation sub-base not included	27.09	2.00	41.00	-	38.48	m	79.48
Access covers and frames							
FACTA (Fabricated Access Cover Trade Association) class: A - 0.5 tonne maximum slow moving wheel load AA - 1.5 tonne maximum slow moving wheel load AAA - 2.5 tonne maximum slow moving wheel load B - 5 tonne maximum slow moving wheel load C - 6.5 tonne maximum slow moving wheel load D - 11 tonne maximum slow moving wheel load							
Access covers and frames; solid top; galvanized; Steelway Brickhouse; "Bristeel"; bedding frame in cement mortar (1:3); cover in grease and sand; clear opening sizes; base size shown in brackets (50 mm depth)							
FACTA AA; single seal							
450 x 450 mm (520 x 520 mm)	75.52	1.50	30.75	-	85.21	nr	115.96
600 x 450 mm (670 x 520 mm)	84.91	1.80	36.90	-	95.68	nr	132.58
600 x 600 mm (670 x 670 mm)	91.52	2.00	41.00	-	104.44	nr	145.44
FACTA AA; double seal							
450 x 450 mm (560 x 560 mm)	129.99	1.50	30.75	-	139.67	nr	170.42
60 x 450 mm (710 x 560 mm)	144.04	1.80	36.90	-	154.80	nr	191.70
600 x 600 mm (710 x 710 mm)	158.82	2.00	41.00	-	171.73	nr	212.73
FACTA B; single seal							
450 x 450 mm (520 x 520 mm)	94.58	1.50	30.75	-	104.26	nr	135.01
600 x 450 mm (670 x 520 mm)	114.01	1.80	36.90	-	124.77	nr	161.67
600 x 600 mm (670 x 670 mm)	134.28	2.00	41.00	-	147.19	nr	188.19

R DISPOSAL SYSTEMS

Item Excluding site overheads and profit	PC £	Labour hours	Labour £	Plant £	Material £	Unit	Total rate £
R12 DRAINAGE BELOW GROUND – cont'd							
Access covers and frames; solid top – cont'd							
FACTA B; double seal							
450 x 450 mm (560 x 560 mm)	151.08	1.50	30.75	-	160.76	nr	**191.51**
600 x 450 mm (710 x 560 mm)	169.58	1.80	36.90	-	180.35	nr	**217.25**
600 x 600 mm (710 x 710 mm)	202.84	2.00	41.00	-	215.76	nr	**256.76**
Access covers and frames; recessed; galvanized; Steelway Brickhouse; "Bripave"; bedding frame in cement mortar (1:3); cover in grease and sand; clear opening sizes; base size shown in brackets (for block depths 50 mm, 65 mm, 80 mm or 100 mm)							
FACTA AA							
450 x 450 mm (562 x 562 mm)	204.74	1.50	30.75	-	214.43	nr	**245.18**
600 x 450 mm (712 x 562 mm)	229.33	1.80	36.90	-	240.09	nr	**276.99**
600 x 600 mm (712 x 712 mm)	239.51	2.00	41.00	-	483.55	nr	**524.55**
750 x 600 mm (862 x 712 mm)	272.94	2.40	49.20	-	537.32	nr	**586.52**
FACTA B							
450 x 450 mm (562 x 562 mm)	205.18	1.50	30.75	-	214.87	nr	**245.62**
600 x 450 mm (712 x 562 mm)	246.50	1.80	36.90	-	257.26	nr	**294.16**
600 x 600 mm (712 x 712 mm)	251.15	2.00	41.00	-	264.07	nr	**305.07**
750 x 600 mm (862 x 712 mm)	303.52	2.40	49.20	-	317.51	nr	**366.71**
FACTA D							
450 x 450 mm (562 x 562 mm)	235.89	1.50	30.75	-	245.58	nr	**276.33**
600 x 450 mm (712 x 562 mm)	253.69	1.80	36.90	-	264.45	nr	**301.35**
600 x 600 mm (712 x 712 mm)	263.73	2.00	41.00	-	276.64	nr	**317.64**
750 x 600 mm (862 x 712 mm)	331.61	2.40	49.20	-	345.60	nr	**394.80**
Extra over manhole frames and covers for filling recessed manhole covers with brick paviors; PC £305.00/1000	12.20	1.00	20.50	-	14.05	m^2	**34.55**
filling recessed manhole covers with vehicular paving blocks; PC £8.72/m^2	8.72	0.75	15.38	-	9.35	m^2	**24.73**
filling recessed manhole covers with concrete paving flags; PC £12.54/m^2	12.54	0.35	7.17	-	12.92	m^2	**20.10**
R13 LAND DRAINAGE							
Land drainage; excavating for drains; by 3 tonne tracked excavator; including disposing spoil to spoil heaps not exceeding 100 m							
Width 225 mm							
depth 450 mm	-	-	-	328.92	-	100 m	**328.92**
depth 600 mm	-	-	-	338.06	-	100 m	**338.06**
depth 700 mm	-	-	-	347.99	-	100 m	**347.99**
depth 900 mm	-	-	-	370.69	-	100 m	**370.69**
depth 1000 mm	-	-	-	383.74	-	100 m	**383.74**
depth 1200 mm	-	-	-	398.17	-	100 m	**398.17**

R DISPOSAL SYSTEMS

Item Excluding site overheads and profit	PC £	Labour hours	Labour £	Plant £	Material £	Unit	Total rate £
Land drainage; excavating for drains; by hand; including disposing spoil to spoil heaps not exceeding 100 m							
Width 150 mm							
depth 450 mm	-	22.03	451.62	-	-	100 m	**451.62**
depth 600 mm	-	29.38	602.29	-	-	100 m	**602.29**
depth 700 mm	-	34.27	702.53	-	-	100 m	**702.53**
depth 900 mm	-	44.06	903.23	-	-	100 m	**903.23**
Width 225 mm							
depth 450 mm	-	33.05	677.52	-	-	100 m	**677.52**
depth 600 mm	-	44.06	903.23	-	-	100 m	**903.23**
depth 700 mm	-	51.41	1053.90	-	-	100 m	**1053.90**
depth 900 mm	-	66.10	1355.05	-	-	100 m	**1355.05**
depth 1000 mm	-	146.88	3011.04	-	-	100 m	**3011.04**
Wavin Plastics Ltd; flexible plastic perforated pipes in trenches (not included); to a minimum depth of 450 mm (couplings not included)							
"OsmaDrain"; flexible plastic perforated pipes in trenches (not included); to a minimum depth of 450 mm (couplings not included)							
80 mm diameter; available in 100 m coil	68.00	2.00	41.00	-	68.00	100 m	**109.00**
100 mm diameter; available in 100 m coil	110.00	2.00	41.00	-	110.00	100 m	**151.00**
160 mm diameter; available in 35 m coil	266.50	2.00	41.00	-	273.16	100 m	**314.16**
"WavinCoil"; plastic pipe junctions							
80 x 80 mm	3.07	0.05	1.02	-	3.07	nr	**4.09**
100 x 100 mm	3.43	0.05	1.02	-	3.43	nr	**4.46**
100 x 60 mm	3.38	0.05	1.02	-	3.38	nr	**4.40**
100 x 80 mm	3.26	0.05	1.02	-	3.26	nr	**4.29**
160 x 160 mm	8.16	0.05	1.02	-	8.16	nr	**9.19**
"WavinCoil"; couplings for flexible pipes							
80 mm diameter	1.08	0.03	0.68	-	1.08	nr	**1.76**
100 mm diameter	1.19	0.03	0.68	-	1.19	nr	**1.87**
160 mm diameter	1.61	0.03	0.68	-	1.61	nr	**2.29**
Market prices of backfilling materials							
Sand	29.70	-	-	-	35.64	m³	-
Gravel rejects	31.50	-	-	-	34.65	m³	-
Topsoil; allowing for 20% settlement	26.00	-	-	-	31.20	m³	-
Land drainage; backfilling trench after laying pipes with gravel rejects or similar; blind filling with ash or sand; topping with 150 mm topsoil from dumps not exceeding 100 m; by machine							
Width 150 mm							
depth 450 mm	-	3.20	65.60	47.01	219.83	100 m	**332.45**
depth 600 mm	-	4.20	86.10	62.68	296.05	100 m	**444.84**
depth 750 mm	-	4.86	99.63	78.36	337.96	100 m	**515.94**
depth 900 mm	-	6.20	127.10	94.03	432.46	100 m	**653.59**
Width 225 mm							
depth 450 mm	-	4.80	98.40	72.09	297.76	100 m	**468.24**
depth 600 mm	-	6.30	129.15	94.03	444.23	100 m	**667.41**
depth 750 mm	-	7.80	159.90	117.50	297.76	100 m	**575.15**
depth 900 mm	-	9.30	190.65	144.17	297.76	100 m	**632.58**

R DISPOSAL SYSTEMS

Item Excluding site overheads and profit	PC £	Labour hours	Labour £	Plant £	Material £	Unit	Total rate £
R13 LAND DRAINAGE – cont'd							
Land drainage; backfilling trench after laying pipes with gravel rejects or similar; blind filling with ash or sand; topping with 150 mm topsoil from dumps not exceeding 100 m; by hand Width 150 mm							
depth 450 mm	-	18.63	381.92	-	194.09	100 m	**576.01**
depth 600 mm	-	24.84	509.22	-	282.01	100 m	**791.24**
depth 750 mm	-	31.05	636.52	-	265.42	100 m	**901.94**
depth 900 mm	-	37.26	763.83	-	418.42	100 m	**1182.25**

PART 6

Fees for Professional Services

LANDSCAPE ARCHITECTS' FEES

The Landscape Institute no longer sets any fee scales for its members. A publication which offers guidance in determining fees for different types of project is available from:

The Landscape Institute
33 Great Portland Street
London
W1W 8QG
Website: www.landscapeinstitute.org

Telephone: 020 7299 4500

The publication is entitled "Engaging a Landscape Consultant - Guidance for Clients on Fees 2002"

The document refers to suggested fee systems on the following basis:

- Time Charged Fee Basis
- Lump Sum Fee Basis
- Percentage Fee Basis
- Retainer Fee Basis

In regard to the 'Percentage Fee Basis' the publication lists various project types and relates them to complexity ratings for works valued at £22,500 and above. The suggested fee scales are dependent on the complexity rating of the project.

The following are samples of the Complexity Categories:

Category 1: Golf Courses, Country Parks and Estates, Planting Schemes

Category 2: Coastal, River and Agricultural Works, Roads, Rural Recreation Schemes

Category 3: Hospital Grounds, Sport Stadia, Urban Offices and Commercial Properties, Housing

Category 4: Domestic/Historic Garden Design, Urban Rehabilitation and Environmental Improvements

Table Showing Suggested Percentage Fees for Various Categories of Complexity of Landscape Design or Consultancy

Project Value £	Complexity			
	1	2	3	4
22,500	15.00%	16.50%	18.00%	21.00%
30,000	13.75	15.00	16.50	19.25
50,000	10.50	12.75	14.00	16.25
100,000	9.50	10.50	11.50	13.50
150,000	8.75	9.50	10.50	12.50
200,000	8.00	8.75	9.75	11.25
300,000	7.50	8.25	9.00	10.50
500,000	6.75	7.50	8.25	9.75
750,000	6.50	7.25	7.75	9.25
1,000,000	6.25	7.00	7.50	8.75
10,000,000	6.25	6.75	7.25	8.25

(extrapolated from the graph/curve chart in the aforementioned publication)

Guide to Stage Payments of Fees, Relevant Fee Basis and Proportion of Fee Applicable to Lump Sum and Percentage Fee Basis. *Details of Preliminary, Standard and Other Services are set out in detail in the Landscape Consultant's Appointment.*

Work Stage		Relevant Fee Basis			Proportion of Fee	
		Time	Lump	%	*Proportion of fee*	Total
Preliminary Services						
A	Inception	✓	✓	n/a	n/a	n/a
B	Feasibility	✓	✓	n/a	n/a	n/a
Standard Services						
C	Outline Proposals	✓	✓	✓	15%	15%
D	Sketch Scheme Proposals	✓	✓	✓	15%	30%
E	Detailed Proposals	✓	✓	✓	15%	45%
FG	Production Information	✓	✓	✓	20%	65%
HJ	Tender Action & Contract Preparation	✓	✓	✓	5%	70%
K	Operations on Site	✓	✓	✓	25%	95%
L	Completion	✓	✓	✓	5%	100%
Other Services		✓	✓	n/a		
Timing of Fee Payments Percentage fees are normally paid at the end of each work stage. Time based fees are normally paid at monthly intervals. Lump sum fees are normally paid at intervals by agreement. Retainer or term commission fees are normally paid in advance, for predetermined periods of service.						

WORKED EXAMPLES OF PERCENTAGE FEE CALCULATIONS

Worked Example 1

Project Type Caravan Site
Services Required To Detailed Proposals - Work Stages C to E
Budget £120,000

Step 1	Decide on Work Type and therefore Complexity Rating - Complexity Rating 2
Step 2	Decide on Services required and Proportion of Fee - To Detailed Proposals, 45%
Step 3	Read off Graph, Complexity Rating 2, the % fee of £120,000 - Graph Fee 9.9%
Step 4	Multiply the Proportion of Fee (45%) by the Graph Fee (9.9%) - Adjusted Fee - 4.46%
Step 5	Calculate the Guide Fee (4.46% of £120,000) - Guide Fee - £5,352
Step 6	Agree fee with Client, complete Memorandum of Agreement & Schedule of Services & Fees

Worked Example 2

Project Type New Housing
Services Required Full Standard Services - Work Stages C to L
Budget £350,000

Step 1	Decide on Work Type and therefore Complexity Rating - Complexity Rating 3
Step 2	Decide on Services required and Proportion of Fee - To Completion, 100%
Step 3	Read off Graph, Complexity Rating 3, the % fee of £350,000 - Graph Fee 8.8%
Step 4	Multiply the Proportion of Fee (100%) by the Graph Fee (8.8%) - Adjusted Fee - 8.8%
Step 5	Calculate the Guide Fee (8.8% of £350,000) - Guide Fee - £30,800
Step 6	Agree fee with Client, complete Memorandum of Agreement & Schedule of Services & Fees

Worked Example 3

Project Type Urban Environmental Improvements
Services Required To Production Information - Work Stages C to G
Budge £1,250,000

Step 1	Decide on Work Type and therefore Complexity Rating - Complexity Rating 4
Step 2	Decide on Services required and Proportion of Fee - To Production Information, 65%
Step 3	Read off Graph, Complexity Rating 4, the % fee of £1,250,000 - Graph Fee 8.6%
Step 4	Multiply the Proportion of Fee (65%) by the Graph Fee (8.6%) - Adjusted Fee - 5.59%
Step 5	Calculate the Guide Fee (5.59% of £1,250,000) - Guide Fee - £69,875
Step 6	Agree fee with Client, complete Memorandum of Agreement & Schedule of Services & Fees

PART 7

Rates of Wages

BUILDING INDUSTRY WAGES

Authorized rates of wages, etc., in the building industry in England, Wales and Scotland agreed by the Building and Civil Engineering Joint Negotiating Committee. Effective from 25 June 2008. Basic pay. Weekly rate based on 39 hours.

Craft Rate   £401.70
Skill Rate 1   £382.98
Skill Rate 2   £368.94
Skill Rate 3   £345.15
Skill Rate 4   £325.65
General Operative £302.25

The Prices for Measured Works used in this book are based upon the commercial wage rates currently used in the landscaping industry and are typical of the London area.
However, it is recognized that some contractors involved principally in soft landscaping and planting works may base their wages on the rates determined by the Agricultural Wages Board (England and Wales) and therefore the following information is given to assist readers to adjust the prices if necessary.

AGRICULTURAL WAGES, ENGLAND AND WALES

Minimum payments agreed by the Agricultural Wages Board for standard 39 hour, five day week. Effective 1st October 2008.

Minimum rates of pay are specified for 6 grades and 4 categories of workers:

       Grade 1 - Initial Grade
       Grade 2 - Standard Worker
       Grade 3 - Lead Worker
       Grade 4 - Craft Grade
       Grade 5 - Supervisory Grade
       Grade 6 - Farm Management Grade
       Full Time Flexible Worker
       Part Time Flexible Worker
       Apprentice
       Trainee

The following rates apply to workers who meet the conditions for the applicable grade who are not Full or Part-Time Flexible Workers.

Grade	Weekly Pay (£ per week)	Hourly Pay (£ per hour)	Overtime (£ per hour)
1 - of compulsory school age	-	2.87	4.31
1 - above compulsory school age	223.86	5.74	8.61
2	244.14	6.26	9.39
Grade	Weekly Pay (£ per week)	Hourly Pay (£ per hour)	Overtime (£ per hour)
3	268.71	6.89	10.34
4	288.21	7.39	11.09
5	304.98	7.82	11.73
6	329.55	8.45	12.68

Weekly rates apply to workers contractually required to work 39 basic hours per week; hourly rates apply to workers contractually required to work less than a 39 basic hours per week.

Overtime rates apply to all overtime hours worked by workers other than Full or Part-Time Flexible Workers -

- more than 39 basic hours in any week
- or more than 8 hours on any day
- or any hours beyond the normal working hours specified in a contract
- or on a Public Holiday
- or on a Sunday (applies only to workers engaged under a contract which started before 1 October 2007)

The rate of holiday pay should be calculated using the following principle:

$$\text{Holiday pay per day} = \frac{\text{Normal or average weekly pay}}{\text{Number of days worked weekly}}$$

Copies of the Agricultural Wages Order 2008 may be obtained from:

The Secretary to the Agricultural Wages Board for England and Wales
Area 7E, 9 Millbank
c/o 17, Smith Square
London SW1P 3JR
Telephone: 020 7238 6523
Fax: 020 7238 6553
Email: agriwages@arp.defra.gsi.gov.uk, or online from www.defra.gov.uk

AGRICULTURAL WAGES, SCOTLAND

New rules relating to agricultural wages and other conditions of service took effect from 1 January 2008. The new rules are contained in the Agricultural Wages (Scotland) Order (No. 56) 2008. The Order contains the detailed legal requirements for the calculations of minimum pay etc, but the Scottish Office has produced 'A Guide for Workers and Employers' which attempts to explain the new rules in simpler terms.

Both 'The Agricultural Wages (Scotland) Order (No. 56) 2008 and 'A Guide for Workers and Employers' may be obtained from:

Scottish Agricultural Wages Board
Pentland House
47 Robb's Loan
Edinburgh EH14 1TY
Telephone: 0131 244 6397
Fax: 0131 244 6551
Email: sawb@scotland.gsi.gov.uk

The Order applies equally to all workers employed in agriculture in Scotland. It makes no distinction between full-time employees, parttime employees and students.

The old classifications of general farm worker, shepherd, stockworker, tractorman and supervisory grade have been removed and replaced by a single category covering all agricultural workers. The one exception to this is that special arrangements have been made for hillshepherds.

Instead of minimum weekly rates of pay, the new Order works on minimum hourly rates of pay (with the exception of hillshepherds). It is very important to note that although wages are to be calculated on an hourly basis, a worker is still entitled to be paid for the full number of hours for which he is contracted (unless he is unavailable for work). In particular, this means that an employer cannot reduce the number of hours for which he is to be paid simply by e.g. sending him home early.

Minimum rates of wages
There are two main pay scales: one for workers who have been with their present employer for not more than 26 weeks and the other for workers who have been with their present employer for more than 26 weeks.

The minimum hourly rates of pay are as follows, effective from 1st October 2008:

All ages of worker	Minimum Hourly Rates (£ per hour)	
	Up to 26 weeks continuous employment	More than 26 weeks continuous employment
Minimum Basic Rate	5.73	6.20
Minimum Overtime Rate	8.59	9.30

Workers who have been with the same employer for more than 26 weeks and hold either:

(a) a Scottish, or National, Vocational Qualification in an agricultural subject at Level III or above, or
(b) an apprenticeship certificate approved by Lantra, NTO (formerly ATB Landbase), or a certificate of acquired experience issued by the ATB Landbase

shall be paid, for each hour worked, an additional sum of not less than £0.94.

Minimum hourly overtime rate
The minimum hourly rate of wages payable to a worker:

(a) for each hour worked in excess of 8 hours on any day, and
(b) for each hour worked in excess of 39 hours in any week

shall be calculated in accordance with the following formula:

M x 1.5 (where M = the minimum hourly rate of pay to which the worker is entitled)

Example of Computation using National Wage Awards - Building Craft and General Operatives
(not used for rate calculations in this book)

Effective from 30 June 2008

			Craft Operatives £	£	General Operatives £	£
Wages at standard basic rate						
Productive time	44.3	wks	401.70	17795.31	302.25	13389.68
Lost time allowance	0.9	wks	401.70	361.53	302.25	272.03
Overtime	0.0	wks	0.00	0.00	0.00	0.00
				18156.84		13661.70
National Working Rules	45.2	wks		-		-
Sick pay	1	wk		-		-
CITB allowance (0.50% of payroll)	1	year		102.43		77.07
Holiday pay	4.2	wks	401.70	1687.14	302.25	1269.45
Public Holiday	1.6	wks	401.70	642.72	302.25	483.60
Travelling allowance (say 30 km)	226	dys	2.62		2.62	
Fare allowance (say 30 km)	226	dys	6.45		6.45	
Employer's contributions to						
EasyBuild Stakeholder Pension (death and accident cover is provided free)	52	wks	5.00	260.00	5.00	260.00
National Insurance (average weekly payment)	48	wks	34.34	1648.32	22.35	1072.80
				22497.45		16824.62
Severance pay and sundry costs	Plus		1.5%	337.46	1.5%	252.37
				22834.91		17076.99
Employer's liability and third-party insurance	Plus		2.0%	456.70	2.0%	341.54
Total cost per annum				23291.61		17418.53
Total cost per hour				**£13.48**		**£10.08**

Notes:

1. Absence due to sickness has been assumed to be for periods not exceeding 3 days for which no payment is due (Working Rule 20.7.3).
2. EasyBuild Stakeholder Pension effective from 1 July 2002. Death and accident benefit cover is provided free of charge. Taken as £5.00/week average as range increased for 2006/9 wage award.
3. All N.I. Payments are at not-contracted out rates applicable from April 2009. National Insurance is paid for 48 complete weeks (52 wks - 4.2 wks) based on employer making regular monthly payments into the Template holiday pay scheme and by doing so the employer achieves National Insurance savings on holiday wages.
4. At the time of preparing the book, wage rates for June 2009 are deemed to be frozen.

PART 8

Daywork and Prime Cost

When work is carried out which cannot be valued in any other way it is customary to assess the value on a cost basis with an allowance to cover overheads and profit. The basis of costing is a matter for agreement between the parties concerned, but definitions of prime cost for the building industry have been prepared and published jointly by the Royal Institution of Chartered Surveyors and the National Federation of Building Trades Employers (now the Construction Confederation) for the convenience of those who wish to use them. These documents are reproduced with the permission of the Royal Institution of Chartered Surveyors, which owns the copyright.

The daywork schedule published by the Civil Engineering Contractors Association is included in the A & B's companion title, *Spons Civil Engineering and Highway Works Price Book*.

For larger Prime Cost contracts the reader is referred to the form of contract issued by the Royal Institute of British Architects.

BUILDING INDUSTRY

DEFINITION OF PRIME COST OF DAYWORK CARRIED OUT UNDER A BUILDING CONTRACT (JUNE 2007 - THIRD EDITION)

This definition of Prime Cost is published by the Royal Institution of Chartered Surveyors and the Construction Confederation, for convenience and for use by people who choose to use it. Members of the Construction Confederation are not in any way debarred from defining Prime Cost and rendering their accounts for work carried out on that basis in any way they choose. Building owners are advised to reach agreement with contractors on the Definition of Prime Cost to be used prior to issuing instructions.

INTRODUCTION

This new edition of the Definition includes two options for dealing with the prime cost of labour:

Option 'A' – Percentage Addition, is based upon the traditional method of pricing labour in daywork, and allows for a percentage addition to be made for incidental costs, overheads and profit, to the prime cost of labour applicable at the time the daywork is carried out.

Option 'B' – All-inclusive Rates, includes not only the prime cost of labour but also includes an allowance for incidental costs, overheads and profit. The all-inclusive rates are deemed to be fixed for the period of the contract. However, where a fluctuating price contract is used, or where the rates in the contract are to be index-linked, the all-inclusive rates shall be adjusted by a suitable index in accordance with the contract conditions.

Model documentation, intended for inclusion in a building contract, is included in Appendix A, which illustrates how the Definition of Prime Cost may be applied in practice.

Example calculations of the Prime Cost of Labour in Daywork are given in Appendix B.

SECTION 1 APPLICATION

1.1   This Definition provides a basis for the valuation of daywork executed under such building contracts as provide for its use.
1.2   It is not applicable in the case of daywork executed after the date of practical completion.
1.3   It is applicable to works carried out incidental to contract work but may not be deemed appropriate for use in 'daywork only' work or work carried out on an 'hourly' basis only, for which the 'Definition of Prime Cost of Building Works of a Jobbing or Maintenance Character' may be more suitable.
1.4   The terms 'contract' and 'contractor' herein shall be read as 'subcontract' and 'subcontractor' as applicable.
1.5   Dayworks are to be calculated by reference to the rate(s) current and prevailing on the day the work is carried out, except where Option 'B' for labour is used which may be adjusted by a suitable index in accordance with the contract conditions.

SECTION 2 COMPOSITION OF TOTAL CHARGES

2.1   The prime cost of daywork comprises the sum of the following costs:

   2.1.1   Labour as defined in Section 3.
   2.1.2   Material and goods as defined in Section 4.
   2.1.3   Plant as defined in Section 5.

2.2   Incidental costs, overheads and profit as defined in Section 6, as provided in the building contract and expressed therein as percentage adjustments are applicable to each of 2.1.1, (Option A for Labour – Section 3) – 2.1.3. NB: If using Option 'B' for the labour element of prime cost in Section 3, incidental costs, overheads and profit are deemed included.

SECTION 3 LABOUR

Option A – Percentage Addition

3.1   The prime cost of labour is defined in 3.5. Incidental costs, overheads and profit should be added as defined in Section 6.
3.2   The standard wage rates, payments and expenses referred to below and the standard working hours referred to in 3.3 are those laid down for the time being in the rules or decisions of the Construction Industry Joint Council (CIJC) and the terms of the Building and Civil Engineering Benefits Scheme (managed by the Building and Civil Engineering Holidays Scheme Management Ltd) applicable to the works, or the rules or decisions or agreements of such body, other than the CIJC, as may be applicable relating to the grade and type of operative concerned at the time when and in the area where the daywork is executed.
3.3   Hourly base rates for labour are computed by dividing the annual prime cost of labour, based upon standard working hours and as defined in 3.5, by the number of standard working hours per annum (see Example 1 on page 465).
3.4   The hourly rates computed in accordance with 3.3 shall be applied in respect of the time spent by operatives directly engaged on daywork, including those operating mechanical plant and transport and erecting and dismantling other plant (unless otherwise expressly provided in the building contract) and handling and distributing the materials and goods used in the daywork.
3.5   The annual prime cost of labour comprises the following:

   (a)   Standard or guaranteed minimum weekly earnings.*
   (b)   All other guaranteed minimum payments (unless included in Section 6). *
   (c)   Differentials or extra payments in respect of skill, responsibility, discomfort, inconvenience or risk (excluding those in respect of supervisory responsibility see 3.6). *
   (d)   Payments in respect of public holidays.
   (e)   Any amounts which may become payable by the Contractor to or in respect of operatives arising from the operation of the rules or decisions referred to in 3.2 which are not provided for in 3.5 (a) to (d) or in Section 6. *

(f) Employer's contributions to industry's annual holiday with pay scheme or payment in lieu thereof.
(g) Employer's contributions to industry's welfare benefits scheme or payment in lieu thereof.
(h) Employer's National Insurance contributions applicable to 3.5 (a) to (g).
(i) Any contribution, levy or tax imposed by statute, payable by the contractor in his capacity as an employer, or compliance with any legislation which has a direct effect on the cost of labour. *

3.6 Differentials or extra payments in respect of supervisory responsibility are excluded from the annual prime cost (see Section 6). The time of supervisory staff such as principals, foremen, gangers, leading hands and the like, when working manually, is admissible under this Section only at the appropriate standard/normal rates for the grade of operative suitable for the operation concerned.

3.7 An example calculation of a typical standard hourly base rate is provided in Example 1 on page 465.

Non-productive Overtime

3.8 * The prime cost for non-productive overtime should be based only on the hourly payments for items marked with an asterisk in 3.5. #

3.9 An example calculation of a typical non-productive overtime rate is précised in Example 2 on page 397.

Option B – All-inclusive Rates

3.10 The prime cost of labour is based on the all-inclusive rates for labour provided for in the building contract. The all-inclusive rates are to include all costs associated with employing the labour including all items listed in 3.5.

3.11 The all-inclusive hourly rates are also to include all costs, fixed and time-related charges, overheads and profit (as defined in Section 6) in connection with labour.

3.12 The all-inclusive hourly rates shall be applied in respect of the time actually spent by the operatives directly engaged on daywork, including those operating mechanical plant and transport and erecting and dismantling other plant (unless otherwise expressly provided in the building contract) and handling and distributing the materials and goods used in the daywork.

3.13 The time of supervisory staff, such as principals, foremen, gangers, leading hands and the like, when working manually, is admissible under this Section only at the appropriate all-inclusive hourly rates for the grade of operative suitable for the operations concerned. Any extra payment in respect of supervisory responsibility is not allowable.

3.14 The all-inclusive rates are deemed to be fixed for the period of the contract. However, where a fluctuating price contract is used, or where the rates in the contract are to be index-linked, the all-inclusive rates shall be adjusted by a suitable index in accordance with the contract conditions.

Non-productive Overtime

3.15 Allowance for non-productive overtime should be made in accordance with the Model Documentation included in Appendix A. #

SECTION 4 MATERIALS AND GOODS

4.1 The prime cost of materials and goods obtained specifically for the daywork is the invoice cost after deducting all trade discounts and any portion of cash discounts in excess of 5%, plus any appropriate handling and delivery charges.

4.2 The prime cost of materials and goods supplied from the Contractor's stock is based upon the current market prices after deducting all trade discounts and any portion of cash discounts in excess of 5%, plus any appropriate handling charges.

4.3 Any Value Added Tax which is treated, or is capable of being treated, as input tax (as defined in the Finance Act, 1972, or any re-enactment or amendment thereof or substitution thereof) by the Contractor is excluded, for the purpose of calculations.

SECTION 5 PLANT

5.1 Unless otherwise stated in the building contract, the prime cost of plant comprises the cost of the following:

(a) Use or hire of mechanical operated plant and transport for the time employed/engaged for the daywork.
(b) Use of nonmechanical plant (excluding nonmechanical hand tools) for the time employed/engaged for the daywork.
(c) Transport/delivery to and from site and erection and dismantling where applicable.
(d) Qualified professional operators (e.g. crane drivers) not employed by the contractor (see 5.5 below).

5.2 Where plant is hired, the prime cost of plant shall be the invoice cost after deducting all trade discounts and any portion of cash discount in excess of 5%.

5.3 Where plant is not hired, the prime cost of plant shall be calculated in accordance with the latest edition of the Royal Institution of Chartered Surveyor's (RICS) Schedule of Basic Plant Charges for Use in Connection with Daywork Under a Building Contract.

5.4 The use of nonmechanical hand tools and of erected scaffolding, staging, trestles or the like is excluded (see Section 6).

5.5 Where hired or other plant is operated by the Contractor's operatives, the operative's time is to be included under Section 3 unless otherwise provided in the contract.

5.6 Any Value Added Tax which is treated, or is capable of being treated, as input tax (as defined by the Finance Act, 1972, or any re-enactment or amendment thereof or substitution therefor) by the Contractor is excluded, for the purposes of calculation.

SECTION 6 INCIDENTAL COSTS, OVERHEADS AND PROFIT

6.1 The percentage adjustments provided in the building contract, which are applicable to each of the totals of Sections 3 (Option A), 4 and 5, include the following:

(a) Head Office charges.
(b) Site staff, including site supervision.
(c) The additional cost of overtime.
(d) Time lost due to inclement weather.
(e) The additional cost of bonuses and all other incentive payments in excess of any guaranteed minimum included in 3.5 (a).
(f) Apprentices' study time.
(g) Subsistence, lodging and periodic allowances.
(h) Fares and travelling allowances.
(i) Sick pay or insurance in respect thereof.
(j) Thirdparty and employers' liability insurance.
(k) Liability in respect of redundancy payments to employees.
(l) Employers' National Insurance contributions not included in Section 3.5.
(m) Tool allowances.
(n) Use and maintenance of non-mechanical hand tools.
(o) Use of erected scaffolding, staging, trestles or the like.
(p) Use of tarpaulins, plastic sheeting or the like, all necessary protective clothing, artificial lighting, safety and welfare facilities, storage and the like that may be available on the site.
(q) Any variation to basic rates required by the Contractor in cases where the building contract provides for the use of a specified schedule of basic plant charges (to the extent that no other provision is made for such variation – see Section 5).
(r) All other liabilities and obligations whatsoever not specifically referred to in this Section nor chargeable under any other Section.
(s) Any variation in welfare/pension payments from industry standard.
(t) Profit (including main contractor's profit as appropriate).

Non-productive Overtime

6.2 When calculating the percentage adjustment for incidental costs, overheads and profit, if the Option A calculation of price cost of labour is prescribed in the contract, it should be borne in mind that not all items listed in 6.1 are necessarily applicable to non-productive overtime. When Option B is prescribed, non-productive overtime should be shown separately in the contract documents as detailed in the Model Documentation in Appendix A

The additional cost of non-productive overtime, where specifically ordered by the Architect/Supervising Officer/Contract Administrator/Employer's Agent, shall only be chargeable on the terms of prior written agreement between the parties to the building contract.

APPENDIX A

Model Documentation for Inclusion in a Building Contract

This model document is included to illustrate how the Definition of Prime Cost may be applied in practice. It does not form part of the Definition. It is, however, in a form agreed between the RICS and the Construction Confederation and its use in this form amended only as required to suit the specific building contract is encouraged.

Where Using Option A for Labour

Dayworks

The Contractor will be paid as defined below for the cost of works carried out as daywork in accordance with the building contract.

For building works, the prime cost of daywork will be calculated in accordance with the latest *Definition of Prime Cost of Daywork carried out under a Building Contract,* (*State edition*_____), published by the Royal Institution of Chartered Surveyors and the Construction Confederation.

For electrical works, the prime cost of daywork will be calculated in accordance with the latest *Definition of Prime Cost of Daywork carried out under an Electrical Contract,* (*State edition*_____), published by the Royal Institution of Chartered Surveyors, the Electrical Contractors' Association and 'SELECT' the Electrical Contractors' Association of Scotland.

For heating and ventilating work etc, the prime cost of daywork will be calculated in accordance with the latest *Definition of Prime Cost of Daywork carried out under a Heating, Ventilating, Air-Conditioning, Refrigeration, Pipework and/or Domestic Engineering Contract,* (*State edition*_____), published by the Royal Institution of Chartered Surveyors and the Heating and Ventilating Contractors' Association.

For plumbing work, the prime cost of daywork will be calculated in accordance with the latest *Definition of Prime Cost of Daywork carried out under a Plumbing Contract,* (*State edition*_____), published by the Royal Institution of Chartered Surveyors, the Association of Plumbing and Heating Contractors and the Scottish and Northern Ireland Plumbing Employers' Confederation.*

* It is anticipated that the 1st Edition of this Definition will be published in 2007. Until such time, reference should be made to the April 1985 formula agreed between the Royal Institution of Chartered Surveyors, the National Association of Plumbing, Heating and Mechanical Services Contractors and the Scottish and Northern Ireland Plumbing Employers' Federation.

Labour

Building Operatives	Provisional Sum	£
Add for Incidental Costs, Overheads and Profit	............%	£
Electrical Operatives	Provisional Sum	£
Add for Incidental Costs, Overheads and Profit	............%	£
Heating and Ventilating Operatives	Provisional Sum	£
Add for Incidental Costs, Overheads and Profit	............%	£
Plumbing Operatives	Provisional Sum	£
Add for Incidental Costs, Overheads and Profit	............%	£

Non-productive Overtime

Building Operatives	Provisional Sum	£
Add for Incidental Costs, Overheads and Profit	............%	£
Electrical Operatives	Provisional Sum	£
Add for Incidental Costs, Overheads and Profit	............%	£
Heating and Ventilating Operatives	Provisional Sum	£
Add for Incidental Costs, Overheads and Profit	............%	£
Plumbing Operatives	Provisional Sum	£
Add for Incidental Costs, Overheads and Profit	............%	£

Where using Option B for Labour

Dayworks

The Contractor will be paid as defined below for the cost of works carried out as daywork in accordance with the building contract.

For building works, the prime cost of daywork will be calculated in accordance with the latest *Definition of Prime Cost of Daywork carried out under a Building Contract*, (*State edition*_____), published by the Royal Institution of Chartered Surveyors and the Construction Confederation.

For electrical works, the prime cost of daywork will be calculated in accordance with the latest *Definition of Prime Cost of Daywork carried out under an Electrical Contract*, (*State edition*_____), published by the Royal Institution of Chartered Surveyors, the Electrical Contractors' Association and 'SELECT' the Electrical Contractors' Association of Scotland.

For heating and ventilating work etc, the prime cost of daywork will be calculated in accordance with the latest *Definition of Prime Cost of Daywork carried out under a Heating, Ventilating, Air-Conditioning, Refrigeration, Pipework and/or Domestic Engineering Contract*, (*State edition*_____), published by the Royal Institution of Chartered Surveyors and the Heating and Ventilating Contractors' Association.

For plumbing work, the prime cost of daywork will be calculated in accordance with the latest *Definition of Prime Cost of Daywork carried out under a Plumbing Contract*, (*State edition*_____), published by the Royal Institution of Chartered Surveyors, the Association of Plumbing and Heating Contractors and the Scottish and Northern Ireland Plumbing Employers' Confederation.**

** *It is anticipated that the 1st Edition of this Definition will be published in 2007. Until such time, reference should be made to the April 1985 formula agreed between the Royal Institution of Chartered Surveyors, the National Association of Plumbing, Heating and Mechanical Services Contractors and the Scottish and Northern Ireland Plumbing employers' Federation.*

Labour

The Contractor must state below the all-inclusive prime cost hourly rates required for labour as defined in Section 3 (Option B) and the core working ours to which they apply.

Core Hours

General Operatives	£............ per hour
Skilled Operatives (all grades)	£............ per hour
Craft Operatives	£............ per hour
Other Grades/Trades:	
………………………………………..	£............ per hour
………………………………………..	£............ per hour
………………………………………..	£............ per hour
………………………………………..	£............ per hour
………………………………………..	£............ per hour

Core hours are ___am to ___pm Monday to Friday (excluding statutory holidays)

Overtime specifically ordered by the Architect/Supervising Officer/Contract Administrator/ Employer's Agent

The non-productive element of overtime should be as defined in the relevant Working Rule Agreement. However, if different, please state below.

Trade	Day	Time	Non-Productive Element (hours)
…………………………………	…………	………to………	…………………………………
…………………………………	…………	………to………	…………………………………
…………………………………	…………	………to………	…………………………………
…………………………………	…………	………to………	…………………………………
…………………………………	…………	………to………	…………………………………
…………………………………	…………	………to………	…………………………………
…………………………………	…………	………to………	…………………………………

Provide the all-inclusive prime cost of labour as defined in Section 3 (Option B)

Productive Hours
[ ] hours (Provisional) General Operatives @ £.........per hour £
[ ] hours (Provisional) General Operatives @ £.........per hour £
[ ] hours (Provisional) General Operatives @ £.........per hour £

Other Grades/Trades:
[ ] hours (Provisional) General Operatives @ £.........per hour £
[ ] hours (Provisional) General Operatives @ £.........per hour £
[ ] hours (Provisional) General Operatives @ £.........per hour £

Non-productive Hours
[ ] hours (Provisional) General Operatives @ £.........per hour £
[ ] hours (Provisional) General Operatives @ £.........per hour £
[ ] hours (Provisional) General Operatives @ £.........per hour £

Other Grades/Trades:
[ ] hours (Provisional) General Operatives @ £.........per hour £
[ ] hours (Provisional) General Operatives @ £.........per hour £
[ ] hours (Provisional) General Operatives @ £.........per hour £

Materials and Goods

Provide for the prime cost of materials and goods
as defined in Section 4 (Provisional)   £ [ ]

Add the percentage addition for incidental costs,
overheads and profit as defined in Section 6   _____%

Plant

Provide for the prime cost of plant hired by the
Contractor as defined in Section 5 (Provisional)   £ [ ]

Add the percentage addition for incidental costs,
overheads and profit as defined in Section 6   _____%

Rates for plant not hired by the Contractor shall be as set out in *The Schedule of Basic Plant Charges for Use in Connection with Daywork Under a Building Contract* published by the Royal Institution of Chartered Surveyors (_____ Edition dated _____)

Provide for the prime cost of plant not hired by the
Contractor, as defined in Section 5 (Provisional)   £ [ ]

Add the percentage addition for incidental costs,
overheads and profit as defined in Section 6   _____%

Daywork Rates - Building Operatives

APPENDIX B

Example Calculations of Prime Cost of Labour in Daywork

Example 1
Example of calculation of typical standard hourly base rate (as defined in Section 3) for CIJC Building Craft operative and General Operative based upon rates applicable from 30th June 2008.

		Rate (£)	Craft Operative	Rate (£)	General Operative
Basic Wages:	46.2 weeks	401.70	£18,558.54	302.25	£13,963.95
Extra Payments:	Where applicable	-	-		-
Subtotal:			£18,558.54		£13,963.95
National Insurance:	12.8% after the first £110.01 per week		£1,724.94		£1,136.83
Holidays with Pay:	226 hours	10.30	£2,327.80	7.75	£1,751.50
Welfare Benefit:	52 stamps	11.00	£572.00	11.00	£572.00
CITB Levy:	0.5% of payroll		£104.43		£78.58
Annual Labour Cost:			**£23,287.71**		**£17,502.86**
Hourly Base Rate:	Divide by 1802 hours		**£12.92**		**£9.71**

Note:

(1) Standard working hours per annum calculated as follows:

52 weeks @ 39 hours	2028
Less	
hours annual holiday	163
hours public holiday	63
Standard working hours per year	1802

(2) It has been assumed that employers who follow the CIJC Working Rules Agreement will match the employee pension contributions (part of welfare benefit) between £3.00 and £10.00 per week. Furthermore it has been assumed that employees have contributed £10.00 per week to the pension scheme and £1.00 per week for life insurance.

(3) It should be noted that all labour costs incurred by the Contractor in his capacity as an employer other than those contained in the hourly base rate, are to be taken into account under Section 6.

(4) The above example is for the convenience of users only and does not form part of the Definition; all the basic costs are subject to re-examination according to the time when and in the area where the daywork is executed.

Example 2
Example of calculation of typical non-productive overtime rate (as defined in section 3) for CIJC Building Craft Operative and General Operative based upon rates applicable 30th June 2008.

		Rate (£)	Craft Operative	Rate (£)	General Operative
Basic Wages:	46.2 weeks	401.70	£18,558.54	302.25	£13,963.95
Extra Payments:	Where applicable	-	-		-
Subtotal:			£18,558.54		£13,963.95
National Insurance:	12.8% after the first £110.01 per week		£1,724.94		£1,136.83
CITB Levy:	0.5% of payroll		£92.79		£69.82
Annual Labour Cost:			£20,376.27		£15,170.60
Hourly Base Rate:	Divide by 1802 hours		£11.31		£8.42

Note:

(1) Standard working hours per annum calculated as follows:

52 weeks @ 39 hours	2028
Less \	
hours annual holiday	163
hours public holiday	63
Standard working hours per year	1802

(2) It should be noted that all labour costs incurred by the Contractor in his capacity as an employer other than those contained in the hourly base rate, are to be taken into account under Section 6.

(3) The above example is for the convenience of users only and does not form part of the Definition; all the basic costs are subject to re-examination according to the time when and in the area where the daywork is executed.

PART 9
Tables and Memoranda

This part of the book contains the following sections:
Conversion Tables, page 468
Formulae, page 479
Earthwork, page 484
Concrete Work, page 491
Masonry, page 500
Timber, page 506
Metal, page 510
Kerbs/Edgings/Channels, page 511
Bases, page 513
Interlocking Brick/Block Roads/Pavings, page 513
Paving and Surfacing, page 514
Sports, page 518
Seeding/Turfing and Planting, page 520
Fencing and Gates, page 523
Drainage, page 528
Drainage Below Ground and Land Drainage, page 531
Piped Supply Systems, page 534
Electrical Supply/Power/Lighting Systems, page 535

CONVERSION TABLES
QUICK REFERENCE CONVERSION TABLES

	Imperial		Metric	
1. LINEAR				
0.039	in	1	mm	25.4
3.281	ft	1	metre	0.305
1.094	yd	1	metre	0.914
2. WEIGHT				
0.020	cwt	1	kg	50.802
0.984	ton	1	tonne	1.016
2.205	lb	1	kg	0.454
3. CAPACITY				
1.760	pint	1	litre	0.568
0.220	gal	1	litre	4.546
4. AREA				
0.002	in²	1	mm²	645.16
10.764	ft²	1	m²	0.093
1.196	yd²	1	m²	0.836
2.471	acre	1	ha	0.405
0.386	mile²	1	km²	2.59
5. VOLUME				
0.061	in³	1	cm³	16.387
35.315	ft³	1	m³	0.028
1.308	yd³	1	m³	0.765
6. POWER				
1.310	HP	1	kW	0.746

CONVERSION FACTORS - METRIC TO IMPERIAL

Multiply Metric	Unit	By	To obtain Imperial	Unit
Length				
kilometre	km	0.6214	statute mile	ml
metre	m	1.0936	yard	yd
centimetre	cm	0.0328	foot	ft
millimetre	mm	0.0394	inch	in
Area				
hectare	ha	2.471	acre	
square kilometre	km²	0.3861	square mile	sq ml
square metre	m²	10.764	square foot	sq ft
square metre	m²	1550	square inch	sq in
square centimetre	cm²	0.155	square inch	sq in
Volume				
cubic metre	m³	35.336	cubic foot	cu ft
cubic metre	m³	1.308	cubic yard	cu yd
cubic centimetre	cm³	0.061	cubic inch	cu in
cubic centimetre	cm³	0.0338	fluid ounce	fl oz
Liquid volume				
litre	l	0.0013	cubic yard	cu yd
litre	l	61.02	cubic inch	cu in
litre	l	0.22	Imperial gallon	gal
litre	l	0.2642	US gallon	US gal
litre	l	1.7596	pint	pt
Mass				
metric tonne	t	0.984	long ton	lg ton
metric tonne	t	1.102	short ton	sh ton
kilogram	kg	2.205	pound, avoirdupois	lb
gram	g or gr	0.0353	ounce, avoirdupois	oz
Unit mass				
kilograms/cubic metre	kg/m³	0.062	pounds/cubic foot	lbs/cu ft
kilograms/cubic metre	kg/m³	1.686	pounds/cubic yard	lbs/cu yd
tonnes/cubic metre	t/m³	1692	pound/cubic yard	lbs/cu yd
kilograms/sq centimetre	kg/cm²	14.225	pounds/square inch	lbs/sq in
kilogram-metre	kg.m	7.233	foot-pound	ft-lb

Multiply Metric	Unit	By	To obtain Imperial	Unit
Force				
meganewton	MN	9.3197	tons force	tonf
kilonewton	kN	225	pounds force	lbf
newton	N	0.225	pounds force	lbf
Pressure and stress				
meganewton per square metre	MN/m²	9.3197	tons force/square foot	tonf/ft²
kilopascal	kPa	0.145	pounds/square inch	psi
bar		14.5	pounds/square inch	psi
kilograme metre	kgm	7.2307	foot-pound	ft-lb
Energy				
kilocalorie	kcal	3.968	British thermal unit	Btu
metric horsepower	CV	0.9863	horsepower	hp
kilowatt	kW	1.341	horsepower	hp
Speed				
kilometres/hour	km/h	0.621	miles/hour	mph

CONVERSION FACTORS - IMPERIAL TO METRIC

Multiply Imperial	Unit	By	To obtain metric	Unit
Length				
statute mile	ml	1.609	kilometre	km
yard	yd	0.9144	metre	m
foot	ft	0.3048	metre	m
inch	in	25.4	millimetre	mm
Area				
acre	acre	0.4047	hectare	ha
square mile	sq ml	2.59	square kilometre	km^2
square foot	sq ft	0.0929	square metre	m^2
square inch	sq in	0.0006	square metre	m^2
square inch	sq in	6.4516	square centimetre	cm^2
Volume				
cubic foot	cu ft	0.0283	cubic metre	m^3
cubic yard	cu yd	0.7645	cubic metre	m^3
cubic inch	cu in	16.387	cubic centimetre	cm^3
fluid ounce	fl oz	29.57	cubic centimetre	cm^3
Liquid volume				
cubic yard	cu yd	764.55	litre	l
cubic inch	cu in	0.0164	litre	l
Imperial gallon	gal	4.5464	litre	l
US gallon	US gal	3.785	litre	l
US gallon	US gal	0.833	Imperial gallon	gal
pint	pt	0.5683	litre	l
Mass				
long ton	lg ton	1.016	metric tonne	tonne
short ton	sh ton	0.907	metric tonne	tonne
pound	lb	0.4536	kilogram	kg
ounce	oz	28.35	gram	g
Unit mass				
pounds/cubic foot	lb/cu ft	16.018	kilograms/cubic metre	kg/m^3
pounds/cubic yard	lb/cu yd	0.5933	kilograms /cubic metre	kg/m^3
pounds/cubic yard	lb/cu yd	0.0006	tonnes/cubic metre	t/m^3
foot-pound	ft-lb	0.1383	kilogram-metre	kg.m

Multiply Imperial	Unit	By	To obtain metric	Unit
Force				
tons force	tonf	0.1073	meganewton	MN
pounds force	lbf	0.0045	kilonewton	kN
pounds force	lbf	4.45	newton	N
Pressure and stress				
pounds/square inch	psi	0.1073	kilogram/sq. centimetre	kg/cm^2
pounds/square inch	psi	6.89	kilopascal	kPa
pounds/square inch	psi	0.0689	bar	
foot-pound	ft-lb	0.1383	kilogram metre	kgm
Energy				
British thermal unit	Btu	0.252	kilocalorie	kcal
horsepower (hp)	hp	1.014	metric horsepower	CV
horsepower (hp)	hp	0.7457	kilowatt	kW
Speed				
miles/hour	mph	1.61	kilometres/hour	km/h

Length

millimetre	mm	1 in	= 25.4 cm	1 mm	=	0.0394 in
centimetre	cm	1 in	= 2.54 cm	1 cm	=	0.3937 in
metre	m	1 ft	= 0.3048 m	1 m	=	3.2808 ft
		1 yd	= 0.9144 m	1 m	=	1.0936 yd
kilometre	km	1 mile	= 1.6093 km	1 km	=	0.6214 mile
	Note:	1 cm	= 10 mm	1 ft	=	12 in
		1 m	= 100 cm	1 yd	=	3 ft
		1 km	= 1,000 m	1 mile	=	1,760 yd

Area

square millimetre	mm²	1 in²	= 645.2 mm²	1 mm²	=	0.0016 in²
square centimetre	cm²	1 in²	= 6.4516 cm²	1 cm²	=	1.1550 in²
square metre	m²	1 ft²	= 0.0929 m²	1 m²	=	10.764 ft²
		1 yd²	= 0.8361 m²	1 m²	=	1.1960 yd²
square kilometre	km²	1 mile²	= 2.590 km²	1 km²	=	0.3861 mile²
	Note:	1 cm²	= 100 m²	1 ft²	=	144 in²
		1 m²	= 10,000 cm²	1 yd²	=	9 ft²
		1 km²	= 100 hectares	1 mile²	=	640 acres
				1 acre	=	4,840 yd²

Volume

cubic centimetre	cm³	1 cm³	= 0.0610 in³	1 in³	=	16.387 cm³
cubic decimetre	dm³	1 dm³	= 0.0353 ft³	1 ft³	=	28.329 dm³
cubic metre	m³	1 m³	= 35.3147 ft³	1 ft³	=	0.0283 m³
		1 m³	= 1.3080 yd³	1 yd³	=	0.7646 m³
litre	L	1 L	= 1.76 pint	1 pint	=	0.5683 L
		1 L	= 2.113 US pt	1 pint	=	0.4733 L
	Note:	1 dm³	= 1,000 cm³	1 ft³	=	1.728 in³
		1 m³	= 1,000 dm³	1 yd³	=	27 ft³
		1 L	= 1 dm³	1 pint	=	20 fl oz
		1 HL	= 100 L	1 gal	=	8 pints

Mass

milligram	mg	1 mg	= 0.0154 grain	1 grain	=	64.935 mg
gram	g	1 g	= 0.0353 oz	1 oz	=	28.35 g
kilogram	kg	1 kg	= 2.2046 lb	1 lb	=	0.4536 kg
tonne	t	1 t	= 0.9842 ton	1 ton	=	1.016 t

Note:
- 1 g = 1,000 mg
- 1 kg = 1000 g
- 1 t = 1,000 kg
- 1 oz = 437.5 grains
- 1 lb = 16 oz
- 1 stone = 14 lb
- 1 cwt = 112 lb
- 1 ton = 20 cwt

Force

newton	N	1 lbf	= 4.448 N	1 kgf	=	9.807 N
kilonewton	kN	1 lbf	= 0.00448 kN	1 ton f	=	9.964 kN
meganewton	MN	100 tonf	= 0.9964 MN			

Pressure and stress

kilonewton per square metre	kN/m²	1 lbf/in²	= 6.895 kN/m²
		1 bar	= 100 kN/m²
meganewton per square metre		1 tonf/ft²	= 107.3 kN/m² = 0.1073 MN/m²
	MN/m²	1 kgf/cm²	= 98.07 kN/m²
		1 lbf/ft²	= 0.04788 kN/m²

Temperature

degree Celsius °C

$$°C = \frac{5 \times (°F - 32)}{9} \qquad °F = \frac{(9 \times °C) + 32}{5}$$

Metric Equivalents		
1 km	=	1000 m
1 m	=	100 cm
1 cm	=	10 mm
1 km²	=	100 ha
1 ha	=	10,000 m²
1 m²	=	10,000 cm²
1 cm²	=	100 mm²
1 m³	=	1,000 litres
1 litre	=	1,000 cm³
1 metric tonne	=	1,000 kg
1 quintal	=	100 kg
1 N	=	0.10197 kg
1 kg	=	1000 g
1 g	=	1000 mg
1 bar	=	14.504 psi
1 cal	=	427 kg.m
1 cal	=	0.0016 cv.h
torque unit	=	0.00116 kw.h
1 CV	=	75 kg.m/s
1 kg/cm²	=	0.97 atmosphere
English Unit Equivalents		
1 mile	=	1760 yd
1 yd	=	3 ft
1 ft	=	12 in
1 sq mile	=	640 acres
1 acre	=	43,560 sq ft
1 sq ft	=	144 sq in
1 cu ft	=	7.48 gal liq
1 gal	=	231 cu in
	=	4 quarts liq
1 quart	=	32 fl oz
1 fl oz	=	1.80 cu in
	=	437.5 grains
1 stone	=	14 lb
1 cwt	=	112 lb
1 sh ton	=	2000 lb
1 lg ton	=	2240 lb
	=	20 cwt
1 lb	=	16 oz, avdp
1 Btu	=	778 ft lb
	=	0.000393 hph
	=	0.000293 kwh
1 hp	=	550 ft-lb/sec
1 atmosph	=	14.7 lb/in²

Power Units		
kW	=	Kilowatt
HP	=	Horsepower
CV	=	Cheval Vapeur (Steam Horsepower)
	=	French designation for Metric Horsepower
PS	=	Pferdestarke (Horsepower)
	=	German designation of Metric Horsepower
1 HP	=	1.014 CV = 1.014 PS
	=	0.7457 kW
1 PS	=	1 CV = .9863 HP
	=	0.7355 kW
1 kW	=	1.341 HP
	=	1.359 CV
	=	1.359 PS

SPEED CONVERSION

km/h	m/min	Mph	fpm
1	16.7	0.6	54.7
2	33.3	1.2	109.4
3	50.0	1.9	164.0
4	66.7	2.5	218.7
5	83.3	3.1	273.4
6	100.0	3.7	328.1
7	116.7	4.3	382.8
8	133.3	5.0	437.4
9	150.0	5.6	492.1
10	166.7	6.2	546.8
11	183.3	6.8	601.5
12	200.0	7.5	656.2
13	216.7	8.1	710.8
14	233.3	8.7	765.5
15	250.0	9.3	820.2
16	266.7	9.9	874.9
17	283.3	10.6	929.6
18	300.0	11.2	984.3
19	316.7	11.8	1038.9
20	333.3	12.4	1093.6
21	350.0	13.0	1148.3
22	366.7	13.7	1203.0
23	383.3	14.3	1257.7
24	400.0	14.9	1312.3
25	416.7	15.5	1367.0
26	433.3	16.2	1421.7
27	450.0	16.8	1476.4
28	466.7	17.4	1531.1
29	483.3	18.0	1585.7
30	500.0	18.6	1640.4
31	516.7	19.3	1695.1
32	533.3	19.9	1749.8
33	550.0	20.5	1804.5
34	566.7	21.1	1859.1
35	583.3	21.7	1913.8
36	600.0	22.4	1968.5
37	616.7	23.0	2023.2
38	633.3	23.6	2077.9
39	650.0	24.2	2132.5
40	666.7	24.9	2187.2

km/h	m/min	Mph	fpm
41	683.3	25.5	2241.9
42	700.0	26.1	2296.6
43	716.7	26.7	2351.3
44	733.3	27.3	2405.9
45	750.0	28.0	2460.6
46	766.7	28.6	2515.3
47	783.3	29.2	2570.0
48	800.0	29.8	2624.7
49	816.7	30.4	2679.4
50	833.3	31.1	2734.0

FORMULAE

Two dimensional figures

Figure	Diagram of figure	Surface area	Perimeter
Square		a^2	$4a$
Rectangle		ab	$2(a+b)$
Triangle		$\frac{1}{2}ch$ $\frac{1}{2}ab\sin C$ $\sqrt{\{s(s-a)(s-b)(s-c)\}}$ where $s=\frac{1}{2}(a+b+c)$	$a+b+c$
Circle		πr^2 $\frac{1}{4}\pi d^2$ where $2r=d$	$2\pi r$ πd
Parallelogram		ah	$2(a+b)$
Trapezium		$\frac{1}{2}h(a+b)$	$a+b+c+d$
Ellipse		Approximately πab	$\pi(a+b)$
Hexagon		$2.6 \times a^2$	

Figure	Diagram of figure	Surface area	Perimeter
Octagon		$4.83 \times a^2$	
Sector of a circle		$\frac{1}{2}rb$ or $\frac{q}{360}\pi r^2$ note $b =$ angle $\frac{q}{360} \times \pi 2r$	
Segment of a circle		$S - T$ where $S =$ area of sector, $T =$ area of triangle	
Bellmouth		$\frac{3}{14} \times r^2$	

Three dimensional figures

Figure	Diagram of figure	Surface area	Perimeter
Cube		$6a^2$	a^3
Cuboid/ rectangular block		$2(ab + ac + bc)$	abc
Prism/ triangular block		$bd + hc + dc + ad$	$\frac{1}{2}hcd$ $\frac{1}{2}ab \sin C \, d$ $d\sqrt{s(s-a)(s-b)(s-c)}$ where $s = \frac{1}{2}(a+b+c)$

Figure	Diagram of figure	Surface area	Perimeter
Cylinder		$2\pi rh + 2\pi r^2$ $\pi dh + \tfrac{1}{2}\pi d^2$	$\pi r^2 h$ $\tfrac{1}{4}\pi d^2 h$
Sphere		$4\pi r^2$	$\tfrac{4}{3}r^3$
Segment of sphere		$2\pi Rh$	$\tfrac{1}{6}\pi h(3r^2 + h^2)$ $\tfrac{1}{3}\pi h^2(3R - H)$
Pyramid		$(a + b)l + ab$	$\tfrac{1}{3}abh$
Frustum of a pyramid		$l(a + b + c + d) + \sqrt{(ab + cd)}$ [rectangular figure only]	$\tfrac{h}{3}(ab + cd + \sqrt{abcd})$

Figure	Diagram of figure	Surface area	Perimeter
Cone		$\pi r l + \pi r^2$ $\frac{1}{2}\pi dh + \frac{1}{4}\pi d^2$	$\frac{1}{3}\pi r^2 h$ $\frac{1}{12}\pi d^2 h$
Frustum of a cone		$\pi r^2 + \pi R^2 + \pi l(R+r)$	$\frac{1}{3}\pi(R^2 + Rr + r^2)$

Formula	Description
Pythagoras theorem	$A^2 = B^2 + C^2$ where A is the hypotenuse of a right-angled triangle and B and C are the two adjacent sides

Simpsons Rule	The Area is divided into an even number of strips of equal width, and therefore has an odd number of ordinates at the division points
	
	$$\text{area} = \frac{S(A + 2B + 4C)}{3}$$
	
	where S = common interval (strip width)
	            A = sum of first and last ordinates
	            B = sum of remaining odd ordinates
	            C = sum of the even ordinates
	
	The Volume can be calculated by the same formula, but by substituting the area of each co-ordinate rather than its length.

Trapezoidal Rule	A given trench is divided into two equal sections, giving three ordinates, the first, the middle and the last.
	
	$$\text{volume} = \frac{S \times (A + B + 2C)}{2}$$
	
	where S = width of the strips
	            A = area of the first section
	            B = area of the last section
	            C = area of the rest of the sections

Prismoidal Rule	A given trench is divided into two equal sections, giving three ordinates, the first, the middle and the last.
	
	$$\text{volume} = \frac{L \times (A + 4B + C)}{6}$$
	
	where L = total length of trench
	            A = area of the first section
	            B = area of the middle section
	            C = area of the last section

EARTHWORK

Weights of Typical Materials Handled by Excavators

The weight of the material is that of the state in its natural bed and includes moisture. Adjustments should be made to allow for loose or compacted states.

Material	kg/m³	lb/cu yd
Adobe	1914	3230
Ashes	610	1030
Asphalt, rock	2400	4050
Basalt	2933	4950
Bauxite: alum ore	2619	4420
Borax	1730	2920
Caliche	1440	2430
Carnotite	2459	4150
Cement	1600	2700
Chalk (hard)	2406	4060
Cinders	759	1280
Clay: dry	1908	3220
Clay: wet	1985	3350
Coal: bituminous	1351	2280
Coke	510	860
Conglomerate	2204	3720
Dolomite	2886	4870
Earth: dry	1796	3030
Earth: moist	1997	3370
Earth: wet	1742	2940
Feldspar	2613	4410
Felsite	2495	4210
Fluorite	3093	5220
Gabbro	3093	5220
Gneiss	2696	4550
Granite	2690	4540
Gravel, dry	1790	3020
Gypsum	2418	4080
Hardcore (consolidated)	1928	120
Lignite broken	1244	2100
Limestone	2596	4380
Magnesite, magnesium ore	2993	5050
Marble	2679	4520
Marl	2216	3740
Peat	700	1180

Material	kg/m³	lb/cu yd
Potash	2193	3700
Pumice	640	1080
Quarry waste	1438	90
Quartz	2584	4360
Rhyolite	2400	4050
Sand: dry	1707	2880
Sand: wet	1831	3090
Sand and gravel - dry	1790	3020
- wet	2092	3530
Sandstone	2412	4070
Schist	2684	4530
Shale	2637	4450
Slag (blast)	2868	4840
Slate	2667	4500
Snow - dry	130	220
- wet	510	860
Taconite	3182	5370
Topsoil	1440	2430
Trachyte	2400	4050
Traprock	2791	4710
Water	1000	62

Transport Capacities

Type of vehicle	Capacity of vehicle	
	Payload	Heaped capacity
Wheelbarrow	150	0.10
1 tonne dumper	1250	1.00
2.5 tonne dumper	4000	2.50
Articulated dump truck (Volvo A20 6x4)	18500	11.00
Articulated dump truck (Volvo A35 6x6)	32000	19.00
Large capacity rear dumper (Euclid R35)	35000	22.00
Large capacity rear dumper (Euclid R85)	85000	50.00

Machine Volumes for Excavating and Filling

Machine type	Cycles per minute	Volume per minute (m^3)
1.5 tonne excavator	1	0.04
	2	0.08
	3	0.12
3 tonne excavator	1	0.13
	2	0.26
	3	0.39
5 tonne excavator	1	0.28
	2	0.56
	3	0.84
7 tonne excavator	1	0.28
	2	0.56
	3	0.84
21 tonne excavator	1	1.21
	2	2.42
	3	3.63
Backhoe loader JCB3CX excavating Rear bucket capacity $0.28m^3$	1	0.28
	2	0.56
	3	0.84
Backhoe loader JCB3CX loading Front bucket capacity $1.00m^3$	1	1.00
	2	2.00

Machine Volumes for Excavating and Filling

Machine type	Loads per hour	Volume per hour (m³)
1 tonne high tip skip loader	5	2.43
Volume 0.485m³	7	3.40
	10	4.85
3 tonne dumper	4	7.60
Max. volume 2.40m³	5	9.50
Available volume 1.9m³	7	13.30
	10	19.00
6 tonne dumper	4	15.08
Max. volume 3.40m³	5	18.85
Available volume 3.77m³	7	26.39
	10	37.70

Bulkage of Soils (After excavation)

Type of soil	Approximate bulking of 1 m³ after excavation
Vegetable soil and loam	25 - 30%
Soft clay	30 - 40%
Stiff clay	10 - 20%
Gravel	20 - 25%
Sand	40 - 50%
Chalk	40 - 50%
Rock, weathered	30 - 40%
Rock, unweathered	50 - 60%

Shrinkage of Materials (On being deposited)

Type of soil	Approximate bulking of 1 m³ after excavation
Clay	10%
Gravel	8%
Gravel and sand	9%
Loam and light sandy soils	12%
Loose vegetable soils	15%

Voids in Material Used as Sub-bases or Beddings

Material	m³ of voids/m³
Alluvium	0.37
River grit	0.29
Quarry sand	0.24
Shingle	0.37
Gravel	0.39
Broken stone	0.45
Broken bricks	0.42

Angles of Repose

Type of soil		Degrees
Clay	- dry	30
	- damp, well drained	45
	- wet	15 - 20
Earth	- dry	30
	- damp	45
Gravel	- moist	48
Sand	- dry or moist	35
	- wet	25
Loam		40

Slopes and Angles

Ratio of base to height	Angle in degrees
5 : 1	11
4 : 1	14
3 : 1	18
2 : 1	27
1½ : 1	34
1 : 1	45
1 : 1½	56
1 : 2	63
1 : 3	72
1 : 4	76
1 : 5	79

Grades (In Degrees and Percents)

Degrees	Percent	Degrees	Percent
1	1.8	24	44.5
2	3.5	25	46.6
3	5.2	26	48.8
4	7.0	27	51.0
5	8.8	28	53.2
6	10.5	29	55.4
7	12.3	30	57.7
8	14.0	31	60.0
9	15.8	32	62.5
10	17.6	33	64.9
11	19.4	34	67.4
12	21.3	35	70.0
13	23.1	36	72.7
14	24.9	37	75.4
15	26.8	38	78.1
16	28.7	39	81.0
17	30.6	40	83.9
18	32.5	41	86.9
19	34.4	42	90.0
20	36.4	43	93.3
21	38.4	44	96.6
22	40.4	45	100.0

Bearing Powers

Ground conditions	Bearing power		
	kg/m²	lb/in²	Metric t/m²
Rock (broken)	483	70	50
Rock (solid)	2,415	350	240
Clay, dry or hard	380	55	40
medium dry	190	27	20
soft or wet	100	14	10
Gravel, cemented	760	110	80
Sand, compacted	380	55	40
clean dry	190	27	20
Swamp and alluvial soils	48	7	5

Earthwork Support

Maximum depth of excavation in various soils without the use of earthwork support:

Ground conditions	Feet (ft)	Metres (m)
Compact soil	12	3.66
Drained loam	6	1.83
Dry sand	1	0.3
Gravelly earth	2	0.61
Ordinary earth	3	0.91
Stiff clay	10	3.05

It is important to note that the above table should only be used as a guide. Each case must be taken on its merits and, as the limited distances given above are approached, careful watch must be kept for the slightest signs of caving in.

CONCRETE WORK

Weights of Concrete and Concrete Elements

Type of material			kg/m³	lb/cu ft
Ordinary concrete (dense aggregates)				
Non-reinforced plain or mass concrete				
Nominal weight			2305	144
Aggregate	- limestone		2162 to 2407	135 to 150
	- gravel		2244 to 2407	140 to 150
	- broken brick		2000 (av)	125 (av)
	- other crushed stone		2326 to 2489	145 to 155
Reinforced concrete				
Nominal weight			2407	150
Reinforcement	- 1%		2305 to 2468	144 to 154
	- 2%		2356 to 2519	147 to 157
	- 4%		2448 to 2703	153 to 163
Special concretes				
Heavy concrete				
Aggregates - barytes, magnetite			3210 (min)	200 (min)
steel shot, punchings			5280	330
Lean mixes				
Dry-lean (gravel aggregate)			2244	140
Soil-cement (normal mix)			1601	100

Weights of Concrete and Concrete Elements - cont'd

Type of material		kg/m² per mm thick	lb/sq ft per inch thick
Ordinary concrete (dense aggregates)			
Solid slabs (floors, walls etc.)			
Thickness:	75 mm or 3 in	184	37.5
	100 mm or 4 in	245	50
	150 mm or 6 in	378	75
	250 mm or 10 in	612	125
	300 mm or 12 in	734	150
Ribbed slabs			
Thickness:	125 mm or 5 in	204	42
	150 mm or 6 in	219	45
	225 mm or 9 in	281	57
	300 mm or 12 in	342	70
Special concretes			
Finishes etc			
	Rendering, screed etc		
	Granolithic, terrazzo	1928 to 2401	10 to 12.5
	Glass-block (hollow) concrete	1734 (approx)	9 (approx)
Prestressed concrete		Weights as for reinforced concrete (upper limits)	
Air-entrained concrete		Weights as for plain or reinforced concrete	

Average Weight of Aggregates

Materials	Voids %	Weight kg/m³
Sand	39	1660
Gravel 10 - 20 mm	45	1440
Gravel 35 - 75 mm	42	1555
Crushed stone	50	1330
Crushed granite (over 15 mm)	50	1345
(n.e. 15 mm)	47	1440
'All-in' ballast	32	1800 - 2000

Material	kg/m³	lb/cu yd
Vermiculite (aggregate)	64-80	108-135
All-in aggregate	1999	125

Applications and mix design
Site mixed concrete

Recommended mix	Class of work suitable for	Cement (kg)	Sand (kg)	Coarse aggregate (kg)	Nr 25 kg bags cement per m³ of combined aggregate
1:3:6	Roughest type of mass concrete such as footings, road haunching over 300 mm thick	208	905	1509	8.30
1:2.5:5	Mass concrete of better class than 1:3:6 such as bases for machinery, walls below ground etc.	249	881	1474	10.00
1:2:4	Most ordinary uses of concrete, such as mass walls above ground, road slabs etc. and general reinforced concrete work	304	889	1431	12.20
1:1.5:3	Watertight floors, pavements and walls, tanks, pits, steps, paths, surface of 2 course roads, reinforced concrete where extra strength is required	371	801	1336	14.90
1:1:2	sWork of thin section such as fence posts and small precast work	511	720	1206	20.40

Applications and mix design
Ready mixed concrete

Application	Designated concrete	Standardized prescribed concrete	Recommended consistence (nominal slump class)
Foundations			
Mass concrete fill or blinding	GEN 1	ST2	S3
Strip footings	GEN 1	ST2	S3
Mass concrete foundations			
Single storey buildings	GEN 1	ST2	S3
Double storey buildings	GEN 3	ST4	S3
Trench Fill foundations			
Single storey buildings	GEN 1	ST2	S4
Double storey buildings	GEN 3	ST4	S4
General applications			
Kerb bedding and haunching	GEN 0	ST1	S1
Drainage works - immediate support	GEN 1	ST2	S1
Other drainage works	GEN 1	ST2	S3
Oversite below suspended slabs	GEN 1	ST2	S3
Floors			
Garage and house floors with no embedded steel	GEN 3	ST4	S2
Wearing surface: Light foot and trolley traffic	RC30	ST4	S2
Wearing surface: General industrial	RC40	N/A	S2
Wearing surface: Heavy industrial	RC50	N/A	S2
Paving			
House drives, domestic parking and external parking	PAV 1	N/A	S2
Heavy-duty external paving	PAV 2	N/A	S2

Prescribed Mixes for Ordinary Structural Concrete

Weights of cement and total dry aggregates in kg to produce approximately one cubic metre of fully compacted concrete together with the percentages by weight of fine aggregate in total dry aggregates.

Conc. grade	Nominal max. size of aggregate (mm)	40		20		14		10	
	Workability	Med.	High	Med.	High	Med.	High	Med.	High
	Limits to slump that may be expected (mm)	50-100	100-150	25-75	75-125	10-50	50-100	10-25	25-50
7	Cement (kg)	180	200	210	230	-	-	-	-
	Total aggregate (kg)	1950	1850	1900	1800	-	-	-	-
	Fine aggregate (%)	30-45	30-45	35-50	35-50	-	-	-	-
10	Cement (kg)	210	230	240	260	-	-	-	-
	Total aggregate (kg)	1900	1850	1850	1800	-	-	-	-
	Fine aggregate (%)	30-45	30-45	35-50	35-50	-	-	-	-
15	Cement (kg)	250	270	280	310	-	-	-	-
	Total aggregate (kg)	1850	1800	1800	1750	-	-	-	-
	Fine aggregate (%)	30-45	30-45	35-50	35-50	-	-	-	-
20	Cement (kg)	300	320	320	350	340	380	360	410
	Total aggregate (kg)	1850	1750	1800	1750	1750	1700	1750	1650
	Sand								
	Zone 1 (%)	35	40	40	45	45	50	50	55
	Zone 2 (%)	30	35	35	40	40	45	45	50
	Zone 3 (%)	30	30	30	35	35	40	40	45
25	Cement (kg)	340	360	360	390	380	420	400	450
	Total aggregate (kg)	1800	1750	1750	1700	1700	1650	1700	1600
	Sand								
	Zone 1 (%)	35	40	40	45	45	50	50	55
	Zone 2 (%)	30	35	35	40	40	45	45	50
	Zone 3 (%)	30	30	30	35	35	40	40	45
30	Cement (kg)	370	390	400	430	430	470	460	510
	Total aggregate (kg)	1750	1700	1700	1650	1700	1600	1650	1550
	Sand								
	Zone 1 (%)	35	40	40	45	45	50	50	55
	Zone 2 (%)	30	35	35	40	40	45	45	50
	Zone 3 (%)	30	30	30	35	35	40	40	45

Weights of Bar Reinforcement

Nominal sizes (mm)	Cross-sectional area (mm^2)	Mass kg/m	Length of bar m/tonne
6	28.27	0.222	4505
8	50.27	0.395	2534
10	78.54	0.617	1622
12	113.1	0.888	1126
16	201.06	1.578	634
20	314.16	2.466	405
25	490.87	3.853	260
32	804.25	6.313	158
40	1265.64	9.865	101
50	1963.5	15.413	65

Weights of Bars at Specific Spacings

Weights of metric bars in kilograms per square metre.

Size (mm)	Spacing of bars in millimetres									
	75	100	125	150	175	200	225	250	275	300
6	2.96	2.220	1.776	1.480	1.27	1.110	0.99	0.89	0.81	0.74
8	5.26	3.95	3.16	2.63	2.26	1.97	1.75	1.58	1.44	1.32
10	8.22	6.17	4.93	4.11	3.52	3.08	2.74	2.47	2.24	2.06
12	11.84	8.88	7.10	5.92	5.07	4.44	3.95	3.55	3.23	2.96
16	21.04	15.78	12.63	10.52	9.02	7.89	7.02	6.31	5.74	5.26
20	32.88	24.66	19.73	16.44	14.09	12.33	10.96	9.87	8.97	8.22
25	51.38	38.53	30.83	25.69	22.02	19.27	17.13	15.41	14.01	12.84
32	84.18	63.13	50.51	42.09	36.08	31.57	28.06	25.25	22.96	21.04
40	131.53	98.65	78.92	65.76	56.37	49.32	43.84	39.46	35.87	32.88
50	205.51	154.13	123.31	102.76	88.08	77.07	68.50	61.65	56.05	51.38

Basic weight of steelwork taken as 7850 kg/m^3
Basic weight of bar reinforcement per metre run = 0.00785 kg/mm^2
The value of PI has been taken as 3.141592654

Fabric Reinforcement

Fabric reference	Longitudinal wires			Cross wires			Mass
	Nominal wire size (mm)	Pitch (mm)	Area (mm/m^2)	Nominal wire size (mm)	Pitch (mm)	Area (mm/m^2)	(kg/m^2)
Square mesh							
A393	10	200	393	10	200	393	6.16
A252	8	200	252	8	200	252	3.95
A193	7	200	193	7	200	193	3.02
A142	6	200	142	6	200	142	2.22
A98	5	200	98	5	200	98	1.54
Structural mesh							
B1131	12	100	1131	8	200	252	10.90
B785	10	100	785	8	200	252	8.14
B503	8	100	503	8	200	252	5.93
B385	7	100	385	7	200	193	4.53
B283	6	100	283	7	200	193	3.73
B196	5	100	196	7	200	193	3.05
Long mesh							
C785	10	100	785	6	400	70.8	6.72
C636	9	100	636	6	400	70.8	5.55
C503	8	100	503	5	400	49.00	4.34
C385	7	100	385	5	400	49.00	3.41
C283	6	100	283	5	400	49.00	2.61
Wrapping mesh							
D98	5	200	98	5	200	98	1.54
D49	2.5	100	49	2.5	100	49	0.77
Stock sheet size	Length 4.8 m		Width 2.4 m		Sheet area 11.52m^2		

Wire

SWG	6g	5g	4g	3g	2g	1g	1/0g	2/0g	3/0g	4/0g	5/0g
Diameter in mm	0.192	0.212	0.232	0.252	0.276	0.300	0.324	0.348	0.372	0.400	0.432
	4.9	5.4	5.9	6.4	7.0	7.6	8.2	8.8	9.5	0.2	1.0
Area in mm^2	0.029	0.035	0.042	0.050	0.060	0.071	0.082	0.095	0.109	0.126	0.146
	19	23	27	32	39	46	53	61	70	81	95

Average Weight (kg/m³) of Steelwork Reinforcement in Concrete for Various Building Elements

	kg/m³ concrete
Substructure	
Pile caps	110 - 150
Tie beams	130 - 170
Ground beams	230 - 330
Bases	90 - 130
Footings	70 - 110
Retaining walls	110 - 150
Superstructure	
Slabs - one way	75 - 125
Slabs - two way	65 - 135
Plate slab	95 - 135
Cantilevered slab	90 - 130
Ribbed floors	80 - 120
Columns	200 - 300
Beams	250 - 350
Stairs	130 - 170
Walls - normal	30 - 70
Walls - wind	50 - 90

Note: For exposed elements add the following %: Walls 50%, Beams 100%, Columns 15%

Formwork Stripping Times – Normal Curing Periods

Conditions under which concrete is maturing	Minimum periods of protection for different types of cement					
	Number of days (where the average surface temperature of the concrete exceeds 10°C during the whole period)			Equivalent maturity (degree hours) calculated as the age of the concrete in hours multiplied by the number of degrees Celsius by which the average surface temperature of the concrete exceeds 10°C		
	Other	SRPC	OPC or RHPC	Other	SRPC	OPC or RHPC
1. Hot weather or drying winds	7	4	3	3500	2000	1500
2. Conditions not covered by 1	4	3	2	2000	1500	1000

KEY
OPC - Ordinary Portland Cement
RHPC - Rapid-hardening Portland cement
SRPC - Sulphate-resisting Portland cement

Minimum period before striking formwork

	Minimum period before striking		
	Surface temperature of concrete		
	16° C	17° C	t° C(0-25)
Vertical formwork to columns, walls and large beams	12 hours	18 hours	300 hours t+10
Soffit formwork to slabs	4 days	6 days	100 days t+10
Props to slabs	10 days	15 days	250 days t+10
Soffit formwork to beams	9 days	14 days	230 days t+10
Props to beams	14 days	21 days	360 days t+10

MASONRY

Weights of Bricks and Blocks

Walls and components of walls	kg/m² per mm thick	lb/sq ft per inch thick
Blockwork		
Hollow clay blocks; average)	1.15	6
Common clay blocks	1.90	10
Brickwork		
Engineering clay bricks	2.30	12
Refractory bricks	1.15	6
Sand-lime (and similar) bricks	2.02	10.5

Weights of Stones

Type of stone	kg/m3	lb/cu ft
Natural stone (solid)		
Granite	2560 to 2927	160 to 183
Limestone - Bathstone	2081	130
- Marble	2723	170
- Portland stone	2244	140
Sandstone	2244 to 2407	140 to 150
Slate	2880	180
Stone rubble (packed)	2244	140

Quantities of Bricks and Mortar

Materials per m² of wall:		
Thickness	No. of Bricks	Mortar m³
Half brick (112.5 mm)	58	0.022
One brick (225 mm)	116	0.055
Cavity, both skins (275 mm)	116	0.045
1.5 brick (337 mm)	174	0.074
Mass brickwork per m³	464	0.36

Mortar Mixes: Quantities of Dry Materials

	Imperial cu yd			Metric m³		
Mix	Cement cwts	Lime cwts	Sand cu yds	Cement tonnes	Lime tonnes	Sand cu m
1:3	7.0	-	1.04	0.54	-	1.10
1:4	6.3	-	1.10	0.40	-	1.20
1:1:6	3.9	1.6	1.10	0.27	0.13	1.10
1:2:9	2.6	2.1	1.10	0.20	0.15	1.20
0:1:3	-	3.3	1.10	-	0.27	1.00

Mortar Mixes for Various Uses

Mix	Use
1:3	Construction designed to withstand heavy loads in all seasons.
1:1:6	Normal construction not designed for heavy loads. Sheltered and moderate conditions in spring and summer. Work above d:p:c - sand, lime bricks, clay blocks, etc.
1:2:9	Internal partitions with blocks which have high drying shrinkage, pumice blocks, etc. any periods.
0:1:3	Hydraulic lime only should be used in this mix and may be used for construction not designed for heavy loads and above d:p:c spring and summer.

Quantities of Bricks and Mortar Required per m² of Walling

Description	Unit	Nr of bricks required	Mortar required (m³)		
			No frogs	Single frogs	Double frogs
Standard bricks **Brick size** **215 x 102.5 x 50 mm**					
half brick wall (103 mm) (103 mm)	m²	72	0.022	0.027	0.032
2 x half brick cavity wall (270 mm)	m²	144	0.044	0.054	0.064
one brick wall (215 mm)	m²	144	0.052	0.064	0.076
one and a half brick wall (328 mm)	m²	216	0.073	0.091	0.108
mass brickwork	m³	576	0.347	0.413	0.480
Brick size **215 x 102.5 x 65 mm**					
half brick wall (103 mm)	m²	58	0.019	0.022	0.026
2 x half brick cavity wall (270 mm)	m²	116	0.038	0.045	0.055
one brick wall (215 mm)	m²	116	0.046	0.055	0.064
one and a half brick wall (328 mm)	m²	174	0.063	0.074	0.088
mass brickwork	m³	464	0.307	0.360	0.413
Metric modular bricks - Perforated **Brick co-ordinating size** **200 x 100 x 75 mm**					
90 mm thick	m²	67	0.016	0.019	
190 mm thick	m²	133	0.042	0.048	
290 mm thick	m²	200	0.068	0.078	
Brick co-ordinating size **200 x 100 x 100 mm**					
90 mm thick	m²	50	0.013	0.016	
190 mm thick	m²	100	0.036	0.041	
290 mm thick	m²	150	0.059	0.067	
Brick co-ordinating size **300 x 100 x 75 mm**					
90 mm thick	m²	33	-	0.015	
Brick co-ordinating size **300 x 100 x 100 mm**					
90 mm thick	m²	44	0.015	0.018	

Note: Assuming 10 mm deep joints

Mortar Required per m² Blockwork (9.88 blocks/m²)

Wall thickness	75	90	100	125	140	190	215
Mortar m³/m²	0.005	0.006	0.007	0.008	0.009	0.013	0.014

Mortar Mixes

Mortar group	Cement: lime: sand	Masonry cement: sand	Cement: sand with plasticizer
1	1 : 0-0.25:3		
2	1 : 0.5 :4-4.5	1 : 2.5-3.5	1 : 3-4
3	1 : 1:5-6	1 : 4-5	1 : 5-6
4	1 : 2:8-9	1 : 5.5-6.5	1 : 7-8
5	1 : 3:10-12	1 : 6.5-7	1 : 8

Group 1: strong inflexible mortar
Group 5: weak but flexible

All mixes within a group are of approximately similar strength.
Frost resistance increases with the use of plasticisers.
Cement:lime:sand mixes give the strongest bond and greatest resistance to rain penetration.
Masonry cement equals ordinary Portland cement plus a fine neutral mineral filler and an air entraining agent.

Calcium Silicate Bricks

Type	Strength	Location
Class 2 crushing strength	14.0N/mm2	not suitable for walls
Class 3	20.5N/mm2	walls above dpc
Class 4	27.5N/mm2	cappings and copings
Class 5	34.5N/mm2	retaining walls
Class 6	41.5N/mm2	walls below ground
Class 7	48.5N/mm2	walls below ground

The Class 7 calcium silicate bricks are therefore equal in strength to Class B bricks.
Calcium silicate bricks are not suitable for DPCs.

Durability of Bricks

FL	Frost resistant with low salt content
FN	Frost resistant with normal salt content
ML	Moderately frost resistant with low salt content
MN	Moderately frost resistant with normal salt content

Brickwork Dimensions

No. of horizontal bricks	Dimensions (mm)	No. of vertical courses	No. of vertical courses
1/2	112.5	1	75
1	225.0	2	150
1 1/2	337.5	3	225
2	450.0	4	300
2 1/2	562.5	5	375
3	675.0	6	450
3 1/2	787.5	7	525
4	900.0	8	600
4 1/2	1012.5	9	675
5	1125.0	10	750
5 1/2	1237.5	11	825
6	1350.0	12	900
6 1/2	1462.5	13	975
7	1575.0	14	1050
7 1/2	1687.5	15	1125
8	1800.0	16	1200
8 1/2	1912.5	17	1275
9	2025.0	18	1350
9 1/2	2137.5	19	1425
10	2250.0	20	1500
20	4500.0	24	1575
40	9000.0	28	2100
50	11250.0	32	2400
60	13500.0	36	2700
75	16875.0	40	3000

Standard Available Block Sizes

Block	Co-ordinating size Length x height (mm)	Work size Length x height (mm)	Thicknesses (work size) (mm)
A	400 x 100	390 x 90	75, 90, 100, 140, 190
	400 x 200	440 x 190	75, 90, 100, 140, 190
	450 x 225	440 x 215	75, 90, 100, 140, 190, 215
B	400 x 100	390 x 90	75, 90, 100, 140, 190
	400 x 200	390 x 190	75, 90, 100, 140, 190
	450 x 200	440 x 190	75, 90, 100, 140, 190, 215
	450 x 225	440 x 215	75, 90, 100, 140, 190, 215
	450 x 300	440 x 290	75, 90, 100, 140, 190, 215
	600 x 200	590 x 190	75, 90, 100, 140, 190, 215
	600 x 225	590 x 215	75, 90, 100, 140, 190, 215
C	400 x 200	390 x 190	60, 75
	450 x 200	440 x 190	60, 75
	450 x 225	440 x 215	60, 75
	450 x 300	440 x 290	60, 75
	600 x 200	590 x 190	60, 75
	600 x 225	590 x 215	60, 75

TIMBER
Weights of Timber

Material	kg/m³	lb/cu ft
General	806 (avg)	50 (avg)
Douglas fir	479	30
Yellow pine, spruce	479	30
Pitch pine	673	42
Larch, elm	561	35
Oak (English)	724 to 959	45 to 60
Teak	643 to 877	40 to 55
Jarrah	959	60
Greenheart	1040 to 1204	65 to 75
Quebracho	1285	80
Material	**kg/m² per mm thickness**	**lb/sq ft per inch thickness**
Wooden boarding and blocks		
Softwood	0.48	2.5
Hardwood	0.76	4
Hardboard	1.06	5.5
Chipboard	0.76	4
Plywood	0.62	3.25
Blockboard	0.48	2.5
Fibreboard	0.29	1.5
Wood-wool	0.58	3
Plasterboard	0.96	5
Weather boarding	0.35	1.8

Conversion Tables (for sawn timber only)

Inches	>	Millimetres	Feet	>	Metres
1		25	1		0.300
2		50	2		0.600
3		75	3		0.900
4		100	4		1.200
5		125	5		1.500
6		150	6		1.800
7		175	7		2.100
8		200	8		2.400
9		225	9		2.700
10		250	10		3.000
11		275	11		3.300
12		300	12		3.600
13		325	13		3.900
14		350	14		4.200
15		375	15		4.500
16		400	16		4.800
17		425	17		5.100
18		450	18		5.400
19		475	19		5.700
20		500	20		6.000
21		525	21		6.300
22		550	22		6.600
23		575	23		6.900
24		600	24		7.200

Planed Softwood

The finished end section size of planed timber is usually 3/16" less than the original size from which it is produced. This however varies slightly dependent upon availability of material and origin of species used.

Standard (timber) to Cubic Metres and Cubic Metres to Standard (timber)

m³	m³/Standards	Standard
4.672	1	0.214
9.344	2	0.428
14.017	3	0.642
18.689	4	0.856
23.361	5	1.070
28.033	6	1.284
32.706	7	1.498
37.378	8	1.712
42.05	9	1.926
46.722	10	2.140
93.445	20	4.281
140.167	30	6.421
186.890	40	8.561
233.612	50	10.702
280.335	60	12.842
327.057	70	14.982
373.779	80	17.122
420.502	90	19.263
467.224	100	21.403

Standards (timber) to Cubic Metres and Cubic Metres to Standards (timber)

1 cu metre	=	35.3148 cu ft	=	0.21403 std
1 cu ft	=	0.028317 cu metres		
1 std	=	4.67227 cu metres		

Basic Sizes of Sawn Softwood Available (cross sectional areas)

Thickness (mm)	Width (mm)								
	75	100	125	150	175	200	225	250	300
16	x	x	x	x					
19	x	x	x	x					
22	x	x	x	x					
25	x	x	x	x	x	x	x	x	x
32	x	x	x	x	x	x	x	x	x
36	x	x	x	x					
38	x	x	x	x	x	x	x		
44	x	x	x	x	x	x	x	x	x
47*	x	x	x	x	x	x	x	x	x
50	x	x	x	x	x	x	x	x	x
63	x	x	x	x	x	x	x		
75		x	x	x	x	x	x	x	x
100		x		x		x		x	x
150				x		x			x
200						x			
250								x	
300									x

* This range of widths for 47 mm thickness will usually be found to be available in construction quality only.

Note: The smaller sizes below 100 mm thick and 250 mm width are normally but not exclusively of European origin. Sizes beyond this are usually of North and South American origin.

Basic Lengths of Sawn Softwood Available (metres)

1.80	2.10	3.00	4.20	5.10	6.00	7.20
	2.40	3.30	4.50	5.40	6.30	
	2.70	3.60	4.80	5.70	6.60	
		3.90			6.90	

Note: Lengths of 6.00 m and over will generally only be available from North American species and may have to be re-cut from larger sizes.

Reductions From Basic Size to Finished Size of Timber By Planing of Two Opposed Faces

Purpose	15 – 35 mm	36 – 100 mm	101 – 150 mm	Over 150 mm
a) constructional timber	3 mm	3 mm	5 mm	6 mm
b) matching interlocking boards	4 mm	4 mm	6 mm	6 mm
c) wood trim not specified in BS 584	5 mm	7 mm	7 mm	9 mm
d) joinery and cabinet work	7 mm	9 mm	11 mm	13 mm

Note: The reduction of width or depth is overall the extreme size and is exclusive of any reduction of the face by the machining of a tongue or lap joints.

METAL

Weights of Metals

Material	kg/m³	lb/cu ft
Metals, steel construction, etc.		
Iron		
- cast	7207	450
- wrought	7687	480
- ore - general	2407	150
- (crushed) Swedish	3682	230
Steel	7854	490
Copper		
- cast	8731	545
- wrought	8945	558
Brass	8497	530
Bronze	8945	558
Aluminium	2774	173
Lead	11322	707
Zinc (rolled)	7140	446
	g/mm² per metre	**lb/sq ft per foot**
Steel bars	7.85	3.4
Structural steelwork	Net weight of member @ 7854 kg/m³	
riveted	+ 10% for cleats, rivets, bolts, etc	
welded	+ 1.25% to 2.5% for welds, etc	
Rolled sections		
beams	+ 2.5%	
stanchions	+ 5% (extra for caps and bases)	
Plate		
web girders	+ 10% for rivets or welds, stiffeners, etc	
	kg/m	**lb/ft**
Steel stairs: industrial type 1 m or 3 ft wide	84	56
Steel tubes 50 mm or 2 in bore	5 to 6	3 to 4
Gas piping 20 mm or 3/4 in	2	1¼

KERBS/EDGINGS/CHANNELS

Precast Concrete Kerbs to BS 7263
Straight kerb units: length from 450 to 915 mm

150 mm high x 125 mm thick		
bullnosed	type BN	
half battered	type HB3	
255 mm high x 125 mm thick		
45° splayed	type SP	
half battered	type HB2	
305 mm high x 150 mm thick		
half battered	type HB1	
Quadrant kerb units		
150 mm high x 305 and 455 mm radius to match	type BN	type QBN
150 mm high x 305 and 455 mm radius to match	type HB2, HB3	type QHB
150 mm high x 305 and 455 mm radius to match	type SP	type QSP
255 mm high x 305 and 455 mm radius to match	type BN	type QBN
255 mm high x 305 and 455 mm radius to match	type HB2, HB3	type QHB
225 mm high x 305 and 455 mm radius to match	type SP	type QSP
Angle kerb units		
305 x 305 x 225 mm high x 125 mm thick		
bullnosed external angle	type XA	
splayed external angle to match type SP	type XA	
bullnosed internal angle	type IA	
splayed internal angle to match type SP	type IA	
Channels		
255 mm wide x 125 mm high flat	type CS1	
150 mm wide x 125 mm high flat type	CS2	
255 mm wide x 125 mm high dished	type CD	

Transition kerb units
from kerb type SP to HB	left handed	type TL
	right handed	type TR
from kerb type BN to HB	left handed	type DL1
	right handed	type DR1
from kerb type BN to SP	left handed	type DL2
	right handed	type DR2

Radial Kerbs and Channels All profiles of kerbs and channels

External radius (mm)	Internal radius (mm)
1000	3000
2000	4500
3000	6000
4500	7500
6000	9000
7500	1050
9000	1200
1050	
1200	

Precast Concrete Edgings

Round top type ER	Flat top type EF	Bullnosed top type EBN
150 x 50 mm	150 x 50 mm	150 x 50 mm
200 x 50	200 x 50	200 x 50
250 x 50	250 x 50	250 x 50

BASES

Cement Bound Material for Bases and Sub-bases

CBM1:	very carefully graded aggregate from 37.5 - 75 ym, with a 7-day strength of 4.5 N/mm2
CBM2:	same range of aggregate as CBM1 but with more tolerance in each size of aggregate with a 7-day strength of 7.0 N/mm2
CBM3:	crushed natural aggregate or blast furnace slag, graded from 37.5 mm - 150 ym for 40 mm aggregate, and from 20 - 75 ym for 20 mm aggregate, with a 7-day strength of 10 N/mm2
CBM4:	crushed natural aggregate or blast furnace slag, graded from 37.5 mm - 150 ym for 40 mm aggregate, and from 20 - 75 ym for 20 mm aggregate, with a 7-day strength of 15 N/mm2

INTERLOCKING BRICK/BLOCK ROADS/PAVINGS

Sizes of Precast Concrete Paving Blocks

Type R blocks	**Type S**
200 x 100 x 60 mm	Any shape within a 295 mm space
200 x 100 x 65	
200 x 100 x 80	
200 x 100 x 100	

Sizes of Clay Brick Pavers
200 x 100 x 50 mm thick
200 x 100 x 65
210 x 105 x 50
210 x 105 x 65
215 x 102.5 x 50
215 x 102.5 x 65

Type PA: 3 kN
Footpaths and pedestrian areas, private driveways, car parks, light vehicle traffic and over-run.

Type PB: 7 kN
Residential roads, lorry parks, factory yards, docks, petrol station forecourts, hardstandings, bus stations.

PAVING AND SURFACING

Weights and Sizes of Paving and Surfacing

Description of item	Size	Quantity per tonne
Paving 50 mm thick	900 x 600 mm	15
Paving 50 mm thick	750 x 600 mm	18
Paving 50 mm thick	600 x 600 mm	23
Paving 50 mm thick	450 x 600 mm	30
Paving 38 mm thick	600 x 600 mm	30
Path edging	914 x 50 x 150 mm	60
Kerb (including radius and tapers)	125 x 254 x 914 mm	15
Kerb (including radius and tapers)	125 x 150 x 914 mm	25
Square channel	125 x 254 x 914 mm	15
Dished channel	125 x 254 x 914 mm	15
Quadrants	300 x 300 x 254 mm	19
Quadrants	450 x 450 x 254 mm	12
Quadrants	300 x 300 x 150 mm	30
Internal angles	300 x 300 x 254 mm	30
Fluted pavement channel	255 x 75 x 914 mm	25
Corner stones	300 x 300 mm	80
Corner stones	360 x 360 mm	60
Cable covers	914 x 175 mm	55
Gulley kerbs	220 x 220 x 150 mm	60
Gulley kerbs	220 x 200 x 75 mm	120

Weights and Sizes of Paving and Surfacing

Material	kg/m³	lb/cu yd
Tarmacadam	2306	3891
Macadam (waterbound)	2563	4325
Vermiculite (aggregate)	64-80	108-135
Terracotta	2114	3568
Cork - compressed	388	24
	kg/m²	lb/sq ft
Clay floor tiles, 12.7 mm	27.3	5.6
Pavement lights	122	25
Damp proof course	5	1
	kg/m² per mm thickness	lb/sq ft per inch thickness
Paving Slabs (stone)	2.3	12
Granite setts	2.88	15
Asphalt	2.30	12
Rubber flooring	1.68	9
Poly-vinylchloride	1.94 (avg)	10 (avg)

Coverage (m²) Per Cubic Metre of Materials Used as Sub-bases or Capping Layers

Consolidated thickness laid in (mm)	Square metre coverage		
	Gravel	Sand	Hardcore
50	15.80	16.50	-
75	10.50	11.00	-
100	7.92	8.20	7.42
125	6.34	6.60	5.90
150	5.28	5.50	4.95
175	-	-	4.23
200	-	-	3.71
225	-	-	3.30
300	-	-	2.47

Approximate Rate of Spreads

Average thickness of course mm	Description	Approximate rate of spread			
		Open Textured		Dense, Medium & Fine Textured	
		kg/m^2	m^2/t	kg/m^2	m^2/t
35	14 mm open textured or dense wearing course	60-75	13-17	70-85	12-14
40	20 mm open textured or dense base course	70-85	12-14	80-100	10-12
45	20 mm open textured or dense base course	80-100	10-12	95-100	9-10
50	20 mm open textured or dense, or 28 mm dense base course	85-110	9-12	110-120	8-9
60	28 mm dense base course, 40 mm open textured of dense base course or 40 mm single course as base course		8-10	130-150	7-8
65	28 mm dense base course, 40 mm open textured or dense base course or 40 mm single course	100-135	7-10	140-160	6-7
75	40 mm single course, 40 mm open textured or dense base course, 40 mm dense roadbase	120-150	7-8	165-185	5-6
100	40 mm dense base course or roadbase	-	-	220-240	4-4.5

Surface Dressing Roads: Coverage (m²) per Tonne of Material

Size in mm	Sand	Granite chips	Gravel	Limestone chips
Sand	168	-	-	-
3	-	148	152	165
6	-	130	133	144
9	-	111	114	123
13	-	85	87	95
19	-	68	71	78

Sizes of Flags

Reference	Nominal size (mm)	Thickness (mm)
A	600 x 450	50 and 63
B	600 x 600	50 and 63
C	600 x 750	50 and 63
D	600 x 900	50 and 63
E	450 x 450	50 and 70 chamfered top surface
F	400 x 400	50 and 65 chamfered top surface
G	300 x 300	50 and 60 chamfered top surface

Sizes of Natural Stone Setts

Width (mm)		Length (mm)		Depth (mm)
100	x	100	x	100
75	x	150 to 250	x	125
75	x	150 to 250	x	150
100	x	150 to 250	x	100
100	x	150 to 250	x	150

SPORTS

Sizes of Sports Areas
Sizes in metres given include clearances

Association football	Senior	114 x 72
	Junior	108 x 58
	International	100 – 110 x 64 – 75
Football	American	Pitch 109.80 x 48.80 Overall 118.94 x 57.94
	Australian Rules	Overall 135 – 185 x 110 – 155
	Canadian	Overall 145.74 x 59.47
	Gaelic	128 – 146.4 x 76.8 – 91.50
Handball		91 – 110 x 55 – 65
Hurling		137 x 82
Rugby	Union pitch	56 x 81
	League pitch	134 x 80
Hockey pitch		100.5 x 61
Men's lacrosse pitch		106 x 61
Women's lacrosse pitch		110 x 60
Target archery ground		150 x 50
Archery (Clout)		7.3m firing area Range 109.728 (Women), 146.304 (Men). 182.88 (Normal range)
400m running track		115.61 bend length x 2 84.39 straight length x 2 Overall 176.91 long x 92.52 wide
Baseball		Overall 60m x 70
Basketball		14.0 x 26.0
Badminton		6.10 x 13.40
Camogie		91 – 110 x 54 – 68
Discus and Hammer		Safety cage 2.74m square Landing area 45 arc (65° safety) 70 m radius
Javelin		Runway 36.5 x 4.27 Landing area 80 - 95 x 48

Jump	High	Running area 38.8 x 19 Landing area 5 x 4
	Long	Runway 45 x 1.22 Landing area 9 x 2.750
	Triple	Runway 45 x 1.22 Landing area 7.3 x 2.75
Korfball		90 x 40
Netball		15.25 x 30.48
Pole Vault		Runway 45 x 1.22 Landing area 5 x 5
Polo		275 x 183
Rounders		Overall 19 x 17
Shot Putt		Base 2.135 dia Landing area 65° arc, 25m radius from base
Shinty		128 –183 x 64 – 91.5
Tennis		Court 23.77 x 10.97 Overall minimum 36.27 x 18.29
Tug–of–war		46 x 5

SEEDING/TURFING AND PLANTING

Topsoil Quality

Topsoil grade	Properties
Premium	Natural topsoil, high fertility, loamy texture, good soil structure, suitable for intensive cultivation.
General Purpose	Natural or manufactured topsoil of lesser quality than Premium, suitable for agriculture or amenity landscape, may need fertilizer or soil structure improvement.
Economy	Selected subsoil, natural mineral deposit such as river silt or greensand. The grade comprises two sub-grades; "Low clay" and "High clay" which is more liable to compaction in handling. This grade is suitable for low- production agricultural land and amenity woodland or conservation planting areas.

Forms of Trees

Standards:	shall be clear with substantially straight stems. Grafted and budded trees shall have no more than a slight bend at the union. Standards shall be designated as Half, Extra light, Light, Standard, Selected standard, Heavy, and Extra heavy.
Sizes of Standards	
Heavy standard	12-14 cm girth x 3.50 to 5.00 m high
Extra Heavy standard	14-16 cm girth x 4.25 to 5.00 m high
Extra Heavy standard	16-18 cm girth x 4.25 to 6.00 m high
Extra Heavy standard	18-20 cm girth x 5.00 to 6.00 m high
Semi-mature trees:	between 6.0 m and 12.0 m tall with a girth of 20 to 75 cm at 1.0 m above ground.
Feathered trees:	shall have a defined upright central leader, with stem furnished with evenly spread and balanced lateral shoots down to or near the ground.
Whips:	shall be without significant feather growth as determined by visual inspection.
Multi-stemmed trees:	shall have two or more main stems at, near, above or below ground.

Seedlings grown from seed and not transplanted shall be specified when ordered for sale as:

1+0 one year old seedling		
2+0 two year old seedling		
1+1 one year seed bed,	one year transplanted	= two year old seedling
1+2 one year seed bed,	two years transplanted	= three year old seedling
2+1 two year seed bed,	one year transplanted	= three year old seedling
1u1 two years seed bed,	undercut after 1 year	= two year old seedling
2u2 four years seed bed,	undercut after 2 years	= four year old seedling

Cuttings

The age of cuttings (plants grown from shoots, stems, or roots of the mother plant) shall be specified when ordered for sale. The height of transplants and undercut seedlings/cuttings (which have been transplanted or undercut at least once) shall be stated in centimetres. The number of growing seasons before and after transplanting or undercutting shall be stated.

0+1	one year cutting
0+2	two year cutting
0+1+1	one year cutting bed, one year transplanted = two year old seedling
0+1+2	one year cutting bed, two years transplanted = three year old seedling

Grass Cutting Capacities in m² per Hour

Speed mph	\multicolumn{13}{c}{Width Of Cut in metres}
	0.5	0.7	1.0	1.2	1.5	1.7	2.0	2.0	2.1	2.5	2.8	3.0	3.4
1.0	724	1127	1529	1931	2334	2736	3138	3219	3380	4023	4506	4828	5472
1.5	1086	1690	2293	2897	3500	4104	4707	4828	5069	6035	6759	7242	8208
2.0	1448	2253	3058	3862	4667	5472	6276	6437	6759	8047	9012	9656	10944
2.5	1811	2816	3822	4828	5834	6840	7846	8047	8449	10058	11265	12070	13679
3.0	2173	3380	4587	5794	7001	8208	9415	9656	10139	12070	13518	14484	16415
3.5	2535	3943	5351	6759	8167	9576	10984	11265	11829	14082	15772	16898	19151
4.0	2897	4506	6115	7725	9334	10944	12553	12875	13518	16093	18025	19312	21887
4.5	3259	5069	6880	8690	10501	12311	14122	14484	15208	18105	20278	21726	24623
5.0	3621	5633	7644	9656	11668	13679	15691	16093	16898	20117	22531	24140	27359
5.5	3983	6196	8409	10622	12834	15047	17260	17703	18588	22128	24784	26554	30095
6.0	4345	6759	9173	11587	14001	16415	18829	19312	20278	24140	27037	28968	32831
6.5	4707	7322	9938	12553	15168	17783	20398	20921	21967	26152	29290	31382	35566
7.0	5069	7886	10702	13518	16335	19151	21967	22531	23657	28163	31543	33796	38302

Number of Plants per m²: For Plants Planted on an Evenly Spaced Grid

Planting distances

mm	0.10	0.15	0.20	0.25	0.35	0.40	0.45	0.50	0.60	0.75	0.90	1.00	1.20	1.50
0.10	100.00	66.67	50.00	40.00	28.57	25.00	22.22	20.00	16.67	13.33	11.11	10.00	8.33	6.67
0.15	66.67	44.44	33.33	26.67	19.05	16.67	14.81	13.33	11.11	8.89	7.41	6.67	5.56	4.44
0.20	50.00	33.33	25.00	20.00	14.29	12.50	11.11	10.00	8.33	6.67	5.56	5.00	4.17	3.33
0.25	40.00	26.67	20.00	16.00	11.43	10.00	8.89	8.00	6.67	5.33	4.44	4.00	3.33	2.67
0.35	28.57	19.05	14.29	11.43	8.16	7.14	6.35	5.71	4.76	3.81	3.17	2.86	2.38	1.90
0.40	25.00	16.67	12.50	10.00	7.14	6.25	5.56	5.00	4.17	3.33	2.78	2.50	2.08	1.67
0.45	22.22	14.81	11.11	8.89	6.35	5.56	4.94	4.44	3.70	2.96	2.47	2.22	1.85	1.48
0.50	20.00	13.33	10.00	8.00	5.71	5.00	4.44	4.00	3.33	2.67	2.22	2.00	1.67	1.33
0.60	16.67	11.11	8.33	6.67	4.76	4.17	3.70	3.33	2.78	2.22	1.85	1.67	1.39	1.11
0.75	13.33	8.89	6.67	5.33	3.81	3.33	2.96	2.67	2.22	1.78	1.48	1.33	1.11	0.89
0.90	11.11	7.41	5.56	4.44	3.17	2.78	2.47	2.22	1.85	1.48	1.23	1.11	0.93	0.74
1.00	10.00	6.67	5.00	4.00	2.86	2.50	2.22	2.00	1.67	1.33	1.11	1.00	0.83	0.67
1.20	8.33	5.56	4.17	3.33	2.38	2.08	1.85	1.67	1.39	1.11	0.93	0.83	0.69	0.56
1.50	6.67	4.44	3.33	2.67	1.90	1.67	1.48	1.33	1.11	0.89	0.74	0.67	0.56	0.44

Grass Clippings Wet: Based on 3.5 m³ /tonne

Annual kg/100 m²	Average 20 cuts kg/100m²	m²/tonne	m²/m³
32.0	1.6	61162.1	214067.3

Nr of cuts	22	20	18	16	12	4
kg/cut	1.45	1.60	1.78	2.00	2.67	8.00
Area capacity of 3 tonne vehicle per load m²	206250	187500	168750	150000	112500	37500
Load m³	100 m² units/m³ of vehicle space					
1	196.4	178.6	160.7	142.9	107.1	35.7
2	392.9	357.1	321.4	285.7	214.3	71.4
3	589.3	535.7	482.1	428.6	321.4	107.1
4	785.7	714.3	642.9	571.4	428.6	142.9
5	982.1	892.9	803.6	714.3	535.7	178.6

Transportation of Trees

To unload large trees a machine with the necessary lifting strength is required. The weight of the trees must therefore be known in advance. The following table gives a rough overview. The additional columns with root ball dimensions and the number of plants per trailer provide additional information for example about preparing planting holes and calculating unloading times.

Girth in cm	Rootball diameter in cm	Ball height in cm	Weight in kg	Numbers of trees per trailer
16 - 18	50 - 60	40	150	100 - 120
18 - 20	60 - 70	40 - 50	200	80 - 100
20 - 25	60 - 70	40 - 50	270	50 - 70
25 - 30	80	50 - 60	350	50
30 - 35	90 - 100	60 - 70	500	12 - 18
35 - 40	100 - 110	60 - 70	650	10 - 15
40 - 45	110 - 120	60 - 70	850	8 - 12
45 - 50	110 - 120	60 - 70	1100	5 - 7
50 - 60	130 - 140	60 - 70	1600	1 - 3
60 - 70	150 - 160	60 - 70	2500	1
70 - 80	180 - 200	70	4000	1
80 - 90	200 - 220	70 - 80	5500	1
90 - 100	230 - 250	80 - 90	7500	1
100 - 120	250 - 270	80 - 90	9500	1

Data supplied by Lorenz von Ehren GmbH
The information in the table is approximate; deviations depend on soil type, genus and weather.

FENCING AND GATES

Types of Preservative

Creosote (tar oil) can be "factory" applied	
	by pressure to BS 144: pts 1&2
	by immersion to BS 144: pt 1
	by hot and cold open tank to BS 144: pts 1&2
Copper/chromium/arsenic (CCA)	by full cell process to BS 4072 pts 1&2
Organic solvent (OS)	by double vacuum (vacvac) to BS 5707 pts 1&3
	by immersion to BS 5057 pts 1&3
Pentachlorophenol (PCP)	by heavy oil double vacuum to BS 5705 pts 2&3
Boron diffusion process (treated with disodium octaborate to BWPA Manual 1986.	

Note: Boron is used on green timber at source and the timber is supplied dry.

Cleft Chestnut Pale Fences

Pales	Pale spacing	Wire lines	
900 mm long	75 mm	2	temporary protection
1050	75 or 100	2	light protective fences
1200	75	3	perimeter fences
1350	75	3	perimeter fences
1500	50	3	narrow perimeter fences
1800	50	3	light security fences

Close-boarded Fences

Close-boarded fences 1.05 to 1.8m high
Type BCR (recessed) or BCM (morticed) with concrete posts 140 x 115 mm tapered and Type BW with timber posts.

Palisade Fences

Wooden palisade fences
Type WPC with concrete posts 140 x 115 mm tapered and Type WPW with timber posts.

For both types of fence:
Height of fence 1050 mm: two rails
Height of fence 1200 mm: two rails
Height of fence 1500 mm: three rails
Height of fence 1650 mm: three rails
Height of fence 1800 mm: three rails

Post and Rail Fences

Wooden post and rail fences
Type MPR 11/3 morticed rails and Type SPR 11/3 nailed rails
Height to top of rail 1100 mm
Rails: three rails 87 mm 38 mm

Type MPR 11/4 morticed rails and Type SPR 11/4 nailed rails
Height to top of rail 1100 mm
Rails: four rails 87 mm 38 mm.

Type MPR 13/4 morticed rails and Type SPR 13/4 nailed rails
Height to top of rail 1300 mm
Rail spacing 250 mm, 250 mm, and 225 mm from top
Rails: four rails 87 mm 38 mm.

Steel Posts

Rolled steel angle iron posts for chain link fencing:

Posts	Fence height	Strut	Straining post
1500 x 40 x 40 x 5 mm	900 mm	1500 x 40 x 40 x 5 mm	1500 x 50 x 50 x 6 mm
1800 x 40 x 40 x 5 mm	1200 mm	1800 x 40 x 40 x 5 mm	1800 x 50 x 50 x 6 mm
2000 x 45 x 45 x 5 mm	1400 mm	2000 x 45 x 45 x 5 mm	2000 x 60 x 60 x 6 mm
2600 x 45 x 45 x 5 mm	1800 mm	2600 x 45 x 45 x 5 mm	2600 x 60 x 60 x 6 mm
3000 x 50 x 50 x 6 mm with arms	1800 mm	2600 x 45 x 45 x 5 mm	3000 x 60 x 60 x 6 mm

Concrete Posts

Concrete posts for chain link fencing:

Posts and straining posts	Fence height	Strut
1570 mm 100 x 100 mm	900 mm	1500 mm x 75 x 75 mm
1870 mm 125 x 125 mm	1200 mm	1830 mm x 100 x 75 mm
2070 mm 125 x 125 mm	1400 mm	1980 mm x 100 x 75 mm
2620 mm 125 x 125 mm	1800 mm	2590 mm x 100 x 85 mm
3040 mm 125 x 125 mm	1800 mm	2590 mm x 100 x 85 mm (with arms)

Rolled Steel Angle Posts

Rolled steel angle posts for rectangular wire mesh (field) fencing

Posts	Fence height	Strut	Straining post
1200 x 40 x 40 x 5 mm	600 mm	1200 x 75 x 75 mm	1350 x 100 x 100 mm
1400 x 40 x 40 x 5 mm	800 mm	1400 x 75 x 75 mm	1550 x 100 x 100 mm
1500 x 40 x 40 x 5 mm	900 mm	1500 x 75 x 75 mm	1650 x 100 x 100 mm
1600 x 40 x 40 x 5 mm	1000 mm	1600 x 75 x 75 mm	1750 x 100 x 100 mm
1750 x 40 x 40 x 5 mm	1150 mm	1750 x 75 x 100 mm	1900 x 125 x 125 mm

Concrete Posts

Concrete posts for rectangular wire mesh (field) fencing

Posts	Fence height	Strut	Straining post
1270 x 100 x 100 mm	600 mm	1200 x 75 x 75 mm	1420 x 100 x 100 mm
1470 x 100 x 100 mm	800 mm	1350 x 75 x 75 mm	1620 x 100 x 100 mm
1570 x 100 x 100 mm	900 mm	1500 x 75 x 75 mm	1720 x 100 x 100 mm
1670 x 100 x 100 mm	600 mm	1650 x 75 x 75 mm	1820 x 100 x 100 mm
1820 x 125 x 125 mm	1150 mm	1830 x 75 x 100 mm	1970 x 125 x 125 mm

Cleft Chestnut Pale Fences
Timber Posts

Timber posts for wire mesh and hexagonal wire netting fences
Round timber for general fences

Posts	Fence height	Strut	Straining post
1300 x 65 mm dia.	600 mm	1200 x 80 mm dia	1450 x 100 mm dia
1500 x 65 mm dia	800 mm	1400 x 80 mm dia	1650 x 100 mm dia
1600 x 65 mm dia.	900 mm	1500 x 80 mm dia	1750 x 100 mm dia
1700 x 65 mm dia.	1050 mm	1600 x 80 mm dia	1850 x 100 mm dia
1800 x 65 mm dia.	1150 mm	1750 x 80 mm dia	2000 x 120 mm dia

Squared timber for general fences

Posts	Fence height	Strut	Straining post
1300 x 75 x 75 mm	600 mm	1200 x 75 x 75 mm	1450 x 100 x 100 mm
1500 x 75 x 75 mm	800 mm	1400 x 75 x 75 mm	1650 x 100 x 100 mm
1600 x 75 x 75 mm	900 mm	1500 x 75 x 75 mm	1750 x 100 x 100 mm
1700 x 75 x 75 mm	1050 mm	1600 x 75 x 75 mm	1850 x 100 x 100 mm
1800 x 75 x 75 mm	1150 mm	1750 x 75 x 75 mm	2000 x 125 x 100 mm

Steel Fences to BS 1722: Part 9: 1992

	Fence height	Top/bottom rails and flat posts	Vertical bars
Light	1000 mm	40 x 10 mm 450 mm in ground	12 mm dia at 115 mm cs
	1200 mm	40 x 10 mm 550 mm in ground	12 mm dia at 115 mm cs
	1400 mm	40 x 10 mm 550 mm in ground	12 mm dia at 115 mm cs
Light	1000 mm	40 x 10 mm 450 mm in ground	16 mm dia at 120 mm cs
	1200 mm	40 x 10 mm 550 mm in ground	16 mm dia at 120 mm cs
	1400 mm	40 x 10 mm 550 mm in ground	16 mm dia at 120 mm cs
Medium	1200 mm	50 x 10 mm 550 mm in ground	20 mm dia at 125 mm cs
	1400 mm	50 x 10 mm 550 mm in ground	20 mm dia at 125 mm cs
	1600 mm	50 x 10 mm 600 mm in ground	22 mm dia at 145 mm cs
	1800 mm	50 x 10 mm 600 mm in ground	22 mm dia at 145 mm cs
Heavy	1600 mm	50 x 10 mm 600 mm in ground	22 mm dia at 145 mm cs
	1800 mm	50 x 10 mm 600 mm in ground	22 mm dia at 145 mm cs
	2000 mm	50 x 10 mm 600 mm in ground	22 mm dia at 145 mm cs
	2200 mm	50 x 10 mm 600 mm in ground	22 mm dia at 145 mm cs

Notes: Mild steel fences: round or square verticals; flat standards and horizontals. Tops of vertical bars may be bow-top, blunt, or pointed. Round or square bar railings.

Timber Field Gates to BS 3470: 1975

Gates made to this standard are designed to open one way only.
All timber gates are 1100 mm high.
Width over stiles 2400, 2700, 3000, 3300, 3600, and 4200 mm.
Gates over 4200 mm should be made in two leaves.

Steel Field Gates to BS 3470: 1975

All steel gates are 1100 mm high.
Heavy duty: width over stiles 2400, 3000, 3600 and 4500 mm
Light duty: width over stiles 2400, 3000, and 3600 mm

Domestic Front Entrance Gates to BS 4092: Part 1: 1966

Metal gates:	Single gates are 900 mm high minimum, 900 mm, 1000 mm and 1100 mm wide

Domestic Front Entrance Gates to BS 4092: Part 2: 1966

Wooden gates:	All rails shall be tenoned into the stiles Single gates are 840 mm high minimum, 801 mm and 1020 mm wide Double gates are 840 mm high minimum, 2130, 2340 and 2640 mm wide

Timber Bridle Gates to BS 5709:1979 (Horse or Hunting Gates)

Gates open one way only	
Minimum width between posts	1525 mm
Minimum height	1100 mm

Timber Kissing Gates to BS 5709:1979

Minimum width	700 mm
Minimum height	1000 mm
Minimum distance between shutting posts	600 mm
Minimum clearance at mid-point	600 mm

Metal Kissing Gates to BS 5709:1979

Sizes are the same as those for timber kissing gates.
Maximum gaps between rails 120 mm.

Categories of Pedestrian Guard Rail to BS 3049:1976

Class A for normal use.
Class B where vandalism is expected.
Class C where crowd pressure is likely.

DRAINAGE

Weights and Dimensions - Vitrified Clay Pipes

Product	Nominal diameter (mm)	Effective length (mm)	BS 65 limits of tolerance		Crushing strength (kN/m)	Weight	
			min (mm)	max (mm)		kg/pipe	kg/m
Supersleve	100	1600	96	105	35.00	14.71	9.19
	150	1750	146	158	35.00	29.24	16.71
Hepsleve	225	1850	221	236	28.00	84.03	45.42
	300	2500	295	313	34.00	193.05	77.22
	150	1500	146	158	22.00	37.04	24.69
Hepseal	225	1750	221	236	28.00	85.47	48.84
	300	2500	295	313	34.00	204.08	81.63
	400	2500	394	414	44.00	357.14	142.86
	450	2500	444	464	44.00	454.55	181.63
	500	2500	494	514	48.00	555.56	222.22
	600	2500	591	615	57.00	796.23	307.69
	700	3000	689	719	67.00	1111.11	370.45
	800	3000	788	822	72.00	1351.35	450.45
Hepline	100	1600	95	107	22.00	14.71	9.19
	150	1750	145	160	22.00	29.24	16.71
	225	1850	219	239	28.00	84.03	45.42
	300	1850	292	317	34.00	142.86	77.22
Hepduct (conduit)	90	1500	-	-	28.00	12.05	8.03
	100	1600	-	-	28.00	14.71	9.19
	125	1750	-	-	28.00	20.73	11.84
	150	1750	-	-	28.00	29.24	16.71
	225	1850	-	-	28.00	84.03	45.42
	300	1850	-	-	34.00	142.86	77.22

Weights and Dimensions - Vitrified Clay Pipes

Nominal internal diameter (mm)	Nominal wall thickness (mm)	Approximate weight kg/m
150	25	45
225	29	71
300	32	122
375	35	162
450	38	191
600	48	317
750	54	454
900	60	616
1200	76	912
1500	89	1458
1800	102	1884
2100	127	2619

Wall thickness, weights and pipe lengths vary, depending on type of pipe required.
The particulars shown above represent a selection of available diameters and are applicable to strength class 1 pipes with flexible rubber ring joints.
Tubes with Ogee joints are also available.

Weights and Dimensions - PVC-u Pipes

	Nominal size	Mean outside diameter (mm)		Wall thickness	Weight
	min	max	(mm)	(kg/m)	
Standard pipes	82.4	82.4	82.7	3.2	1.2
	110.0	110.0	110.4	3.2	1.6
	160.0	160.0	160.6	4.1	3.0
	200.0	200.0	200.6	4.9	4.6
	250.0	250.0	250.7	6.1	7.2
Perforated pipes heavy grade	As above	As above	As above	As above	As above
thin wall	82.4	82.4	82.7	1.7	-
	110.0	110.0	110.4	2.2	-
	160.0	160.0	160.6	3.2	-

Width of Trenches Required for Various Diameters of Pipes

Pipe diameter (mm)	Trench n.e. 1.5 m deep (mm)	Trench over 1.5 m deep (mm)
n.e. 100	450	600
100-150	500	650
150-225	600	750
225-300	650	800
300-400	750	900
400-450	900	1050
450-600	1100	1300

DRAINAGE BELOW GROUND AND LAND DRAINAGE

Flow of Water Which Can Be Carried by Various Sizes of Pipe

Clay or concrete pipes

Pipe size	Gradient of pipeline							
	1:10	1:20	1:30	1:40	1:50	1:60	1:80	1:100
	Flow in litres per second							
DN 100 15.0	8.5	6.8	5.8	5.2	4.7	4.0	3.5	
DN 150 28.0	19.0	16.0	14.0	12.0	11.0	9.1	8.0	
DN 225 140.0	95.0	76.0	66.0	58.0	53.0	46.0	40.0	

Plastic pipes

Pipe size	Gradient of pipeline							
	1:10	1:20	1:30	1:40	1:50	1:60	1:80	1:100
	Flow in litres per second							
82.4 mm i/dia	12.0	8.5	6.8	5.8	5.2	4.7	4.0	3.5
110 mm i/dia	28.0	19.0	16.0	14.0	12.0	11.0	9.1	8.0
160 mm i/dia	76.0	53.0	43.0	37.0	33.0	29.0	25.0	22.0
200 mm i/dia	140.0	95.0	76.0	66.0	58.0	53.0	46.0	40.0

Vitrified (Perforated) Clay Pipes and Fittings to BS En 295-5 1994

Length not specified		
75 mm bore	250 mm bore	600 mm bore
100	300	700
125	350	800
150	400	1000
200	450	1200
225	500	

Precast Concrete Pipes: Prestressed Non-pressure Pipes and Fittings: Flexible Joints to BS 5911: Pt. 103: 1994

Rationalized metric nominal sizes: 450, 500	
Length:	500 - 1000 by 100 increments
	1000 - 2200 by 200 increments
	2200 - 2800 by 300 increments
Angles: length:	450 - 600 angles 45, 22.5, 11.25°
	600 or more angles 22.5, 11.25°

Precast Concrete Pipes: Un-reinforced and Circular Manholes and Soakaways to BS 5911: Pt. 200: 1994

Nominal sizes:	
Shafts:	675, 900 mm
Chambers:	900, 1050, 1200, 1350, 1500, 1800, 2100, 2400, 2700, 3000 mm.
Large chambers:	To have either tapered reducing rings or a flat reducing slab in order to accept the standard cover.
Ring depths:	1. 300 - 1200 mm by 300 mm increments except for bottom slab and rings below cover slab, these are by 150 mm increments.
	2. 250 - 1000 mm by 250 mm increments except for bottom slab and rings below cover slab, these are by 125 mm increments.
Access hole:	750 x 750 mm for DN 1050 chamber
	1200 x 675 mm for DN 1350 chamber

Calculation of Soakaway Depth

The following formula determines the depth of concrete ring soakaway that would be required for draining given amounts of water.

$$h = \frac{4ar}{3\pi D^2}$$

h = depth of the chamber below the invert pipe
A = The area to be drained
r = The hourly rate of rainfall (50 mm per hour)
π = pi
D = internal diameter of the soakaway

This table shows the depth of chambers in each ring size which would be required to contain the volume of water specified. These allow a recommended storage capacity of 1/3 (one third of the hourly rain fall figure).

Table Showing Required Depth of Concrete Ring Chambers in Metres

Area m^2	50	100	150	200	300	400	500
Ring size							
0.9	1.31	2.62	3.93	5.24	7.86	10.48	13.10
1.1	0.96	1.92	2.89	3.85	5.77	7.70	9.62
1.2	0.74	1.47	2.21	2.95	4.42	5.89	7.37
1.4	0.58	1.16	1.75	2.33	3.49	4.66	5.82
1.5	0.47	0.94	1.41	1.89	2.83	3.77	4.72
1.8	0.33	0.65	0.98	1.31	1.96	2.62	3.27
2.1	0.24	0.48	0.72	0.96	1.44	1.92	2.41
2.4	0.18	0.37	0.55	0.74	1.11	1.47	1.84
2.7	0.15	0.29	0.44	0.58	0.87	1.16	1.46
3.0	0.12	0.24	0.35	0.47	0.71	0.94	1.18

Precast Concrete Inspection Chambers and Gullies to BS 5911: Part. 230: 1994

Nominal sizes:	375 diameter, 750, 900 mm deep 450 diameter, 750, 900, 1050, 1200 mm deep

Depths:	from the top for trapped or un-trapped units:
	    centre of outlet 300 mm
	    invert (bottom) of the outlet pipe 400 mm

Depth of water seal for trapped gullies:	85 mm, rodding eye int. diam. 100 mm
Cover slab:	65 mm min.

Ductile Iron Pipes to BS En 598: 1995
Type K9 with flexible joints should be used for surface water drainage. 5500 mm or 8000 mm long

80 mm bore	400 mm bore	1000 mm bore
100	450	1100
150	500	1200
200	600	1400
250	700	1600
300	800	
350	900	

Bedding Flexible Pipes: PVC-u Or Ductile Iron

Type 1 =	100 mm fill below pipe, 300 mm above pipe: single size material
Type 2 =	100 mm fill below pipe, 300 mm above pipe: single size or graded material
Type 3 =	100 mm fill below pipe, 75 mm above pipe with concrete protective slab over
Type 4 =	100 mm fill below pipe, fill laid level with top of pipe
Type 5 =	200 mm fill below pipe, fill laid level with top of pipe
Concrete =	25 mm sand blinding to bottom of trench, pipe supported on chocks, 100 mm concrete under the pipe, 150 mm concrete over the pipe.

Bedding Rigid Pipes: Clay or Concrete
(for vitrified clay pipes the manufacturer should be consulted)

Class D:	Pipe laid on natural ground with cut-outs for joints, soil screened to remove stones over 40 mm and returned over pipe to 150 m min depth. Suitable for firm ground with trenches trimmed by hand.
Class N:	Pipe laid on 50 mm granular material of graded aggregate to Table 4 of BS 882, or 10 mm aggregate to Table 6 of BS 882, or as dug light soil (not clay) screened to remove stones over 10 mm. Suitable for machine dug trenches.
Class B:	As Class N, but with granular bedding extending half way up the pipe diameter.
Class F:	Pipe laid on 100 mm granular fill to BS 882 below pipe, minimum 150 mm granular fill above pipe: single size material. Suitable for machine dug trenches.
Class A:	Concrete 100 mm thick under the pipe extending half way up the pipe, backfilled with the appropriate class of fill. Used where there is only a very shallow fall to the drain. Class A bedding allows the pipes to be laid to an exact gradient.
Concrete surround:	25 mm sand blinding to bottom of trench, pipe supported on chocks, 100 mm concrete under the pipe, 150 mm concrete over the pipe. It is preferable to bed pipes under slabs or wall in granular material.

PIPED SUPPLY SYSTEMS
Identification of Service Tubes From Utility to Dwellings

Utility	Colour	Size	Depth
British Telecom	grey	54 mm od	450 mm
Electricity	black	38 mm od	450 mm
Gas	yellow	42 mm od rigid 60 mm od convoluted	450 mm
Water	may be blue	(normally untubed)	750 mm

ELECTRICAL SUPPLY/POWER/LIGHTING SYSTEMS
Electrical Insulation Class En 60.598 BS 4533

Class 1:	luminaires comply with class 1 (I) earthed electrical requirements
Class 2:	luminaires comply with class 2 (II) double insulated electrical requirements
Class 3:	luminaires comply with class 3 (III) electrical requirements

Protection to Light Fittings
BS EN 60529:1992 Classification for degrees of protection provided by enclosures.(IP Code - International or ingress Protection)

1st characteristic: against ingress of solid foreign objects		
The figure	2	indicates that fingers cannot enter
	3	that a 2.5 mm diameter probe cannot enter
	4	that a 1.0 mm diameter probe cannot enter
	5	the fitting is dust proof (no dust around live parts)
	6	the fitting is dust tight (no dust entry)
2nd characteristic: ingress of water with harmful effects		
The figure	0	indicates unprotected
	1	vertically dripping water cannot enter
	2	water dripping 15° (tilt) cannot enter
	3	spraying water cannot enter
	4	splashing water cannot enter
	5	jetting water cannot enter
	6	powerful jetting water cannot enter
	7	proof against temporary immersion
	8	proof against continuous immersion
Optional additional codes:		A-D protects against access to hazardous parts
	H	High voltage apparatus
	M	fitting was in motion during water test
	S	fitting was static during water test
	W	protects against weather
Marking code arrangement:		(example) IPX5S = IP (International or Ingress Protection) X (denotes omission of first characteristic); 5 = jetting; S = static during water test.

ESSENTIAL READING FROM TAYLOR AND FRANCIS

Design for Outdoor Recreation
Second Edition

Simon Bell

A manual for planners, designers and managers of outdoor recreation destinations, this book works through the processes of design and provides the tools to find the most appropriate balance between visitor needs and the capacity of the landscape.

A range of different aspects are covered including car parking, information signing, hiking, waterside activities, wildlife watching and camping.

This second edition incorporates new examples from overseas, including Australia, New Zealand, Japan and Eastern Europe as well as focusing on more current issues such as accessibility and the changing demands for recreational use.

July 2008: 276x219: 272pp
Pb: 978-0-415-44172-8 **£44.99**

To Order: Tel: +44 (0) 1235 400524 **Fax:** +44 (0) 1235 400525
or Post: Taylor and Francis Customer Services,
Bookpoint Ltd, Unit T1, 200 Milton Park, Abingdon, Oxon, OX14 4TA UK
Email: book.orders@tandf.co.uk

**For a complete listing of all our titles visit:
www.tandf.co.uk**

Taylor & Francis
Taylor & Francis Group

Index

Access, restriction 65
Access Covers And Frames 300, 444
Aeration 246
   of turfed areas 59
Aggregates, bound 78
Agricultural, drainage 67
Amsterdam Soil 215
Anchoring Systems 109-110
Aquatic
   herbicides 89
   plants 73
   planting 230, 421
Artificial, sports pitches 185
Artificial Grass 47, 405
Asbestos Survey 102
Asphalt 383
   paving 162-164
   roads 162-164
Automatic, irrigation 72

Backfilling 308
Backfilling Trenches 447
Ballstop, fencing 64
Bark
   paving 46
   play area 47
   surfaces 399
Barriers 268-269
Bases 159, 380-382
   in situ concrete 381
   paving 338
   street furniture 270
Bedding 57, 344, 425
   maintenance 249, 424
Benches 65, 273-276
   stone 273-276
Bins
   dog waste 271-273
   grit 273
   Litter 272
Blocks
   concrete 135
   edgings 157, 377
   paving 168-169, 339
   walling 337, 368–371
Bollards 65, 279-281

columns 328
lighted 327
precinct lanterns 329
reproduction 328
reproduction brackets 328
reproduction suspensions 328
street 328
Boreholes 313
Bound Aggregates 78, 165, 385
Bowling Green 47
   irrigation system 47
Brick
   copings 135
   edgings 157, 377
   paving 170-173, 339, 390
   walling 337, 368-371
Brick/Block Walling 131
Bricks
   engineering 132
   facing 133-134
   market prices 131
   reclaimed 134
Building Fabric Survey 102
Buildings
   cast stone 100
   ornamental timber 100
Bulb Planting 229, 420

Cast Stone, buildings 100
Cast Stone Walling 137
Cattle Grid 64, 74
CDA 234
Cellular Soakaways 302
Cement, market prices 131
Chain Link, fencing 61, 63
Channel, edgings 157
Channel Drains 298, 347, 443
Channels 153-159, 439
   precast 153
Chemical
   applications 82
   landscape 82
Chemicals 234
Chestnut
   cleft paling 61
   pale fencing 62

Chestnut Fencing 209, 408
Cladding
  cast stone 141
  yorkstone 141
Clay
  channels 291
  drain 67
  gully 66
  pipes 293
Clearance 332, 358
Closeboard Fencing 255
Cobble Paving 46, 177-178, 181, 395
Cobbles, concrete 395
Combined Herbicides 86
Compost 197, 401
Concrete
  blocks 135
  block paving 388
  fencing 62
  gullies 66, 294
  in situ 125, 336, 354-367
  mixes 125-126
  paving 394
  plain in situ 126
  reinforced 365
  reinforced in situ 128
  roads 160
Conservation
  metal railings 104
  of historic railings 63
Contact Herbicides 85
Containers 276-277
Control
  insect 90
  moss 90
  of vegetation 59
Conveyor Belt 97
Copings 370
  brick 135
  cast stone 138
  precast 139
Cover Slabs 302
Crane, hire 98
Crib Walls 105
Cricket
  outfield 48
  square 48
Cultivating 124
Cultivation 195-197, 209-211, 363, 400, 408
Cutting, grass 59

Cycle
  stand 65
  track 48
Cycle Holders 277-278

Damp Proof Courses 139
  mastic asphalt 142
Damp Proofing 373
Decking, timber 74, 140, 372
Deer, fencing 62
Demolition 102, 332, 353
Deterrent, paving 177
Disposal 117-118, 121-122, 358, 361
Ditching 310-311
Documents, tender and contract 94
Drainage 347, 438
  agricultural 67
  below ground 289
  boards 299
  linear 66-67
  subsoil 68
  surface water 66
Dry Stone Walling 136
Dumpers 116, 357

Earthwork Support 152, 291
Edge Maintenance 247
Edge Restraints 171
Edges, granite sett 76-77
Edgings 153-159, 338, 375-380
  block 376
  brick 376
  channel 378
  concrete 375
  granite 156
  granite setts 351
  metal 379-380, 386
  polythene 350
  precast 153
  sett 376
  timber 387
  vinyl 350
  yorkstone 179
Electrical Cable 329
Embankments 108, 110
Erosion, control 52
Erosion Control, mats 113
Estate Fencing 264
Excavation 118-120, 334, 359-360
  and filling 115

for land drainage 304-307
  pathways 165
  trenches 152, 291-292, 440
Excavators 357
Expansion Joints 161
Extensive Green Roofs 145, 147

Felling Trees 116
Fencing 428-436
  ball stop 263
  ballstop 64
  boundary 257, 431
  chain link 61, 63
  chestnut pale 62
  close-boarded 255
  concrete 62, 263, 435
  deer 62
  forestry 62
  intruder guards 261
  oak 62
  palisade 256
  panels 255
  post and rail 256
  protective 61, 254, 428
  rabbit 259, 433
  rabbit-stop 62
  razor wire 262
  security 63, 96, 261
  site 61
  steel bar 261
  strained wire and mesh 259
  temporary 61, 96, 428
  timber 61, 64, 254, 428
  tubular 260
  weld mesh 260
  windbreak 263, 435
  wire 61
Fencing And Gates 254
Fertilizers 231
  market prices 188, 208
Field, gate 64
Filling 334, 361, 387
  to make up levels 120-121
Fin Drains 298
Flagpoles 286
Flexible Sheet Materials 113
Flexible Sheet Tanking 143
Footpath Gravels 388
Footpaths 46, 167-168
Forestry, fencing 62

Forestry Planting 233
Formwork 366
  plain vertical 129
Formwork for In Situ Concrete 129
Foundations, kerbs 153-159, 375
Fountains, wall 320-321
Fraise Mowing 76, 123
French Drains 308-310
Fungicides 90, 242

Gabions 357
Games Equipment 283-286
Gates 435
  field 64
  hardwood 267
  kissing 268
  sliding 268
  softwood 267
  stock 64
  tubular steel 268
Grading 211, 362
  operations 123
  surface 123
Granite
  edgings 156
  paving 180
  sett edges 76-77, 351
  walls 135
Granite Sett Paving 178, 396
Granite Slab Paving 396
Granular Pavings 387
Granular Sub-Bases 159
Grass
  artificial 405
  cutting 59
  reinforcement 110
Grass Concrete, paving 181-182
Grass Cutting 243-346
Grass Seed, market prices 190-193
Grassblocks 169-170, 389
Gravel
  paths 46
  paving 165
  roads 165
Gravel Paving 385
Green Roof Systems 76, 144-148
Groundcover Planting 420
Guard Rails 268
Gullies 294-296, 441
  clay 66, 348

concrete 66, 294
PVC-u 66, 295, 348
vitrified clay 294

Hardcore, bases 159
Harrowing 245
Hedges 225, 346, 418
  cutting 426
  maintenance 250
  planting 53
Herbaceous Planting 229, 420
Herbicides
  application 234
  aquatic 89
  combined 86
  contact 85
  market prices 235-242
  residual 85
  selective 86-88
  total 83
  woody weed 88
Hire, crane 98
Hit And Miss, rail 62

In Situ Concrete 125-130, 336, 354-367
  bases 380-382
  formwork 129
  paving 160-161
  reinforced 128
  reinforcement for 129
  roads 160-161
  steps 78
Indian, paving 180
Indian Stone Paving 340
Insect Control 90
Inspection Chambers 66, 289, 348, 439
Intensive Green Roofs 144, 147
Intercepting Traps 291, 439
Interlocking Paving 168
Intruder Guards 261
Inverted Waterproofing Systems 146
Irrigation 249, 313-316, 426
  automatic 72
  landscape 72
Irrigation System, bowling green 47
Irrigations, leaky pipe 72

Japanese Knotweed 251-254
  control 89
  management plan 91
  treatment 91-92
Jogging Track 47

Kerb Drains 296
Kerbs 153-159, 338, 375
  foundations 153-159
  precast 153
  stone 155-156
Kissing Gates 437

Lake Construction 318
Lake Liners 317
Lakes 317
Lakes And Ponds 73
Land Drainage 301, 446
Landfill Tax 93
Landscape
  chemicals 82
  irrigation 72
Landscape Maintenance 234, 422
Laser Surface Grading 76
Leaf Clearance 246, 423
Leaky Pipe Irrigation 72, 315
Leca 299
Lifebuoy Stations 273
Lifting Turf 201, 342, 357
Lighting 323-330
  bulkhead fittings 323
  floodlights 324
  lamps 323
  lighted bollards 327
  low level 327
  recessed 325
  spotlights 325
  underwater 326
Linear Drainage 66-67, 296, 347, 443
Liners, lakes and ponds 317
Litter Clearance 247

Macadam 383
  paving 162-164
  roads 162-164
Machine Selection Table 115, 357
Maintenance 234, 422
  bedding 249
  beds 248, 424
  edges 247, 423
  grass cutting 346
  hedges 250, 346, 426
  irrigation 249, 426

leaves 423
litter 247, 423
of grassed areas 59
of planted area 59
planting 421
post planting 59
pruning shrubs 247
shrub beds 346
shrubs 424
spraying 427
tree guards 247, 423
tree stakes 423
turf 203-206, 243-345, 406-407, 423
watering 249
Maintenance Operations 232
Manholes 66, 290, 347, 438
Market Prices
bricks 131
cement 131
fertilizers 190-193, 208
grass seed 190-193
herbicides 235-242
landscape chemicals 83
mulch 207
native species 81
of materials 131
sand 131
sports surfacing materials 184
topsoil 116
wild flora seed 193-195
Market Prices of, surfacing materials 399
Mastic Asphalt, tanking 142
Materials
flexible sheet 113
market prices 131
Metal, edgings 158-159
Metal Estate Fencing 264
Method Statements 94
Mortar 368
mixes 131
specialized 184
Moss Control 90, 243
Movement of Materials 410
Mowing, fraise 123
Muck Away, see disposal 358
Mulch Mats 231
Mulching Materials, market prices 207

Native Species Planting 230
Natural Stone, paving 178

Operations after Planting 421
Ornamental
pools 320
shrub beds 60
stone structures 101
timber buildings 100
Outdoor Seats 273
Outfield
cricket 48
cycle track 48

Painting 150, 374
Paling, cleft chestnut 61
Path Preparation 385
Paths
brick paved 46
gravel 46
Paving
bark 46
beach cobble 46
block 339, 388
brick 339, 390
bricks 170-173
cobble 177-178, 181, 398
concrete setts 178
deterrent 177
granite 180
granite setts 178, 396
grassblock 388
Indian stone 180, 340, 398
natural stone 178
precast concrete 174-178
quartzite 397
reconstituted yorkstone 178
resin 165, 351
setts 179-180
slab 174
SUDS 79-80
sustainable 79
yorkstone 179-180, 340, 397
Pavings
granular 167
permeable 173
sustainable 173
Pedestrian, guard rail 64
Pedestrian Deterrent Paving 177
Pergolas 93, 288
Permeable, paving 173
Pervious Surfacing Materials 80
Picnic Bench 276

Pier
    balls and bases 137
    caps 137
Pier Caps, precast 139
Piers, cast stone 137
Pipe Laying 66, 307, 347
Piped Water 313
Pipes
    clay 293, 441
    perforated 447
    PVC-u 293, 441
Pitches
    football 48
    hockey 48
    lacrosse 48
    maintenance 205
    rugby league 48
    rugby union 48
    shinty 48
    sports 48
Plant Containers 276-277
Planters
    planting 58
    street 65
Planting 81, 207, 406-407, 420
    bulbs 420
    groundcovers 420
    hedge 53
    hedges 418
    herbaceous 420
    native species 81
    planters 58
    shrubs 343, 418
    tree 53, 55
    trees 343
Plants, aquatic 73
Play, area 47
Play Equipment 283-286
Play Structures 285
Playgrounds 47, 186-187, 283-286, 399
Polo Ground 48
Ponds 317
Pools, ornamental 320
Prefabricated Buildings/Structures 100
Preparation, site 116
Programmes 95
Pruning Shrubs 247

Q20 Title 77
Quarry Tiles 149

Quartzite Paving 397

Rabbit Netting 259, 433
Rabbit-Stop, fencing 62
Rail
    hit and miss 62
    trip 64
Railings 63, 264
    conservation of historic railings 63
Railway Sleeper Walls 111-112, 355
Rainwater Harvesting 316
Recessed Access Covers 300
Reconstituted Yorkstone Pavings 178
Reinforcement 367
    bars 129-130
    fabric 130
    for in situ concrete 129
    grass 110
    to bases 381
Rendering 142, 149
Reservoir Board 144
Residual Herbicides 85
Resin Paving 165, 351
Resin Surfacing 385
Restoration of Railings 150
Retaining Walls 105-108, 138
    gabions 357
    sleeper 355
    timber 355
    timber log 111
Roads, in situ concrete 381
Rockery, stone 136
Rolling 195, 201, 246
Root Barriers 92, 144

Safety Surfaces 48, 399
Sand, market prices 131
Sand Banding 304
Sand Grooving 304
Sand Slitting 304
Sandpits 286
Scarifying 245
Seating 65
Seats 273-376
Security Fencing 63, 96, 261
Sedum Mat 148
Seedbed Preparation 198
Seedbeds 402
Seeding 51, 188, 199-201, 342, 400-403
    market prices 188

Selective Herbicides 86-88
Semi Mature Trees 222
Sett Edgings 156, 351, 377
Setting out 96
Setting out Planting 227
Setts
   concrete 395
   granite 178
Shotblasting 150
Shrub, ornamental beds 60
Shrub Bed Maintenance 346
Shrub Beds 346
Shrub Maintenance 424
Shrub Planting 227-229, 343, 418
Shrubs, Groundcovers and Bulbs 56
Signage 278
Silt Pits 289
Site
   clearance 117, 332, 358
   fencing 61
   preparation 116
   protection 209, 408
Skips 361
Slab Paving 174, 393
Sleeper
   retaining walls 355
   walls 111-112
Slides 284
Slot Drains 297, 347
Soakaways 71, 348
   cellular 303
   concrete 301-303
Soil Additives 401
Soil Stabilization 105, 355
Sport 399
Sports Drainage 304
Sports Equipment 283, 287
Sports Pitches 48
   artificial 185
   natural 185
Sports Surfaces 184-186
Sportsgrounds 48
Spraying Herbicides 234
Spreading, topsoil 195
Stabilization
   soil 105
   surface 109
Standing Stones 269
Step Irons 290
Steps 78, 183

in situ concrete 78
Stiles 268, 437
Stock Gate 64
Stone
   ornamental stone structures 101
   paving 340, 395
   rockery 136
   structures 100
   walls 135-136
Street Furniture 269
Structural Soil 215
Structures 93
   ornamental stone 101
   stone 100
Stump Grinding 116
Sub-bases 159
Subsoil Drainage 68
SUDS, paving 79-80
Surface
   applications 197, 212-214
   grading 76, 123
   preparation 49-50, 342
   stabilization 109
   treatment 124
Surface Water Drainage 66
Surfacing
   macadam 383
   pervious surfacing materials 80
Surveys
   asbestos 102
   building fabric 102
   underground services 102
Sustainable Paving 79, 173
Swimming Pools 320
   natural 319
Swings 284
Systems
   anchoring 109-110
   green roof 76
   temporary roadway 98

Tanking 373
Temporary
   accommodation 96
   fencing 61, 96
   protective fencing 61
   roadway system 98
   storage 96
Tendering
   costs 95

documents 94
material costs 95
Tennis Courts 186
Tiling 149
Timber
　decking 74, 140, 372
　edgings 158, 399
　fencing 61, 64
　log walls 355
　ornamental timber buildings 100
　retaining walls 111
Toilets 97
Top Dressing 246
Topiary 226
Topsoil 400, 404, 408
　spreading 195
Topsoil Improvements 409
Topsoil to Turf Beds 201
Total Herbicides 83
Transformers 329
Treatment of Japanese Knotweed 91-92
Treatments, surface 124
Tree
　anchors 412
　felling 116
　grilles 282
　pits 410-411
　pit additives 412
　pit mulching 413
　planting 53, 55
　protection 224-225
　planting labours 413
　soil 215
　staking 412
Tree Planting 214-225, 343, 410, 416-417
　Airpot containers 221
　anchors 215
　bare root 220, 414
　containerized 219, 415
　labours 218
　mature 416
　pit additives 215
　pit irrigation 217
　pit mulching 218
　root balled 219
　root barriers 217

rootballed 415
semi-mature 222, 416
staking 215
Trellis 265-266, 434
Trenches 152, 291-292, 335, 440
Trip Rails 64, 262
Turf 80
　aeration 246, 423
　artificial 203
　industrially grown 80
　maintenance 203-206, 243-245, 406-407
　reinforced 203
Turfbed Preparation 404
Turfing 51, 342, 404-405
　preparation 201-203
Type 1, bases 160

Underground Services Survey 102

Vapour Barriers 144-145
Vegetation, control 59
Vehicle Crash Barriers 270

Walls
　block 337, 368-371
　brick 337, 368-371
　brick/block 131
　crib 105
　dry stone 136
　granite 135
　natural stone rubble 135
　retaining 105-108, 138
　sleeper 111-112
　willow walling 113
Warm Roof Waterproofing 146
Water, features 73
Waterfall Construction 319
Watering 249, 344, 426
Waterproofing 142-143, 373
Wild Flora Seed, market prices 193-195
Willow Walling 113
Wood Chip Surfaces 168, 388
Woody Weed Herbicides 88

Yorkstone Paving 340, 396-397

ESSENTIAL READING FROM TAYLOR AND FRANCIS

Rethinking Landscape
A Critical Reader

Ian H. Thompson

This unusually wide-ranging critical tool in the field of landscape architecture provides extensive excerpted materials from, and detailed critical perspectives on, standard and neglected texts from the 18th century to the present day. Considering the aesthetic, social, cultural and environmental foundations of our thinking about landscape this book explores the key writings which shaped the field in its emergence and maturity. Uniquely the book also includes original materials drawn from philosophical, ethical and political writings.

Selected Contents:

Part 1: Pluralism
Part 2: Aesthetics
Part 3: The Social Mission
Part 4: Ecology
Part 5: Some other Perspectives
Part 6: Conclusions and Suggestions

2008: 246x174: 288pp
Hb: 978-0-415-42463-9 **£85.00**
Pb: 978-0-415-42464-6 **£24.99**

To Order: Tel: +44 (0) 1235 400524 **Fax:** +44 (0) 1235 400525
or Post: Taylor and Francis Customer Services,
Bookpoint Ltd, Unit T1, 200 Milton Park, Abingdon, Oxon, OX14 4TA UK
Email: book.orders@tandf.co.uk

For a complete listing of all our titles visit:
www.tandf.co.uk

Taylor & Francis
Taylor & Francis Group

Software and eBook Single-User Licence Agreement

We welcome you as a user of this Spon Price Book Software and eBook and hope that you find it a useful and valuable tool. Please read this document carefully. **This is a legal agreement** between you (hereinafter referred to as the "Licensee") and Taylor and Francis Books Ltd. (the "Publisher"), which defines the terms under which you may use the Product. **By breaking the seal and opening the document inside the back cover of the book containing the access code you agree to these terms and conditions outlined herein. If you do not agree to these terms you must return the Product to your supplier intact, with the seal on the document unbroken.**

1. **Definition of the Product**
   The product which is the subject of this Agreement, *Spon's External Works and Landscape Price Book 2010* Software and eBook (the "Product") consists of:
   1.1 Underlying data comprised in the product (the "Data")
   1.2 A compilation of the Data (the "Database")
   1.3 Software (the "Software") for accessing and using the Database
   1.4 An electronic book containing the data in the price book (the "eBook")

2. **Commencement and licence**
   2.1 This Agreement commences upon the breaking open of the document containing the access code by the Licensee (the "Commencement Date").
   2.2 This is a licence agreement (the "Agreement") for the use of the Product by the Licensee, and not an agreement for sale.
   2.3 The Publisher licenses the Licensee on a non-exclusive and non-transferable basis to use the Product on condition that the Licensee complies with this Agreement. The Licensee acknowledges that it is only permitted to use the Product in accordance with this Agreement.

3. **Multiple use**
   For more than one user or for a wide area network or consortium, use is only permissible with the purchase from the Publisher of a multiple-user licence and adherence to the terms and conditions of that licence.

4. **Installation and Use**
   4.1 The Licensee may provide access to the Product for individual study in the following manner: The Licensee may install the Product on a secure local area network on a single site for use by one user.
   4.2 The Licensee shall be responsible for installing the Product and for the effectiveness of such installation.
   4.3 Text from the Product may be incorporated in a coursepack. Such use is only permissible with the express permission of the Publisher in writing and requires the payment of the appropriate fee as specified by the Publisher and signature of a separate licence agreement.
   4.4 The Product is a free addition to the book and no technical support will be provided.

5. **Permitted Activities**
   5.1 The Licensee shall be entitled:
      5.1.1 to use the Product for its own internal purposes;
      5.1.2 to download onto electronic, magnetic, optical or similar storage medium reasonable portions of the Database provided that the purpose of the Licensee is to undertake internal research or study and provided that such storage is temporary;
   5.2 The Licensee acknowledges that its rights to use the Product are strictly set out in this Agreement, and all other uses (whether expressly mentioned in Clause 6 below or not) are prohibited.

6. **Prohibited Activities**
   The following are prohibited without the express permission of the Publisher:
   6.1 The commercial exploitation of any part of the Product.
   6.2 The rental, loan, (free or for money or money's worth) or hire purchase of this product, save with the express consent of the Publisher.
   6.3 Any activity which raises the reasonable prospect of impeding the Publisher's ability or opportunities to market the Product.
   6.4 Any networking, physical or electronic distribution or dissemination of the product save as expressly permitted by this Agreement.
   6.5 Any reverse engineering, decompilation, disassembly or other alteration of the Product save in accordance with applicable national laws.
   6.6 The right to create any derivative product or service from the Product save as expressly provided for in this Agreement.
   6.7 Any alteration, amendment, modification or deletion from the Product, whether for the purposes of error correction or otherwise.

7. General Responsibilities of the License

7.1 The Licensee will take all reasonable steps to ensure that the Product is used in accordance with the terms and conditions of this Agreement.

7.2 The Licensee acknowledges that damages may not be a sufficient remedy for the Publisher in the event of breach of this Agreement by the Licensee, and that an injunction may be appropriate.

7.3 The Licensee undertakes to keep the Product safe and to use its best endeavours to ensure that the product does not fall into the hands of third parties, whether as a result of theft or otherwise.

7.4 Where information of a confidential nature relating to the product of the business affairs of the Publisher comes into the possession of the Licensee pursuant to this Agreement (or otherwise), the Licensee agrees to use such information solely for the purposes of this Agreement, and under no circumstances to disclose any element of the information to any third party save strictly as permitted under this Agreement. For the avoidance of doubt, the Licensee's obligations under this sub-clause 7.4 shall survive the termination of this Agreement.

8. Warrant and Liability

8.1 The Publisher warrants that it has the authority to enter into this agreement and that it has secured all rights and permissions necessary to enable the Licensee to use the Product in accordance with this Agreement.

8.2 The Publisher warrants that the Product as supplied on the Commencement Date shall be free of defects in materials and workmanship, and undertakes to replace any defective Product within 28 days of notice of such defect being received provided such notice is received within 30 days of such supply. As an alternative to replacement, the Publisher agrees fully to refund the Licensee in such circumstances, if the Licensee so requests, provided that the Licensee returns this copy of *Spon's External Works and Landscape Price Book 2010* to the Publisher. The provisions of this sub-clause 8.2 do not apply where the defect results from an accident or from misuse of the product by the Licensee.

8.3 Sub-clause 8.2 sets out the sole and exclusive remedy of the Licensee in relation to defects in the Product.

8.4 The Publisher and the Licensee acknowledge that the Publisher supplies the Product on an "as is" basis. The Publisher gives no warranties:
   - 8.4.1 that the Product satisfies the individual requirements of the Licensee; or
   - 8.4.2 that the Product is otherwise fit for the Licensee's purpose; or
   - 8.4.3 that the Data are accurate or complete or free of errors or omissions; or
   - 8.4.4 that the Product is compatible with the Licensee's hardware equipment and software operating environment.

8.5 The Publisher hereby disclaims all warranties and conditions, express or implied, which are not stated above.

8.6 Nothing in this Clause 8 limits the Publisher's liability to the Licensee in the event of death or personal injury resulting from the Publisher's negligence.

8.7 The Publisher hereby excludes liability for loss of revenue, reputation, business, profits, or for indirect or consequential losses, irrespective of whether the Publisher was advised by the Licensee of the potential of such losses.

8.8 The Licensee acknowledges the merit of independently verifying Data prior to taking any decisions of material significance (commercial or otherwise) based on such data. It is agreed that the Publisher shall not be liable for any losses which result from the Licensee placing reliance on the Data or on the Database, under any circumstances.

8.9 Subject to sub-clause 8.6 above, the Publisher's liability under this Agreement shall be limited to the purchase price.

9. Intellectual Property Rights

9.1 Nothing in this Agreement affects the ownership of copyright or other intellectual property rights in the Data, the Database of the Software.

9.2 The Licensee agrees to display the Publishers' copyright notice in the manner described in the Product.

9.3 The Licensee hereby agrees to abide by copyright and similar notice requirements required by the Publisher, details of which are as follows:

"© 2010 Taylor & Francis. All rights reserved. All materials in *Spon's External Works and Landscape Price Book 2010* are copyright protected. All rights reserved. No such materials may be used, displayed, modified, adapted, distributed, transmitted, transferred, published or otherwise reproduced in any form or by any means now or hereafter developed other than strictly in accordance with the terms of the licence agreement enclosed with *Spon's External Works and Landscape Price Book 2010*. However, text and images may be printed and copied for research and private study within the preset program limitations. Please note the copyright notice above, and that any text or images printed or copied must credit the source."

9.4 This Product contains material proprietary to and copyedited by the Publisher and others. Except for the licence granted herein, all rights, title and interest in the Product, in all languages, formats and media throughout the world, including copyrights therein, are and remain the property of the Publisher or other copyright holders identified in the Product.

10. Non-assignment

This Agreement and the licence contained within it may not be assigned to any other person or entity without the written consent of the Publisher.

11. Termination and Consequences of Termination.

11.1 The Publisher shall have the right to terminate this Agreement if:

11.1.1 the Licensee is in material breach of this Agreement and fails to remedy such breach (where capable of remedy) within 14 days of a written notice from the Publisher requiring it to do so; or

11.1.2 the Licensee becomes insolvent, becomes subject to receivership, liquidation or similar external administration; or

11.1.3 the Licensee ceases to operate in business.

11.2 The Licensee shall have the right to terminate this Agreement for any reason upon two month's written notice. The Licensee shall not be entitled to any refund for payments made under this Agreement prior to termination under this sub-clause 11.2.

11.3 Termination by either of the parties is without prejudice to any other rights or remedies under the general law to which they may be entitled, or which survive such termination (including rights of the Publisher under sub-clause 7.4 above).

11.4 Upon termination of this Agreement, or expiry of its terms, the Licensee must destroy all copies and any back up copies of the product or part thereof.

12. General

12.1 **Compliance with export provisions**

The Publisher hereby agrees to comply fully with all relevant export laws and regulations of the United Kingdom to ensure that the Product is not exported, directly or indirectly, in violation of English law.

12.2 **Force majeure**

The parties accept no responsibility for breaches of this Agreement occurring as a result of circumstances beyond their control.

12.3 **No waiver**

Any failure or delay by either party to exercise or enforce any right conferred by this Agreement shall not be deemed to be a waiver of such right.

12.4 **Entire agreement**

This Agreement represents the entire agreement between the Publisher and the Licensee concerning the Product. The terms of this Agreement supersede all prior purchase orders, written terms and conditions, written or verbal representations, advertising or statements relating in any way to the Product.

12.5 **Severability**

If any provision of this Agreement is found to be invalid or unenforceable by a court of law of competent jurisdiction, such a finding shall not affect the other provisions of this Agreement and all provisions of this Agreement unaffected by such a finding shall remain in full force and effect.

12.6 **Variations**

This agreement may only be varied in writing by means of variation signed in writing by both parties.

12.7 **Notices**

All notices to be delivered to: Spon's Price Books, Taylor & Francis Books Ltd., 2 Park Square, Milton Park, Abingdon, Oxfordshire, OX14 4RN, UK.

12.8 **Governing law**

This Agreement is governed by English law and the parties hereby agree that any dispute arising under this Agreement shall be subject to the jurisdiction of the English courts.

If you have any queries about the terms of this licence, please contact:

Spon's Price Books
Taylor & Francis Books Ltd.
2 Park Square, Milton Park, Abingdon, Oxfordshire, OX14 4RN
Tel: +44 (0) 20 7017 6672
Fax: +44 (0) 20 7017 6702
www.tandfbuiltenvironment.com

Spon Press
Taylor & Francis Group

Multiple-user use of the Spon Press Software and eBook

To buy a licence to install your Spon Press Price Book Software and eBook on a secure local area network or a wide area network, and for the supply of network key files, for an agreed number of users please contact:

Spon's Price Books
Taylor & Francis Books Ltd.
2 Park Square, Milton Park, Abingdon, Oxfordshire, OX14 4RN
Tel: +44 (0) 207 017 6672
Fax: +44 (0) 207 017 6072
www.pricebooks.co.uk

Number of users	Licence cost
2–5	£450
6–10	£915
11–20	£1400
21–30	£2150
31–50	£4200
51–75	£5900
76–100	£7100
Over 100	Please contact Spon for details

Software Installation and Use Instructions

System requirements

Minimum

- Pentium processor
- 256 MB of RAM
- 20 MB available hard disk space
- Microsoft Windows 98/2000/NT/ME/XP/Vista
- SVGA screen
- Internet connection

Recommended

- Intel 466 MHz processor
- 512 MB of RAM (1,024MB for Vista)
- 100 MB available hard disk space
- Microsoft Windows XP/Vista
- XVGA screen
- Broadband Internet connection

Microsoft® is a registered trademark and Windows™ is a trademark of the Microsoft Corporation.

Installation

Spon's External Works and Landscape Price Book 2010 Electronic Version is supplied solely by internet download. No CD-ROM is supplied.

In your internet browser type in www.pricebooks.co.uk/downloads (or as a fallback www.ebookstore.tandf.co.uk/supplements/pricebook) and follow the instructions on screen. Then type in the unique access code which is sealed inside the back cover of this book.

When the access code is successfully validated, click on the download links to download and then save the files.

Please note: you will only be allowed one download of these files and onto one computer.

A folder called *PriceBook* will be added to your desktop, which will need to be unzipped. Then click on the application called *install*.

Use

- The installation process will create a folder containing the price book program links as well as a program icon on your desktop.
- Double click the icon (from the folder or desktop) installed by the Setup program.
- Follow the instructions on screen.

Technical Support

Support for the installation is provided on www.ebookstore.tandf.co.uk/supplements/helpdesk.asp

The *Electronic Version* is a free addition to the book. For help with the running of the software please visit www.pricebooks.co.uk

The software used in *Spon's External Works and Landscape Price Book 2010 Electronic Version* is furnished under a single user licence agreement. The software may be used only in accordance with the terms of the licence agreement, unless the additional multi-user licence agreement is purchased from Spon Press, 2 Park Square, Milton Park, Abingdon, Oxon OX14 4RN Tel: +44 (0) 20 7017 6000

© COPYRIGHT ALL RIGHTS RESERVED

Free Updates

with three easy steps...

1. Register today on www.pricebooks.co.uk/updates

2. We'll alert you by email when new updates are posted on our website

3. Then go to www.pricebooks.co.uk/updates
   and download the update.

All four Spon Price Books – *Architects' and Builders'*, *Civil Engineering and Highway Works*, *External Works and Landscape* and *Mechanical and Electrical Services* – are supported by an updating service. Three updates are loaded on our website during the year, in November, February and May. Each gives details of changes in prices of materials, wage rates and other significant items, with regional price level adjustments for Northern Ireland, Scotland and Wales and regions of England. The updates terminate with the publication of the next annual edition.

As a purchaser of a Spon Price Book you are entitled to this updating service for the 2010 edition – free of charge. Simply register via the website www.pricebooks.co.uk/updates and we will send you an email when each update becomes available.

If you haven't got internet access we can supply the updates by an alternative means. Please write to us for details: Spon Price Book Updates, Spon Press Marketing Department, 2 Park Square, Milton Park, Abingdon, Oxfordshire, OX14 4RN.

Find out more about Spon books
Visit www.sponpress.com for more details.